Mechanical Engineering Series

Series Editor

Francis A. Kulacki, Department of Mechanical Engineering, University of Minnesota, Minneapolis, MN, USA

The Mechanical Engineering Series presents advanced level treatment of topics on the cutting edge of mechanical engineering. Designed for use by students, researchers and practicing engineers, the series presents modern developments in mechanical engineering and its innovative applications in applied mechanics, bio-engineering, dynamic systems and control, energy, energy conversion and energy systems, fluid mechanics and fluid machinery, heat and mass transfer, manufacturing science and technology, mechanical design, mechanics of materials, micro- and nano-science technology, thermal physics, tribology, and vibration and acoustics. The series features graduate-level texts, professional books, and research monographs in key engineering science concentrations.

More information about this series at http://www.springer.com/series/1161

Sameer Khandekar • K. Muralidhar

Drop Dynamics and Dropwise Condensation on Textured Surfaces

Springer

Sameer Khandekar
Department of Mechanical Engineering
Indian Institute of Technology Kanpur
Kanpur, Uttar Pradesh, India

K. Muralidhar
Department of Mechanical Engineering
Indian Institute of Technology Kanpur
Kanpur, Uttar Pradesh, India

ISSN 0941-5122 ISSN 2192-063X (electronic)
Mechanical Engineering Series
ISBN 978-3-030-48463-7 ISBN 978-3-030-48461-3 (eBook)
https://doi.org/10.1007/978-3-030-48461-3

This Springer imprint is published by the registered company Springer Nature Switzerland AG
The registered company address is: Gewerbestrasse 11, 6330 Cham, Switzerland

Preface

This book is an expanded form of our previous monograph, **Khandekar S. and Muralidhar K., *Dropwise Condensation on Inclined Textured Surfaces*, Springer, 2013**. Here, we had described mathematical modeling of dropwise condensation over inclined textured surfaces, followed by simulations and comparison with experiments. The model factored in several details of the overall cyclic process but approximated processes at the scale of individual drops. In the last five years, drop level dynamics and spreading over hydrophobic surfaces have been extensively studied. These aspects can now be incorporated in the condensation model with greater generality. The present monograph has been written from this perspective.

Dropwise condensation is frequently encountered in the power, process, and materials industry and is an efficient route to heat transfer. Drops are also formed during condensation in distillation devices that work with diverse fluids ranging from water to liquid metal. Design of such equipment requires careful understanding of the condensation cycle starting from the birth of nuclei, followed by clusters, direct growth, and coalescence all the way to instability and fall-off. These steps of birth, growth and fall-off repeat over the surface, making the process periodic in the time domain.

Apart from the dropwise condensation model, additional topics covered include drop shape determination under static conditions, a fundamental study of drop spreading in sessile and pendant configurations, and the details of the coalescence phenomena. These are subsequently embedded in the condensation model and their consequences are examined. Since the mathematical model is spread over multiple scales of length and time, simulation is often computationally intensive. A parallelization approach to simulation of the condensation model is then presented.

Special topics covered in this book include three-phase contact line modeling, surface preparation techniques, fundamentals of evaporation and evaporation rates of a single liquid drop, and measurement of heat transfer coefficient during large-scale condensation of water vapor in the dropwise and filmwise modes.

Texts and review articles on dropwise condensation have focused on the phenomenology itself and reported experimental data. Models reported in the literature are primarily heuristic and empirical in nature. We are not aware of any text that describes a unified model of the complete dropwise condensation process over textured surfaces. The present book includes an explicit treatment of drop shapes, drop coalescence, effect of hydrophobicity on transport characteristics, contact angle hysteresis, and drop instability. The elaborate and generic model allows for progressive improvement and suggests possible experiments needed for validation.

Most chapters have been written by the authors in collaboration with their students. Their names have been included in the respective chapters.

We hope that this significantly improved text meets the expectations of design engineers, analysts, and researchers working in areas related to phase change phenomena and related heat transfer devices.

Kanpur, India

Sameer Khandekar
K. Muralidhar

Permissions

Several figures have been adapted by the authors from their previous journal articles, with permission. A list of these publications is given below.

1. Somwanshi, P., Muralidhar, K., & Khandekar, S. (2018). Coalescence dynamics of sessile and pendant liquid drops placed on a hydrophobic surface. *Physics of Fluids*, *30*, 092103.
2. Somwanshi, P., Muralidhar, K., & Khandekar, S. (2018). Dropwise condensation patterns of bismuth formed on horizontal and vertical surfaces. *International Journal of Heat and Mass Transfer*, *122*, 1024–1039.
3. Mistry, A., & Muralidhar, K. (2018). Spreading of a pendant liquid drop underneath a textured substrate. *Physics of Fluids*, *30*, 042104.
4. Singh, S.K., Gogna, M., Muralidhar, K., & Khandekar, S. (2017). Combined effect of substrate wettability and thermal properties on evaporation dynamics of a sessile droplet. *Interfacial Phenomena and Heat Transfer*, *5*, 321–335.
5. Singh, S.K., Yadav, M., Khandekar, S., & Muralidhar, K. (2017). Estimation of time-dependent wall heat flux from single thermocouple data. *International Journal of Thermal Sciences*, *115*, 1–15.
6. Gunjan, M. R., Somwanshi, P., Agarwal, A., Khandekar, S., & Muralidhar, K. (2015). Recoil of drops during coalescence on super-hydrophobic surfaces. *Interfacial Phenomena and Heat Transfer*, *3*, 1–18.
7. Sikarwar, B.S., Muralidhar, K., & Khandekar, S. (2015). Dropwise condensation of metal vapour underneath inclined textured substrates. *Interfacial Phenomena and Heat Transfer* (Kutateladze special issue), *3*, 85–113.
8. Mistry, A., & Muralidhar, K. (2015). Axisymmetric model of drop spreading on a horizontal surface. *Physics of Fluids*, *27*, 092103.
9. Sikarwar, B.S., Muralidhar, K., & Khandekar S. (2013). Effect of drop shape on heat transfer during dropwise condensation underneath inclined surfaces. *Interfacial Phenomena and Heat Transfer*, *1*, 339–356.

10. Bhutani, G., Muralidhar, K., & Khandekar, S. (2013). Determination of apparent contact angle and shape of a static pendant drop on a physically textured inclined surface. *Interfacial Phenomena and Heat Transfer, 1*, 29–49.
11. Sikarwar, B.S., Khandekar, S., & Muralidhar, K. (2013). Simulation of flow and heat transfer in a drop sliding underneath a hydrophobic surface. *International Journal of Heat and Mass Transfer, 57*, 786–811.
12. Sikarwar, B.S., Khandekar, S., & Muralidhar, K. (2013). Mathematical modeling of dropwise condensation on textured surfaces. *Sadhana, 38*(6), 1–37.
13. Sikarwar, B., Battoo, N.K., Khandekar, S., & Muralidhar, K. (2011). Dropwise condensation underneath chemically textured surfaces: simulation and experiments. *ASME Journal of Heat Transfer, 133*, 021501-1–21501-15.
14. Sikarwar, B.S., Khandekar, S., Agrawal, S., Kumar, S., & Muralidhar, K. (2012). Dropwise condensation studies on multiple scales. *Heat Transfer Engineering, 33*(4–5), 301–341.
15. Bansal, G.D., Khandekar, S., & Muralidhar, K. (2009). Measurement of heat transfer during dropwise condensation of water on polyethylene. *Nanoscale and Microscale Thermophysical Engineering, 13*(3), 184–201.
16. Chatterjee, S., Bhattacharya, S., Maurya, S.K., Srinivasan. V., Khare, K. Khandekar, S., (2017). Surface wettability of an atomically heterogeneous system and the resulting intermolecular forces. *Europhysics Letters, 118*, 68006 (1–7).
17. Singh S.K., Khandekar, S., Pratap, D. and Ramakrishna, S.A.,(2013). Wetting dynamics and evaporation of sessile droplets on nano-porous alumina surfaces. *Colloids and Surfaces A: Physicochemical and Engineering Aspects, 432*, 71–81.

The authors gratefully acknowledge permissions granted by the publishers for using the following figures in the respective chapters.

Figure 1.26 from APS, NY

Figures 2.1–2.3; 2.9–2.14 from APS, NY

Figures 3.13–3.23; 3.25–3.26 from APS, NY

Figures 10.1–10.9 from Elsevier Publishers, NY Figures 12.6 and 12.7 from Elsevier Publishers, NY

Permissions for certain figures in Chaps. 5, 6, 9, 11, and 14 were obtained during the publication of the earlier monograph: **Khandekar S and Muralidhar K., Dropwise Condensation on Inclined Textured Surfaces, Springer, 2013**.

Abstract

Dropwise condensation is a heterogeneous phase change process in which vapor condenses in the form of discrete liquid drops on or underneath a cold substrate. The heat transfer coefficient of dropwise condensation can be up to an order higher than film condensation and mixed-mode condensation, particularly with low conductivity liquids. Therefore, it is of considerable interest in applications such as thermal power plants and condensing equipment. It is also of interest in the material enrichment of large molecular weight liquids. Dropwise condensation is a complex process, involving drop formation at the atomic scale, growth of drops by direct condensation, coalescence of drops, drop instability and movement, followed by fresh nucleation. Hence, the dropwise condensation process is hierarchical in the sense that it occurs over a wide range of length and timescales. In addition, it depends on the thermophysical properties of the condensing fluid, physico-chemical and thermal properties of the cold substrate, orientation of the cold substrate, surface texture, subcooling, thermodynamic saturation conditions, and the presence of non-condensable gases.

Against this background, a mathematical model of dropwise condensation process underneath an inclined surface is presented in this monograph. The model includes formation of drops at the atomic scale, growth by direct condensation, coalescence, gravitational instability including slide-off and fall-off, followed by fresh nucleation of liquid droplets. The stability criterion is developed as a force balance equation at the level of a drop. Transport parameters of a sliding drop are determined using a CFD model and presented in the form of correlations. Performing the simulation of the complete cycle of dropwise condensation, the spatio-temporal distribution of drops is obtained. Consequently, quantities such as instantaneous condensation pattern, area of coverage, wall friction, and heat transfer rates, as well as important time- and area-averaged wall heat fluxes are determined. The simulated condensation patterns are compared against experimentally recorded images. The model is also validated against wall heat fluxes reported in the literature. While applicable for a wide range of fluids such as water and liquid metals, the model is seen to be sensitive to surface texture, inclination, and saturation conditions.

Dropwise condensation is a multiscale process in the sense that small-scale processes impact the macroscopically observed drop distribution and heat transfer rates. The smallest drop that is nucleated over the substrate arises from a balance of internal pressure and drop curvature over the thermodynamic phase diagram. Subsequently, it grows by direct condensation of vapor over the liquid-vapor interface. In this connection, static considerations of a liquid drop resting over a textured surface are discussed in the text. The governing equation arising from force balance is the Young-Laplace equation with surface characteristics specified in terms of the contact angle distribution around the three-phase contact line of the drop. Methods of solving this equation, mainly for the drop shape, are discussed. The contact line and the contact angle are shown to be important factors in fixing the drop shape.

For a given volume of the drop and prescribed surface properties and inclination, the drop attains an equilibrium shape. The shape tends to become a part of a sphere for increasing levels of hydrophobicity and diminishing contact hysteresis. Any departure from equilibrium leads to drop spreading, namely a process in which fluid velocity arises from non-equilibrium conditions. Spreading itself is a multiscale process in time during which droplet oscillations may be observed, but is ultimately followed by a viscous dissipation-controlled asymptotic decay of the transient. The spreading process is intricate and is rich in details. In the text, a contact line model that is free of modeling approximations is presented. It is shown to match experiments quite well.

The coalescence model adopted in most studies approximates the nuances of the overall merger process by requiring that it occurs instantaneously, moving from one equilibrium shape to the next. Preliminary experiments highlight the subtleties of the process and reveal complex flow patterns including oscillations of the free surface and large instantaneous wall shear stresses and wall heat fluxes. Specifically, drop coalescence commences when two droplets approach each other and contact either at the three-phase contact line or above the surface, depending on the degree of hydrophobicity of the surface. A tiny liquid bridge is immediately formed, induced by the Van der Waals forces. The coalescence process gets initiated by the extra surface energy released in the process. The difference in internal pressure between the two drops drives fluid motion. Immediately afterward, the coalescence process is limited by viscous and inertia forces. In water, free surface oscillation can last ~20 to 40 ms, depending on the size of droplets, substrate orientation, and thermophysical properties. Long-term relaxation can occur over 40–100 ms. Rapid transients in the early stage of coalescence will induce large shear stresses over the substrate, further accompanied by enhancement of the wall heat fluxes. To address this issue, experiments on coalescence of sessile and pendant drops were carried out on a hydrophobic surface. The phenomenon was imaged using a high-speed camera. The length, velocity, and timescales of coalescence were determined and compared with analytical estimates. This data was subsequently integrated with the dropwise condensation model to incorporate finite time effects in the simulation.

Since the simulation spans the size of the nuclei (in nm) to the size of the largest drop (in mm) over a substrate (10–100 mm), the computational effort involved is enormous. Simulation is facilitated by running the model on a computing machine with a parallel architecture.

The extended dropwise condensation model has been simulated with water and the condensation characteristics compared with bismuth. Differences arise mainly from those in thermophysical properties. Relative to water, liquid bismuth has a higher density, higher thermal conductivity, higher surface tension with vapor, higher viscosity but a smaller latent heat of vaporization, thus generating several contrasts with it. These differences are discussed at length.

Dropwise condensation can only be supported on non-wetting surfaces. Thus, the methods of surface engineering that will produce hydrophobicity on metallic and non-metallic surfaces are of great interest. These techniques are further discussed.

As the driving temperature difference for the process is small, experimental measurement of heat transfer coefficient in dropwise condensation is a challenging task. Heat fluxes show substantial spatial and temporal variation during a dropwise condensation process and their measurements are of substantial importance. Two approaches are discussed: one, the spatial distribution of wall heat flux under a condensing drop using liquid crystal thermography and two, the spatially averaged but temporally resolved heat flux using an inverse heat transfer formulation.

With a variety of additional details provided, the text should be of interest to researchers working not only on dropwise condensation, but generally to all those studying interfacial phenomena and heat transfer.

Acknowledgements

We thank our institute, the Indian Institute of Technology Kanpur for providing an enriching environment to carry out research and enough time for collecting and organizing the material in the form of a monograph. Students and faculty colleagues of the Department of Mechanical Engineering provided the much-needed inspiration. We thank Professor Frank A. Kulacki, Editor, Mechanical Engineering Series (Springer) for his continuous encouragement over the years. Part of the research contained in this text was financially supported by the Board of Research in Nuclear Sciences, Mumbai (India) and the Science and Engineering Research Board, New Delhi (India). The funding agencies are gratefully acknowledged. Finally, we thank our respective families for their patience and abundant support.

Contents

Part II Modeling Dropwise Condensation

Contributors

Smita Agrawal The Dow Chemical Company, Lake Jackson, TX, USA

Nirmal K. Battoo Department of Mechanical Engineering, Indian Institute of Technology Kanpur, Kanpur, India

Abhinav Bhanawat Department of Mechanical Engineering, Indian Institute of Technology Kanpur, Kanpur, India

Sudeep Bhattacharjee Department of Mechanical Engineering, Indian Institute of Technology Kanpur, Kanpur, India

Manish Bhendura Department of Mechanical Engineering, Indian Institute of Technology Kanpur, Kanpur, India

Gaurav Bhutani School of Engineering, Indian Institute of Technology Mandi, Mandi, India

Sanghamitro Chatterjee Department of Mechanical Engineering, Indian Institute of Technology Bombay, Mumbai, India

Mohit Gogna Department of Mechanical Engineering, Indian Institute of Technology Kanpur, Kanpur, India

Sameer Khandekar Department of Mechanical Engineering, Indian Institute of Technology Kanpur, Kanpur, India

Sumeet Kumar Department of Mechanical Engineering, Massachusetts Institute of Technology, Cambridge, MA, USA

Aashutosh Mistry Chemical Sciences and Engineering Division, Argonne National Laboratory, Lemont, IL, USA

K. Muralidhar Department of Mechanical Engineering, Indian Institute of Technology Kanpur, Kanpur, India

Maneesh Punetha Department of Mechanical Engineering, Indian Institute of Technology Kanpur, Kanpur, India

Basant S. Sikarwar Department of Mechanical Engineering, Amity University, Noida, India

Sachin Kumar Singh Department of Mechanical and Industrial Engineering, University of Chicago, Chicago, IL, USA

Praveen Somwanshi Department of Mechanical Engineering, Indian Institute of Technology Kanpur, Kanpur, India

Mahesh K. Yadav Department of Mechanical Engineering, Punjab Engineering College, Chandigarh, India

Part I
Statics, Spreading, Coalescence

Chapter 1
Droplet Statics

Gaurav Bhutani, K. Muralidhar, and Sameer Khandekar

Nomenclature

a	Radius of a spherical drop (m)
c	Constant
d	Drop width (m)
f	Drop shape function
$\widetilde{f}_i$	Force acting on plane Γ (N)
$\widehat{f}$	A unit vector in the tangential direction
g	3D drop shape function
$\mathbf{g}$	Acceleration due to gravity (m s^{-2})
l	Coordinate along the three-phase contact line (m)
l_{sl}	Slip length (m)
$\widehat{\mathbf{n}}$	Unit normal vector
p	Hydrostatic pressure (Pa)
r	Roughness measure (m)
r_{ij}	Distance between spheres in MD simulations (m)
$t_i^{(\widehat{\mathbf{n}})}$	Traction vector for a cutting plane in direction $\widehat{\mathbf{n}}$
u_{sl}	Slip velocity (m s^{-1})
u_{cl}	Contact line speed (m s^{-1})
x	x-coordinate (m)
y	y-coordinate (m)
z	z-coordinate (m)
A	Interfacial contact area (m^2)
Bo	Bond number
C	Constant
Ca	Capillary number
F	Helmholtz free energy (J)
N	Number of moles
P	Thermodynamic pressure (Pa)

S. Khandekar, K. Muralidhar, *Drop Dynamics and Dropwise Condensation on Textured Surfaces*, Mechanical Engineering Series,
https://doi.org/10.1007/978-3-030-48461-3_1

R	Radius (m)
S	Entropy (J K^{-1})
T	Temperature (K)
U	Contact line speed (m s^{-1})
V	Volume (m^3)
int	Interface
l	Liquid
s	Solid
v	Vapor
α	Range of contact angle (rad)
β	Substrate plate inclination angle (rad)
δ_{ij}	Kronecker delta
λ_{ij}	Dispersion and core interaction between beads i and j
$\widetilde{\epsilon}_{ij}$	Potential parameter
κ_i	Local curvature in i^{th} direction
μ	Viscosity (Pa s)
ρ	Density (kg m^{-3})
θ	Contact angle (rad)
θ^*	Effective contact angle (rad)
θ_d	Dynamic contact angle (rad)
θ_e	Equilibrium contact angle (rad)
θ_{max}	Maximum contact angle (rad)
θ_{min}	Minimum contact angle (rad)
θ_A	Advancing angle (rad)
θ_R	Receding angle (rad)
θ_Y	Young's equilibrium contact angle (rad)
σ	Interfacial tension (N m^{-1})
$\widetilde{\sigma}_{ij}$	Potential parameter
ϕ	Azimuthal angle (rad)
ϕ_1, ϕ_2	Fractional surface area occupied by a drop
ψ	Angle of the tangent drawn over the drop surface (rad)
Γ	Area of cutting plane (m^2)
Σ_{ij}	Cauchy stress tensor (N m^{-2})
2D	Two-dimensional
3D	Three-dimensional
CAH	Contact angle hysteresis
CB	Cassie-Baxter
CFD	Computational fluid dynamics
CVD	Chemical vapor deposition
DCA	Dynamic contact angle
DWC	Dropwise condensation
IK	Irving and Kirkwood
MD	Molecular dynamics
ODE	Ordinary differential equation

RHS	Right hand side
RMS	Root mean square
SAFT	Statistical associating fluid theory
W	Wenzel
IVP	Initial value problem
RK4	Fourth-order Runge-Kutta method

1.1 Wettability

A small amount of liquid ethanol spreads out to form a thin film when placed on the underside of flat aluminum substrate; however, a small quantity of water deposited on the same surface forms a discrete drop, as shown in Fig. 1.1. Depending on the surface energy of the solid-liquid contact, liquids with weak affinity for a solid wall will collect themselves into near-spherical drops. Those with a high affinity for the solid will spread to form a film to maximize liquid-solid contact area. This affinity of liquids for solids is referred to as wettability. Figure 1.1 shows a bead-shaped liquid drop in contact with a solid and surrounded by gas; or it can also be vapor. The solid surface is taken to be locally flat and is idealized as perfectly homogeneous in terms of the affinity for the deposited liquid. The overall shape of the liquid drop is near-spherical and, to a naked eye, meets the surface along the periphery at a finite angle. It is also called an *apparent* angle since the contact region may have a variety of local heterogeneities.

A liquid drop in static equilibrium is usually the final configuration attained through a series of intermediate states during which a flow field prevails within the liquid body. The drop is then said to spread over the surface. For example, one may place a spherical drop over the surface using a syringe, or let a drop fall over the surface from a given height. In both instances, velocities are set up within the drop, which decay in time till the drop finally attains a static shape. Thus, drop spreading invariably precedes the equilibrium configuration, whether on a horizontal or an inclined surface.

During spreading, the footprint radius of the drop may exceed the equilibrium value, to be followed by retraction. More generally, for low viscosity liquids such as

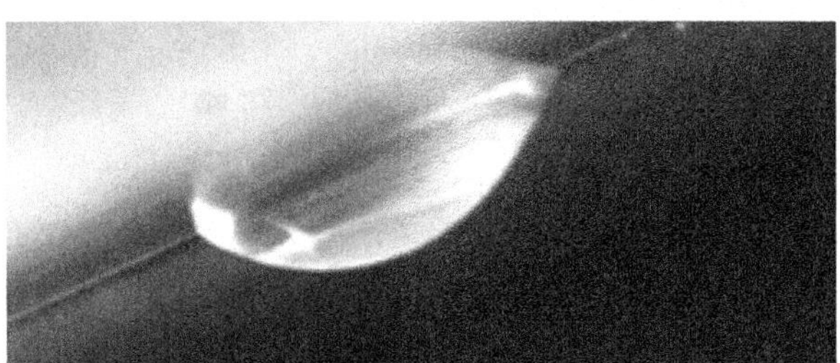

Fig. 1.1 Water drop on the underside of an aluminum substrate

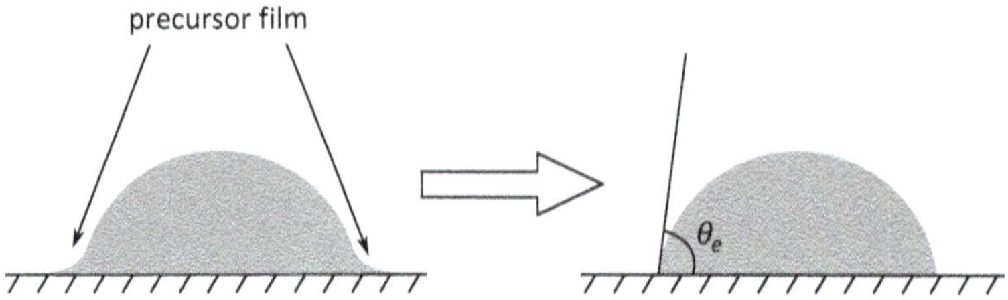

Fig. 1.2 A liquid drop with a precursor film approximated as one with a finite equilibrium contact angle

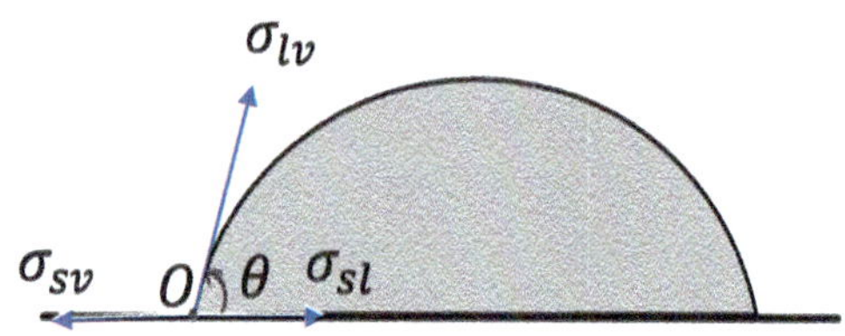

Fig. 1.3 Schematic drawing of a liquid drop on an idealized substrate showing interfacial tensions acting along the contact line

water, the footprint radius may oscillate relative to the equilibrium value. All such motions are collectively referred to as *spreading*.

The process of spreading is expected to create a thin film of liquid that is adsorbed by the surface and is termed the *precursor layer*. Experiments show the layer thickness to be at the nanometer scale and is not visible in a normal image, as in Fig. 1.1. The precursor film also makes the contact angle indeterminate, if the image near the contact line is recorded with high magnification. Quantification of the drop shape along with the precursor film is often a challenge. In this chapter, the drop is taken to sit on the substrate with a definite footprint radius and the precursor film is ignored (Fig. 1.2).

1.2 Contact Angle

The wettability of a liquid relative to the substrate placed in a given environment of another fluid (gas or liquid) is quantified using the *equilibrium contact angle* θ_e. In the present study, the environment surrounding the drop can either be vapor or air. Equilibrium contact angle is defined as the angle between the liquid-vapor interface and the substrate under equilibrium conditions; it is also referred as *contact angle* in short. It is measured along the gas-liquid-solid interface, namely the point of intersection point O, in a cross-sectional view, as shown in Fig. 1.3 where all the three phases meet. In view of the precursor film, the contact angle becomes indeterminate if the drop is viewed under increasing magnification. For this reason, this angle is also referred as the *apparent contact angle* that one would see without the precursor film at the scale of the drop.

When the drop is in motion, the contact angle at the three-phase contact line will depart from the equilibrium value and is called the *dynamic* contact angle, or, the *dynamic apparent* contact angle.

As θ_e decreases, a fixed quantity of liquid will spread to a greater extent over the surface. In the limit of θ_e approaching 0°, the liquid spreads over the entire available surface forming a thin film of zero thickness. The other limit is reached on a superhydrophobic surface when θ_e approaches 180° and the drop shape becomes truly spherical. Surfaces that have equilibrium contact angles closer to 0° are classified as hydrophilic; closer to 180° are hydrophobic. Such high contact angles are difficult to realize and hydrophobic surfaces with an equilibrium contact angle closer to 150° are more common. Tentatively, a value of 90° may be used to demarcate hydrophilic and hydrophobic surfaces, with those exceeding 150° being usually classified as superhydrophobic.

Figure 1.3 shows a three-phase system in equilibrium with a liquid-vapor interface, a liquid-solid interface, and a solid-vapor interface. The interfacial tensions associated with these three interfaces are σ_{lv}, σ_{sl}, and σ_{sv}, respectively. At static equilibrium, a force balance at point O in the horizontal direction requires

$$\sigma_{\mathrm{sv}} = \sigma_{\mathrm{sl}} + \sigma_{\mathrm{lv}} \cos\theta_e \tag{1.1}$$

The vertical component of interfacial tension $\sigma_{\mathrm{lv}} \sin \theta_e$ will be balanced by a vertical reaction at the solid surface along with the atmospheric pressure and the weight of the drop. Equation (1.1) is also referred to as the *Young's equation* (Pozrikidis 1997).

Young's equation can also be derived from thermodynamic considerations. Following Carey (2008, Eq. 2.16), consider that the interfaces are actually interfacial regions of small but finite thickness. We specifically consider a perturbation which results in a differential change in the positions of the interfacial surfaces, while the temperature and the volume of the individual phases in the system are constant. The resulting change in the total Helmholtz free energy F is equal to the sum of the changes in the bulk phases and the interfacial regions given as

$$\mathrm{d}F = \mathrm{d}F_{\mathrm{int}}^{\mathrm{lv}} + \mathrm{d}F_{\mathrm{int}}^{\mathrm{sl}} + \mathrm{d}F_{\mathrm{int}}^{\mathrm{sv}} + \mathrm{d}F_{\mathrm{v}} + \mathrm{d}F_{\mathrm{l}} + \mathrm{d}F_{\mathrm{s}} \tag{1.2}$$

Here, the first three terms on the RHS represent the interfacial contribution to the free energy, and the last three terms correspond to bulk phase properties. As shown by Carey (2008),

$$\mathrm{d}F_{\mathrm{int}}^{ab} = -S\mathrm{d}T + \rho \mathrm{d}N^{ab} + \sigma_{ab}\mathrm{d}A_{ab} \tag{1.3}$$

where S denotes entropy, T is temperature, ρ is density, N is the number of moles, and A is the interfacial contact area between phases a and b. All properties are defined for a finite-thickness interface between these phases. Although Eq. (1.3) was originally developed specifically for a fluid-fluid interface, the arguments used in its development are equally valid for solid-fluid interfaces. Hence for the bulk phases (Carey 2008),

$$dF = -SdT + \rho dN - PdV \tag{1.4}$$

Using $dT = 0$ (isothermal system) and $dV = 0$ (incompressible media) and substituting Eqs. (1.3) and (1.4) in Eq. (1.2), we get

$$\begin{aligned} dF = \rho dN_v + \rho dN_l + \rho dN_s + \rho dN^{lv} + \rho dN^{sl} + \rho dN^{sv} + \sigma_{lv} dA_{lv} + \sigma_{sl} dA_{sl} \\ + \sigma_{sv} dA_{sv} \end{aligned} \tag{1.5}$$

In deriving this result, density ρ is set to be equal in the bulk and interfacial regions. Since the total number of moles in the system is unchanged, we must have

$$dN = dN_v + dN_l + dN_s + dN^{lv} + dN^{sl} + dN^{sv} = 0 \tag{1.6}$$

Substituting Eq. (1.6) in Eq. (1.5) yields,

$$dF = \sigma_{lv} dA_{lv} + \sigma_{sl} dA_{sl} + \sigma_{sv} dA_{sv} \tag{1.7}$$

A shift in the interface that increases the area of the solid-liquid interface would result in an equal reduction in the solid-vapor interface area. This requires

$$dA_{sv} = -dA_{sl} \tag{1.8}$$

If it is further assumed that the liquid-vapor interface is a part of a sphere (*spherical cap*, then it can be shown from geometric considerations that

$$dA_{lv} = dA_{sl} \cos\theta \tag{1.9}$$

Substituting Eqs. (1.8) and (1.9) into Eq. (1.7) yields,

$$\frac{\partial F}{\partial A_{sl}} = \sigma_{lv} \cos\theta + \sigma_{sl} - \sigma_{sv} \tag{1.10}$$

This relation provides the rate of change of total free energy of the system with changing A_{sl} for a constant drop volume and constant temperature. At equilibrium, F must be a minimum with respect to A_{sl}, that is, $\partial F/\partial A_{sl} = 0$. In addition, at equilibrium $\theta = \theta_e$. Hence Eq. (1.10), when rearranged, yields the Young's equation (Eq. 1.1).

This thermodynamic derivation of Young's equation ignores certain aspects of real systems. Young's equation is valid when there is no adsorbed liquid film on the solid surface, namely the precursor film. Therefore, for an experimental realization, one needs to follow a stringent cleaning technique. In addition, it is assumed in the

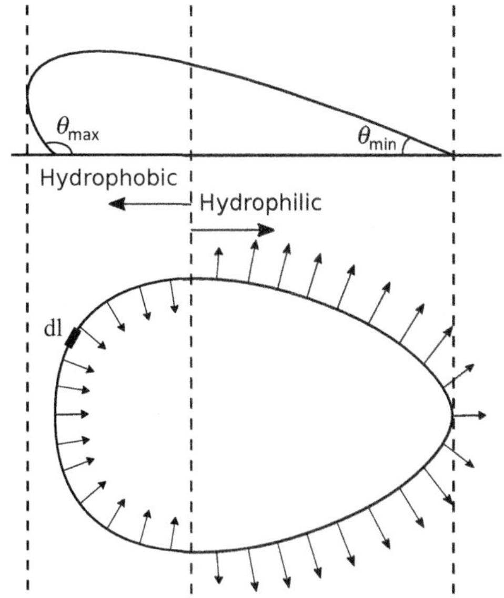

Fig. 1.4 Liquid-vapor interfacial force on a three-phase contact line

development presented above that the interfacial tensions are constant everywhere along their respective interfaces. However, arising from accumulated impurities or a higher evaporation rate, the interfacial tension near the contact line may be different from the value far from the contact line. In addition, the solid-vapor (or solid-gas) surface tension is not readily measurable. Hence, Young's equation cannot be experimentally validated. However, inferences based on the Young's equation have stood the test of time, and Eq. (1.1) is taken to be applicable for clean nonvolatile liquid drops under static conditions.

On a chemically and physically horizontal homogeneous surface, the equilibrium contact angle will be spatially uniform, at least over the scale of the footprint radius. The drop shape is then expected to be symmetric with respect to the vertical midplane. On an inclined surface or one with a wettability gradient, the drop shape is unsymmetrical and the footprint is non-circular. For a three-dimensional drop, contact angle distribution along the contact line (coordinate l) can be derived from theory or from experiments and used to calculate the resultant tangential force on the liquid mass. Figure 1.4 is an example of a sessile drop placed on a surface with inclination or surface inhomogeneities. The infinitesimal force due to liquid-vapor interfacial tension has a component on the substrate plane given by $\sigma_{\mathrm{lv}}\mathrm{d}l \cos \theta$; a collective sum of these incremental forces will pull the drop towards the hydrophilic side. Under static conditions, gravity and pressure will provide restoring forces for balance.

A *goniometer* (Fig. 1.5) is used for measuring contact angle and drop shape on wetting as well as non-wetting surfaces using photography. The recorded image is analyzed using digital image processing tools. The central idea here is to estimate the contact angle by drawing a local tangent around the three-phase contact line.

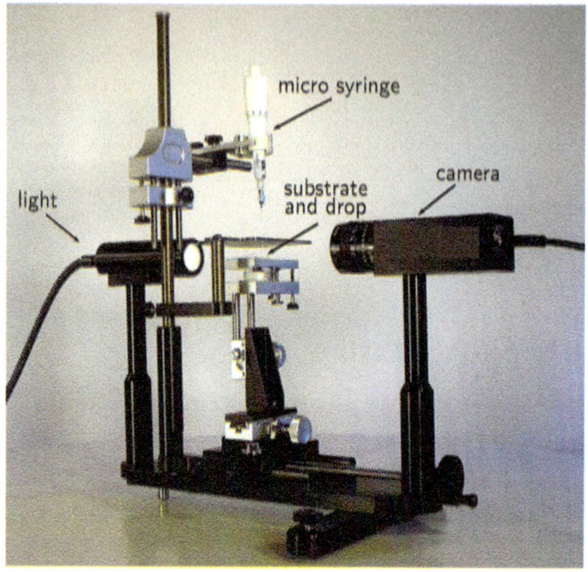

Fig. 1.5 Goniometer used for contact angle and drop shape analysis. (Credits: By Ramehart—Own work, CC BY 3.0, https://commons.wikimedia.org/w/index.php?curid=9508700)

However, there can be considerable uncertainty in such measurements due to the limitation of the tangent method. A more robust alternative approach for measurement of contact angle consistent with the overall shape of the drop is discussed in the later part of this chapter.

Most naturally occurring surfaces are hydrophilic, particularly for common liquids such as water. An extreme example is mercury which is hydrophobic on practically all surfaces. The present discussion relates to water (by default) and many of the observations may carry over to aqueous solution and possibly glycerin.

The process of making a naturally occurring surface or hydrophilic or hydrophobic by design is referred to as *texturing*. Creating a hydrophobic surface may involve depositing a low surface energy chemical layer or coating over the surface that has low affinity for the liquid and the process is called *chemical texturing*. Patterning the surface with nano- or hierarchical pillars is an example of *physical texturing*. In both instances, the liquid drop can be made to acquire a high enough contact angle under equilibrium conditions. Similar procedures are available for creating a hydrophilic surface. Apart from surface engineering, contact angle also depends on the thermophysical properties of the liquid phase and those of the fluid medium constituting the environment. The above discussion is applicable under nominally room temperature conditions for nonvolatile liquids. Higher temperatures or gradients will introduce additional phenomena such as evaporation, buoyancy, and Marangoni convection, thus leading to non-equilibrium conditions and variations in the contact angle. On the other hand, the equilibrium contact angle is practically independent of the drop volume, provided one is in the surface tension-dominated regime. This is equivalent to the statement that the Bond number that scales gravity with surface tension is small (Bo < 1).

In the presence of external body forces including gravity and an electric field, the shape of the drop may be altered and so will the contact angle. External impurities, either in the liquid medium or the surface, alter surface tension and will have an effect on the contact angle as well. Thus, the data on equilibrium contact angle is presented under clearly defined reference conditions in terms of the fluid phases and the surface characteristics of the bounding substrate.

1.3 Surface Texturing

Texturing refers to altering the wetting properties of a natural surface, making it hydrophilic or hydrophobic. In the present discussion on droplets, hydrophobicity is the main property of interest.

Synthetic substrates with surface energy gradients create contact angle gradients and can generate selective movement of liquid drops in microfluidic devices, even in the absence of plate inclination. Mobilizing drops in a controlled manner in dropwise condensation is another application. Controlling surface texture with a view towards realizing a given equilibrium contact angle is discussed here.

The wetting behavior on any flat surface is fixed by its chemical composition. Wettability can be altered by treating it chemically by grafting or adsorbing molecules with wetting characteristics of their own. Molecules that encourage hydrophobicity are often long chain organic monoloayers, octyl-decyltrichloro-silane ($C_{18}H_{37}C_{13}Si$), for example. However, physical texturing can also be used for the modification of contact angle. Physical texturing can create surface roughness and render the surface hydrophilic. On the other hand, patterned surfaces at the nanoscale have been shown to make the surface hydrophobic. A class of surfaces with mildly disordered or strongly disordered roughness distribution, and even fractal surfaces, have been shown to be hydrophobic. A combination of chemical and physical texturing can also be adopted in practice. As a rule, chemically textured surfaces tend to age rapidly due to surface leaching while physically patterned surfaces are comparatively long lasting and industrially relevant. The latter is also expected to be expensive in comparison to chemical coatings that can be deposited using simple chemical vapor deposition (CVD) techniques.

1.3.1 Cassie-Baxter Model

Wettability of a surface can be chemically altered by grafting molecules with desired wetting characteristics. Let a chemically heterogeneous surface be made up of two species (Fig. 1.6), each characterized by its own equilibrium contact angle—θ_1 and θ_2, respectively. Let the fractional areas occupied by each of these species be ϕ_1 and

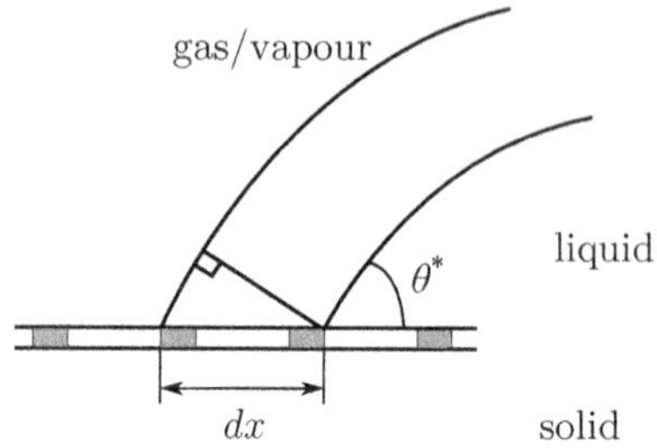

Fig. 1.6 Edge of a drop placed on a chemically textured surface

ϕ_2, respectively, while $\phi_1 + \phi_2 = 1$. In the present discussion, the individual patches are assumed to be small compared to the footprint diameter of the drop.

The energy variation, dF, associated with small displacement dx is

$$\mathrm{d}F = (\sigma_{\mathrm{sl}} - \sigma_{\mathrm{sv}})_1 \phi_1 \, \mathrm{d}x + (\sigma_{\mathrm{sl}} - \sigma_{\mathrm{sv}})_2 \phi_2 \, \mathrm{d}x + \sigma_{\mathrm{lv}} \, \mathrm{d}x \cos\theta^* \tag{1.11}$$

where θ^* denotes the effective (apparent) contact angle that one would see on the macroscale. The minimum energy principle, together with Young's relation applied to each area, leads to the Cassie-Baxter relation (Carey 2008)

$$\cos\theta^* = \phi_1 \cos\theta_1 + \phi_2 \cos\theta_2 \tag{1.12}$$

Therefore, the effective or Cassie-Baxter contact angle θ^* is restricted to the interval [θ_1, θ_2] and is given by an average involving the contact angles characteristic of each constituent, while the average is applied to the cosines of these angles.

The effects of chemical texturing on contact angle can be seen during surface cleaning. Alcohol, which is commonly used for surface cleaning, sticks to the surface resulting in a reduced contact angle; vacuum drying helps remove the traces of alcohol (Bhutani et al. 2013).

1.3.2 *Wenzel's Model*

Physical texturing is more robust than adsorbed chemical layers, which have a tendency to leach away under shear stresses generated by fluid motion. One of the first attempts at understanding the influence of roughness on wetting was due to Wenzel (Carey 2008). The effective contact angle θ^* on a rough, but chemically homogeneous surface, will be estimated in this section when the local contact angle is given by the Young's equation. The roughness scale is taken to be small as compared to the footprint diameter of the drop.

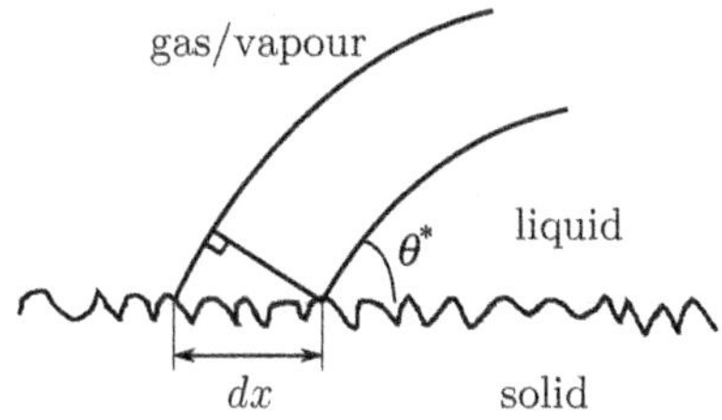

Fig. 1.7 Wenzel's contact angle, θ^*, on a physically rough surface

Similar to the discussion in Sect. 1.3.1, the effective contact angle θ^* can be evaluated by considering a small displacement dx of the line of contact in a direction parallel to the surface, as shown in Fig. 1.7. Under these circumstances, the surface energy changes by an amount dF, which can be written per unit length of the line of contact as

$$\mathrm{d}F = r(\sigma_{\mathrm{sl}} - \sigma_{\mathrm{sv}})\mathrm{d}x + \sigma_{\mathrm{lv}}\,\mathrm{d}x\cos\theta^* \tag{1.13}$$

where r is the roughness measure (such as rms roughness) of the solid surface. At equilibrium, the requirement of minimum surface energy leads to the expression

$$\frac{\partial F}{\partial x} = 0$$

For a smooth surface, $r = 1$ and the Young's equation is recovered. In contrast, for $r > 1$, the equilibrium condition leads to the Wenzel's relation

$$\cos\theta^* = r\cos\theta_{\mathrm{Y}} \tag{1.14}$$

where θ_{Y} represents the Young's equilibrium contact angle.

Wenzel's relation embodies two types of behavior

1. If $\theta_{\mathrm{Y}} < 90°$ (hydrophilic), then $\theta^* < \theta_{\mathrm{Y}}$ (more hydrophilic) since $r > 1$.
2. Similarly if $\theta_{\mathrm{Y}} > 90°$, then $\theta^* > \theta_{\mathrm{Y}}$ (more hydrophobic).

Surface roughness, therefore, magnifies the underlying wetting properties. Since the roughness r is not bounded, there should exist a threshold value r^* beyond which wetting becomes either total (or zero) depending on the sign of $\cos\theta_{\mathrm{Y}}$; this limiting value is

$$r^* = \frac{1}{\cos\theta_{\mathrm{Y}}} \tag{1.15}$$

1.3.3 Superhydrophobicity

When the equilibrium contact angle of a liquid on a solid becomes very high, for example, greater than ~150°, the surface is termed as superhydrophobic. Both chemically and physically textured surfaces may display superhydrophobicity. Lotus leaf is an example of a naturally occurring superhydrophobic surface relative to water drops. It arises from physical patterning that realizes hierarchical roughness patterns over the surface.

For hydrophobic solids ($\theta > \pi/2$), the dry solid has a surface energy lower than the wet one, i.e., $\sigma_{sv} < \sigma_{sl}$, as seen from Young's equation. Thus, the net surface energy can be lowered (and the effective contact angle maximized) if air is trapped below the drop. The drop, in such a case, rests jointly on a mix of air and solid, with the effective contact angle calculated using Cassie-Baxter theory; hence, $\theta^* \in [\pi/2, \pi]$. Such air pockets are observed in plant leaves, which have a roughness of the order of 25–35 μm. Applying the Cassie-Baxter theory leads to

$$\begin{aligned} \cos\theta^* &= \phi\cos\theta + (1-\phi)\cos\pi \\ &= -1 + \phi(\cos\theta + 1) \end{aligned} \tag{1.16}$$

where ϕ is the fraction of the drop footprint that is in contact with the solid surface. The above equation implies that total drying cannot be achieved by texturing a surface; $\theta^* = \pi$ implies either $\phi = 0$, which is not physically feasible, or $\theta = \pi$, in which case it is not necessary to modify the surface to make it hydrophobic.

Thus, it is evident from Eq. (1.16) that a hydrophilic solid surface can be rendered superhydrophobic provided that the physical pattern allows trapping of air pockets below the drop. Experiments have shown that ϕ values could be as large as 20% resulting in superhydrophobicity. Although the textured surface has been treated as heterogeneous in the discussion until now, it can also be modeled as a rough surface resulting in a Wenzel description. The two effects—surface heterogeneity and roughness, compete and the drop *chooses* the state which is thermodynamically more stable. The Wenzel state is preferred if

$$\cos\theta_Y > \frac{\phi - 1}{r - \phi} \tag{1.17}$$

and the Cassie-Baxter state, otherwise. This relation can easily be obtained by comparing the free energy change (per unit displacement, i.e., dF/dx) associated with the two states in Eqs. (1.11) and (1.13), for a given intermediate effective contact angle θ^*.

The following section discusses transitions between the Cassie-Baxter and Wenzel states in some more detail.

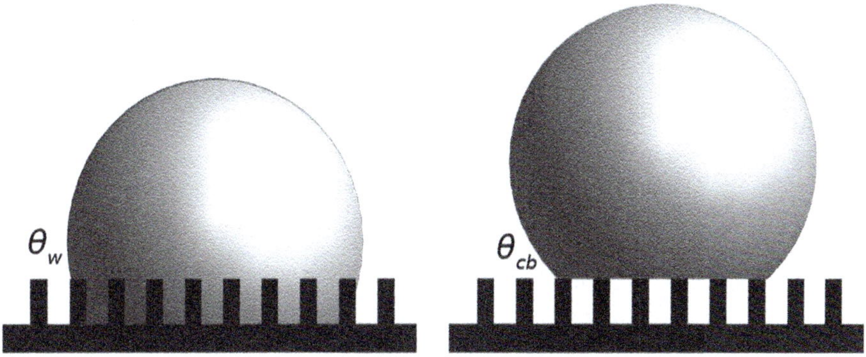

Fig. 1.8 Schematic that shows a drop making Wenzel (left) and Cassie-Baxter (right) contact, respectively, with a rough surface

Fig. 1.9 This schematic shows the energy barrier between a metastable Cassie-Baxter state and a more stable Wenzel state

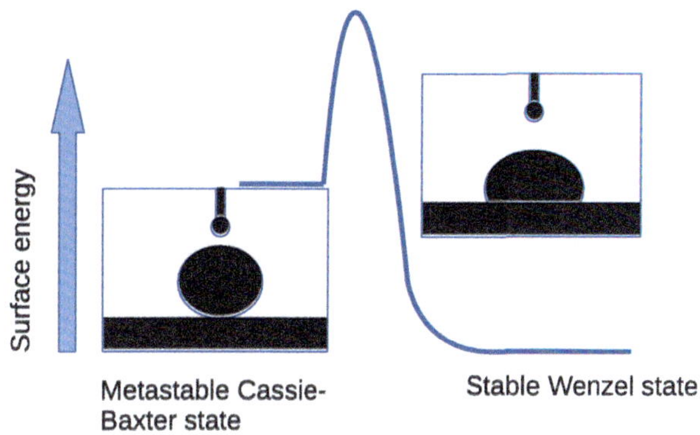

1.3.4 *Wetting Transitions*

A droplet can sit on a solid surface in two distinct configurations or states, as shown in Fig. 1.8. It is said to be in Wenzel state when it conforms with the topography; Wenzel's equation, Eq. (1.14), can be used to compute the apparent contact angle in this case. On the other hand, the drop can sit on the air pockets formed by the protrusions, touching only the spikes on the solid surface, which is referred to as the Cassie-Baxter (or Fakir) state.

As shown in the previous section, the preference for one state over the other is governed by thermodynamic stability, Eq. (1.17). This equation includes θ_Y, r, and ϕ. However, both states—Cassie-Baxter and Wenzel can be stable; these correspond to local and global energy minima and referred to as the metastable (CB) and stable (W) states, respectively. One such situation in depicted in Fig. 1.9, clearly showing the energy barrier preventing a spontaneous transition from the metastable state (CB) to the stable (W).

1.4 Contact Angle Hysteresis

Real surfaces, which are often defaced by defects that are either *chemical* (stains and blotches) or *physical* (surface irregularities), show a contact angle different from the Young's contact angle θ_Y, as described in the previous sections. The modified contact angle is known as the effective contact angle θ^*, which can correspond to the Wenzel or the Cassie-Baxter states depending on thermodynamic (mechanical, thermal, and chemical) equilibrium. In addition to modifying the contact angle, texturing also leads to non-uniqueness, in the sense of distributing it over the contact perimeter.

The equilibrium contact angle is an idealization realized on a perfectly horizontal surface, chemically and physically homogeneous, free of vibrations and air currents as well as thermal inhomogeneities when the drop is placed over it at an infinitesimally small speed. In an experiment, the drop is introduced over the surface by a syringe and initially the liquid region has a velocity distribution associated with it. The equilibrium shape is then acquired by the drop after the passage of a considerably long time when the velocities have become sufficiently small. For such a drop, its shape is symmetric about the vertical coordinate and the contact angle measured around the periphery of the three-phase contact line is practically constant, except for minor variations related to surface and ambient non-uniformities, if present. Under a variety of external perturbations, the drop shape may become unsymmetrical, the contact angle distribution is non-uniform, and a unique equilibrium contact angle is no longer useful. Specific instances where such a possibility exists are outlined:

1. A thin sheet that is partially wetting is immersed in the liquid and is pulled upward at uniform speed in its own plane. The liquid film attached to the sheet drains below and the contact angle seen at the meniscus is less than the equilibrium value. This angle is often referred as the *receding* contact angle. In contrast, a similar sheet moving into a liquid bath will see a build-up of liquid mass at the three-phase contact line and a higher contact angle, called the *advancing* contact angle. The difference between the advancing and receding angles is called the *contact angle hysteresis* (CAH). At low enough speed, it is a property shared by the surface, the liquid, and the environment, analogous to the equilibrium contact angle itself.
2. Hysteresis is also to be seen when the surface carrying a static drop is slowly inclined. In the limiting case of the drop just sliding over the surface, the drop deformation is the highest. The contact angle on the front side of the drop tending to lead its motion is the advancing angle, the one on the rear side is the receding angle, and the difference is usually termed as the contact angle hysteresis. The contact angle around the three-phase contact line will vary from a maximum of the advancing angle to a minimum of the receding angle.
3. If we inflate a drop by adding liquid to it (Fig 1.10a), the contact angle θ can exceed θ_Y without the line of contact moving at all. Eventually, θ reaches a threshold value θ_A beyond which the line of contact finally does move; θ_A is

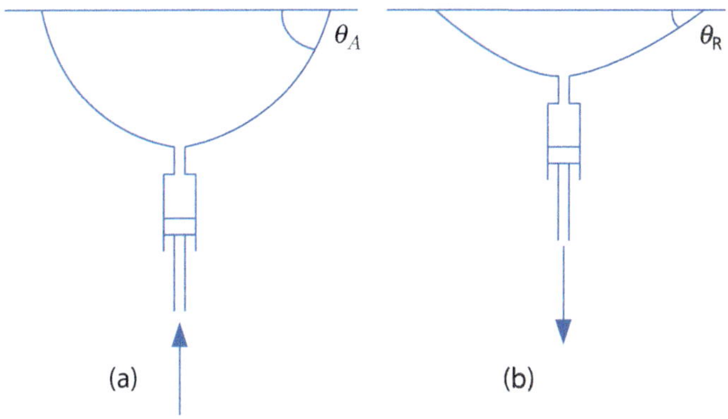

Fig. 1.10 (**a**) Advancing angle when the drop is inflated and (**b**) receding angle when the drop is deflated

referred to as the *advancing angle*. Likewise, when deflating a drop (Fig 1.10b), θ can decrease down to a limiting value θ_{R}, known as the receding angle; the line of contact shifts when θ tries to go below θ_{R}. Generally speaking, it can be said that observed contact angle (θ) depends on the nature of surface "preparation" or its roughness. Contact angle hysteresis is quantified by the difference between the limiting angles θ_{A} and θ_{R}, i.e., CAH $= \theta_{\mathrm{A}} - \theta_{\mathrm{R}}$.

Hysteresis is also noticed when the substrate on which a pendant or sessile drop is deposited is inclined. For a horizontal and (macroscopically[1]) homogeneous solid substrate, a three-dimensional drop forms a circular footprint on the base, with its bounding circle being the three-phase contact line. As the substrate is inclined, a change in the direction of body force, i.e., gravity in the present case, results in a net tangential[2] force on the drop, which creates a shift in the liquid mass, and a tendency for the drop to slide. However, up to a certain plate inclination, CAH allows a change in contact angle along the contact line, which continues to stay circular, keeping the drop in equilibrium. When the contact line, despite external influences, is fixed, the drop is said to be *pinned*.

For an inclined substrate, contact angle is a function of the (circular) contact line, with the contact angle varying between a maximum and a minimum value, denoted by $\theta_{\max}$ and $\theta_{\min}$, respectively. As long as $\theta_{\max} < \theta_{\mathrm{A}}$ and $\theta_{\min} > \theta_{\mathrm{R}}$ the drop stays pinned, i.e., $\theta \in [\theta_{\min}, \theta_{\max}] \subset (\theta_{\mathrm{R}}, \theta_{\mathrm{A}})$. If $\theta_{\max}$ becomes equal to θ_{A} or $\theta_{\min}$ becomes equal to θ_{R}, but not both, the contact line cannot stay circular anymore and moves until a new equilibrium shape is achieved. The drop is now in a state of

[1]Macroscopic homogeneity is different from microscopic homogeneity in the present context. Macroscopic homogeneity simply means that the effective contact angle—Cassie-Baxter/Wenzel—is the same throughout the surface and, therefore, the contact line is circular. Microscopic homogeneity may result in a Young's contact angle throughout.

[2]Tangent and normal directions are defined with respect to the substrate here.

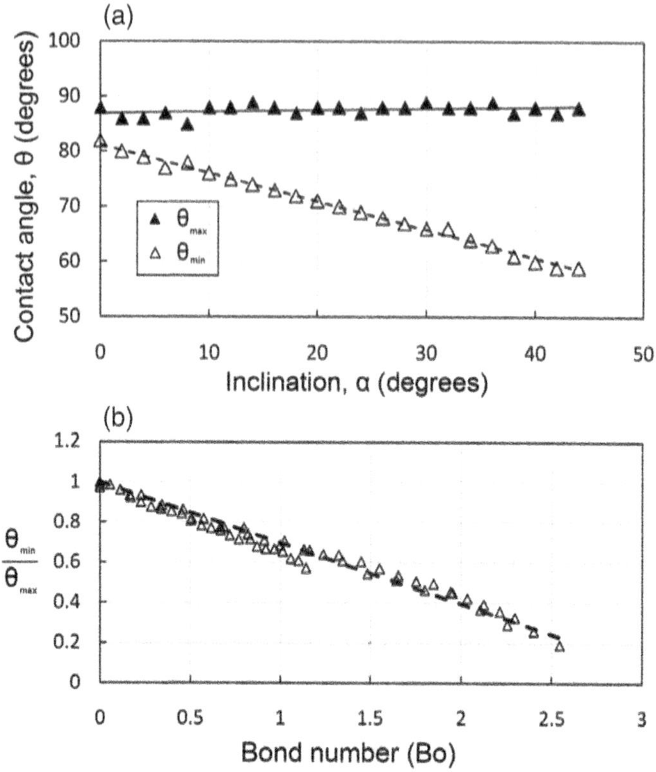

Fig. 1.11 (**a**) Contact angle data with substrate inclination for a 5 μl pendant glycerin drop on aluminum substrate; (**b**) Parameterized contact angle with Bond number glycerin drops of varying volumes and surface inclinations. (Figure adapted from Bhutani et al. 2013)

quasi-equilibrium, with only one point on the contact line pinned. When the plate inclination reaches a limiting value such that the limiting hysteresis (CAH) is achieved, i.e., when $\theta_{\max} = \theta_A$ and $\theta_{\min} = \theta_R$, the drop becomes unstable and flows down the incline.

Contact angle hysteresis can be used to characterize pinning; small CAH values will not allow much variation in the contact angle and the drop will move at small plate inclinations. Superhydrophobic surfaces with contact angles in excess of 150° typically have CAH $< 5°$, which is what leads to beads of water rolling down readily on a lotus leaf.

When a syringe is used to deposit pendant or sessile drops on horizontal surfaces, the measured contact angle, with the syringe in position, is the advancing contact angle. When the syringe is detached, the drop will equilibrate and the contact angle will approach the equilibrium value. As soon as the plate in inclined, the advancing end in the direction of inclination starts to move along with the other parts of the contact line, except for the receding point, which typically remains pinned (Bhutani

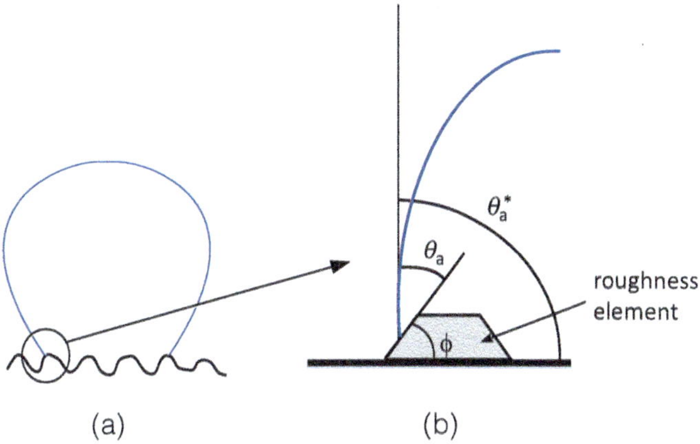

Fig. 1.12 Effect of surface roughness on drop shape close to the substrate. (**a**) Drop resting on a rough surface; (**b**) magnified view of contact line depicting local contact angle θ_a and the *apparent contact angle* $\theta_a^*(=\phi+\theta_a)$

et al. 2013). This results in constant θ_{max} values and a linearly varying θ_{min} as a function of increasing plate inclination, see Fig. 1.11a. Larger volume drops show a higher sensitivity in the change of θ_{min} with inclination angle. Brown et al. (1980) and ElSherbini and Jacobi (2004a, b) have discussed the effect of inclination on CAH, with Bhutani et al. (2013) particularly focusing on inclined pendant drops. The following general trends can now be summarized. The contact angle data for various plate inclinations and drop volumes collapses onto a single line, as shown in Fig. 1.11b. Here, θ_{min} (normalized with the advancing angle) is plotted as a function of the *Bond number* (Annapragada et al. 2012; Bhutani et al. 2013).

Real surfaces are prone to imperfections in composition and smoothness, resulting from manufacturing and related processing operations, making the surface heterogeneous on a small length scale. In addition, ultra-clean surfaces are highly reactive and stray molecules may get adsorbed on the substrate, modifying the interfacial properties. For a liquid drop sitting on (or under) a surface, the following factors contribute to hysteresis: surface heterogeneity, surface roughness, and adsorbed impurities.

Understanding the effects of asperities—engineered and natural, on contact line pinning, is important to understand CAH (Kalinin et al. 2009). The effect of surface roughness on hysteresis can be explained through a schematic drawing, as in Fig. 1.12. Drop contact line may encounter a local surface asperity inclined at an angle ϕ to the surface normal. The local drop curvature changes to accommodate the asperity. Although the local contact angle between the liquid drop and the solid surface stays equal to the equilibrium value, the *apparent contact angle* measured by an observer will appear as $\theta_Y + \phi$ since the angles are measured with respect to the average plane defining the surface. This is the origin of hysteresis of contact angle on rough surfaces.

1.5 Determination of Equilibrium Shapes

Knowledge of shapes of drops is useful in condensation applications for the determination of their growth rates and wall heat fluxes. Drop shape analysis is also essential in other microfluidic applications. When an inverse approach is adopted for a robust determination of contact angle from experimental images, computation of the equilibrium drop shape is an essential intermediate step. This section discusses the mechanistic and variational approaches to drop shape determination. Pendant and sessile drops for horizontal and inclined substrates are analyzed. An inverse approach, which helps in the measurement of equilibrium contact angles, is subsequently discussed.

To understand the derivation of the equilibrium shapes of static drops using principles of statics, an expression for the jump in traction across a liquid-gas interface needs to be developed.

The traction vector at a point P is defined $t_i^{(\widehat{\mathbf{n}})} = \mathrm{d}\widetilde{f}_i/\mathrm{d}\Gamma$, where the derivative is evaluated on a hypothetical plane cutting through P with a unit normal vector $\widehat{\mathbf{n}}$, as shown in Fig. 1.13. Vector $\mathrm{d}\widetilde{f}_i$ represents the force acting on an infinitesimal area $\mathrm{d}\Gamma$ around P. With the symmetric Cauchy stress tensor Σ_{ij} defined as $\Sigma_{ij} = t_j^{(\widehat{e_i})}$, the traction vector on an arbitrary plane can be written as $t_i^{(\widehat{\mathbf{n}})} = \Sigma_{ji} n_j$. Here, the Einstein index notation is used for compact representation of the equations.

For a homogeneous fluid, the tractions are equal and opposite at a point for any directions $\widehat{\mathbf{n}}$ and $-\widehat{\mathbf{n}}$, i.e., $t_i^{(\widehat{\mathbf{n}})} = -t_i^{(-\widehat{\mathbf{n}})}$. However, for an interface an unbalanced tangential force arises from chemical disparity between the phases involved and cannot be neglected. In the present context, the tangential force is that due to surface tension, given as $\sigma_{12}\Delta s\widehat{f}$, where $\widehat{f}$ represents a unit vector in the tangential direction, as shown in Fig. 1.14, $\Delta s = \Delta x$ or Δy depending on the interface boundary under

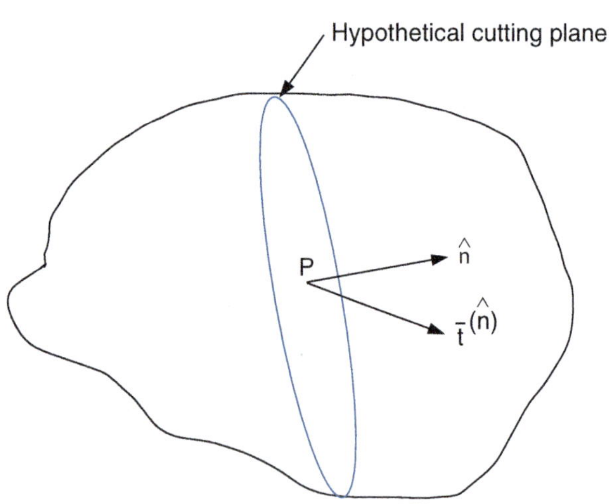

Fig. 1.13 Traction vector on a plane cutting through point P

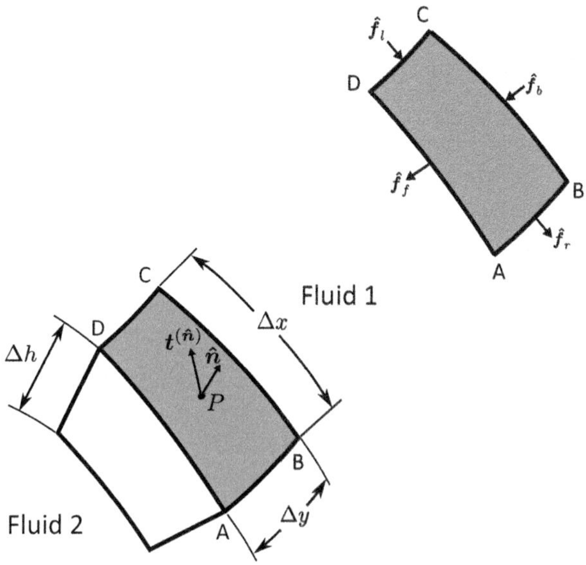

Fig. 1.14 A three-dimensional infinitesimal drop element of thickness Δh with the interface *ABCD* separating two fluids—1 and 2. Here, $\widehat{\mathbf{n}}$ is a normal from Fluid 2 to Fluid 1, with traction vector $\widehat{\mathbf{t}}^{(\widehat{\mathbf{n}})}$ drawn at point *P* on the interface. Further, $\widehat{f}$ vectors denote tangential unit vectors drawn over the interface

consideration, and σ_{12} is the interfacial tension between fluids 1 and 2; for simplicity, it is subsequently represented as σ and is a property shared by fluids 1 and 2. Figure 1.14 shows an element of thickiness Δh centered around an interface between fluids 1 and 2 with the unit normal $\widehat{\mathbf{n}}$ pointing from Fluid 2 to Fluid 1.

Force equilibrium on the scale of the drop translates to one applicable for an infinitesimal interface of area $\Delta\Gamma$ ($=\Delta x \Delta y$) and is

$$t_i^{(\widehat{\mathbf{n}})(1)} \Delta\Gamma + t_i^{(-\widehat{\mathbf{n}})(2)} \Delta\Gamma + \int_{\Delta l} \sigma f_i \mathrm{d}l = 0 \tag{1.18}$$

Here, superscripts (1) and (2) denote Fluid 1 and Fluid 2, respectively, Δl is the boundary arc length of the interface, and f_i is the indicial representation of unit tangential vector $\widehat{f}$. For the interface shown in Fig. 1.14, the equilibrium equation simplifies to

$$\left(\Sigma_{ji}^{(1)} - \Sigma_{ji}^{(2)}\right) n_j \Delta x \Delta y + \sigma_\mathrm{r} \Delta y f_{\mathrm{r},i} - \sigma_\mathrm{l} \Delta y f_{\mathrm{l},i} + \sigma_\mathrm{f} \Delta x f_{\mathrm{f},i} - \sigma_\mathrm{b} \Delta x f_{\mathrm{b},i} = 0 \tag{1.19}$$

Subscripts r, l, f, and b represent the boundaries on the right, left, front, and back, respectively. Rearrangement of interfacial tension terms results in

$$\left(\Sigma_{ji}^{(1)} - \Sigma_{ji}^{(2)}\right) n_j = -\left(\frac{\sigma_\mathrm{r} f_{\mathrm{r},i} - \sigma_\mathrm{l} f_{\mathrm{l},i}}{\Delta x}\right) - \left(\frac{\sigma_\mathrm{f} f_{\mathrm{f},i} - \sigma_\mathrm{b} f_{\mathrm{b},i}}{\Delta y}\right) \tag{1.20}$$

In the limit $\Delta\Gamma \to 0$,

$$\left(\Sigma_{ji}^{(1)} - \Sigma_{ji}^{(2)}\right)n_j = -\frac{\mathrm{d}(\sigma f_i)}{\mathrm{d}x} - \frac{\mathrm{d}(\sigma f_i)}{\mathrm{d}y} \\ = -\sigma\frac{\mathrm{d}f_i}{\mathrm{d}x} - \frac{\mathrm{d}\sigma}{\mathrm{d}x}f_i - \sigma\frac{\mathrm{d}f_i}{\mathrm{d}y} - \frac{\mathrm{d}\sigma}{\mathrm{d}y}f_i \tag{1.21}$$

The terms due to the change in interfacial tension are collectively referred to as Marangoni traction and act in the tangential direction. However, there is no contribution from these terms to the traction jump if the interfacial tension is uniform (and constant) over the interface, as considered in the present discussion. From geometric considerations, it is easy to show that $\mathrm{d}f_i/\mathrm{d}x = -n_i\kappa_x$ and , where κ_x and κ_y are the local curvatures in x and y directions, respectively. Using these relations, Eq. (1.21) is simplified further as

$$\left(\Sigma_{ji}^{(1)} - \Sigma_{ji}^{(2)}\right)n_j = \sigma\left(\kappa_x + \kappa_y\right)n_i = \sigma(2\kappa_\mathrm{m})n_i \tag{1.22}$$

Here, the mean curvature is defined as $\kappa_\mathrm{m} \equiv (1/2)(\kappa_x + \kappa_y)$. The mean curvature, however, can be calculated using curvatures in any two orthogonal directions in a plane perpendicular to the normal $\hat{\mathbf{n}}$. In the absence of a tangential contribution from an inhomogeneous σ (namely, Marangoni effects), the traction discontinuity only has a component normal to the interface, as represented in Eq. (1.22).

The above derivation (Eq. 1.22) can also be presented using Stokes' theorem applied on a finite interface, instead of an infinitesimal interface, as pursued here.

For stationary fluids, Pascal's law states that the Cauchy stress tensor is isotropic, i.e., $\Sigma_{ij} = -p\delta_{ij}$, where p is hydrostatic pressure and δ_{ij} is the Kronecker delta symbol. Substituting this expression for Σ_{ij} in Eq. (1.22) yields

$$p^{(1)} - p^{(2)} = -\sigma(2\kappa_\mathrm{m}) \tag{1.23}$$

which is also known as Laplace's law; it links the pressure jump across an interface to the interfacial tension and curvature. If the curvature is positive, namely the unit normal is outward from Fluid 2 into Fluid 1, then, the right hand side of Laplace's law is negative and $p^{(1)} < p^{(2)}$. Thus, in a liquid drop sitting on a hydrophobic surface in air, the liquid side pressure is greater than the gas side pressure, to an extent that is determined by the surface curvature. Small drops have a high curvature and hence a large internal pressure relative to the atmospheric pressure. Increasing drop volume will diminish this pressure difference. The one-to-one correspondence between (high/low) interface curvature and (high/low) pressure difference holds under static conditions. However, it can be stretched to dynamic situations as well when fluid velocities are moderate to small. In this manner, drop visualization can become a tool for bulk pressure measurement.

The mean curvature can also be expressed as a reciprocal of the mean radius, i.e.,

$$\kappa_{\mathrm{m}} = \frac{1}{R_{\mathrm{m}}} = \frac{1}{2}\left(\frac{1}{R_x} + \frac{1}{R_y}\right) \tag{1.24}$$

Hydrostatic pressure variation is

$$p = \rho g_i x_i + c \tag{1.25}$$

where c is a constant. Using Eq. (1.25) in Eq. (1.23), results in the *Young-Laplace* equation

$$2\kappa_{\mathrm{m}} = -\frac{\Delta\rho g_i x_i}{\sigma} + C \tag{1.26}$$

where $\Delta\rho \equiv \rho_1 - \rho_2$. In addition, C is a constant with dimensions of inverse length, which cannot be prescribed in advance. Mostly, it should be found as a part of the solution. It can be related to the drop volume. This equation is solved in conjunction with a given equilibrium contact angle boundary condition specified at the contact line. For a low Bond number, gravitational effects can be neglected, resulting in a spherical drop, i.e., constant curvature, as seen from Eq. (1.26). For a liquid drop in a gaseous environment, $\Delta\rho$ is practically equal to ρ_{l}, the liquid density.

We introduce three approaches to solving the Young-Laplace equation: solution of second-order ODE using Runge-Kutta method (method 1), solving a parametric form of the Young-Laplace equation (method 2), and a variational (or energy minimization) approach using the *Surface Evolver* code (method 3). The three methods are discussed in the following sections for the solution of 2D horizontal pendant drops, 2D inclined drops, and 3D non-symmetric drops, respectively.

1.5.1 Method 1. Two-Dimensional Horizontal Pendant Drop

The determination of the shape of a liquid drop from the Young-Laplace equation is first demonstrated here. For definiteness, the drop is in a pendant configuration and taken to be two dimensional (like a cut cylinder) as shown in Fig. 1.15. The mean curvature in Young-Laplace equation can be represented in coordinate-free form as

$$2\kappa_{\mathrm{m}} = n_{i,i} \tag{1.27}$$

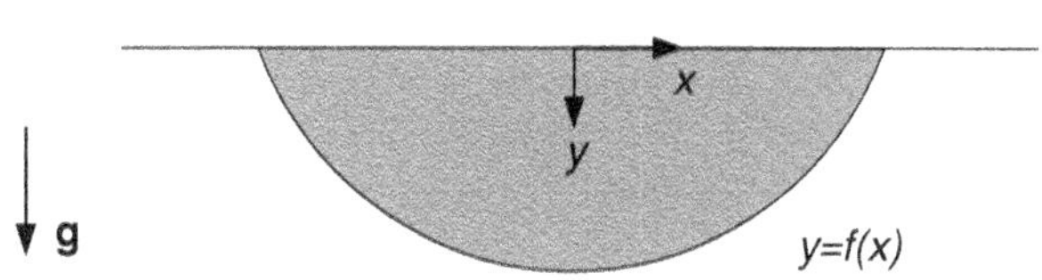

Fig. 1.15 Coordinate system for drop shape analysis

As per the index notation, the notation $n_{i,j}$ is the derivative of the ith component of n with respect to the jth coordinate. Thus, $n_{i,i}$ denotes the divergence of n. The mean curvature in two dimensions has been derived in terms of the drop shape function $y = f(x)$ in the literature (Pozrikidis 1997) and is given as

$$2\kappa_{\mathrm{m}} = -\left(\frac{f''}{\left(1+f'^2\right)^{3/2}}\right) \tag{1.28}$$

Substituting Eq. (1.28) in Eq. (1.26),

$$-\frac{f''}{\left(1+f'^2\right)^{3/2}} = -\frac{\Delta\rho g_i x_i}{\sigma} + C \tag{1.29}$$

which, for the coordinate system in Fig. 1.15 simplifies to

$$-\frac{f''}{\left(1+f'^2\right)^{3/2}} = -\frac{\Delta\rho g y}{\sigma} + C \tag{1.30}$$

where $\mathbf{g} = (0, g)$, and the coordinate system $\mathbf{x} = (x, y)$ has the y-coordinate pointing in the direction of gravity. While the present discussion is for a pendant drop, sessile drops can be similarly treated by writing the gravity vector as $\mathbf{g} = (0, -g)$. Returning to the pendant drop, the governing equation for the drop shape $y = f(x)$ can be rearranged to read

$$\frac{\mathrm{d}^2 y}{\mathrm{d}x^2} = -\left(1+\left(\frac{\mathrm{d}y}{\mathrm{d}x}\right)^2\right)^{\frac{3}{2}}\left(-\frac{\Delta\rho g y}{\sigma} + C\right) \tag{1.31}$$

Equation (1.31) is the governing equation for predicting the shape of a two-dimensional pendant drop. It is solved with boundary conditions as

$$y(x = R) = 0 \tag{1.32}$$

$$\frac{\mathrm{d}y}{\mathrm{d}x}(x = R) = -\tan\left(\theta_{\mathrm{e}}\right) \tag{1.33}$$

where R is the footprint radius of the drop and θ_{e} is the equilibrium contact angle.

Referring to Eqs. (1.31)–(1.33), we see that there are several unknown quantities. These must be prescribed to attain closure in the calculations. Experiments will provide data on the equilibrium contact angle (θ_{e}) for the chosen substrate as well as R, the footprint radius. The constant parameter C relates to internal pressure developed in the drop. It can be found with reference to the drop volume which should be a prescribed quantity in the calculation.

Since the drop on a horizontal surface is symmetric, the volume constraint in two dimensions can be applied as

$$V = 2\int_{0}^{y_{\max}} x\mathrm{d}y \tag{1.34}$$

The drop shape function $y = f(x)$ is now determined by solving the second-order ODE (Eq. 1.31) with the two initial conditions at $x = R$ by first assuming the value of the constant parameter C. The parameter is iterated upon till the volume constraint (Eq. 1.34) is satisfied. In this context, the computational steps can be presented in the form of the following algorithm.

Algorithm

1. Guess the value of C. A good estimate for an initial value of C is obtained by assuming the drop shape to be a section of a sphere of radius a whose volume is equal to the volume of drop. The contact angle of the drop is also equal to the contact angle of the spherical section. Hence, $C = 1/a$ is a good initial guess.

2. Integrate Eq. (1.31) as an Initial Value Problem (IVP) from $x = R$ to $x = 0$ with the initial conditions stated as Eqs. (1.32) and (1.33). The step size will be prescribed in the x-coordinate; for example, it can be 0.1–1% of the footprint radius R. A fourth-order Runge-Kutta (RK4) method can be used to solve the ordinary differential equation.

3. Check for the volume constraint Eq. (1.34). If it is not satisfied, we seek a correction in the parameter C. A Newton-Raphson method can be used to improve on the guessed value.

4. Look for convergence in the parameter C to within, for example, 0.1%. Else, return to Step 2 and continue iterations until the volume constraint is satisfied.

Figure 1.16 shows the plot of a (3D axisymmetric) horizontal pendant drop shape with a contact angle of 80° and 5 μl volume; simulated drop shape is compared to an experimental profile on an aluminum substrate (Bhutani 2007); for 3D drops, the volume integral calculation must be performed over circular disks. A Matlab code for 2D horizontal drops shape calculation is available at https://github.com/gbhutani/Drop-Dynamics-and-Dropwise-Condensation-on-Textured-Surfaces. Algorithms described later in this chapter are also placed here.

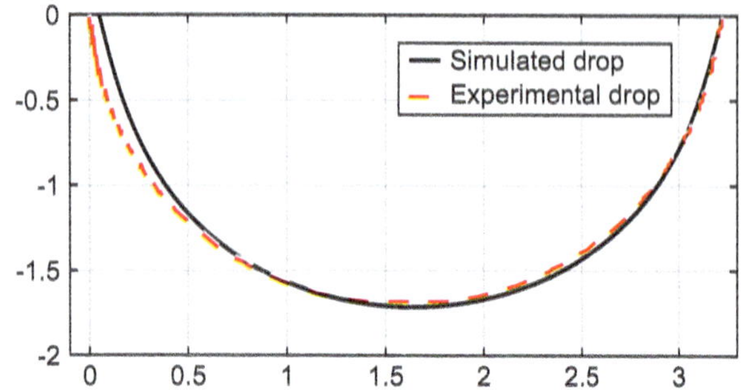

Fig. 1.16 Horizontal pendant drop shape of glycerine determined from the Young-Laplace equation and comparison with experiments; drop volume is 5 μl and equilibrium contact angle is 80°

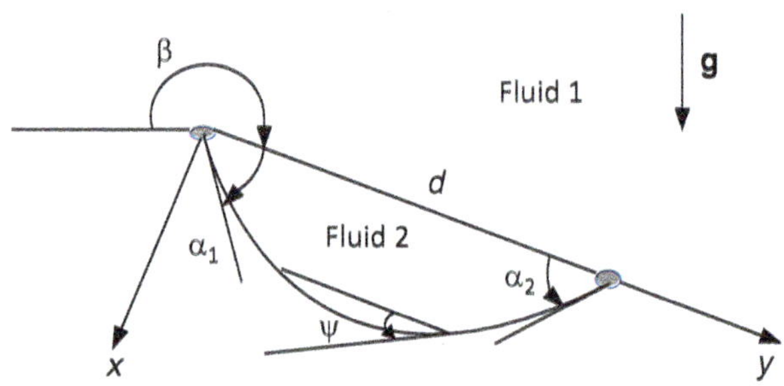

Fig. 1.17 Inclined pendant drop

1.5.2 *Method 2. Two-Dimensional Inclined Pendant Drop*

We proceed to finding the shape of the drop placed over an inclined surface. Figure 1.17 shows an inclined pendant drop, with a plate inclination angle of β. Using $\mathbf{g} = (-g\cos\beta, -g\sin\beta)$ in Eq. (1.26), the following equation for an inclined drop is obtained as

$$\frac{d^2y}{dx^2} = -\left(1 + \left(\frac{dy}{dx}\right)^2\right)^{\frac{3}{2}} \left(-\frac{\Delta\rho}{\sigma}(-gx\cos\beta - gy\sin\beta) + C\right) \tag{1.35}$$

The above-governing equation can be parameterized in terms of ψ, which is the angle of the tangent drawn over the drop surface varying between the contact angles $-\alpha_1$ and α_2 on rear and forward halves of the drop, respectively (Pozrikidis 1997). The parametric form of the governing equation is

$$\frac{dx}{d\psi} = \frac{\sin\psi}{Q} \tag{1.36a}$$

$$\frac{\mathrm{d}y}{\mathrm{d}\psi} = -\frac{\cos\psi}{Q} \tag{1.36b}$$

where

$$Q = -\frac{(x\cos\beta + y\sin\beta)\Delta\rho g}{\sigma} - C \tag{1.37}$$

Equations (1.36a) and (1.36b) are solved with boundary conditions

$$x(-\alpha_1) = 0 \tag{1.38a}$$

$$y(-\alpha_1) = 0 \tag{1.38b}$$

$$x(\alpha_2) = 0 \tag{1.38c}$$

and the volume constraint

$$V = \int_0^d x\mathrm{d}y \tag{1.39}$$

Algorithm

1. Solve for an equivalent horizontal pendant drop, as discussed in the previous section. This gives the drop width d and an initial guess for C.

2. Guess values for α_1 and α_2.

3. Integrate Eqs. (1.36a) and (1.36b) as IVP from $\psi = -\alpha_1$ to $\psi = \alpha_2$.

4. Check for the following three constraints: volume constraint, $x(\alpha_2) = 0$ and $y(\alpha_2) - d = 0$. Improve the value of C, α_1, and α_2 until the three constraints are satisfied.

Simulated drop shapes for inclined pendant drops with fixed advancing angle and pinned contact line are presented in Fig. 1.18. Also see Pozrikidis (1997).

Axisymmetric 3D drop shapes can also be predicted using a very similar parametric approach. Therefore, 2D and axisymmetric (near-spherical) drop shapes can be predicted by solving the Young-Laplace equation. For non-symmetric 3D cases (e.g., inclined pendant drops) the Young-Laplace equation can be derived as follows:

For a 3D drop surface given by the function $z = g(x, y)$,

$$2\kappa_{\mathrm{m}} = -\frac{\left(1 + g_y^2\right)g_{xx} - g_x g_y g_{xy} + \left(1 + g_x^2\right)g_{yy}}{\left(1 + g_x^2 + g_y^2\right)^{3/2}} \tag{1.40}$$

which has been derived using the geometric relation

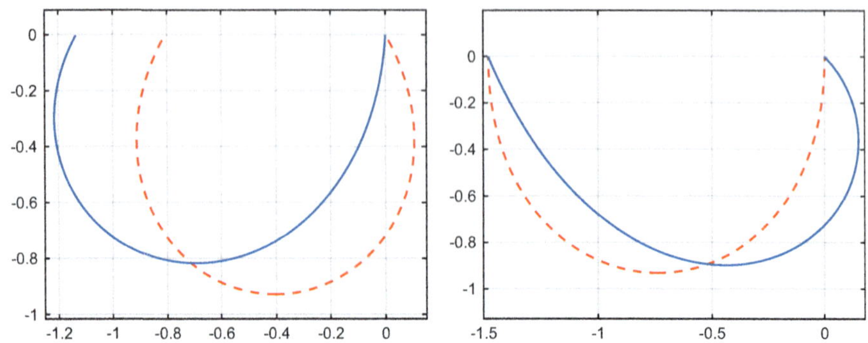

(a) Fixed advancing angle (b) Pinned drops

Fig. 1.18 Simulated inclined pendant drops. (**a**) Capillary length = 0.1 mm, drop volume (area in 2D) = 0.008 m^2, inclination angle from the positive x-axis (β) = 225°, equilibrium contact angle = 120°, Bo = 0.74. (**b**) Capillary length = 0.1, equilibrium contact angle = 90°, inclination angle (β) = 90°, Bo = 1.84. SI units used

$$2\kappa_{\mathrm{m}} = n_{i,i} \tag{1.41}$$

The normal direction can be found as $n_i = G_{,i}/\sqrt{G_{,j}G_{,j}}$ with $G = z - g(x, y)$.

This equation is difficult to solve iteratively. An alternative approach based on the variational principle is preferred, also known as the method 3 in this text. Santos and White (2011) were pioneers in the use of such an energy-minimization approach for predicting non-axisymmetric drop shapes of sessile drops. These calculations have been numerically carried out using a gradient descent method and is available in the form of an open source code called *Surface Evolver.* We have adopted this approach for pendant drops; see Bhutani et al. (2013). The details of the variational method are discussed next.

1.5.3 Variational Approach

An initial volume of fluid in the shape of a cube is placed on the substrate; the total energy of the system, which is a sum of interfacial and gravitational potential energy, is then minimized sequentially using a gradient descent method available in the open-source code *Surface Evolver.*

Surface Evolver is a code written in C programming language that can be used for predicting shapes of surfaces which depend on surface tension and other energies, subject to constraints. A surface in *Surface Evolver* is partitioned into triangles, also known as 2-simplices. Starting from a user-defined surface, Surface Evolver proceeds towards a minimal energy surface using a gradient descent method. The total energy in Surface Evolver is written as a sum of surface tension, gravitational potential energy, squared mean curvature, user-defined surface integrals, and knot

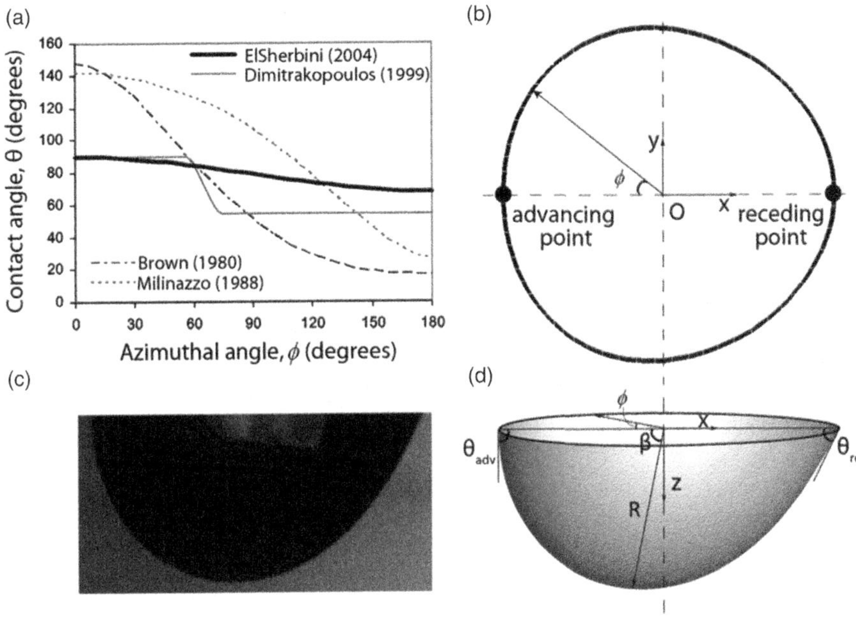

Fig. 1.19 Inclined pendant drop analysis. (**a**) Contact angle variation as a function of the azimuthal angle; (**b**) base contour of a 3D drop; (**c**) experimentally recorded image of a pendant drop; and (**d**) numerical simulation of a 3D pendant drop (polar coordinates R, β, ϕ, advancing angle θ_{adv}, receding angle θ_{rec}). (Figure adapted from Bhutani et al. 2013)

energies, as specified in Brakke (1992). In constraints, *Surface Evolver* can include arbitrary topology, volume constraints, boundary constraints, boundary contact angles, prescribed mean curvature, gravity, and constraints expressed as surface integrals.

The interfacial energy of the solid-liquid contact in a 3D pendant drop has to be specified as an input in the form of contact angle θ over the contact line as a function of the azimuthal angle ϕ. The choice of this function is based on empirical fitting functions proposed in the literature (Dimitrakopoulos and Higdon 1999; ElSherbini and Jacobi 2004a, b), which are shown in Fig. 1.19a. One such polynomial fitting function, proposed by ElSherbini and Jacobi (2004a, b), can be written as

$$\theta = 2\frac{\theta_{\max} - \theta_{\min}}{\pi^3}\phi^3 - 3\frac{\theta_{\max} - \theta_{\min}}{\pi^2}\phi^2 + \theta_{\max} \tag{1.42}$$

where $\theta_{\max}$ and $\theta_{\min}$ are the maximum and minimum contact angles, respectively, specified as inputs to the model. This fitting function has been used in the literature for predicting sessile Annapragada et al. 2012) and pendant (Bhutani et al. 2013) drop shapes. Since the three-phase contact line evolves with iterations, the drop base center O has to be recomputed for the correct calculation of the azimuthal angle. The final base contour for one such case is shown in Fig. 1.19b; a clear deviation from

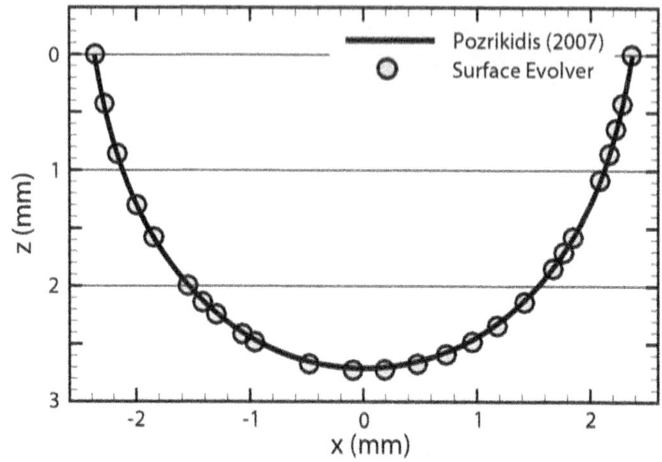

Fig. 1.20 Verification of the *Surface Evolver* code to predict three-dimensional pendant drop shapes. Three-dimensional shape of a 30 μl pendant glycerin drop with a contact angle of 80° simulated using the *Surface Evolver* code is compared with the solution of the axisymmetric Young-Laplace equation using the parametric form suggested by Pozrikidis (1997), as discussed in Sect. 1.5. (Figure adapted from Bhutani et al. 2013)

circular contact line can be noticed. A verification of the variational method for drop shape prediction is shown in Fig. 1.20: the results are compared with the solution of the Young-Laplace equation for an axisymmetric 3D drop. Algorithm for the use of *Surface Evolver* for the determination of 3D drop shapes is as follows:

1. Define a cube of liquid with an initial volume equal to the required drop volume V_D.
2. Specify the substrate surface inclination, volume constraint, and physical parameters.
3. Specify the interfacial energy on the solid-liquid contact plane $z = 0$ using the contact angle variation of Eq. (1.42), with θ_{max} and θ_{min}, as inputs.
4. Specify the gravitational potential energy of the liquid as a function of the plate inclination.
5. Use the gradient descent method in Surface Evolver to approach the new 3D drop shape.

The Surface Evolver code for the above algorithm is available at the link given in Sect. 1.5.1.

1.5.4 Inverse Method for Estimating Contact Angles

Standard contact angle analyzer softwares draw a tangent at the three-phase contact point on the 2D drop image and back calculate the contact angle. This approach can result in significant error as the computed slope is sensitive to the choice of the neighboring pixels. In addition, convergence in the contact angle with increasing image magnification is not guaranteed since newer features of the surface and the precursor film continuously get revealed. An inverse method, which minimizes error between a section of computed drop shape to the measured shape, can provide a robust estimate for equilibrium contact angle over a horizontal surface and advancing and receding angles over inclined surfaces (Bhutani et al. 2013). Moreover, the computed non-axisymmetric 3D drop also provides information about drop profile in other azimuthal planes along with the drop footprint, both of which are quantities of interest.

The numerical algorithm for the inverse technique proceeds as follows:

1. Guess θ_{max} and θ_{min}; a good starting guess can be the measured contact angles using tangent method.

2. *Forward problem*: Obtain the 3D drop shape using energy minimization method in *Surface Evolver* for the above contact angles, given plate inclination angle and fluid properties.

3. Extract the 2D drop shape from the vertical midplane.

4. *Inverse problem*: Calculate the RMS error between the experimental and numerical profiles. Improve the guess for θ_{max} and θ_{min} using an optimization technique. A clear global minimum can be seen in Fig. 1.21, which justifies the search of this minimum.

5. Repeat steps 2–4 to minimize the RMS error obtained in the previous step.

6. The optimum θ_{max} and θ_{min}, corresponding to the minimum RMS error, are reported as the maximum and minimum contact angles.

Figure 1.21 shows a contour plot of the error function with respect to θ_{max} and θ_{min}. Error in this plot is the ratio of the RMS error between the numerical and experimental drop profiles and drop diameter for the plate inclination. Drop diameter is defined as the maximum drop span along the direction of plate inclination. A global minimum exists, as clearly seen from the contours and forms the basis of the inverse method for measuring the contact angles. The error is most often below 1% and represents an acceptable match between experimental and numerical drop shapes.

The codes for the above implementation are included in the link of Sect. 1.5.1.

The contact angle measured from drop images using the tangent method often involves a significant level of uncertainty. The inverse technique overcomes this by creating 3D drop shapes from a numerical model and fitting 2D sections that are extracted from it to the experimental drop profile. The error associated with the fitting is small and has a very small contribution from the solid-liquid-vapor interface. Therefore, the inverse contact angles calculated by the authors are robust for predicting not only the drop shapes but also in providing better estimates of the contact angles compared to the angles estimated from the tangent method.

It is important to obtain the shape of the three-phase contact line along with the knowledge of contact angles. The droplet footprint changes as the plate is inclined. The drop contours can be extracted in the $z = 0$ plane from the 3D shapes of the non-axisymmetric pendant drops obtained using the variational method using *Surface Evolver*. The base contour is initially circular for a horizontal drop and gets

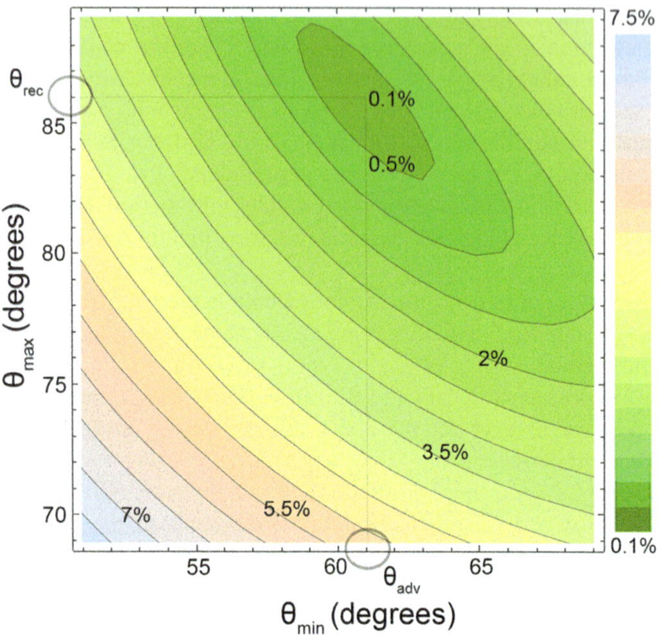

Fig. 1.21 Error contours for a 15 μl pendant glycerin drop on an aluminum substrate inclined at 16° with the horizontal; error has been presented as a ratio of the absolute difference in the drop shapes and its diameter. A clear error minimum is to be seen. (Figure adapted from Bhutani et al. 2013)

distorted as the substrate is inclined. Figure 1.22 presents the base contours of a 15 μl pendant glycerine drop on an aluminum substrate of 1.45 μm RMS roughness. Unlike previous approximations of the drop being pinned, the figure clearly shows that the contact line is not pinned everywhere but only at the receding point. As the inclination increases, the liquid mass starts shifting towards the advancing point and the resulting shapes reported in Fig. 1.22 are obtained. These contours for inclined pendant drops are qualitatively similar to the experimental images of ElSherbini and Jacobi (2004a, b) although the reference is for a sessile drop. Considering the experimental complexity in measuring the three-phase contact line for inclined pendant drops, numerical simulations offer a convenient alternative. The 3D numerical drop shapes predicted using *Surface Evolver* are shown in Fig. 1.23.

Figure 1.24 presents the evolution of overall drop shape with plate inclination. Experimental data is shown to be in good match with simulations of the drop shape on inclined surfaces. Simulation details related to the figure are available in the previous work of the authors Bhutani et al. (2013). The corresponding experimental drop images are shown in Fig. 1.25.

The maximum and minimum contact angles attained by a drop on an inclined surface just before the onset of sliding instability are often equal to the *advancing* and *receding* angles; their difference is the contact angle hysteresis. These can be

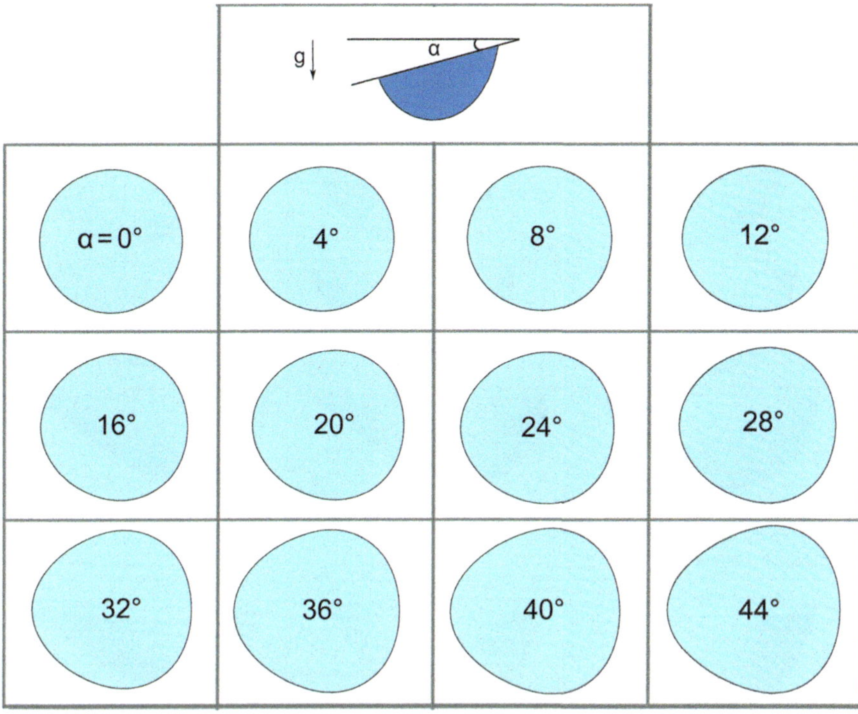

Fig. 1.22 Evolution of the shape of the contact line with plate inclination for a 15 μl pendant glycerin drop under an aluminum substrate. (Figure adapted from Bhutani et al. 2013)

found independently for sessile and pendant drops. These limiting angles are characteristic of the surface relative to the chosen liquid and the overall environment.

1.6 Dynamic Contact Angle and Contact Line Motion

A drop can acquire contact angles different from the equilibrium value when in a state of motion. These are *dynamic contact angles* and often seen to be a function of the velocity and the liquid properties. Specifically, they depend on the capillary number

$$\mathrm{Ca} = \frac{\mu U}{\sigma} \tag{1.43}$$

Here, μ is the liquid viscosity, σ is the coefficient of surface tension, and U is the contact line speed. In imaging experiments, contact line speed may be determined by examining the changes taking place in the drop footprint over a sequence of images.

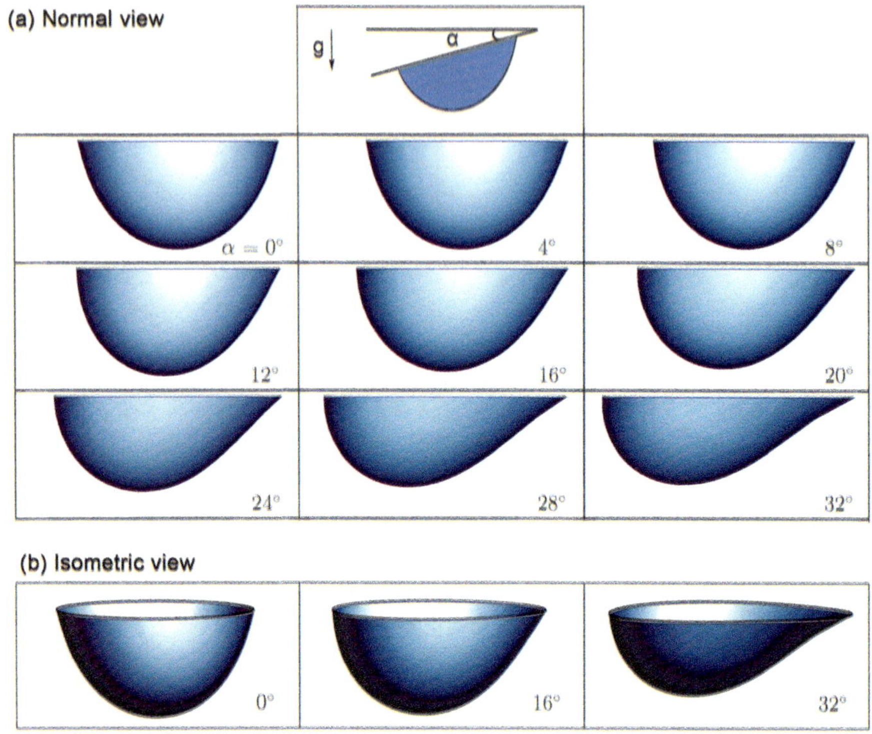

Fig. 1.23 (**a**) Numerical simulation of a 30 μl pendant glycerine drop with a uniform horizontal equilibrium contact angle of 30° on an aluminum substrate of 1.46 μm RMS roughness for various plate inclinations α and (**b**) isometric view of the simulated drops for three plate inclinations. (Figure adapted from Bhutani et al. 2013)

Dynamic contact angles (represented as $\langle\theta\rangle$ or θ_{d}) will also depend on the medium surrounding the droplet liquid, but the discussion here is for a gaseous environment.

Experiments show the dynamic contact angle (DCA) to be a function of equilibrium contact angle and the capillary number. Based on the experimental data recorded under various conditions of drop motion, empirical models connecting DCA to the capillary number are available. In addition, they have been predicted using hydrodynamic and molecular kinetic considerations. The first focuses on viscous dissipation near the contact line region. The molecular kinetic theory underscores the dependence of contact line velocity on the behavior of molecules in the contact line zone. Hybrid models have also been proposed recently (Zhu et al. 2016).

The dynamic contact angle is significant because it occurs when the no-slip condition is violated at the solid-liquid interface. The resulting stress singularity at the three-phase interface boundary should be accounted for in the mathematical

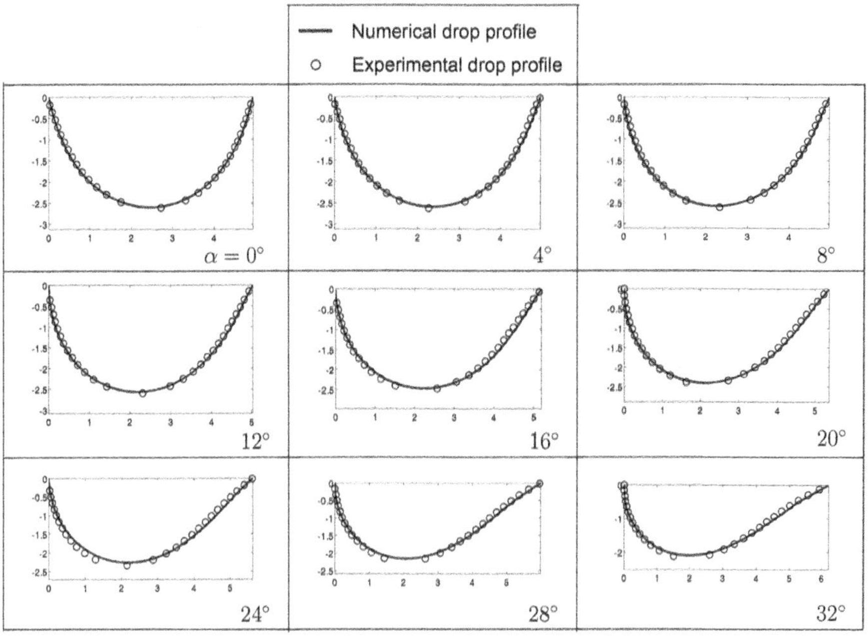

Fig. 1.24 Evolution of the contour of the drop shape with plate inclination for a 30 μl pendant glycerine drop with a uniform horizontal equilibrium contact angle of 30° under an aluminum substrate; numerical drop shape is compared to experiments. (Figure adapted from Bhutani et al. 2013)

modeling of fluid flow. The Navier slip model is an example of such a treatment wherein the slip velocity u_{sl} is prescribed as a boundary condition

$$u_{\mathrm{sl}} = l_{\mathrm{sl}} \frac{\partial u}{\partial n} \tag{1.44}$$

Here, l_{sl} is called a slip length. This model assumes the existence of a virtual no-slip point that is at a distance l_{sl} below the solid boundary with a linear velocity profile in the virtual zone. Slip lengths have been tabulated in the literature and may be of the order of a few microns in commonly prepared hydrophobic surfaces. A difference in the dynamic contact angle $\langle\theta\rangle$ and the equilibrium contact angle θ_{e} can be considered as a potential driving the contact line, as in Tanner's law,

$$u_{\mathrm{cl}} = A_{\mathrm{cl}}(\langle\theta\rangle - \theta_{\mathrm{e}})^{n} \tag{1.45}$$

where u_{cl} is the contact line speed, and A_{cl} and n are parameters that best fit the measurement data. Molecular dynamics simulations aid in the calculation of the dynamic contact angle $\langle\theta\rangle$, which represents an average of local microscopic contact angle values.

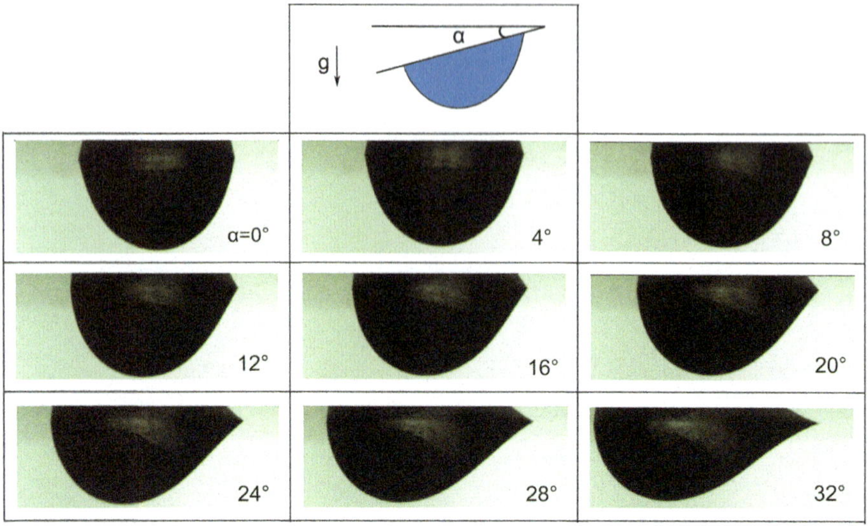

Fig. 1.25 Experimental drop images for a 30 μl pendant glycerine drop with a uniform horizontal equilibrium contact angle of 30° under an aluminum substrate. (Figure adapted from Bhutani et al. 2013)

1.7 Microscopic Modeling of Drop Shapes

Molecular dynamics (MD) presents a first-principles approach to microscopic modeling of interfaces (Razavi et al. 2014). The dynamics of contact line can be predicted using MD since it can capture the microscale mechanism responsible for the motion (Smith et al. 2018). This is not possible using the continuum approach, which accepts the dynamics predicted using the MD approach as an input. In addition, other "macroscale" quantities such as liquid-vapor interfaces, surface tension, and dynamic contact angles can also be predicted using molecular dynamics. Wall effects such as stacking and stick-slip, which are not possible to predict using continuum modeling, can also be obtained using MD models.

Continuum models in the bulk can be coupled to MD at the interface, with the overall behavior of the system evolving together. Such an approach has been described by Smith et al. (2018) who studied superspreading of a drop in the presence of surfactant. In molecular dynamics, Newton's second law of equation is solved for interacting molecules or a collection of molecules known as "beads," with periodicity boundary conditions. Use of these beads allows handling larger time and length scales in the problem, making the method computationally more tractable.

Smith et al. (2018) demonstrated the use of SAFT (Statistical Associating Fluid Theory) class of models in their MD simulations. A SAFT-γ model, using Mie potential (which belongs the Lennard-Jones type), was employed, resulting in a good agreement between MD predictions and experiments. The Mie potential between beads i and j is given as

$$\phi_{ij} = c\widetilde{\epsilon}_{ij}\left[\left(\frac{\widetilde{\sigma}_{ij}}{r_{ij}}\right)^{\lambda_{ij}^{r}} - \left(\frac{\widetilde{\sigma}_{ij}}{r_{ij}}\right)^{\lambda_{ij}^{a}}\right] \tag{1.46}$$

where r_{ij} is the distance between spheres, λ_{ij}^{a} represents dispersion interaction between beads i, and j, λ_{ij}^{r} considers core interaction between the beads and c is a function of λ_{ij}^{a} and λ_{ij}^{r} (Smith et al. 2018). The potential parameters $\widetilde{\sigma}_{ij}$ and $\widetilde{\epsilon}_{ij}$ have to be tuned to create an agreement of the thermophysical properties with experiments, in addition to a dependence on fluid and solid material. For instance, $\widetilde{\epsilon}$ can be tuned to the surface wettability in this model. Spherical beads consisting of two water molecules were used in the simulations by Smith et al. (2018). Wall effects in MD can be modeled in two ways—employing a Mie potential between fluid and solid beads or using a separate harmonic potential for wall adhesion.

The fluctuating molecular data from MD has to be averaged to convert it to thermophysical quantities such as viscosity, surface tension, and contact angles. The average values can then be used in continuum models, as in Eqs. (1.44) and (1.45) for prediction of slip length and dynamic contact angle, respectively.

The liquid-vapor interface is defined as a layer of certain thickness, such as $1.5\widetilde{\sigma}_{ij}$, around the liquid surface cluster. The interfacial surface can be constructed by fitting a function through the centers of molecules in the interfacial region; this fitting function can be represented as a Fourier series or sum of polynomials. Surface tension can also be obtained from the MD data employing a mechanistic approach. The Kirkwood and Buff method (Kirkwood and Buff 1949) helps describe surface tension by integrating a difference of normal and tangential stress components, which can be obtained using the Irving and Kirkwood (IK) method (Irving and Kirkwood 1950). It provides an expression for the stress tensor in terms of fluctuating MD data—velocities and intermolecular forces; for details, see Smith et al. (2018).

The molecular behavior of the contact line can be modeled using a Langevin equation with parameters obtained as statistics from MD simulations. The use of Langevin equation is allowed since the molecular fluctuations are Gaussian and the autocorrelation of the fluctuations follows an exponential decay. The prediction of dynamic contact angle obtained as a solution to the Langevin equation can be incorporated into continuum models for drop motion, as shown in Fig. 1.26. The MD simulation data for liquid bridge under top layer shear is presented in the figure. Figure 1.26a plots parameterized wall sliding speed resulting in a straight line fit between the parameterized speed and the difference in dynamic and equilibrium contact angles. Figure 1.26b shows evolution of the solution with time obtained using a computational fluid dynamics (CFD) model. The dynamic contact angle is seen to approach equilibrium contact angle corresponding to four different values of ϵ_{wall}. The corresponding contact line speeds are also shown in the inset.

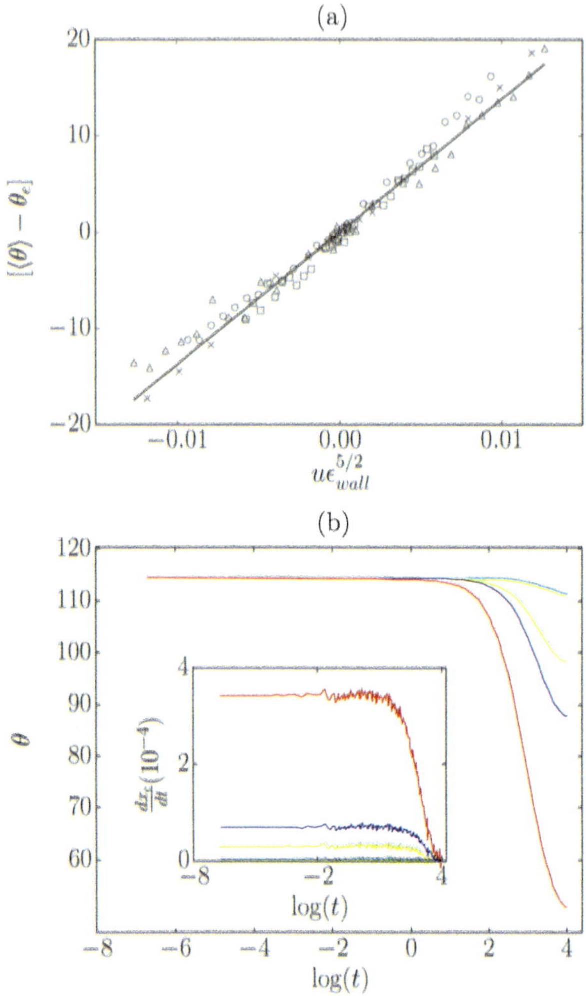

Fig. 1.26 MD simulation data for liquid bridge under top layer shear. (**a**) Wall sliding speed has been parameterized using the Mie potential parameter ϵ_{wall}. The data collapses into a line validating Tanner's law. (**b**) Solution evolution with time obtained using the CFD model is plotted. The dynamic contact angle is seen to approach equilibrium contact angles corresponding to four different values of ϵ_{wall}. The corresponding contact line speeds are also shown in the inset. (Figure taken with permission from Smith et al. 2018)

Droplet shape evolution to equilibrium can be predicted using the MD approach. If the number of MD beads is smaller than 65,000, the contact angle may show dependence on drop size due to line tension effects. This renders MD simulations expensive for droplets larger than the nanoscale.

Liquid-vapor interfaces, surface tension, dynamic contact line, and contact angles can be predicted using molecular dynamics simulations, as discussed in this section. Reduced models such as the Langevin equation can be derived from MD data, which can be incorporated into the continuum description of microdrop dynamics. However, considering the computational cost associated with these simulations, the field of MD is still in its infancy, the growth of which depends on the improvement in distributed computational infrastructure.

1.8 Closure

Fundamental ideas of wettability, equilibrium contact angles, contact angle hysteresis, and wetting transitions have been described in this chapter. Physical and chemical methods to control wettability are outlined. Drop shapes that can be realized on physically textured surfaces, specifically the Cassie-Baxter and Wenzel models, along with superhydrophobicity and transition between the wetting regimes are discussed. The Young-Laplace equation as the equation of statics is derived. A computational approach to deriving the drop shape by solving the Young-Laplace equation is presented. An equivalent approach based on energy minimization is also described. The dependence of drop shapes on various physical parameters, such as fluid density, interfacial tensions, drop volume, substrate texturing, and gravity, can be explored in this framework. An inverse method for the measurement of equilibrium contact angle from image data is proposed. Initial results obtained by the authors on the laboratory scale are presented. The notion of a dynamic contact angle for a drop moving on a textured surface is briefly described. The chapter concludes with recent trends in microscopic modeling of gas-liquid interfaces using molecular dynamic simulation.

Chapter 2
Spreading of Sessile and Pendant Drops on Partially Wetting Surfaces

Aashutosh Mistry and K. Muralidhar

Nomenclature

a, b	Parameters for molecular kinetic theory-based contact line velocity expression
C	Thermal heat capacity, J/kg K
CS	Control surface; suffix *free* for free surface; *contact* for contact line
CV	Control volume
E, E_k, E_g, E_s, E_c	Total energy with components–kinetic, gravitational, free surface, and contact line (dimensionless)
$\boldsymbol{f_B}$	Body force vector, N/m^3
f_C	Contact line force, N/m
f	Dimensionless force with components related to shear, viscous normal stress, and pressure
g	Acceleration due to gravity, m/s^2
H	Height of a disk element measured from the substrate, m
k	Thermal conductivity, W/m K
ℓ	Appropriate length scale, m; also capillary length, m
ℓ_{slip}	Slip length, m
L	Representative macroscopic length scale, m
M	Number of disks for discretization
$\hat{\boldsymbol{n}}$	Direction normal to the liquid-air interface
p, P	Pressure field inside the drop; characteristic value, N/m^2
r, z	Radial and axial coordinates, m
r_c	Contact radius, m
R	Radius referred to a disk element, m
u, w; $\boldsymbol{u}$	r- and z-direction velocity components, m/s; velocity vector, m/s

S. Khandekar, K. Muralidhar, *Drop Dynamics and Dropwise Condensation on Textured Surfaces*, Mechanical Engineering Series,
https://doi.org/10.1007/978-3-030-48461-3_2

u_0	Intrinsic velocity scale, m/s
U, W	Characteristic velocity, m/s
U_c	Contact line velocity, m/s
U_p, W_p	Discretized velocity components at the centroid of the chosen disk, m/s
V_{drop}; V_{disk}	Volume of drop; volume of a disk element, m^3
V_0	Impact velocity, m/s
t; Δt	Time variable; time step in numerical simulation, s
t	Unit tangent vector drawn over the contact line
T, ΔT	Temperature, temperature difference between hot and cold boundaries, K
$q_{footprint}$	Heat flux averaged over the drop footprint, W/m^2

Dimensionless Quantities

Bo	Bond number
Ca	Capillary number
Oh	Ohnesorge number
Pr	Prandtl number

Greek Letters

α	Exponent of time in the spreading process
κ	Droplet curvature, m^{-1}
$\phi = \frac{T-T_L}{T_H-T_L}$	Dimensionless temperature
ρ	Droplet density, kg/m^3
σ	Surface tension, N/m; suffix s for surface-air, ls for liquid-solid, and lv (or l) for liquid-vapor components
Σ	Surface-to-sensible energy ratio
τ_μ, τ_κ	Viscous and conduction timescales, s
ω	Rotational speed, s^{-1}
μ	Dynamic viscosity, kg/m s
θ	Instantaneous, also called, dynamic contact angle, °
θ_{eq}	Equilibrium contact angle, °

2.1 Introduction

Spreading of liquid drops (De Gennes et al. 2013; Yarin 2006) is widely seen in nature, for example, raindrops over leaves. Several process applications rely on spreading dynamics, e.g., inkjet printing, thermal sprays, and soldering in electronics. An application of interest to this monograph is dropwise condensation. The characteristic features of drop spreading originate from surface tension at the liquid-gas interface and the unbalanced forces at the three-phase contact line. Drop spreading is either neglected or approximated and is often treated as a limiting step for various measurements such as equilibrium contact angle estimation. However, spreading dynamics exhibit rich flow features which are prone to misinterpretation. For example, late time relaxation of viscous liquid drops to their equilibrium configuration proceeds in a slow quasi-steady fashion (Eddi et al. 2013). In turn, the drop shape would be apparently static but be far from equilibrium, necessary for an accurate estimation of the static apparent contact angle. Mechanistic understanding of spreading is also essential to engineering applications. For example, the spreading timescale in soldering should be larger than solidification to ensure that localized spots neither form larger patches nor consume unnecessary material. Given environmental concerns, lead-based alloys are being phased out of soldering practice. The new class of materials should be selected in such a way that they exhibit the desired solidification range with limited spreading.

Despite the importance of drop spreading in applications, its detailed understanding has been elusive. Experimental probing in the past could not resolve short-time features, but the new generation high frame-rate cameras are providing additional insight. High-speed visualization (Eddi et al. 2013; Biance et al. 2004; Bird et al. 2008; Winkels et al. 2012) has revealed temporal dynamics at µs–ms timescales with imaging speeds of ~10,000 fps or better. On the other hand, theoretical explorations (Tanner 1979; Braun et al. 1995; Shikhmurzaev 1997; Legendre and Maglio 2013) are impaired given the difficulties of contact line motion and large free surface deformation. As a consequence, a piecemeal interpretation over characteristic timescales is available, but questions such as wall shear stress fluctuations at the drop footprint, exchanges among distinct energy forms, conjugate interactions in heat transfer, and subcritical instability of a pendant drop configuration have remained largely unanswered. This chapter presents an account of the theoretical investigations pursued by the authors to understand dynamical transitions in the spreading of liquid drops (Mistry 2014; Mistry and Muralidhar 2015a, b; 2018).

In a strict sense (Yarin 2006; Mistry and Muralidhar 2015a), drop spreading refers to spontaneous wetting of the solid surface by a liquid drop gently placed underneath or over it. A drop sitting over the solid substrate is called sessile, while the one hanging below is in the pendant state. Figure 2.1 presents a schematic drawing of a part of the sessile drop with the coordinate system shown. Spreading discussed in this chapter differs from drop impact, where the drop approaches the solid surface with a fixed velocity, or drop sliding, where external forces (e.g., gravity on an inclined surface, centrifugal force on a rotating stage, and locomotion provided by an electric field) are responsible for the motion of the three-phase contact line. Such discrimination outlines the physical differences in the associated flow fields. Impact and sliding have prescribed velocity scales, namely the velocity of approach, V_0 for drop impact, $\sqrt{g\ell}$ for gravitational sliding and $\omega\ell$ for centrifugal sliding. Here, g is gravitational acceleration, ω is rotational speed of the surface, and ℓ is the relevant length scale. Drop spreading, on the other hand, is devoid of definite

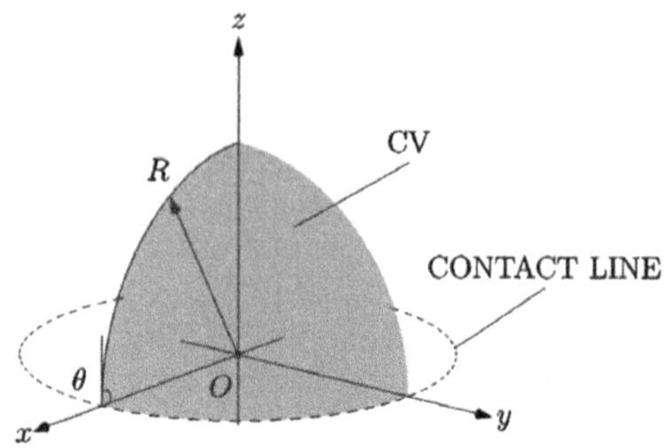

Fig. 2.1 Schematic of a spreading drop. A symmetric portion of the drop volume is shown. CV refers to control volume. Contact line is explicitly shown. Angle θ discerns the instantaneous (dynamic) contact angle during spreading

velocity scales. An intrinsic velocity, $u_0 \sim \sqrt{\sigma/\rho\ell}$, based on fluid properties is a representative scale. Here, σ is the surface tension at the liquid-air interface (or an interface formed between liquid and vapor or gas), and ρ is the density of the liquid drop. Such differences in the driving force create distinct flow regimes. In addition, the absence of an external impetus makes drop spreading quite sensitive to a variety of influences, some of which are described below.

1. Singularity of Initial Contact. When a small drop is gently made to touch the solid surface, it initially forms a point contact. The contact line force, which is responsible for spreading, is proportional to the perimeter of the contact line. Hence, spreading cannot begin unless the point contact expands to a tiny patch with a finite radius. Given the finite time nature of this transition, actual spreading commences from the time instant when the contact patch is formed. Small tangents at the contact make it very difficult to visualize this early transition from a side view. A line of sight perpendicular to the spreading plane is required (Eddi et al. 2013; Winkels et al. 2012) but is optically not amenable for opaque liquids or for severely refracting drops if spreading is observed from above and for non-transparent substrates when spreading is observed from below.
2. Initial Conditions. The initial spreading transients are sensitive to slight perturbations in the manner in which the drop is placed over (or under) the substrate. Specifically, the initial velocity of placement and the degree of angular asymmetry (Lo et al. 2017) are important factors. It is not enough to carry out high-speed visualization but also ensure representative spreading onset. Two common configurations exist for drop-solid contact. Either the drop is slowly formed at the end of a needle in the proximity of the surface of interest (Bird et al. 2008) or it is first placed on a superhydrophobic surface and the test substrate is gradually raised or lowered to commence spreading (Biance et al. 2004). Both strategies exhibit uncertainties that generate a non-universal spreading behavior and departure from theoretical considerations. In other experiments of drop impact, spreading may be examined in the limit of impact velocity approaching zero. However, the contact time of patch formation is unspecified and is a source of uncertainty. Thus, pure spreading experiments suffer from high sensitivity to initial conditions.
3. Contact Line Singularity. In the continuum limit, liquids fulfill the no-slip condition (Leal 2007) at their interface with solid surfaces. However, the line-

boundary formed at the gas-liquid-solid contact, namely the three-phase contact line (Fig. 2.1), violates no-slip, and moves in response to the local force imbalance. Gas here refers to the phase and discussion carries over when vapor or air is present. The drop footprint is thus associated with zero velocity in the interior and non-zero velocity at the periphery (Dussan 1976), becoming zero once again in the gas phase outside the drop. In a simulation framework, contact line motion is empirically prescribed (Snoeijer and Andreotti 2013; Sui et al. 2014). Representative contact line descriptions are summarized in Table 2.1.

Enough resolution of the initial contact singularity, namely the zero contact radius, is an experimental challenge. On the other hand, the singularity related to contact line motion is a major theoretical bottleneck but is routinely observed in experiments. It is a source of mismatch between simulation and experiments and influences analysis and interpretation of the spreading dynamics. A survey of the literature in Table 2.1 shows the following. The motion of the contact line is described by assuming a separation of length scales in the sense that it is localized, with dependence on the immediate flow field surrounding it. The dependence may be heuristically developed or from first principles using a molecular dynamics approach or an equivalent theory. The local information is blended with the continuum scale governing equations applicable to the entire fluid domain. Scale separation may be considered a stringent assumption, unlike other flow configurations such as turbulence and transport in porous media, where scale separation is readily apparent and hence, is justified (Kundu and Cohen 2004).

The assumption of separated scales, length or time, implies universality in contact line motion. This behavior has not been experimentally observed (Snoeijer and Andreotti 2013; Sui et al. 2014). Intuitively, contact line motion is a result of unbalanced forces. The Navier-Stokes equations connect the force resultant with acceleration. They should, in principle, be able to predict contact line motion similar to the time evolution of the gas-liquid interface. Such a paradigm consistently predicts contact line motion across a range of conditions and liquid properties, without resorting to any tunable parameter. This approach has been adopted in the subsequent sections of this chapter.

We first present integral expressions of motion for a control volume that includes the body of the liquid drop, liquid-solid contact surface, gas-liquid free surface, and a moving contact line. These expressions are further simplified for a spreading drop assuming the flow field to be axisymmetric. The partial differential equations in a Lagrangian framework are transformed to a set of algebraic expressions in a mixed continuous-discrete fashion, where field variations in the radial direction are analytically expressed, while the drop is segmented into cylindrical disks along the axial direction. The framework developed is numerically simulated and predicts the spreading of sessile and pendant drops of water and glycerin over wetting and partially wetting chemically textured surfaces. Rough and (micro/nano) pillared surfaces are not within the scope of the study. The particular choice of fluids is in part due to their high surface tension coefficient but widely different viscosities that consequently reveal distinct spreading dynamics. Additionally, high-speed

Table 2.1 Representative models for contact line motion. The Lagrangian treatment proposed by the authors is not included here

Description	Features	Remarks
Navier slip (Huh and Scriven 1971; Lauga et al. 2007)	Contact line velocity scales with the local shear perpendicular to the solid surface: $U_c \sim \ell_{slip} \frac{\partial u}{\partial n}$	Slip length, ℓ_{slip}, is a tunable parameter that changes with the flow regime.
Molecular kinetic theory (Blake 2006; De Coninck and Blake 2008)	A molecular description; contact line motion is treated as an adsorption-desorption reaction with an associated activation energy barrier: $U_c \sim a \sinh \{b(\cos\theta - \cos\theta_{eq})\}$, where a and b are physical properties characteristic of the three-phase zone (TPZ), and θ_{eq} is the equilibrium contact angle.	Properties a and b are treated as tunable parameters, rather than being estimated through measurements.
Cox–Voinov law (Voinov 1976; Cox 1986, 1998)	Navier slip model is used to resolve flow structure in the vicinity of the contact line; given a small length scale of the TPZ, lubrication approximation is adopted. $\widehat{g}(\theta) = \widehat{g}(\theta_{eq}) + \text{Ca} \cdot \ln\left(\frac{L}{\ell_{slip}}\right)$, where $\widehat{g}(\theta)$ is a function of contact angle, $\text{Ca} = \mu U_c/\sigma$ is the capillary number representing dimensionless contact line velocity, and L is a representative macroscopic length scale.	Choice of length scales is highly subjective.
Rolling contact (Dussan 1976)	The liquid-liquid interface is assumed to gradually convert to solid-liquid contact near the advancing contact line (and *vice versa* for a receding contact).	No straightforward theoretical implementation is reported.
Upscaling (Zhang et al. 2014)	Molecular scale simulations are carried out, where the discrete nature of three-phase contact is simulated and upscaled to continuum calculations.	Given the large length scale disparity, molecular details are coarse-grained, resulting in, at best, qualitative fidelity.
Precursor film (Ludviksson and Lightfoot 1968)	A thin liquid film is assumed to be adsorbed on the solid; hence, contact line moves on a liquid surface, rather than pristine solid.	Could be applicable only to receding lines for nearly wetting liquids though some experiments suggest applicability to an advancing contact line.
Evaporation–condensation (Wayner 1993)	Contact line motion is due to evaporation–condensation phase changes in the vicinity of the contact line.	Contact line motion is observed even when liquids are far away from boiling temperatures.

visualization data is available in the recent literature for these liquids (Eddi et al. 2013; Biance et al. 2004; Bird et al. 2008; Winkels et al. 2012).

Theoretical predictions are compared against the published experimental data. Spreading dynamics is also discussed on an energy landscape, where energy components are followed in time. Pendant drop spreading is examined with a specific emphasis on gravitational instability wherein the drop may detach before reaching equilibrium. Preliminary results on heat transfer characteristics of spreading drops are presented. The chapter closes with a discussion on open questions and ongoing research initiatives.

2.2 Mathematical Modeling

Newton's second law of motion for an arbitrarily selected liquid volume (Leal 2007) can be mathematically expressed (CV $\equiv$ control volume, CS $\equiv$ control surface, CL $\equiv$ contact line)

$$\frac{\mathrm{d}}{\mathrm{d}t}\left(\int_{CV}\rho \boldsymbol{u}\,\mathrm{d}V\right) = -\int_{CS} p\widehat{\boldsymbol{n}}\,\mathrm{d}S + \int_{CS}\mu\nabla \boldsymbol{u}\cdot\widehat{\boldsymbol{n}}\,\mathrm{d}S + \int_{CV}\boldsymbol{f_B}\,\mathrm{d}V + \int_{CL}\boldsymbol{f_C}\,\mathrm{d}L \tag{2.1}$$

The left side term quantifies material acceleration of the fluid volume, while the right side comprises various forms of forces acting on the control volume. As notation, $\boldsymbol{u}$ is velocity, p pressure, μ viscosity, $\widehat{\boldsymbol{n}}$ unit surface normal, $\boldsymbol{f_B}$ body force per unit volume, and $\boldsymbol{f_C}$ the line force per unit length at the contact line. For a generalized volume element, as in Fig. 2.1, the solid-liquid contact area is denoted as $\mathrm{CS}_{\mathrm{contant}}$, the gas-liquid surface as $\mathrm{CS}_{\mathrm{free}}$. Since gas density and viscosity are much smaller than those of the liquid, their interface is classified as free and hence stresses at the free surface can be explicitly substituted from the liquid side. The relevant form of the expression for stresses at the free surface is developed as follows. Here, surface tension gradients (Leal 2007) arising from temperature gradients or those in surfactant concentration (Israelachvili 2011) are neglected. With $\boldsymbol{u}_{\|} = \boldsymbol{u} - (\boldsymbol{u}\cdot\widehat{\boldsymbol{n}})\widehat{\boldsymbol{n}}$, the velocity component parallel to the free surface, continuity of shear stresses at the gas-liquid interface yields

$$\mu\left.\frac{\partial \boldsymbol{u}_{\|}}{\partial n}\right|_{\mathrm{drop}} = \mu\left.\frac{\partial \boldsymbol{u}_{\|}}{\partial n}\right|_{\mathrm{air}} \Rightarrow \left.\frac{\partial \boldsymbol{u}_{\|}}{\partial n}\right|_{\mathrm{drop}} = \frac{\mu_{\mathrm{air}}}{\mu_{\mathrm{drop}}}\cdot\left.\frac{\partial \boldsymbol{u}_{\|}}{\partial n}\right|_{\mathrm{air}} \approx 0 \quad \text{since } \mu_{\mathrm{air}} \ll \mu_{\mathrm{drop}}$$

With $\boldsymbol{u}_{\perp} = (\boldsymbol{u}\cdot\widehat{\boldsymbol{n}})\widehat{\boldsymbol{n}}$, the velocity component normal to the free surface, and κ the interface curvature, the normal stress continuity across the gas-liquid interface simplifies to

Table 2.2 Fluid properties at 20 °C, 1 atm pressure (White 2015)

Properties	Water	Glycerin	Air
Density, ρ (kg/m^3)	998	1260	1.20
Viscosity, μ (Pa s)	10^{-3}	1.49	1.8×10^{-5}
Surface tension[a], σ (mN/m)	72.8	63.3	–
Capillary length[b], ℓ_c (mm)	~2.73	~2.26	–

[a]Relative to air as the surrounding fluid
[b]For acceleration due to gravity, $g = 9.81$ m/s^2

$$-p\widehat{\boldsymbol{n}} + \mu\frac{\partial \boldsymbol{u}_\perp}{\partial n}\bigg|_{\text{drop}} = -p\widehat{\boldsymbol{n}} + \mu\frac{\partial \boldsymbol{u}_\perp}{\partial n}\bigg|_{\text{air}} + \sigma\kappa\widehat{\boldsymbol{n}},$$
$$\text{i.e.} \quad p_{\text{drop}} \approx p_{\text{air}} + \sigma\kappa$$

Since interface deformation will occur at modest rates, the viscous contribution to the normal stress is ignored. Air density being several orders of magnitude smaller than the liquid and the drop itself being spatially small, the gravitational pressure gradient in air is negligible and its pressure is taken to be spatially constant. The constant value is set to zero in the present development. Thus, at the gas-liquid interface,

$$p_{\text{drop}} \approx \sigma\kappa \quad \text{since } \rho_{\text{air}} \ll \rho_{\text{drop}}$$

The assumptions related to density and viscosity ratios of gas-to-liquid are justified based on physical properties presented in Table 2.2. Assigning free surface conditions and including the gravitational contribution within the drop as $\boldsymbol{f_B} = \rho\boldsymbol{g}$, Eq. (2.1) is

$$\frac{\mathrm{d}}{\mathrm{d}t}\left(\int_{\text{CV}} \rho\boldsymbol{u}\,\mathrm{d}V\right) = -\int_{\text{CS-CS}_{\text{free}}} p\widehat{\boldsymbol{n}}\,\mathrm{d}S - \int_{\text{CS}_{\text{free}}} \sigma\kappa\widehat{\boldsymbol{n}}\,\mathrm{d}S + \int_{\text{CS-CS}_{\text{free}}} \mu\nabla\boldsymbol{u}\cdot\widehat{\boldsymbol{n}}\,\mathrm{d}S + \int_{\text{CV}} \rho\boldsymbol{g}\,\mathrm{d}V + \int_{\text{CL}} \boldsymbol{f}_C\,\mathrm{d}L \tag{2.2}$$

The direction of the gravity vector is chosen to represent a sessile or a pendant configuration. Mass conservation for an incompressible liquid drop is

$$\int_{\text{CS}} \boldsymbol{u}\cdot\widehat{\boldsymbol{n}}\,\mathrm{d}S = 0 \tag{2.3}$$

The drop-solid contact, $\text{CS}_{\text{contact}}$, is impermeable to fluid flow, i.e., $\boldsymbol{u}\cdot\widehat{\boldsymbol{n}} = 0$. Additionally, evaporation–condensation is absent, and no mass flow takes place across the gas-liquid free surface; hence, over $\mathrm{CS}_{\text{free}}$ $\boldsymbol{u} \cdot \widehat{\boldsymbol{n}} = 0$. On the other hand, the free surface moves with the local fluid velocity, a result to be used to follow its temporal evolution. Accounting for these boundary conditions, the mass conservation equation simplifies to

$$\int\limits_{\mathrm{CS}-\mathrm{CS}_{\text{contact}}-\mathrm{CS}_{\text{free}}} \boldsymbol{u} \cdot \widehat{\boldsymbol{n}}\, \mathrm{d}S = 0 \tag{2.4}$$

Equations (2.2) and (2.4) govern flow dynamics of a liquid volume that is bounded by solid-liquid contact, free surface, and the three-phase contact line.

Fluid inertia contained on the left side of Eq. (2.2) includes total acceleration, namely the temporal and spatial contributions. In the present context, drop volume in its entirety is to be determined as a function of time. Hence, a Lagrangian treatment (Leal 2007; Kundu and Cohen 2004) is more suitable than the Eulerian and acceleration is represented here as a total derivative. In the following discussion, the analytical form of the contact line force, $\boldsymbol{f}_C$, and the resulting governing equations in an axisymmetric coordinate system are presented.

2.2.1 Unbalanced Force at the Contact Line

For a liquid drop in equilibrium, the contact line is also in equilibrium, requiring the force components to be in balance. In terms of the equilibrium contact angle θ_{eq}, the force balance of the component forms of surface tension in the vicinity of the contact line is expressed

$$\sigma_{\text{l}} \cos \theta_{\text{eq}} + \sigma_{\text{ls}} = \sigma_{\text{s}} \tag{2.5}$$

Here σ_{l} is surface tension of the liquid-air interface, while σ_{s} and σ_{ls} refer to surface energies of solid-air and liquid-solid interfaces, respectively. In the literature, σ_{l} and σ are often interchangeably used. Equation (2.5) is known as the Young's equation (Israelachvili 2011). Changes in the corresponding interfacial area scale with surface energy changes, in the sense that a greater amount of energy is stored when the surface is stretched. Each of these terms represents conservative surface forces, and in turn, represent energy storage modes. The equilibrium contact angle, as defined by Eq. (2.5), is a property shared by the liquid-gas-solid combination; the outer gas environment could be another liquid as well. Given the equivalence of force and energy for conservative forces in equilibrium, Young's equation may be presented from either standpoint (Israelachvili 2011; De Gennes et al. 2013). Under spreading conditions, the instantaneous contact angle departs from the equilibrium value and $\theta \neq \theta_{\text{eq}}$. This inequality is representative of an unbalanced force; in terms of energy, it indicates a drive towards a lower energy state. Let $\widehat{\boldsymbol{t}}$ be the unit vector

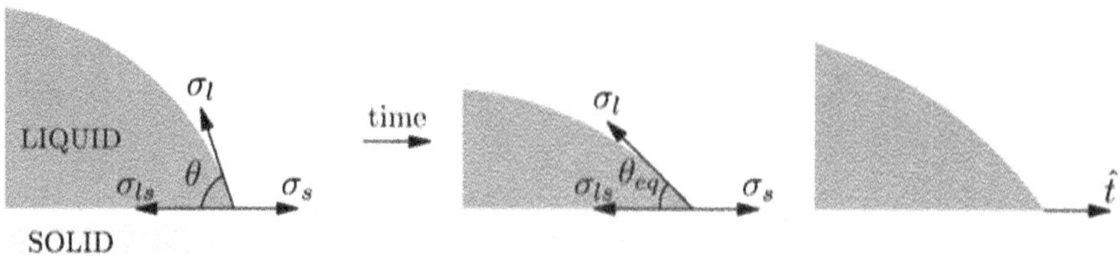

Fig. 2.2 A three-phase contact line is unbalanced when the contact angle departs from its equilibrium value. This imbalance results in a line force perpendicular to the contact line in the plane of solid-fluid contact. A unit vector $\hat{t}$ is sketched for clarity

perpendicular to the contact line, pointing away from the liquid, in the liquid-solid contact plane (Fig. 2.2). The unbalanced force can be estimated as

$$f_C = \{\sigma_s - \sigma_{ls} - \sigma_l \cos\theta\}\hat{t} = \sigma_l(\cos\theta_{eq} - \cos\theta)\hat{t} = \sigma(\cos\theta_{eq} - \cos\theta)\hat{t} \quad (2.6)$$

The unbalanced force in Eq. (2.6) is to be interpreted as force per unit length of the contact line and has the units of N/m. Symbols in bold refer to vector quantities.

2.2.2 Axisymmetric Modeling of Drop Spreading

Given the radial symmetry of drop spreading, the governing equations (Eqs. 2.1 and 2.3) can be further simplified in a cylindrical coordinate system.

Radial momentum:

$$\rho\frac{\mathrm{d}}{\mathrm{d}t}\left(\int_z\int_r u2\pi r\mathrm{d}r\mathrm{d}z\right) = -\int_z\int_r \frac{\partial p}{\partial r}2\pi r\mathrm{d}r\mathrm{d}z + \mu\int_z\int_r\left(\frac{1}{r}\frac{\partial}{\partial r}\left(r\frac{\partial u}{\partial r}\right) - \frac{u}{r^2}\right)2\pi r\mathrm{d}r\mathrm{d}z$$
$$+\mu\int_z\int_r\frac{\partial^2 u}{\partial z^2}2\pi r\mathrm{d}r\mathrm{d}z + \int_0^{2\pi r_c}\sigma(\cos\theta_{eq} - \cos\theta)\mathrm{d}s \quad (2.7)$$

Axial momentum:

$$\rho\frac{\mathrm{d}}{\mathrm{d}t}\left(\int_z\int_r w2\pi r\mathrm{d}r\mathrm{d}z\right) = -\int_z\int_r \frac{\partial p}{\partial r}2\pi r\mathrm{d}r\mathrm{d}z + \mu\int_z\int_r\frac{1}{r}\frac{\partial}{\partial r}\left(r\frac{\partial w}{\partial r}\right)2\pi r\mathrm{d}r\mathrm{d}z$$
$$+\mu\int_z\int_r\frac{\partial^2 w}{\partial z^2}2\pi r\mathrm{d}r\mathrm{d}z \mp \int_z\int_r\rho g2\pi r\mathrm{d}r\mathrm{d}z \quad (2.8)$$

Volume conservation:

$$\int_z \int_r \left(\frac{1}{r} \frac{\partial}{\partial r}(ru) + \frac{\partial w}{\partial z} \right) 2\pi r dr dz = 0 \tag{2.9}$$

Here, the symbols used are as follows: radial velocity component is u, axial velocity w, radial coordinate r, axial coordinate z, arc coordinate along the contact path s, and radius of the contact patch over the base r_c. The contact line lies in the $z = 0$ plane, where Eq. (2.7) is applicable. The contact line force (Eq. 2.6) acts along the radial direction. Gravity appears in the axial momentum (Eq. 2.8), with the sign convention of being negative for sessile and positive for the pendant configuration. The simplifications made earlier to derive Eqs. (2.2) and (2.4) from Eqs. (2.1) and (2.3) are applied while deriving the radial analytical profiles and boundary conditions for discretization in the z-direction.

2.3 Numerical Implementation

Computational difficulties in simulating free surface flows with surface tension effects arise due to three interrelated weaknesses in numerical implementation (Tryggvason et al. 2011). These are seen in drop spreading simulations as well. Specifically:

1. Smearing in the velocity field, i.e., inexact volume conservation, in the presence of large density and viscosity ratios (typically realized in drop spreading; Table 2.2).
2. Error in pressure estimation with strong surface tension causes an artifact in the velocity field–commonly referred to as spurious currents.
3. Curvature estimation for a continuously evolving interface can become erroneous, further intensifying spurious currents.

Given the (body-fitted) Lagrangian treatment adopted in the present study, the volume is conserved in an analytical sense and smearing is entirely avoided. The accuracy in the pressure solution arises from employing identical discretization operators for pressure and curvature terms. Lack of smearing has an immediate impact on improving the specification of geometry. Extra care is to be exercised for curvature estimation since it represents the second derivative of the interface coordinates. Discrete differentiation based on a finite difference method is the primary source of curvature inaccuracy. In the present study, an interpolation-based local analytical reconstruction of the interface is proposed. Given the analytical form, analytical differentiation is carried out to obtain accurate curvatures. An additional concern for the three-phase contact line is a reliable calculation of the dynamic contact angle. Specific geometrical expressions that provide a consistent estimation of the time-dependent contact angle are discussed by the authors elsewhere (Mistry and Muralidhar 2015, 2018).

Table 2.3 The radial dependence of flow fields subject to symmetry and free surface constraints (Mistry and Muralidhar, 2018)

Functional form	Constraint at disk center, $r = 0$	Constraint at disk surface, $r = R$	Representative value at $r = R/2$
$u(r) = \frac{4U}{3}\left(2\left(\frac{r}{R}\right) - \left(\frac{r}{R}\right)^2\right)$	$u = 0$	$\frac{\partial u}{\partial r} = 0$	$u = U$
$w(r) = W$	$\frac{\partial w}{\partial r} = 0$	$\frac{\partial w}{\partial r} = 0$	$w = W$
$p(r) = \left(\frac{4P-\sigma\kappa}{3}\right) - \left(\frac{4P-4\sigma\kappa}{3}\right)\left(\frac{r}{R}\right)^2$	$\frac{\partial p}{\partial r} = 0$	$p = \sigma\kappa$	$p = P$

The axisymmetric nature of drop spreading helps in discretizing the drop into M equi-volume disks (Mistry and Muralidhar 2015), i.e.,

$$V_{\text{disk}} = \frac{V_{\text{drop}}}{M}$$

Radial distribution of velocity and pressure are analytically constructed in such a way that constraints at the drop axis ($r = 0$) and the free surface ($r = R$) are identically satisfied. Based on this discussion, the specific form of the constraints and appropriate functional forms are summarized in Table 2.3. Here, R is the radius of each discretized disk. In general, radii of contiguous disks are unequal.

The analytical expressions from Table 2.3 are substituted next in Eqs. (2.7)–(2.9) for evaluating the integral expressions over each of the disks. The contact line force appears only for the lowermost disk that is in contact with the solid surface. After rearrangement, the spatially discretized governing equations, complementary to Eqs. (2.7)–(2.9) read

$$\rho\frac{\mathrm{d}U_P}{\mathrm{d}t} = \frac{8}{5}\left(\frac{P_P - \sigma\kappa_P}{R_P}\right) - \frac{18}{5}\mu\frac{U_P}{R_P^2} + \mu\frac{R_{uf}^2}{R_P^2 H_P(H_P + H_U)}(a_{UU}U_U - a_{PU}U_P) - \mu\frac{R_{lf}^2}{R_P^2 H_P(H_P + H_L)}(a_{PL}U_P - a_{LL}U_L) + \frac{9}{5}\frac{\sigma r_c}{R_P^2 H_P}\left(\cos\theta_{\text{eq}} - \cos\theta\right) \tag{2.10}$$

$$\rho\frac{\mathrm{d}W_P}{\mathrm{d}t} = -\frac{2R_{uf}^2}{3R_P^2 H_P}P_{uf} + \frac{2R_{lf}^2}{3R_P^2 H_P}P_{lf} - \frac{3R_P^2 - 2R_{uf}^2}{3R_P^2 H_P}\sigma\kappa_{uf} + \frac{3R_P^2 - 2R_{lf}^2}{3R_P^2 H_P}\sigma\kappa_{lf} + \mu\frac{2R_{uf}^2}{R_P^2 H_P(H_P + H_U)}(W_U - W_P) - \mu\frac{2R_{lf}^2}{R_P^2 H_P(H_P + H_L)}(W_P - W_L) \mp \rho g \tag{2.11}$$

$$\frac{8U_P}{3R_P} + \frac{1}{H_P}\left(W_{uf} - W_{lf}\right) = 0 \tag{2.12}$$

Here, indices P, U, and L, respectively, refer to the current disk, the one above P and the one below. Notations uf and lf correspond to values at the disk-disk interfaces. Evaluation of the radial derivatives in Eqs. (2.7)–(2.9) is straightforward after substituting for analytical expressions from Table 2.3. The discontinuous geometry of adjoining disks necessitates additional considerations while estimating the axial derivatives. In general, adjoining disks have unequal radii, even when all disks are of equal volume. Hence, centroids of adjoining disks do not align vertically and standard difference operator, e.g., $\partial y/\partial x \approx (y_{i+1} - y_i)/\Delta x$, is erroneous. To circumvent this time-varying non-orthogonality of the curve passing through centroids of the disks, coordinate transformation is carried out so that orthogonality is ensured in the transformed domain where difference operator becomes accurate. Mathematically, such a transformation can be represented as

$$\begin{Bmatrix} \partial/\partial r \\ \partial/\partial z \end{Bmatrix} = \begin{bmatrix} \dfrac{\partial r'}{\partial r} & \dfrac{\partial z'}{\partial r} \\ \dfrac{\partial r'}{\partial z} & \dfrac{\partial z'}{\partial z} \end{bmatrix} \begin{Bmatrix} \partial/\partial r' \\ \partial/\partial z' \end{Bmatrix} = \begin{bmatrix} 1/R & 0 \\ -\dfrac{r'}{R}\dfrac{\mathrm{d}R}{\mathrm{d}z} & 1 \end{bmatrix} \begin{Bmatrix} \partial/\partial r' \\ \partial/\partial z' \end{Bmatrix} \tag{2.13}$$

$$\text{where } r' = r/R, \quad z' = z$$

Here (r, z) represent physical coordinates, while (r', z') refer to the transformed domain. Furthermore, expressions for u, w, p fields at the disk-disk interface (f) are obtained such that their axial derivatives are continuous, i.e.,

$$\left.\frac{\partial u}{\partial z}\right|_{f^+} = \left.\frac{\partial u}{\partial z}\right|_{f^-} \quad \text{for } r \in [0, R_f] \tag{2.14}$$

Similar conditions are satisfied by the axial velocity, w, and pressure, p. In addition, these fields vary along the radial coordinate. Hence, Eq. (2.14) is to be satisfied at every radial location at the disk interface. The radial dimension of the interface relates to adjoining disks

$$R_f = \frac{R/H_1 + R_2/H_2}{\frac{1}{H_1} + \frac{1}{H_2}} \tag{2.15}$$

Here, the distance between the centroid and the interface is chosen as weights. Once the interface fields are reconstructed, the axial integrals in Eqs. (2.7)–(2.9) can be evaluated, leading to Eqs. (2.10)–(2.12). Details are provided by the authors elsewhere (Mistry and Muralidhar 2018).

Local evolution of the free surface between the liquid drop and air is captured in terms of changes in each of the disk radii

$$\frac{\mathrm{d}R_P}{\mathrm{d}t} = \frac{4}{3}U_P \tag{2.16}$$

Since the volume of each of the disks is constant with time, new radii R_P provide new thicknesses H_P of the disk elements. Further geometrical attributes, namely curvatures, κ_P, contact line radius, r_c, and dynamic contact angle, θ, are consistently

Table 2.4 Boundary conditions along the axial direction to close the discretized form of the governing equations

Field	Drop-substrate boundary	Free surface
Radial velocity, u	No-slip, $u = 0$	Zero shear, $\frac{\partial u}{\partial z} = 0$
Axial velocity, w	No flow, $w = 0$	Zero shear, $\frac{\partial w}{\partial z} = 0$
Pressure, p	Force balance, $\frac{\partial p}{\partial z} = \mp \rho g$ (−) for sessile and (+) for pendant	Curvature pressure, $p = \sigma\kappa$

obtained from the (R_P, H_P) data (Mistry and Muralidhar 2018). Given the explicit discretization along the axial direction, boundary conditions are required to close the system of equations (Table 2.4).

2.3.1 Non-dimensionalization of Governing Equations

The governing equations (Eqs. 2.10–2.16) contain the relevant fluid properties as parameters. To develop universal results, these can be expressed in dimensionless form. Non-dimensionalization can be carried out using the representative length scale, $\ell = (3V_{\text{drop}}/4\pi)^{1/3}$, i.e., the radius of an equivalent spherical drop and velocity scale, $u_0 = \sqrt{\sigma/\rho\ell}$ to obtain dimensionless numbers that demarcate spreading regimes. Two such numbers that arise are

$$\text{Ohnesorge number}: \quad \text{Oh} = \frac{\mu}{\sqrt{\rho\sigma\ell}} \tag{2.17}$$

$$\text{Bond number}: \quad \text{Bo} = \frac{\rho g \ell^2}{\sigma} \tag{2.18}$$

The length scale definition based on drop volume ensures that the dynamics of equal volume drops is mapped on to the same (Bo, Oh) space. For a particular liquid, the two numbers are not independent and are connected through the drop volume. In other words, when comparing spreading across liquid drops of differing sizes, Oh and Bo change concurrently. As drop volume is increased, Bo increases quadratically with the length scale while Oh decreases with an inverse square root dependence. Physically, such a trend amounts to a dominant body force and reduced viscous forces for larger drops.

The length scale ℓ and velocity scale u_0 reproduce other relevant scales, namely timescale (ℓ/u_0), pressure scale (σ/ℓ), and curvature-scale $(1/\ell)$. Dimensionless forms of the governing equations are now stated as

$$\frac{\mathrm{d}U_P}{\mathrm{d}t} = \frac{8}{5}\left(\frac{P_P - \kappa_P}{R_P}\right) - \frac{18}{5}\mathrm{Oh}\frac{U_P}{R_P^2} + \mathrm{Oh}\frac{R_{uf}^2}{R_P^2 H_P (H_P + H_U)}(a_{UU}U_U - a_{PU}U_P)$$
$$- \mathrm{Oh}\frac{R_{lf}^2}{R_P^2 H_P (H_P + H_L)}(a_{PL}U_P - a_{LL}U_L) + \frac{9}{5}\frac{r_c}{R_P^2 H_P}(\cos\theta_{eq} - \cos\theta) \tag{2.19}$$

$$\frac{\mathrm{d}W_P}{\mathrm{d}t} = -\frac{2R_{uf}^2}{3R_P^2 H_P}P_{uf} + \frac{2R_{lf}^2}{3R_P^2 H_P}P_{lf} - \frac{3R_P^2 - 2R_{uf}^2}{3R_P^2 H_P}\kappa_{uf} + \frac{3R_P^2 - 2R_{lf}^2}{3R_P^2 H_P}\kappa_{lf}$$
$$+ \mathrm{Oh}\frac{2R_{uf}^2}{R_P^2 H_P (H_P + H_U)}(W_U - W_P) - \mathrm{Oh}\frac{2R_{lf}^2}{R_P^2 H_P (H_P + H_L)}(W_P - W_L) \mp \mathrm{Bo} \tag{2.20}$$

$$\frac{8U_P}{3R_P} + \frac{1}{H_P}(W_{uf} - W_{lf}) = 0 \tag{2.21}$$

$$\frac{\mathrm{d}R_P}{\mathrm{d}t} = \frac{4}{3}U_P \tag{2.22}$$

In the present set of equations, each variable is dimensionless. As stated earlier, the contact line force in Eq. (2.19) is only present for the disk in contact with the solid substrate. Axial boundary conditions in Table 2.4 can be appropriately translated to their dimensional form.

2.3.2 *Pressure-Velocity Coupling and the Solution Algorithm*

The governing equations (Eqs. 2.19–2.22) represent a set of four equations in four variables, namely velocities (u, w), pressure, and interface shape. However, there is no explicit expression for pressure calculation. The pressure field at the new time instant is to be found in such a way that volume conservation, i.e., mass balance, is ensured. The disconnect between the pressure variable and the mass balance equation, making it necessary to determine the pressure in an indirect manner is typical of simulations involving incompressible fluid flow. Following the SIMPLE strategy adopted in the finite volume method, a predictor-corrector approach is used in the present study to transform the continuity equation (Eq. 2.21) to an equivalent pressure update relation. Mathematical expressions arising from this approach are described in the previous work of the authors (Mistry and Muralidhar 2015, 2018) and are summarized below.

Predictor step:

$$
\begin{aligned}
\frac{\tilde{U}_P - U_P^p}{\Delta t} &= -\frac{18}{5}\mathrm{Oh}\frac{\tilde{U}_P}{R_P^2} + \mathrm{Oh}\frac{R_{uf}^2}{R_P^2 H_P(H_P+H_U)}(a_{UU}\tilde{U}_U - a_{PU}\tilde{U}_P) \\
&- \mathrm{Oh}\frac{R_{lf}^2}{R_P^2 H_P(H_P+H_L)}(a_{PL}\tilde{U}_P - a_{LL}\tilde{U}_L) + \frac{9}{5}\frac{r_c}{R_P^2 H_P}(\cos\theta_{eq} - \cos\theta)
\end{aligned} \tag{2.23}
$$

$$
\begin{aligned}
\frac{\tilde{W}_P - W_P^p}{\Delta t} &= \mathrm{Oh}\frac{2R_{uf}^2}{R_P^2 H_P(H_P+H_U)}(\tilde{W}_U - \tilde{W}_P) \\
&- \mathrm{Oh}\frac{2R_{lf}^2}{R_P^2 H_P(H_P+H_L)}(\tilde{W}_P - \tilde{W}_L) \mp \mathrm{Bo}
\end{aligned} \tag{2.24}
$$

Pressure relation:

$$
\begin{aligned}
&P_P^{p+1}\left(\frac{64\Delta t}{15R_P^2}+\frac{4\Delta t}{3H_P^2}\right) - P_U^{p+1}\left(\frac{2\Delta t}{3H_P H_U}\right) - P_L^{p+1}\left(\frac{2\Delta t}{3H_P H_L}\right) = -\left(\frac{8\tilde{U}_P}{3R_P} + \left(\frac{\tilde{W}_{uf}-\tilde{W}_{lf}}{H_P}\right)\right) \\
&+\kappa_P^{p+1}\left(\frac{64\Delta t}{15R_P^2}-\frac{2\Delta t}{3H_P^2}\right) + \kappa_U^{p+1}\left(\frac{\Delta t}{3H_P H_U}\right) + \kappa_L^{p+1}\left(\frac{\Delta t}{3H_P H_L}\right) \\
&+\kappa_{uf}^{p+1}\left(\frac{\Delta t}{H_P}\left(\frac{1}{H_P}-\frac{1}{H_U}\right)\right) - \kappa_{lf}^{p+1}\left(\frac{\Delta t}{H_P}\left(\frac{1}{H_L}-\frac{1}{H_P}\right)\right)
\end{aligned} \tag{2.25}
$$

Corrector step:

$$
\frac{U_P^{p+1} - \tilde{U}_P}{\Delta t} = \frac{8}{5}\left(\frac{P_P^{p+1} - \kappa_P^{p+1}}{R_P}\right) \tag{2.26}
$$

$$
\begin{aligned}
\frac{W_P^{p+1} - \tilde{W}_P}{\Delta t} &= -\left(\frac{2R_{uf}^2}{3R_P^2 H_P}\right)P_{uf}^{p+1} + \left(\frac{2R_{lf}^2}{3R_P^2 H_P}\right)P_{lf}^{p+1} \\
&- \left(\frac{3R_P^2 - 2R_{uf}^2}{3R_P^2 H_P}\right)\kappa_{uf}^{p+1} + \left(\frac{3R_P^2 - 2R_{lf}^2}{3R_P^2 H_P}\right)\kappa_{lf}^{p+1}
\end{aligned} \tag{2.27}
$$

Here, superscripts p and p + 1 identify the current time step, where the solution is known and the next time step, respectively. Velocities $\left(\tilde{U}, \tilde{W}\right)$ account for contributions from viscous stresses, body force, and contact line imbalance and are determined from the predictor step, namely Eqs. (2.23) and (2.24). These velocities are used to obtain an appropriate pressure field (Eq. 2.25) that leads to a divergence-free velocity field during the corrector step, Eqs. (2.26) and (2.27). The divergence-free velocities are used to obtain new geometry of the gas-liquid interface

$$\frac{dR_P^{p+1}}{dt} = \frac{4}{3} U_P^{p+1} \tag{2.28}$$

Given the continuous nature of drop deformation, implicit time discretization is adopted in the present study. Iterative calculations are thus carried out within a time step as both geometry and flow field evolve in time. The sequence of computational steps can be enumerated as follows:

i. Initialize velocity field to zero, provide values of the dimensionless quantities Oh and Bo, equilibrium contact angle, and prescribe the initial geometry as a full sphere.
ii. Discretize the drop into a specified number of disks and evaluate geometrical properties.
iii. Predict the velocity field using Eqs. (2.23)–(2.24).
iv. Estimate new pressures compatible with a divergence-free velocity field using Eq. (2.25).
v. Modify velocities $\left(\widetilde{U}, \widetilde{W}\right)$ with the new pressure field Eqs. (2.26) and (2.27).
vi. Update geometry based on the divergence-free velocity Eq. (2.28).
vii. Compute revised geometrical estimates including curvature.
viii. If the differences between the successive estimates of the new velocities (U^{p+1}, W^{p+1}) and new geometry fall within the specified tolerance, proceed to step (ix); else, repeat steps (iii) to (vii) with the most recent geometry estimates. A convergence criterion of less than 10^{-10} in the root mean square of each of the quantities being tracked is recommended.
ix. Advance time and compute flow and geometry at the next time step by repeating steps (iii) to (viii) till steady-state in the form of an invariant drop shape or insignificant velocities is reached.

Since the axial direction alone is explicitly discretized, the matrix system from Eqs. (2.23)–(2.25) is tridiagonal and can be solved directly using Thomas' algorithm (Hoffman and Frankel 2001).

2.4 Validation

Late time spreading of small drops is determined from competing viscous and capillary forces. Such dynamics is well-studied (Tanner 1979) and is characterized by a power-law dependence of spreading radius on time, i.e., $r_c \sim t^{\alpha}$ with an exponent $\alpha \approx 0.1$. Recent experiments shed light on early time fast transients (Eddi et al. 2013; Biance et al. 2004; Bird et al. 2008; Winkels et al. 2012). The initial spreading regime is characterized by a much higher exponent of $\alpha \sim 0.5$. Figure 2.3 presents simulated trends based on the Lagrangian formulation discussed in this chapter. Experimentally observed trends are present here for both water and glycerin drops. The high viscosity of glycerin (Table 2.2) results in differences of timescale as compared to water. The late time spreading of glycerin departs, to some extent, from the Tanner limit of $\alpha \sim 0.1$ in Fig. 2.3. This observation has also been recorded in the literature (Eddi et al. 2013).

Experiments with water drops on surfaces with well-defined wettability reveal that the spreading history comprises three regimes: wettability-independent rapid spreading at a short time ($\alpha \sim 0.5$); intermediate regime, where the spreading exponent scales inversely with equilibrium contact angle, i.e., slower spreading for larger equilibrium angles; and late time approach to equilibrium as per Tanner's law. On a fully wetting surface, spreading may continue for a long time. On surfaces with non-zero contact angles, however, spreading ceases in finite time. The present set of calculations, described in the previous work of the authors (Mistry and Muralidhar (2015a, b, 2018) capture these features and is shown in Fig. 2.4. The mechanistic origins of such characteristics are discussed in the following sections.

The sensitivity of the pure spreading experiments to the details of initial conditions and related perturbations make comparisons with theory an involved prospect. The convention is to match statistical measures such as spreading exponents between simulation and experiments. Figures 2.5 and 2.6 compare theory and experiments in the limit of zero impact velocity. Given the finite velocity of drop impact, the time instant of contact can be fixed without ambiguity. Figure 2.5 compares the drop shape evolution for drops of water and transformer oil. Figure 2.6 extracts dynamic contact angle transients for the three different liquids studied. The low viscosity of water makes it particularly prone to uncertainties in the initial contact, and as a consequence Fig. 2.6a exhibits a poor match, though the oscillatory trend is faithfully resolved. High viscosity oils demonstrate a much better correlation in time (Fig. 2.6b, c). The experiments in the literature (Arkhipov et al. 2015) were carried out with three different liquid drops spreading on acrylic glass. Besides the difference in fluid properties, equilibrium contact angles vary. Calculations presented in

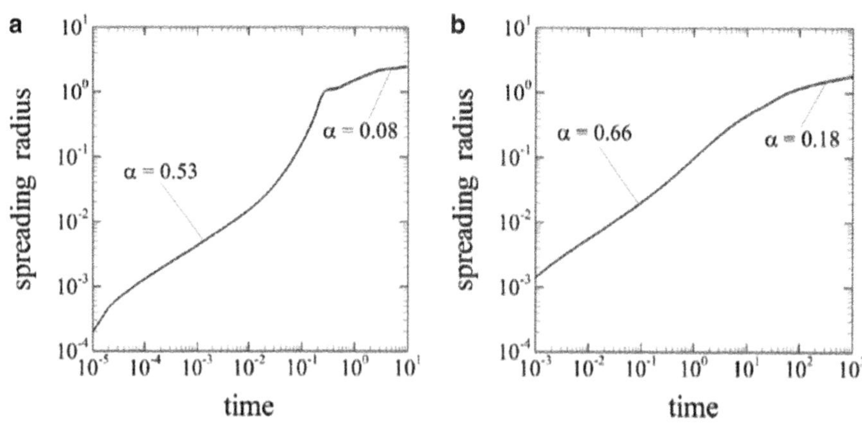

Fig. 2.3 Time evolution of spreading radius for sessile liquid drops of 1% capillary volume (**a**) water and (**b**) glycerin. Bo = 1 for capillary volume, and Bo = 0.05 for 1% capillary volume (Mistry and Muralidhar 2015)

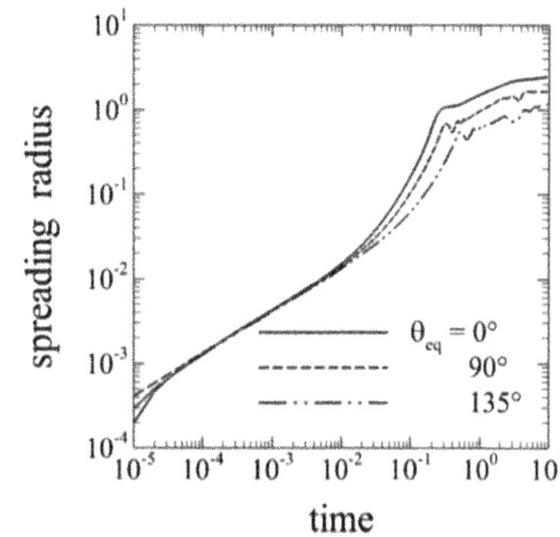

Fig. 2.4 Effect of the substrate, i.e., equilibrium contact angle on spreading dynamics as characterized on a spreading radius versus time plot. Three different wetting conditions are shown in terms of equilibrium contact angles 0° (complete wetting), 90° and 135° (hydrophobic) (Mistry and Muralidhar 2015)

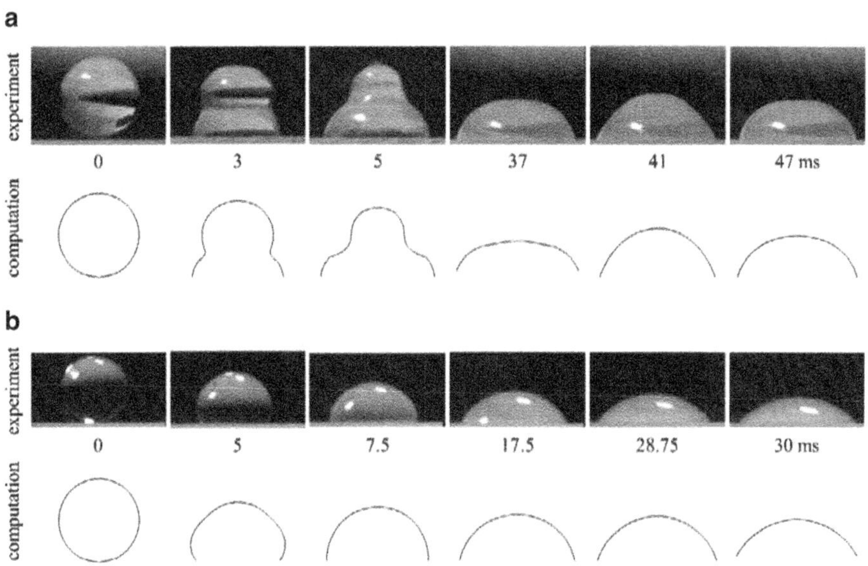

Fig. 2.5 Comparison of drop shape evolution during slow velocity impact of (**a**) water and (**b**) transformer oil drops. The time of impact is fixed as $t = 0$. Experimental imaging and computational predictions are shown (Arkhipov et al. 2015)

Figs. 2.5 and 2.6 account for these differences in the thermophysical and surface properties.

Given the discrete nature of the calculations, temporal and spatial refinements, i.e., Δt (time step) and M (number of equal volume disks) are fixed based on mesh independence tests. The time axis is appropriately shifted to account for the initial contact singularity.

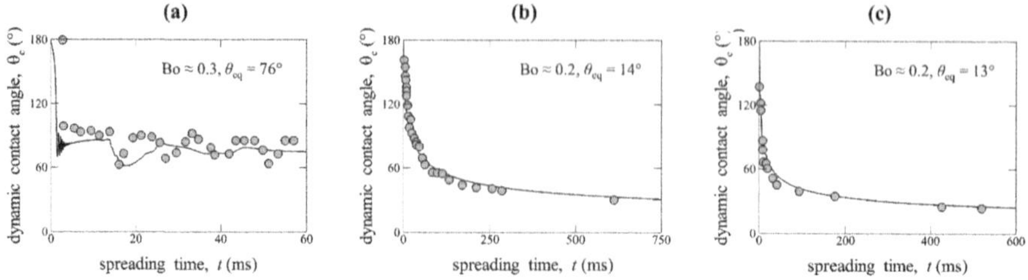

Fig. 2.6 Spreading of (**a**) water, (**b**) silicone oil, and (**c**) transformer oil on acrylic glass, characterized in terms of dynamic contact angle evolution. The drops approach substrates with small but finite velocities to be consistent with the experimental setting (○ Experimental results from Arkhipov *et al.* (2015) *J. Engg. Phys. Thermophysics*, **88** (1), pp. 42–51 compared with the solid line of the present work.)

The strength of the Lagrangian approach adopted in the present work is in its ability to predict spreading characteristics of very different liquid drops and surface wettability, without resorting to any tunable parameter. In addition, there is no need to impose the contact line velocity which emerges as an outcome of the simulation. These data arise naturally from the generalized force imbalance prescribed along the contact line.

2.5 Results and Discussion

The initial capillary force imbalance at the three-phase contact line is responsible for the initiation of spreading. Water and glycerin drops have distinct spreading characteristics, given their vast viscosity differences. A proper comparison across various liquids should be in terms of the dimensionless volume, i.e., identical Bond numbers (Eq. 2.18). Table 2.5 compares the physical dimensions of water and glycerin drops for equal Bond numbers. The corresponding Ohnesorge numbers, Oh, presented in Table 2.5, identify the flow regimes. A large value of Oh indicates dominant viscous resistance to the evolving flow field.

Drop shapes are reconstructed from (R_P, H_P) data available for the individual disks at each time instant. Differences in drop shape evolution for water and glycerin are qualitatively apparent from the data in Fig. 2.7. Viscous effects are strong in glycerin as seen from the monotonic trend of relaxation to equilibrium from the initial spherical shape. In comparison, a water drop shows inertial oscillations during its approach to equilibrium. Unlike glycerin drops (Fig. 2.7c, d) which maintain a nearly spherical shape even as they spread, water drops initially deform strongly close to the solid surface, while maintaining a near-spherical shape away from it. Gradually, the near-wall deformation grows and engulfs the entire drop, as it spreads with the passage of time. Such observations can be correlated to the viscous timescale, $\tau_\mu \sim \ell^2\rho/\mu$. Viscous timescale refers to the time interval associated with the propagation of shear in velocity. The viscosity of water is much lower than

Table 2.5 Representative length scale and Ohnesorge numbers for water and glycerin drops (Mistry and Muralidhar 2015)

	Water			Glycerin		
Bo	ℓ (mm)	Volume (μl)	Oh	ℓ (mm)	Volume (μl)	Oh
1	2.73	85	2.3×10^{-3}	2.26	49	3.5
0.22	1.27	8.5	3.3×10^{-3}	1.05	4.9	5.1
0.05	0.59	0.85	4.8×10^{-3}	0.49	0.49	7.4

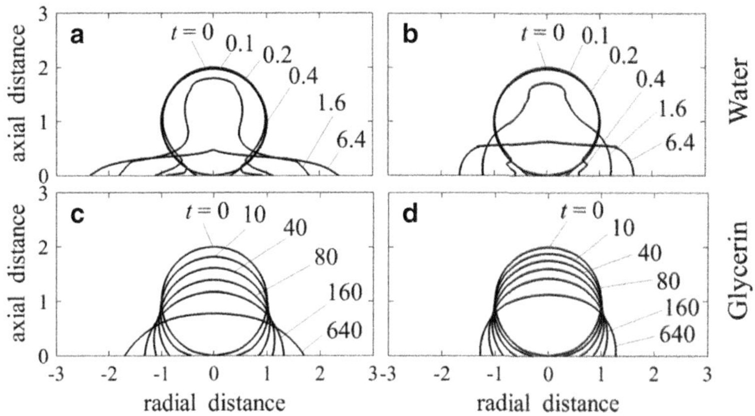

Fig. 2.7 Simulated drop shape evolution for (**a**, **b**) water and (**c**, **d**) glycerin drop of 1% capillary volume (Bo = 0.05) at two different wetting conditions (**a**, **c**) $\theta_{eq} = 0°$ and (**b**, **d**) $\theta_{eq} = 90°$

glycerin but their densities and length scales are comparable for a given Bond number; Tables 2.2 and 2.5. As a result, $\tau_{\mu,\text{ water}} \gg \tau_{\mu,\text{ glycerin}}$. Hence, as the fluid layer closest to solid responds to contact line imbalance (Eq. 2.6), it does not immediately drag subsequent layers of water above. Such behavior is also observed for other levels of surface wettability (Fig. 2.7a vs b). Geometrical differences in Fig. 2.7a, b also arise from differences in equilibrium shapes that the drops tend to approach.

The deformation process of low viscosity water drops is qualitatively sketched in Fig. 2.8. At short times, fluid layers close to the contact line experience spreading given the contact angle imbalance ($\theta \neq \theta_{eq}$). This leads to concave (negative) curvature close to the contact line, forming a neck, as shown in Fig. 2.8b. The drop maintains a spherical shape away from the contact line. The spherical shape has convex (positive) curvature. The curvature gradients contribute to the pressure gradient within and along the free surface and cause fluid flow from high to low pressure locations. This flow will diminish curvature gradients if the contact line were to be pinned over the substrate. However, over time, contact line motion pulls other liquid sheets above the neck of the drop sustaining curvature gradients over the free surface. Pressure gradients continue to drive flow towards the contact line, displacing it further over the surface (Fig. 2.8). Such synergistic contributions

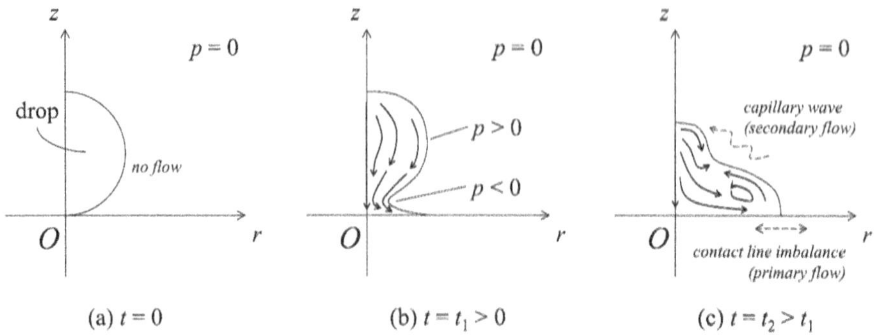

Fig. 2.8 Sketches of drop shape evolution for low viscosity drops. Three distinct times are shown. (**a**) Initial drop just before spreading onset; (**b**) at short times, drop deforms close to the solid surface while retaining near-spherical shape elsewhere; (**c**) at larger times, capillary wave has spread throughout the drop

drive spreading of a low viscosity liquid drop. The onset of this regime of negative curvature is marked by the time when the contact radius becomes comparable to the drop size, i.e., $r_c \sim \ell$.

With the passage of time, concave curvature is observed towards the apex of the drop (Fig. 2.8c). In view of the low pressure prevailing at a site of concave curvature, flow takes place at other instants from the apex towards the concave location. Migration of the site of concave curvature from the contact line to the apex has the appearance of a wave and is termed *capillary wave* propagation (Bird et al. 2008; Winkels et al. 2012) (Fig. 2.8). It is also responsible for oscillatory velocities attained by the liquid. During this time interval, contact line motion and the overall flow field are predominantly governed by contact angle imbalance (Fig. 2.8). In turn, spreading is wettability-dependent, as shown in Fig. 2.4 for intermediate time periods. Beyond this time interval, the liquid body loses kinetic energy by viscous dissipation in the bulk and at the contact line. This is the capillary-viscosity regime for which Tanner's law is applicable. Thus, there are two mechanisms driving fluid motion: contact angle imbalance (Eq. 2.6) and capillary wave propagation. For large drops, gravity additionally promotes spreading.

These three regimes for drop spreading for water and glycerin are highlighted in terms of the contact angle evolution in Fig. 2.9. The short time, wettability-independent regime has a nearly spherical drop shape ($\theta \sim 180°$) and the contact radius, r_c, increases rapidly (compare Fig. 2.4 for spreading radius and Fig. 2.9 for the contact angle). Drop shape further deforms locally, gives rise to capillary waves and flow oscillations in water which manifest as time-dependent fluctuations in the dynamic contact angle. At late times, contact angles gradually diminish towards the equilibrium value. On the contrary, for a glycerin drop, viscosity affects spreading from the beginning. It retards the speed of drop shape evolution and inhibits the formation of capillary waves related to negative curvature and hence, inertial oscillations. This difference between water and glycerin is apparent in the dynamic contact angle variation with time in Fig. 2.9.

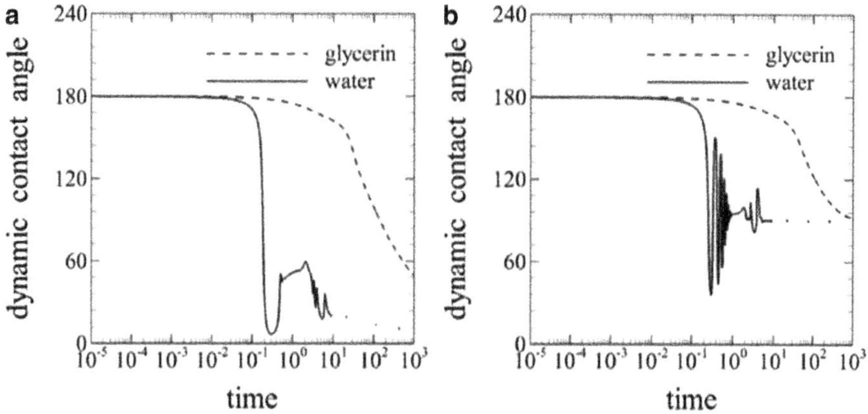

Fig. 2.9 Time evolution of (dynamic) contact angle for water and glycerin drops of 1% capillary volume (Bo $= 0.05$) for two distinct wetting conditions (**a**) $\theta_{eq} = 0°$ and (**b**) $\theta_{eq} = 90°$. Corresponding drop shape evolutions are shown in Fig. 2.7

Spreading under partial wetting conditions ($\theta_{eq} = 90°$, Fig. 2.9b) follows the general trend of their fully wetting counterparts. While water shows oscillations, for glycerin, the late stage relaxation is comparatively slower, given the smaller driving force related to a smaller contact angle imbalance. For the drop spreading on a fully wetting surface (Fig. 2.9a), contact angle imbalance is always of constant sign since $\theta \geq \theta_{eq}$. Flow is generated due to free surface relaxation in response to the curvature disparity and is responsible for the oscillations. For a partially wetting surface, the dynamic contact angle may overshoot the equilibrium value, i.e., $\theta \lesseqgtr \theta_{eq}$, causing intermittent sign reversal of the contact line force and is a source of heightened oscillations. The viscous resistance does not completely materialize until the flow field is established throughout the drop leading to Tanner's law regime for subsequent times.

A phase plot of contact line velocity with the dynamic contact angle is shown in Figs. 2.10 and 2.11. Both water and glycerin are considered. In Fig. 2.10, the phase plot shows a clear dependence on the equilibrium contact angle for both the fluids. The phase plot covers a larger area for spreading on a fully wetting surface, indicating larger amplitude of oscillations. In Fig. 2.11, the phase plot is presented for various Bond numbers. The effect of the Bond number is seen to be quite minor as the drop volumes studied here fall in the capillary dominated flow. These trends are in conformity with contact line models (as in Table 2.1) wherein the dynamic contact angle is shown to be a function of the contact line velocity and equilibrium contact angle but not the Bond number. For larger drops (Bo > 1), gravity becomes the dominant spreading influence and contact line motion is expected to be qualitatively different.

For incomplete wetting, the contact line force (Eq. 2.6) shows intermittent sign reversal, causing a mixed advancing-receding motion of the contact line in water

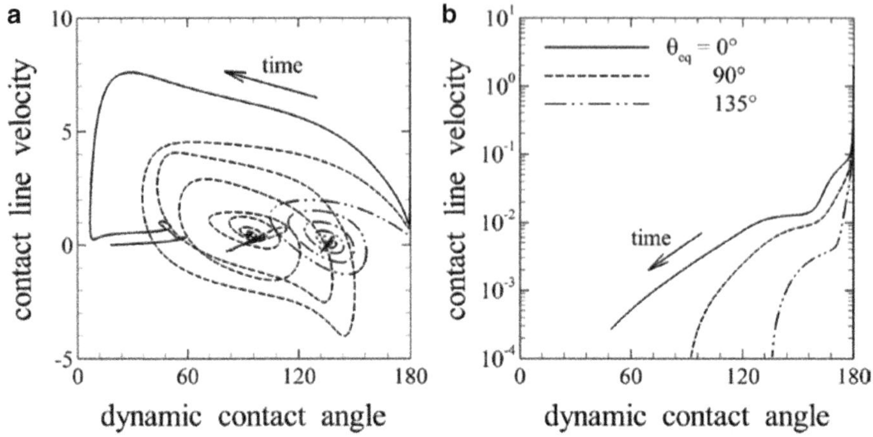

Fig. 2.10 Non-universal relations between contact line velocity and instantaneous (dynamics) contact angle for (**a**) water and (**b**) glycerin drops of 1% capillary volume (Bo = 0.05) for different wetting conditions. The data neither collapse across different fluids nor for different equilibrium angles even for identical drop volumes, thus underscoring the absence of universality commonly assumed

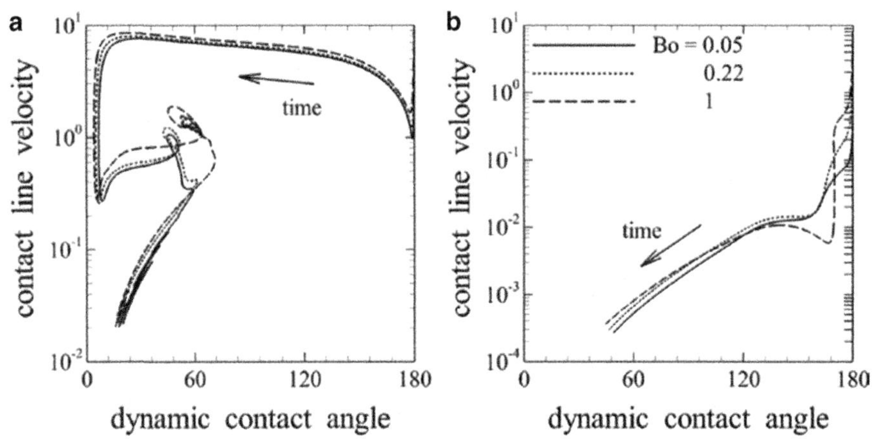

Fig. 2.11 Contact line velocity *versus* dynamic contact angle for (**a**) water and (**b**) glycerin drops of three different volumes–capillary volume (Bo = 1), 10% capillary volume (Bo = 0.22) and 1% capillary volume (Bo = 0.05). Spreading is shown in the fully wetting limit, in turn, contact line velocities are positive even for the water drops

(Fig. 2.10a). Contact line velocities are oscillatory but positive for a fully wetting surface. Oscillations are practically absent in glycerin. The range of velocities acquired by water is smaller at higher contact angles while the range is less sensitive to contact angles in glycerin. In both fluids, the time required to reach equilibrium diminishes with increasing contact angle.

Though the Bond number is less important in the phase plots of Fig. 2.11, a departure from universality is seen for a Bond number of unity (the dashed line in Fig. 2.11a, b). This is expected because, for large Bond numbers, the drop volume is greater and inertial oscillations are expected to be highly pronounced.

2.5.1 Spreading of Pendant Drops and Energy Landscape

Spreading dynamics of pendant water and glycerin drops is summarized in Figs. 2.12 and 2.13, respectively. Wetting conditions for the two liquids are varied to facilitate comparison. For small drops (compared to the capillary volume), gravity has a vanishing contribution and sessile and pendant configurations show nearly identical spreading rates. The spreading trends of a pendant drop in terms of the evolution of footprint radius (Figs. 2.12a and 2.13a), dynamic contact angle (Figs. 2.12b and 2.13b), and contact line velocity conform to earlier observations on sessile drops (Figs. 2.3, 2.4, 2.9, and 2.11). For identical drop volumes, sessile drops evolve to an equilibrium shape that is flat at the apex, while pendant drops are vertically elongated relative to a spherical shape for a given equilibrium contact angle. Equivalently, spreading radii of the footprint follow the relation: sessile > spherical cap > pendant (Mistry and Muralidhar 2018).

Energetically, spreading drops pose an interesting question. For a given equilibrium angle, the initial and final configurations are both static, being only a function of the drop volume, namely Bond number. The final equilibrium shape attained is computable from the Young-Laplace equation. Thus, equal Bond number drops on surfaces with identical equilibrium contact angles will have identical shapes, namely energies, in dimensionless form. While the initial and final energies are independent of the hydrodynamic properties of the liquid, the spreading dynamics strongly depends on the choice of the liquid (Figs. 2.12 and 2.13). As a consequence, each liquid may adopt an independent path on an energy landscape, some of which may result in instability. This situation is realized in pendant drops, as discussed below.

To quantify the energy transitions in an isothermal drop, Navier-Stokes equations (Eq. 2.2) are used to derive a mechanical energy equation (Kundu and Cohen 2004). The mechanical energy balance classifies conservative and non-conservative forces as energy storage and dissipation modes, respectively. It is of the form

$$\frac{\mathrm{d}E}{\mathrm{d}t} = \mathrm{Oh} \int\limits_{\mathrm{TPZ}} (\boldsymbol{u} \cdot \nabla \boldsymbol{u}) \cdot \widehat{\boldsymbol{n}}\, \mathrm{d}S - \mathrm{Oh} \int\limits_{\mathrm{CV}} (\nabla \boldsymbol{u} : \nabla \boldsymbol{u}) \mathrm{d}V \tag{2.29}$$

where total energy E has the following components:

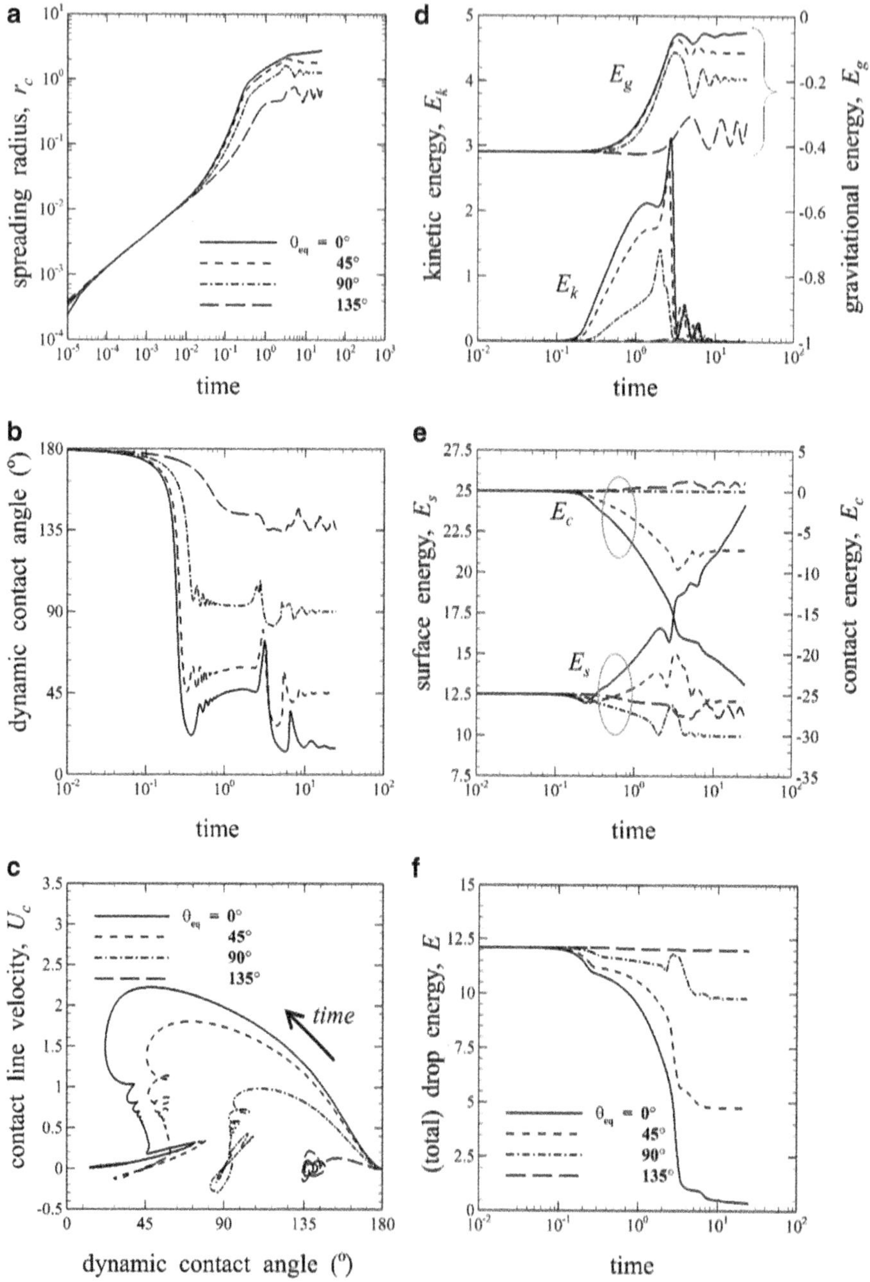

Fig. 2.12 Spreading of pendant water drops of identical volumes (Bo = 0.10) and different substrate wettability: (**a**) spreading radius, (**b**) contact angle, (**c**) contact line velocity, (**d**) bulk energies, (**e**) interfacial energies, and (**f**) total energy evolution over time. Wetting conditions are characterized in terms of equilibrium contact angles, $\theta_{eq} = 0°, 45°, 90°$, and $135°$

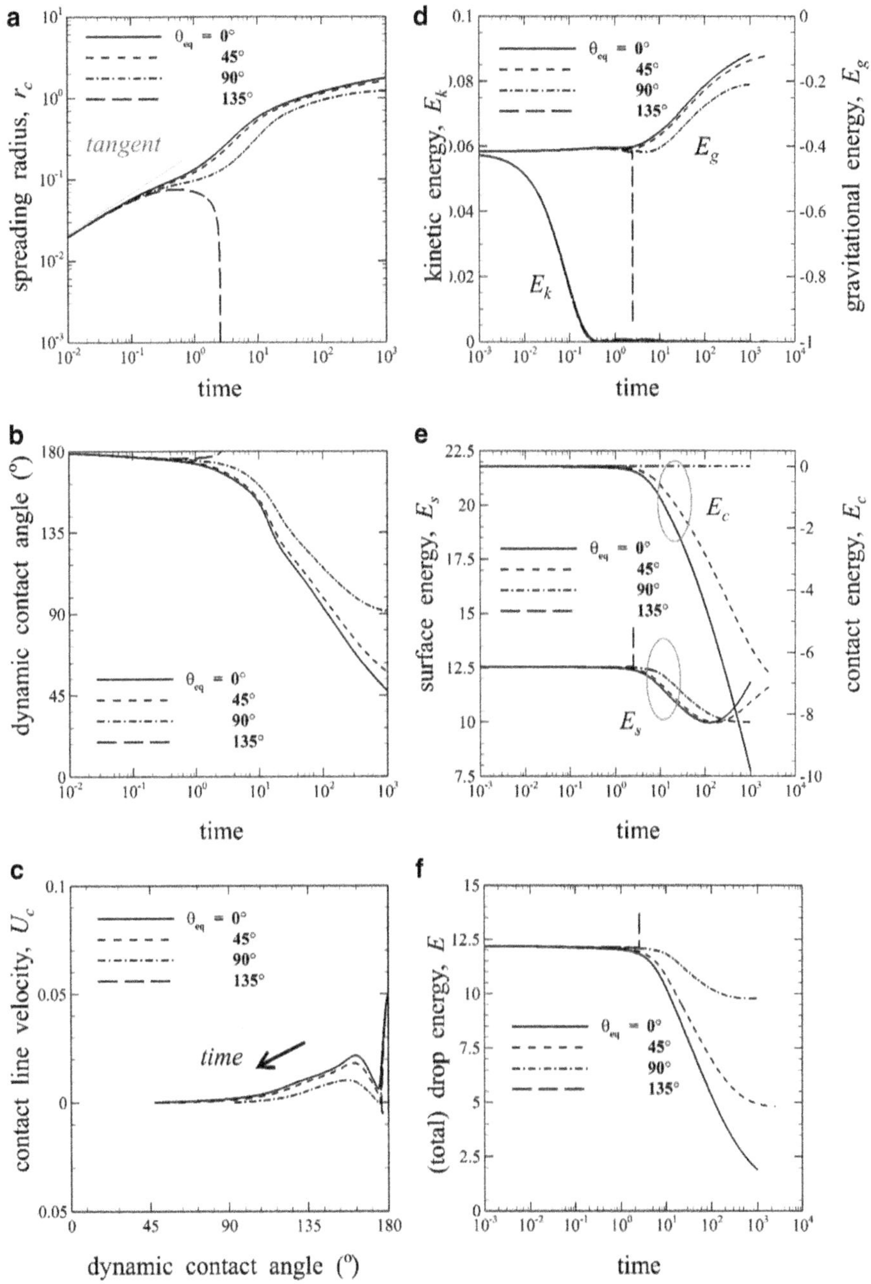

Fig. 2.13 Spreading of pendant glycerin drops of identical volumes (Bo = 0.10) and different substrate wettability: (**a**) spreading radius, (**b**) contact angle, (**c**) contact line velocity, (**d**) bulk energies, (**e**) interfacial energies, and (**f**) total energy evolution over time. Wetting conditions are characterized in terms of equilibrium contact angles, $\theta_{eq} = 0°, 45°, 90°$, and 135°

$$\text{Kinetic energy}: \quad E_{\mathrm{k}} = \int\limits_{\mathrm{CV}} \frac{1}{2} \boldsymbol{u} \cdot \boldsymbol{u} \, \mathrm{d}V \tag{2.30}$$

$$\text{Gravitational potential energy}: \quad E_{\mathrm{g}} = \mathrm{Bo} \int\limits_{\mathrm{CV}} (\pm z) \, \mathrm{d}V, \quad \text{where sessile}(+), \text{pendant}(-) \tag{2.31}$$

$$\text{Free-surface energy}: \quad E_{\mathrm{s}} = \int\limits_{\mathrm{CS_{free}}} \mathrm{d}S \tag{2.32}$$

$$\text{Energy of the contact patch}: \quad E_{\mathrm{c}} = \int\limits_{\mathrm{CS_{contact}}} \left(-\cos\theta_{\mathrm{eq}}\right) \mathrm{d}S = \left(-\cos\theta_{\mathrm{eq}}\right) \pi r_{\mathrm{c}}^2 \tag{2.33}$$

Kinetic energy, E_{k}, and gravitational energy, E_{g}, are for the bulk of the drop, evaluated over the entire control volume. Free surface energy, E_{s}, and contact line energy, E_{c} require interfacial considerations. For small Bond numbers, relative interfacial energy contributions are expected to be significant. Energy balance (Eq. 2.29) suggests two dissipation mechanisms: dissipation close to the three-phase contact (the first term on the right) and viscous dissipation (the second term). Initially, the drop is static, the contact area with the substrate is zero, and total energy is composed of $E_{\mathrm{g}} = \pm \mathrm{Bo} \cdot \ell$ and $E_{\mathrm{s}} = 4\pi\ell^2$ while it is made up of E_{g}, E_{s} and E_{c} in the final equilibrium state. Spreading is spontaneous since $E_{\text{initial}} > E_{\text{final}}$, the difference being accounted for by various modes of dissipation during drop movement.

First, compare the total energy evolution of drops of water (Fig. 2.12f) and glycerin (Fig. 2.13f) on surfaces of different wettability, namely the equilibrium contact angle. Despite the endpoints being equilibrium states, their temporal evolution is markedly different. Glycerin drop on a hydrophobic surface with $\theta_{\mathrm{eq}} = 135°$ falls off due to the gravitational instability; an aspect discussed later in this section. During spreading but before fall-off, the energy response of glycerin is comparatively monotonic, relative to water. The kinetic energy of the glycerin drop (Fig. 2.13d) is smaller than water (Fig. 2.12d), making inertia insignificant in glycerin. As the pendant drop spreads, its centroid moves towards the substrate and the gravitational energy becomes less negative; see Figs. 2.12d and 2.13d. Surface energy varies with the drop shape (Figs. 2.12e and 2.13e) while the contact energy relates to spreading radius transients (Figs. 2.12e and 2.13e), with a sign that depends on the equilibrium contact angle. These components jointly give rise to the total energy trends seen in Figs. 2.12f and 2.13f. For the water drop, dominant energy reduction is during the intermediate stage where wettability-dependent spreading is observed. On the other hand, energy loss in glycerin is spread over the entire spreading duration. Here, the short-time relaxation to $r_c \sim 0.1\ell$ takes place at nearly constant energies (Fig. 2.13a, f). Water drops start dissipating energy much earlier and constant energy spreading is observed for early relaxation times when $r_c \sim 10^{-2}\ell$ (Fig. 2.12a, f). This is because viscosity of water is low and necessitates dissipation over a wider time interval.

Despite the small size of the liquid drops, shear stresses generated during spreading are important since they can result in wear and tear of the chemical coating (Arkhipov et al. 2015). In contrast, physically textured or pillared surfaces are less prone to damage and are long-lived. The drop spreading characteristics on such surfaces are, however, expected to be quite different from a chemically coated surface. For the present model, stress variations are examined at the drop footprint in Fig. 2.14 for the flow fields discussed in Figs. 2.12 and 2.13. Three distinct force components are at play here: viscous shear stress, viscous contribution to normal stress, and pressure. Each of their contributions is integrated over the drop footprint as per the following dimensionless relations:

$$\text{Shear stress}: \quad f_{\text{shear}} = \frac{4\text{Oh}}{r_c^2} \int_0^{r_c} \left.\frac{\partial u}{\partial z}\right|_{z=0} r\,dr \tag{2.34}$$

$$\text{Normal stress}: \quad f_{\text{normal}} = \frac{4\text{Oh}}{r_c^2} \int_0^{r_c} \left.\frac{\partial w}{\partial z}\right|_{z=0} r\,dr \tag{2.35}$$

$$\text{Pressure}: \quad f_{\text{pressure}} = \frac{4}{r_c^2} \int_0^{r_c} p|_{z=0}\, r\,dr \tag{2.36}$$

Shear stresses (Fig. 2.14a) in water show oscillations, in response to those in the contact line motion. For a partially wetting surface, velocities are smaller as in Fig. 2.12d and the corresponding shear stresses are also smaller. The normal stress variations in Fig. 2.14b can also be explained by reference to velocity distribution. Strong vertically directed flow is generated when the contact line relaxes in response to curvature gradients by supplying flow from the apex to the region near the contact line. Such a flow structure is characteristic of drops spreading over surfaces of intermediate-to-high wettability.

When the interface curvature turns concave near the contact line, wall pressure variation will show a steep minimum (Fig. 2.14c). For a non-zero equilibrium contact angle, drop shapes are convex (approaching spherical) in the long run, leading to positive wall pressures. For a fully wetting surface, curvatures go to zero with time and so do the wall pressures over the footprint.

Structural differences in flow behavior between water and glycerin give rise to visible differences in forces over the footprint (Fig. 2.14a–c vs d–f). Unlike water, shear stresses in glycerin (Fig. 2.14d) are devoid of oscillations. The relaxation of the initial contact line singularity causes stresses at the short times that are independent

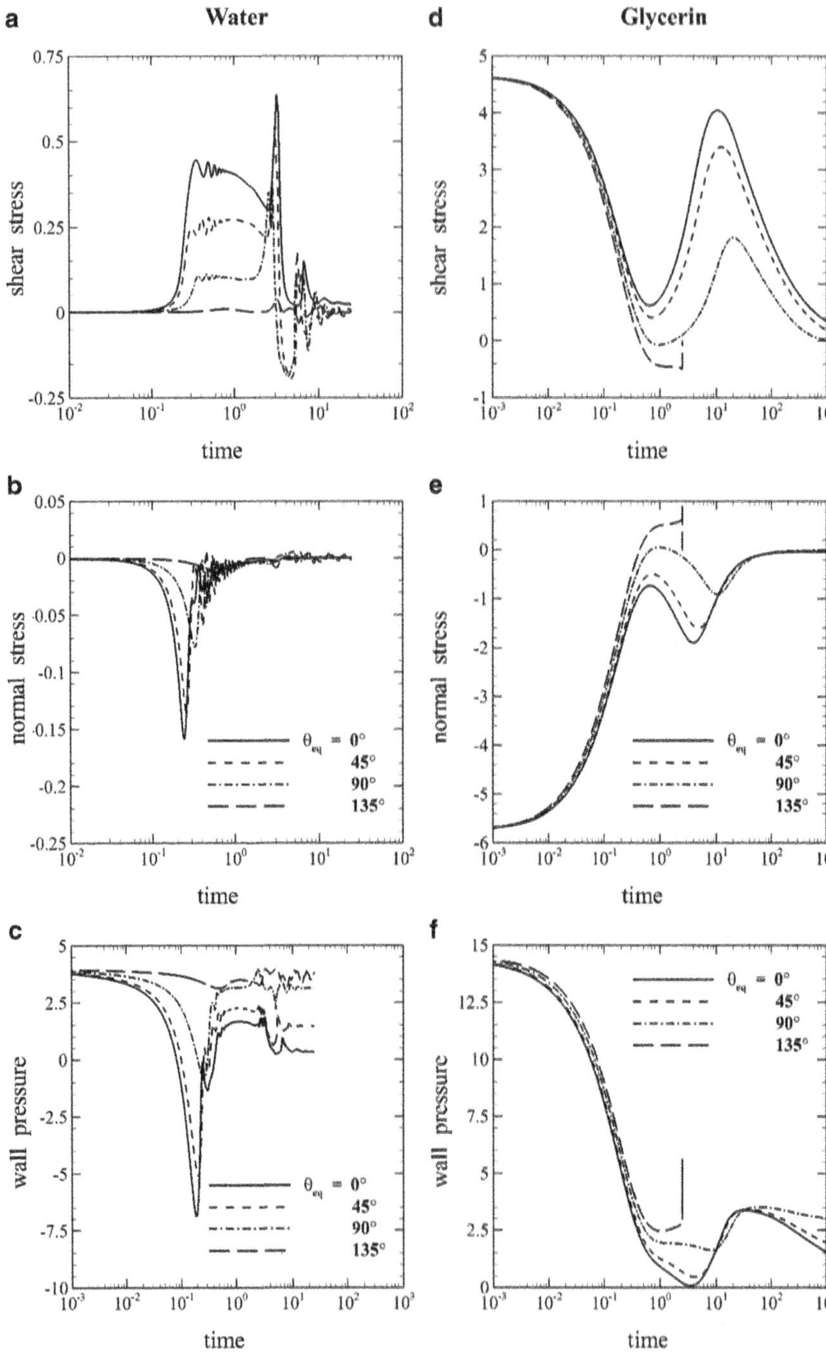

Fig. 2.14 Time evolution of forces at the footprint during spreading of pendant (**a–c**) water and (**d–f**) glycerin drops. These trends complement the spreading dynamics reported in Figs. 2.12 and 2.13. (**a**, **d**) are shear stresses; (**b**, **e**) are normal stresses; and (**c**, **f**) are wall pressure over the drop footprint

of surface wettability (Fig. 2.14d–e). In water (Fig. 2.14a, b), this relaxation is predominantly due to contact line motion and the remainder of the drop is comparatively unperturbed, while in glycerin strong viscous coupling sets up flow field even during early time. Consequently, glycerin exhibits non-zero stresses at short times (Fig. 2.14c, d). The intermediate peak represents the spreading due to contact line force (Eq. 2.6), where high viscosity of glycerin limits the velocity magnitude. As a result, stresses correlate with departure from equilibrium, namely wettability. Wall pressure at small times results from a combination of pressure jump at the curved surface and gravitational head, approaching equilibrium gradually at later times (Fig. 2.14f). In contrast, the initial pressure in water (Fig. 2.14c) diminishes rapidly due to localized contact line relaxation. In general, a liquid drop of low viscosity will display positive as well as negative free surface curvature, and spread with an oscillatory wall force field. As a consequence, it is a source of fatigue loading of the coated surface.

2.5.2 Gravitational Instability of Pendant Drops

For a sessile drop, gravity acts towards the surface and assists spreading, while it opposes spreading of a pendant configuration. Hence, the pendant drop could detach before achieving a stable shape. As discussed before, the equilibrium state is defined by the drop volume and is thermodynamically independent of the initial drop shape. For the pendant drop, the limiting equilibrium state refers to drop weight being balanced by the vertical component of the contact line force ($\sim 2\pi r_c \cdot \sin\theta_{eq}$). Equivalently, for a prescribed contact angle, one finds the largest drop volume possible for equilibrium. Such force balance analysis amounts to solving the Young-Laplace equation (Chap. 1). Figure 2.15a presents the solution in terms of dimensionless volume, i.e., the critical Bond number Bo^*. As the contact angle decreases, the drop footprint becomes larger. On the other hand, $\sin\theta_{eq}$ is a maximum at 90° and decreases for both hydrophilic ($\theta_{eq} \to 0°$) and hydrophobic surfaces ($\theta_{eq} \to 180°$). These effects jointly determine the drop volume, wherein for large angles the reduction in footprint as well as vertical component of contact line force exhibit reduced stability. For partially wetting surfaces (specifically, $\theta_{eq} \leq 75°$), the increase in footprint appears to counter the reduction in the vertical component (i.e., $\sin\theta_{eq}$), causing large critical pendant drop volumes. In the pendant mode, the gravitational pull is seen to elongate the drop relative to the spherical shape of smaller equilibrated drops (Fig. 2.15b).

The force balance calculations in Fig. 2.15 report static stability and do not account for temporal evolution of the drop shape from the initial to the final condition in response to fluid velocities. When the unsteady Navier-Stokes Eqs. (2.7)–(2.9) are solved to examine the equilibration of liquid drop from the initial spherical shape, not all the drops in Fig. 2.15a attain the equilibrium shapes; this trend is shown in Fig. 2.16a. Fluid velocities can develop dynamic pressure variations at the footprint that destabilize the approach to equilibrium. The stability

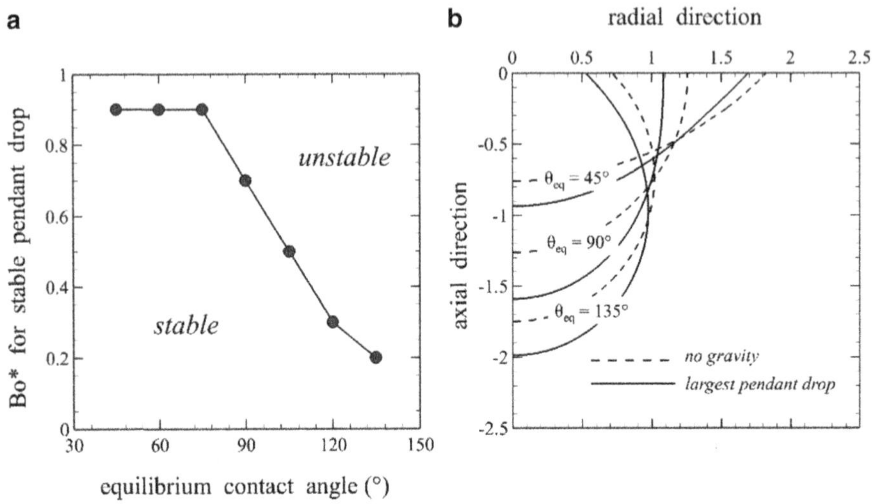

Fig. 2.15 Equilibrium calculations discern (**a**) dimensionless volumes (critical Bond number, Bo^*), for pendant drops hanging from substrates of different wettability. (**b**) Gravity elongates the drop shapes as is revealed from comparing shapes without gravity (spherical) and largest stable drop

predictions accounting for the transients are referred to as dynamic stability. Given the role of fluid velocities in dynamic stability, both the fluid (Fig. 2.16a) as well as initial conditions (Fig. 2.16b) alter the stability bounds. In contrast, static stability in Fig. 2.15 is only dependent on the equilibrium contact angle. Despite oscillations observed during spreading, water drops are consistently more stable than glycerin over a wide range of parameters. This result is related to the region of negative curvature formed near the contact line of a water drop. The resulting flow from the apex of the drop adds to its stability under dynamic conditions. The high viscosity of glycerin inhibits large drop deformation at the free surface, where a larger pressure drop aids the possibility of detachment from the surface. Thus, contact line oscillations stabilize the pendant water drops as compared to the slow spreading glycerin (Fig. 2.16a). When the initial drop is close to the equilibrium shape, velocities associated with spreading are weak, resulting in better stability (Fig. 2.16b). For drops far from equilibrium, larger initial velocities and longer spreading timescales result in additional energy dissipated, thus making them increasingly prone to premature onset of instability. Interestingly, the sluggish response of glycerin does not allow spherical drops to elongate and balance weight, showing an inferior stability even when the initial contact angle $\theta_0 \sim \theta_{eq}$.

Figure 2.17 presents the time evolution of drop shapes during unstable spreading, as determined from the generalized Navier-Stokes equations. The initial non-zero spreading radius, $r_c \sim 0.1$ refers to a small but finite contact patch for $\theta_0 = 175°$. At early times, contact line imbalance will promote spreading. As time progresses, the cumulative effect of gravitational pull will outweigh the stabilizing influence of the spreading contact line, and drop detachment is observed. The detachment

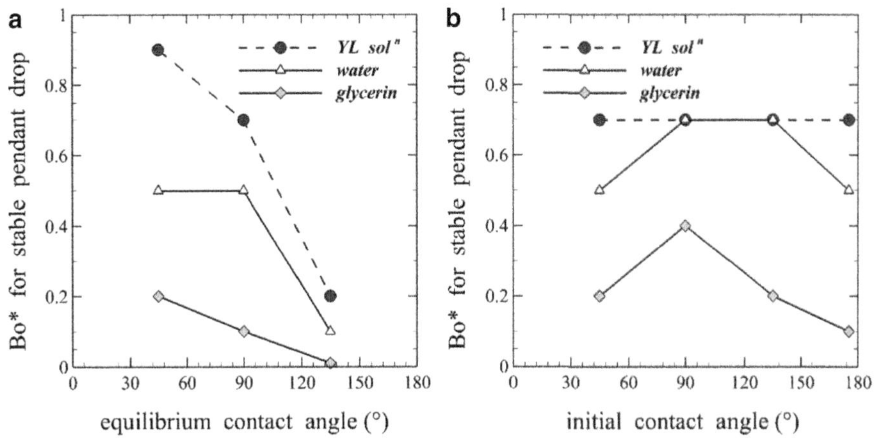

Fig. 2.16 Comparing static and dynamic stabilities for pendant drops. Static stability is estimated from equilibrium calculations (Fig. 2.15), while dynamic stability accounts for the spreading dynamics, thus accounting for flow associated instabilities. Role of (**a**) equilibrium wetting and (**b**) initial contact angle on dynamic stability

sequence is characteristic of the choice of the liquid. For low viscosity water drops contact line spreads (Fig. 2.17a) while the bulk motion responds to the gravitational force (Fig. 2.17b). Once gravity dominates, contact line motion is arrested and the drop detaches from the surface. On the contrary, domineering viscous effects in glycerin do not let the contact line relax (Fig. 2.17c). As the gravitational influence increases, the contact patch becomes smaller and marks the onset of drop detachment (Fig. 2.17d). Hence, drop volumes Bo $=$ 0.6 for water and Bo $=$ 0.2 for glycerin, discussed here qualify for statically stable but dynamically unstable drops. In contrast, static calculations show that the largest stable drop volume for both water and glycerin is Bo ~ 0.7 for $\theta_{eq} = 90°$ (Fig. 2.16a).

2.5.3 Convection in Spreading Drops

The previous discussion revealed rich flow patterns during the spreading of liquid drops. An associated outcome is heat transfer augmentation between the liquid phase and the solid surface due to the presence of strong velocity fields. Building upon the fluid flow formulation (Eq. 2.2) and mechanical energy balance (Eq. 2.29), a conjugate energy balance model can be constructed to examine non-isothermal interactions. A generalized balance for thermal interactions in a fluid volume containing the three-phase contact line and a free surface has the form

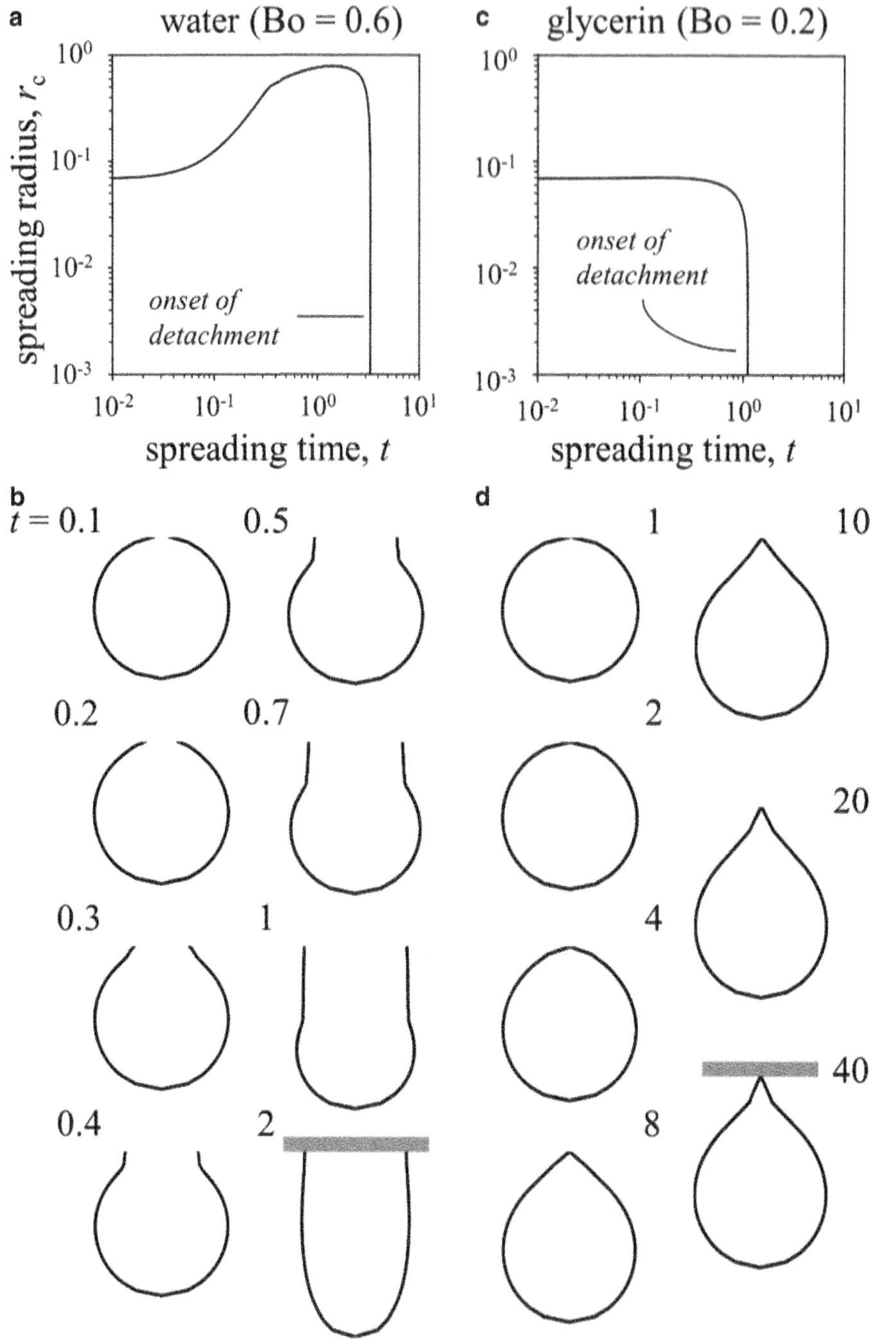

Fig. 2.17 Representative spatiotemporal features during instability growth in (**a**, **b**) water and (**c**, **d**) glycerin drops. The drops relax from an initial spherical shape ($\theta_0 = 175^\circ$) on a neutrally wetting surface, i.e., $\theta_{eq} = 90^\circ$. Drop volumes are chosen such that they are statically stable but dynamically unstable based on Fig. 2.16a. (**a**, **c**) Spreading radius evolution near detachment; (**b**, **d**) instantaneous dynamical drop shapes. The substrate is schematically shown in the final instances

$$\frac{\mathrm{d}}{\mathrm{d}t}\left(\int\limits_{\mathrm{CV}} \rho C T\,\mathrm{d}V\right) = \int\limits_{\mathrm{CS}} k\nabla T\cdot\hat{\boldsymbol{n}}\,\mathrm{d}S - \mu\int\limits_{\mathrm{TPZ}} (\boldsymbol{u}\cdot\nabla\boldsymbol{u})\cdot\hat{\boldsymbol{n}}\,\mathrm{d}S + \mu\int\limits_{\mathrm{CV}} (\nabla\boldsymbol{u}:\nabla\boldsymbol{u})\mathrm{d}V \tag{2.37}$$

The above expression follows from the first law of thermodynamics. The conservative energy forms (e.g., potential energy) have been accounted for by subtracting the mechanical energy equation (Eq. 2.29) from the total energy equation. The resulting thermal energy balance quantifies temperature changes as a result of heat transfer across the boundary (the first term on the right) and possible dissipative mechanisms that decrease mechanical energy and increase the thermal content. Here, dissipation is taken to occur at the three-phase contact line as well as viscous dissipation in the bulk of the drop.

Let the relevant temperature range be $\Delta T = T_{\mathrm{H}} - T_{\mathrm{C}}$. The temperature scale relates to the spreading of a heated drop over a cold substrate. The initial liquid temperature is T_{H} and the surface temperature is T_{C}, maintained constant for all time. One can form the dimensionless temperature

$$\phi = \frac{T - T_{\mathrm{L}}}{T_{\mathrm{H}} - T_{\mathrm{L}}}$$

Using previously defined hydrodynamic scales, Eq. (2.37) is expressed in dimensionless form

$$\frac{\mathrm{d}}{\mathrm{d}t}\left(\int\limits_{\mathrm{CV}} \phi\,\mathrm{d}V\right) = \frac{\mathrm{Oh}}{\mathrm{Pr}}\int\limits_{\mathrm{CS}} \nabla\phi\cdot\hat{\boldsymbol{n}}\,\mathrm{d}S - \sum\cdot\,\mathrm{Oh}\int\limits_{\mathrm{TPZ}} (\boldsymbol{u}\cdot\nabla\boldsymbol{u})\cdot\hat{\boldsymbol{n}}\,\mathrm{d}S + \sum\cdot\,\mathrm{Oh}\int\limits_{\mathrm{CV}} (\nabla\boldsymbol{u}:\nabla\boldsymbol{u})\mathrm{d}V \tag{2.38}$$

with two additional dimensionless parameters

$$\text{Prandtl number}: \quad \mathrm{Pr} = \frac{\mu C}{k} \tag{2.39}$$

$$\text{Surface-to-sensible energy ratio}: \quad \sum = \frac{\sigma}{\rho C \Delta T \ell} \tag{2.40}$$

Here, C is specific heat and k is thermal conductivity. Prandtl number (Leal 2007) compares momentum transport by viscosity to diffusive heat transfer by heat conduction. Analogous to the viscous timescale $\tau_\mu = \ell^2\rho/\mu$, one may define a conduction timescale $\tau_\kappa = \ell^2\rho C/k$, to assess the time required to detect temperature rise at a distance ℓ. Thus, it follows that Prandtl number may also be interpreted by

$$\mathrm{Pr} = \frac{\tau_\kappa}{\tau_\mu}$$

If $\mathrm{Pr} \ll 1$, conduction is faster than viscous dynamics and thermal transients may be neglected. Conversely, if $\mathrm{Pr} \gg 1$, thermal transients will persist long after the drop has reached hydrodynamic equilibrium.

The second dimensionless parameter, Σ, scales the surface energy $\tilde{\mathcal{O}}(\sigma/\ell)$ against sensible thermal energy $\tilde{\mathcal{O}}(\rho C\Delta T)$. In a drop spreading application (with the appropriate length scales and fluid properties), even a small temperature difference of $\Delta T \sim 1$ K yields a surface-to-sensible energy ratio $\Sigma\ \tilde{}\,\mathcal{O}(10^{-3} - 10^{-5})$ (Table 2.6). Hence, hydrodynamic dissipation is much smaller than heat transfer in the presence of temperature gradients. Following this observation, the energy equation (Eq. 2.38) can be further simplified

$$\frac{\mathrm{d}}{\mathrm{d}t}\left(\int_{\mathrm{CV}} \phi\,\mathrm{d}V\right) \approx \frac{\mathrm{Oh}}{\mathrm{Pr}}\int_{\mathrm{CS}} \nabla\phi\cdot\hat{\boldsymbol{n}}\,\mathrm{d}S \tag{2.41}$$

Note that the Lagrangian formulation adopted here accounts implicitly for advective heat transfer since the control volume moves in time. The velocity terms are absent from the governing equations, but they determine the deformation of the Lagrangian control volume.

Based on the continuity of heat flux at the free surface, one can write

$$-k\frac{\partial T}{\partial n}\bigg|_{\mathrm{drop}} = -k\frac{\partial T}{\partial n}\bigg|_{\mathrm{air}} \quad \text{leading to} \quad \frac{\partial T}{\partial n}\bigg|_{\mathrm{drop}} = \frac{k_{\mathrm{air}}}{k_{\mathrm{drop}}}\cdot\frac{\partial T}{\partial n}\bigg|_{\mathrm{air}}$$

The conductivity ratio $k_{\mathrm{air}}/k_{\mathrm{drop}} \sim \mathcal{O}(0.1)$ (Table 2.6) and is larger than the viscosity ratio. In other words, a free surface has a negligible hydrodynamic contribution to the liquid movement from the gas phase but is thermally coupled. The approximation of negligible hydrodynamic coupling at the free surface implies that flow in air has little impact on liquid flow. This does not, however, preclude the possibility of air movement establishing a convective boundary layer at the liquid drop-air interface.

In view of the above discussion, temperature gradient in air cannot be approximated in terms of diffusion length or conduction in a semi-infinite domain. This is because advective contributions in air require the velocity field that has not been estimated in the present discussion (Bergman et al. 2011). Alternatively, we may consider the following two extreme boundary conditions: (1) adiabatic free surface and (2) isothermal free surface. The former indicates an absence of interaction between air and the liquid. The latter will be realized in the limit of very strong

Table 2.6 Thermophysical fluid properties and dimensionless parameters at 27 °C, 1 atm pressure (Bergman et al. 2011)

Properties	Water	Glycerin	Air
Density, ρ (kg/m^3)	997	1259.9	1.20
Viscosity, μ (Pa s)	8.6×10^{-4}	0.8	1.8×10^{-5}
Surface tension, σ (mN/m)[a]	72	63	–
Capillary length, ℓ_c (mm)[b]	~2.71	~2.26	–
Ohnesorge number, Oh[c]	1.9×10^{-3}	1.89	–
Heat capacity, C (J/kg K)	4200	2400	1000
Thermal conductivity, k (W/m K)	0.6	0.3	2.6×10^{-2}
Prandtl number, Pr	6.02	6400	0.69
Surface-to-sensible energy ratio, Σ[c]	6.34×10^{-6}	9.23×10^{-6}	–

[a]Against air
[b]$g = 9.81$ m/s^2
[c]At capillary volume

convective motion of air. Note that the present treatment does not account for Marangoni (thermocapillary) convection that is caused by the temperature dependence of surface tension (Leal 2007). Hence, heat transfer enhancement over conduction arises purely from the spreading motion of the drop.

The temperature equation (Eq. 2.38) is simplified to a set of algebraic equations using the mixed analytical-discrete treatment outlined for the mass and momentum equations, leading to the form

$$\frac{\mathrm{d}}{\mathrm{d}t}\left(\int_z\int_r \phi\, 2\pi r \mathrm{d}r \mathrm{d}z\right) = \frac{\mathrm{Oh}}{\mathrm{Pr}}\int_z\int_r \left(\frac{1}{r}\frac{\partial}{\partial r}\left(r\frac{\partial \phi}{\partial r}\right) + \frac{\partial^2 \phi}{\partial z^2}\right) 2\pi r \mathrm{d}r \mathrm{d}z \tag{2.42}$$

Two sets of discretized equations emerge for the limiting boundary conditions corresponding to isothermal and adiabatic free surface (Mistry and Muralidhar 2015b).

Figure 2.18 presents non-isothermal spreading of sessile water and glycerin drops of capillary volume (Bo = 1). The evolution of spreading radius (Fig. 2.18a) will remain unchanged unless thermocapillary effects are brought in. Figure 2.18b, c shows that the temperature evolution characteristics of the two liquids are markedly different. For water, thermal interactions manifest at large times, while glycerin experiences thermal effects earlier in the wettability-dependent spreading regime and later. Rapid spreading precedes temperature decay in water (Fig. 2.18b). On the other hand, the cooling trend of a glycerin drop takes place concurrently with the flow (Fig. 2.18c). Both adiabatic and isothermal limits of free surface demonstrate such a response, with an isothermal free surface showing temperature reduction at an earlier time instant. For the adiabatic free surface, heat exchange takes place at the drop footprint, while an isothermal free surface represents an additional heat exchange boundary that facilitates faster cooling of the body of liquid.

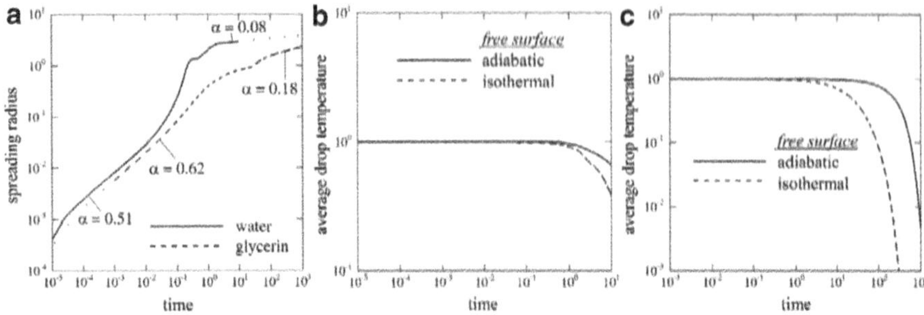

Fig. 2.18 Non-isothermal spreading of capillary volume drops (Bo = 1) of water and glycerin over a fully wetting surface. Temporal evolution of (**a**) spreading radius, (**b**) average temperature of water, and (**c**) glycerin drops. Adiabatic and isothermal conditions refer to those at the free surface

If the goal is heat exchange at the solid surface, the isothermal free surface is less suitable since it represents heat loss that decreases the temperature gradient available for heat transfer between the surface and the drop. For a quantitative comparison, the surface-averaged wall heat flux is computed as a measure of heat exchange effectiveness over the drop footprint by

$$q_{\text{footprint}} = \frac{2}{r_c^2} \int_0^{r_c} \left. \frac{\partial \phi}{\partial z} \right|_{z=0} r\,\mathrm{d}r \tag{2.43}$$

Figure 2.19 presents the wall heat flux evolution during spreading of (a) water and (b) glycerin drops for two different wetting conditions. As discussed earlier, wall heat flux is smaller with an isothermal free surface. This effect is more pronounced for glycerin (Fig. 2.19b), where heat transfer and spreading take place concurrently as compared to spreading of water drops (Fig. 2.19a), where spreading transients have quickly died out. Since incomplete spreading over a partly wetting surface takes place with a mixed advancing-receding motion of the contact line, heat flux transients in water exhibit fluctuations. Such fluctuations enhance the time-averaged content of heat transfer between the liquid and the substrate. However, slower spreading velocities in glycerin render incomplete wetting less effective (Fig. 2.19b).

2.6 Future Directions

The authors propose to explore the contact line model of the present study (Eq. 2.6) with others in the literature in terms of the spreading dynamics. Chemically and physically textured (pillared) surfaces show distinct behavior and require careful analysis. An important new direction is non-isothermal spreading that includes the thermocapillary effect. The resulting secondary flow enhances heat transfer over and above forced convection. It also differentiates the hot drop spreading on a cold

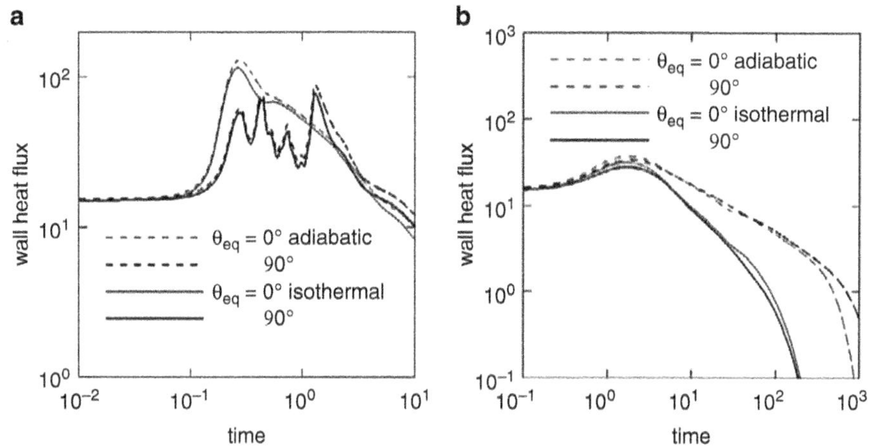

Fig. 2.19 Evolution of wall heat flux over the course of spreading for (**a**) water and (**b**) glycerin drops. Two different wetting conditions, $\theta_{eq} = 0°$ and $90°$, and two distinct free surface thermal conditions are examined. Complementary trends are shown in Fig. 2.18. Lighter curves represent complete wetting conditions

substrate from cold drops spreading over a hot one. Drop spreading with phase change is an important application. Solidification of spreading drops, whether in a soldering process or during ice formation from raindrops are relevant examples. Similarly, evaporation of sessile and pendant drops with spreading, the role of particulates and suspensions, and the importance of rheology are possible directions of future study. Beyond axisymmetric drop spreading, there are numerous flow scenarios where contact lines are central to the flow process. Coalescence of liquid drops over a solid substrate is an example. Incorporation of the contact line treatment in terms of an unbalanced force in computational fluid dynamics simulations is a prerequisite to understanding such flow fields.

2.7 Closure

Transport interactions during spreading are complex, given the continuously evolving nature of the free surface, but also the uncertainty associated with the contact line motion. A rigorous theoretical formulation has been described in the present study that leverages the Lagrangian treatment of fluid motion to consistently predict contact line motion and overall spreading dynamics without resorting to subscale modeling or parameter tuning. The strength of such a formalism is demonstrated through comparison against various experiments and spreading regimes for sessile and pendant drops. Two distinct drop fluids-water and glycerin are examined, given their representative spreading characteristics. Additionally, the substrate effect is studied in terms of its wettability.

Spreading of water is oscillatory while glycerin is monotonic, given its high viscosity, oscillations are greater for a fully wetting surface since the initial contact angle imbalance is also higher. The relationship between the dynamic contact angle and contact line velocity is relatively independent of the Bond number though it depends strongly on the equilibrium contact angle. Pendant drops become statically unstable at a critical size whose upper bound is fixed by the Young-Laplace equation. They are prone to detachment below the critical size if spreading dynamics are factored in. This tendency is more common in glycerin than in the water, where oscillations are seen to stabilize the drop.

The flow details during spreading are examined through mechanical energy balance in terms of the energy storage modes and dissipative mechanisms. Energy loss is the highest for spreading over a fully wetting surface and is smaller over partially wetting surfaces. Additionally, components of fluid stresses at the drop footprint are followed in time to assess their severity to cause coating damage.

Preliminary results are presented for non-isothermal drop spreading and enhancement in the wall heat flux arising from the velocity field. Based on the differences in flow and heat transfer timescales, spreading could considerably precede heat exchange (as in water drops) or both could simultaneously take place (as in glycerin). Oscillations in water translate into those in the wall heat flux while in glycerin, heat fluxes show a monotonic variation, progressively diminishing to zero.

Chapter 3
Coalescence Dynamics of Drops over a Hydrophobic Surface

Praveen Somwanshi, K. Muralidhar, and Sameer Khandekar

Nomenclature

A	Gas-liquid interface area (m^2)
E	Surface energy (J)
E^*	Excess surface energy function (−)
E_{CL}	Contact line dissipation (W)
E_g	Change of potential energy with time (W)
E_k	Change of kinetic energy rate with time (W)
E_s	Surface energy rate of a coalesced drop (W)
E_{vis}	Viscous dissipation (W)
f	Roughness factor (−)
$\overrightarrow{g}$	Acceleration due to gravity (m/s^2)
m	Spring stiffness (N/m)
m	Mass (kg)
M, N	Number of images and number of pixels in each image (−)
N_p	Number of pixels at the interface of a combined drop (−)
p	Pressure (N/m^2)
r	Radius of the drop (m)
R	Characteristic length (m)
$R_{avg} = 1/RC$	Average instantaneous radius of the combined drop (m)
R_b	Radius of the combined drop footprint (m)
(RC)	Radius of curvature averaged over the air-water interface (m^{-1})
t	Time (s); suffix IS is inertia-surface tension; IV is inertia-viscous; VS is viscous-surface tension
U	Velocity scale (m/s)
u_c	x-component of centroid velocity (m/s)
u_c^*	Non-dimensional x-component of centroid velocity (−)

S. Khandekar, K. Muralidhar, *Drop Dynamics and Dropwise Condensation on Textured Surfaces*, Mechanical Engineering Series,
https://doi.org/10.1007/978-3-030-48461-3_3

$u_{\mathrm{res}} = \sqrt{u_{\mathrm{c}}^2 + v_{\mathrm{c}}^2}$	Resultant velocity of the centroid (m/s)
V, $V_{\mathrm{comb\ drop}}$	Volume and volume of the combined drop (m^3)
v_{c}	y-component of centroid velocity (m/s)
v_{c}^*	Non-dimensional x-component of centroid velocity (–)
w_i	Area function for the ith pixel (–)
x_{c}, y_{c}	x- and y-coordinates of centroid (m); time-average is indicated by an overbar
x_{c}^*, y_{c}^*	Non-dimensional x- and y-coordinates of centroid (–);

Dimensionless Quantities

Bo	Bond number, $\frac{\rho g R^2}{\sigma}$
We	Weber number, $\frac{\rho U^2 R}{\sigma}$
Fr	Froude number, $\frac{U^2}{gR}$
Re	Reynolds number, $\frac{\rho U R}{\mu}$
Oh	Ohnesorge number, $\frac{\mu}{\sqrt{\rho R \sigma}}$

Greek Symbols

$\dot{\gamma}$	Shear rate (s^{-1})
γ^*	Non-dimensional hear rate (–)
θ	Contact angle (°)
μ, ν	Dynamic and kinematic viscosity (Pa s; m^2/s)
ρ	Fluid density (kg/m^3)
σ	Coefficient of surface tension (N/m)
τ	Dimensionless time (–)
ζ	Damping ratio (–)

Subscripts

1	Drop placed below (–)
2	Drop placed above (–)
3	Combined drop after coalescence (–)
c	Critical (–)
CB	Cassie-Baxter state (–)
eq	Equilibrium (–)
lg	Liquid-gas interface (–)
sg	Solid-gas interface (–)
sl	Solid-liquid interface (–)
W	Wenzel state (–)

3.1 Introduction

Individual liquid drops tend to take on a spherical shape when suspended in an undisturbed gaseous environment. Their equilibrium shape depends on the extent of surface wettability when placed on a solid substrate. For both conditions, the drop shape is nearly spherical, characterized by an equivalent curvature that depends on the drop volume and determines the magnitude of the internal pressure. When two drops composed of the same liquid, but unequal volumes are brought in contact, their internal pressures are distinct, with the smaller drop having a greater internal pressure. In atmosphere, the drops coalesce spontaneously with flow taking place from the smaller to the larger drop. In the intermediate time period, fluid velocities will displace the combined interface of the two drops; complex drop shapes are encountered in this regime. Since the drops are not externally driven, the velocities generated will diminish over a longer time frame owing to viscous dissipation. For equal volumes, the region of initial contact between the drops deforms till a continuous liquid bridge is formed, creating a single combined liquid entity. The internal pressures of the two equal volume drops are comparable. However, the bridge has concave curvature and is a region of negative pressure. The resulting pressure difference is responsible for fluid motion, as the drops merge with great vigor on the intermediate timescale, followed by slow viscous dissipation in the longer time period. The variation of the drop shape, and hence the interface curvature during the coalescence process is equivalent to a gradient in the surface tension. Such a force can stretch the drop and tear it many ways. For this reason, the consequence of coalescence is a variety of shapes acquired by the liquid mass, which ultimately evolves towards a static equilibrium configuration. The distinct timescales corresponding to bridge formation, interface displacement, and viscous dissipation suggest that coalescence is a multiscale process with inherent complexity.

As discussed in Chap. 2 on drop spreading, the presence of a partially or fully wetting substrate creates contact line motion with its own dissipation characteristics and additional scales of time and velocity.

Coalescence of two small liquid drops can take place under different configurations. For example, coalescence details of two contacting liquid drops independently located away from a solid surface have been reported in the literature, in the context of atomization and spray interactions. One independently suspended drop can also coalesce on another drop resting on a solid surface. Adjacent drops placed on a hydrophobic surface can coalesce, in pendant or sessile mode, as observed during the dropwise condensation process. During such a coalescence process, the three-phase contact line moves over the surface as the footprint of the coalesced drop evolves in time. Length and timescales due to release and dissipation of excess surface energy, and an additional velocity scale of the combined drop are quantities of interest.

While coalescence is realized in several applications such as microfluidics, the present work is motivated by model development for dropwise condensation of vapor on a cooler hydrophobic substrate. Sustained and continued occurrence of coalescing drops on such superhydrophobic surfaces leads to enhanced shear rates as well as local wall heat fluxes, thereby strongly affecting the life of the substrate or the promoter coating on it.

3.2 Coalescence of Liquid Drops

Coalescence characteristics of drops are configuration specific. The presence of a solid substrate substantially alters coalescence scales. The characteristics of coalescing drops in the sessile orientation are different from the pendant. The configuration of two contacting liquid drops away from a solid surface is first considered in this section.

Eggers (1998) studied the coalescence of two liquid spheres, as shown in Fig. 3.1 using theory as well as numerical simulation. The surface area of the coalesced spheres was seen to decrease continuously with time. Rapid motion of the bridge was observed due to large values of surface curvature present at the region of contact of the two spheres. The bridge radius initially scaled linearly with time as $R_b \sim t$.

Eggers et al. (1999) further studied theoretically the coalescence of liquid drops in the configuration of Fig. 3.1. The early-time evolution of coalescence of drops showed dependence on the local curvature of the initial contact. The surface at the bridge of coalescing drops showed large deformations because of large negative pressures in the narrow region. The magnitude of deformation scaled with the height of the bridge. Coalescence of low viscosity drops in the early stage was described by the Stokes equations. Air bubble was found to be trapped between coalescing drops just prior to coalescence. The tension in the neck was responsible for pulling a trapped air bubble. On theoretical grounds, the bridge radius scaled over a longer time period as $R_b \sim t \ln t$.

Wu et al. (2004) studied two merging drops of the same liquid, generated at the tip of a syringe in a pendant–sessile configuration in air (Fig. 3.2). The fluids considered were deionized water, methanol, and a mixture of deionized water and glycerol. The initial growth of the bridge radius was seen to scale with the relaxation time as $t^{0.5}$ with a constant of proportionality depending on fluid properties. The constant of proportionality for methanol was seen to be 20% higher than water and water-glycerol mixture.

Aarts et al. (2005) studied the regimes of coalescence of two drops of the same liquid kept one above the other; the liquids considered were water and silicone oil (Fig. 3.2). Coalescence was observed to be either inertial or viscous, as revealed in

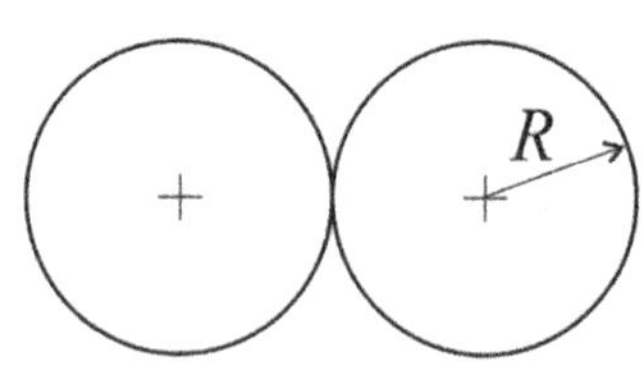

Fig. 3.1 Orientation of coalescing liquid drops placed adjacent to each other away from a solid surface

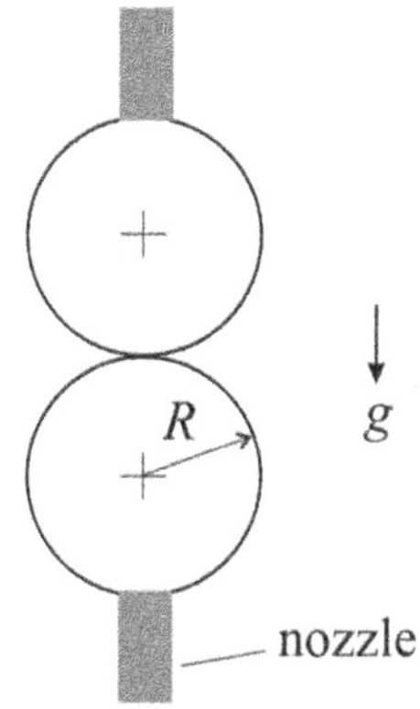

Fig. 3.2 Orientation of coalescing liquid drops placed vertically over each other and away from a solid surface in an experimental setup

the ensuing velocity and timescales. The occurrence of a regime was fixed by the Reynolds number, defined here as Re $= \rho\sigma R_b/\mu^2$, where R_b is the bridge radius. When the regime was inertial, the bridge radius scaled as $R_b = C\sqrt{t}$; constant C was found to be equal to 1.14 and 1.24 for water ($\mu = 1$ mPa s) and silicone oil ($\mu = 5$ mPa s), respectively. Cross-over of regimes was observed at Re = 1. Inertial and viscous coalescence were observed for silicone oil of higher viscosity ($\mu = 50$ mPa s), where cross-over was seen at Re = 1.5. The bridge radius during viscous coalescence exhibited linear scaling, namely $R_b = Ct$. The opening speed of the bridge formation was set by the capillary velocity, and the experiments showed the absence of logarithmic dependence with time.

Thoroddsen et al. (2005) studied the speed of coalescence of a combination of pendant and sessile drops of the same liquid, kept one above the other, using ultra-high-speed imaging (Fig. 3.2). Flow dynamics depended on Reynolds number $\mathrm{Re}_v = \rho\sigma R_d/\mu^2$ and $\mathrm{Re}_\sigma = \sqrt{\rho\sigma R_d}/\mu$. Here, R_d is the radius of the drop, and Re is based on the average radius of curvature and a velocity scale formed by surface tension coefficient and viscosity. The ratio of vertical spacing between the nozzles and the diameter of the drop had strong influence on the coalescence speed showing linear variation. The coalescence speed was seen to slow down because of the pinned boundaries near the neck region. The maximum velocity was seen immediately after the first contact. The diameter of a contact region just before coalescence was seen to be of a definite size of around ~128 μm and contracted with a speed of around ~10 ms^{-1}. The size of the contact region was dependent on the speed of approach of the coalescing drops. Reduction in the coalescence speed arose from the contraction of the air-disk near the neck region.

Thoroddsen et al. (2007) studied coalescence of two miscible liquid drops such as water and ethanol, with one drop above the other (Fig. 3.2). The one with lower surface tension was seen to drive coalescence. Speed of coalescence was proportional to the square root of surface tension. A thin layer of liquid having a lower surface tension coefficient was pulled along the free surface by Marangoni stresses to cover the neck region of the coalescing drops. The speed of coalescence decreased due to reduction in the effective surface tension that pulled the neck. However, it was unaffected by the difference in dynamic viscosities of the coalescing drops. Velocity and timescales in the inertial regime were respectively represented as $v_\sigma = \sqrt{\sigma/\rho R_\mathrm{d}}$ and $t_\sigma = \sqrt{\rho R_\mathrm{n}^3/\sigma}$. In the viscous regime, velocity scaled as $v_\mu = \sigma/\mu$; in this discussion, R_d is the radius of the drop and R_n is the radius of the nozzle.

Xing et al. (2007) studied numerically the coalescence of two drops approaching each other with negligible velocity (Fig. 3.1). A bridge formed between the coalescing drops and its radius R_b scaled with time as $t^{0.5}$. The coalescence speed was found to be directly proportional to the surface tension coefficient.

Zhang et al. (2009) studied the effect of volume ratio on the formation of satellite drops during coalescence of water drops of unequal volumes resting on a steel nozzle as in Fig. 3.2. The capillary length scale $l = \sqrt{\sigma/\rho g}$ and capillary-inertial velocity scale $v_\sigma = \sqrt{\sigma/\rho R_\mathrm{sd}}$ were adopted for scaling. Here, σ is surface tension, ρ is density, and R_sd is the radius of the small drop. A strong capillary wave was produced due to formation of a neck and traveled in the upward direction through the drop kept above. The capillary wave converged at the top, creating protrusions that detached from the surface to form a satellite drop. Capillary wave was seen to carry momentum upward and transfer it towards the center of the neck. Below a critical size ratio of 1.55, satellite drops were not seen in experiments.

Paulsen et al. (2011) studied viscous to inertial cross-over during coalescence of liquid drops of water, NaCl solution, and glycerin on two vertically aligned Teflon nozzles (Fig. 3.2). One drop was held fixed, while the other one could grow using a syringe pump, until the two coalesced under ambient conditions. The scaling law for the length scale was seen to be $R_\mathrm{b} \propto t$ in the viscous regime and $R_\mathrm{b} \propto t^{0.5}$ in the inviscid regime. The authors observed logarithmic correction to the scaling law in the viscous regime, where the ratio of the neck radius to the drop radius was smaller than 0.03. The definition of Reynolds number based on neck width was observed to be more appropriate than the neck radius. The authors estimated the cross-over time $\tau_\mathrm{c} = \mu^2/R_\mathrm{b}^6\sqrt{A_\mathrm{b}/\rho\sigma^3}$ as 2 μs, where A_b is the bridge surface area. This estimate was greater than their theoretical estimate of 0.7 ns. In addition, the cross-over time in experiments scaled as $\tau_\mathrm{c} \propto \mu^2$ for low viscosity fluids and $\tau_\mathrm{c} \propto \mu^3$ for highly viscous fluids.

Eiswirth et al. (2012) studied experimentally and numerically the binary droplet-droplet coalescence event of toluene drops grown adjacent to each other rising in water in the configuration of Fig. 3.1. The growth of the liquid bridge was in the inertial regime. The bridge radius scaled with time as $t^{0.5}$. Mixing was not seen within the first 40 ms after the onset of coalescence of drops of equal volume. Internal pressure, and hence, internal energy of a smaller drop was larger than the larger drop. The liquid jet was directed from smaller to the larger drops during coalescence of unequal drops.

Paulsen et al. (2012) noted three distinct regimes during coalescence of two pendant drops of silicone oil, namely inertial, inertia-limited viscous and the Stokes' regimes (Fig. 3.3). The regime of coalescence was demarcated by the Ohnesorge number with the inertia-limited viscous regime for Oh = 1 and the Stokes' regime

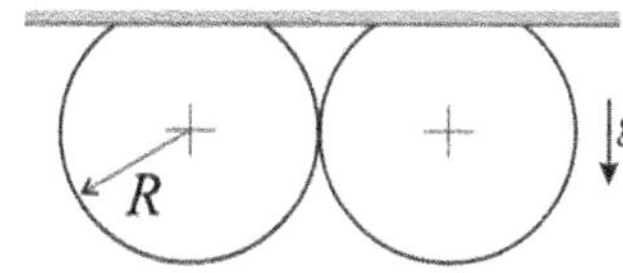

Fig. 3.3 Orientation of coalescing liquid drops placed adjacent to each other in an experimental setup

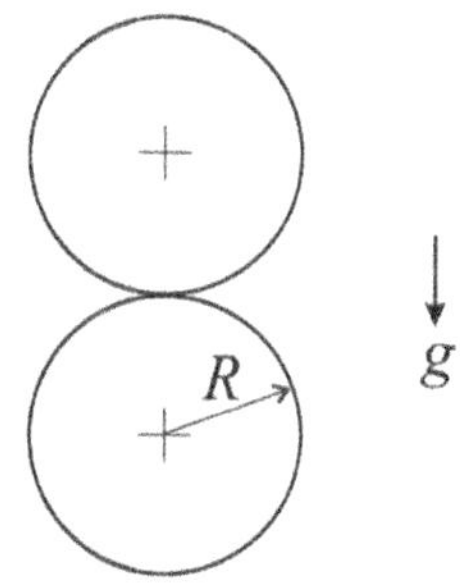

Fig. 3.4 Orientation of coalescing liquid drops placed vertically one above the other

for Oh > 1. The scaling law in the inertia-limited viscous regime was seen to be $R_b \propto t$.

Sprittles and Shikhmurzaev (2012) modeled coalescence of drops in a vertical orientation and compared the data with existing mathematical models and experiments (Fig. 3.4). Unlike bridge formation, the authors suggested that the initial coalescence mechanism of two freely suspended drops involves trapping of the free surface within, followed by a gradual disappearance of the internal interface. The authors showed that each mechanism was accompanied by its own scaling law. The initial motion was slower than what was predicted by the existing models because of gradual evolution of the contact angle. The scaling law showed a trend of $R_b \sim t \ln t$ in the inertial regime. Experiments confirmed the formation of air bubbles trapped internally during coalescence.

Most references on coalescence is concerned with the merging process in free space, while a few deal with coalescence occurring over a textured surface. We examine this literature in the following section.

3.3 Coalescence over Textured and Untextured Surfaces

Coalescence characteristics are significantly altered in the presence of a solid surface. The solid surface introduces a three-phase contact line for the coalescing drops, where the no-slip condition is violated. In mathematical terms, it is a site of singularity in shear stresses. This section discusses the coalescence characteristics of liquid drops in the presence of a substrate in terms of length, velocity, and timescales reported in the literature.

Menchaca-Rocha et al. (2001) studied coalescence of sessile mercury drops on a glass substrate, by bringing another drop near an already placed drop (Fig. 3.5). In experiments, the droplet was seen to be flatter both due to wetting and gravity but not

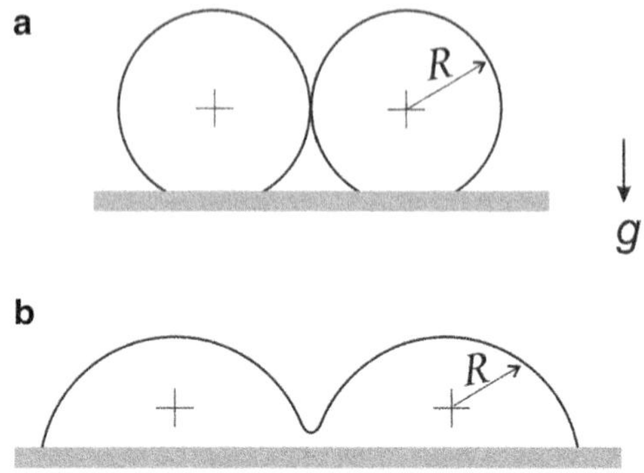

Fig. 3.5 Orientation of coalescing liquid drops placed adjacent to each other and over a solid surface: (**a**) hydrophobic and (**b**) hydrophilic surface

in the numerical simulation. Energy damping introduced due to wetting and gravity was seen to delay coalescence motion in experiments compared to simulation. However, capillary waves showed their presence in both experiments and simulations. Originating from the contact line singularity, capillary waves were seen to reach the extreme end of the drop, traveling through its interior. Energy dissipation in the early stages was larger in experiments than in simulation. In turn, larger amplitude capillary waves were seen in the simulation. The base radius of the combined drop scaled as $t^{0.5}$ in experiments as well as simulation.

Andrieu et al. (2002) studied the time required for complete coalescence of sessile water drops over a plane surface for contact angles of 30° and 53° (Fig. 3.5). Drops were imaged in the plan view and the footprint of the merged drops was followed in time. The authors reported bridge formation over a timescale of a few milliseconds, followed by a long transient, lasting a few min, in proportion to the base radius. The relaxation time was seen to be smaller for smaller contact angles because of pinning of the three-phase contact line on the surface defects. The timescale associated with relaxation was several orders greater than the bulk hydrodynamics. Dissipation at the three-phase contact line was suggested as the mechanism responsible for the long-lived transience of the coalesced drop.

Narhe et al. (2004) compared coalescence induced in sessile drops deposited by a syringe, as opposed to drops grown by direct condensation (Fig. 3.5). These authors observed three stages of coalescence, namely formation of a liquid bridge, subsequent evolution of the bridge and slow growth to attain the final equilibrium shape. Syringe deposition generated large oscillations within the merging drops. The driving force of contact line motion was related to the kinetic energy of the drop generated during coalescence. The drop surface was seen to pull the contact line at each oscillation, thereby accelerating its motion. The relaxation time of coalescence induced by syringe deposition was larger by a factor of 10–100 compared to direct condensation. The relaxation velocity was proportional to the restoring force defined by the variation of the surface energy. The contact line motion was unaffected by changes in the rate of condensation. Free surface oscillations were identified as the reason behind a greater degree of dissipation, leading to a shorter relaxation time for coalescence caused by syringe deposition.

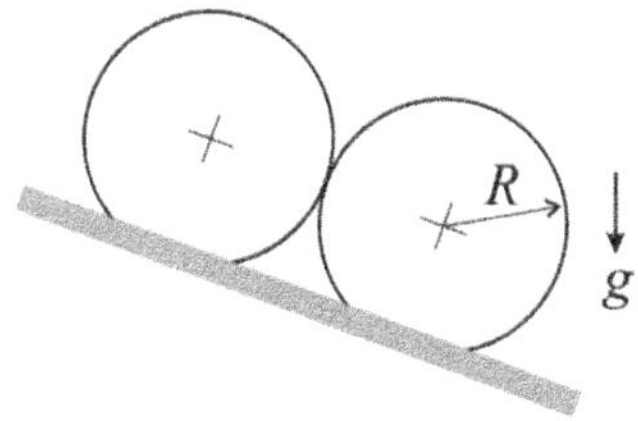

Fig. 3.6 Orientation of two coalescing liquid drops placed over an inclined surface

Narhe et al. (2005a, b) studied the effect of initial conditions and surface properties on the coalescence of water drops by either direct condensation or syringe deposition, sitting on a solid surface (Fig. 3.5). Contact line dissipation was characterized by a suitable coefficient that was seen to be large because of defects on the surface. Stick and slip were experienced by the contact line which showed a large contact line dissipation coefficient. The relaxation rate showed linear dependence on contact angle. Syringe deposition introduced strong oscillations in the coalescing drops. The relaxation time was directly proportional to the combined drop volume.

Ristenpart et al. (2006) studied the coalescence of adjacent silicone oil drops sitting over a substrate (Fig. 3.5). The footprint of the coalesced drop evolved over a longer time to attain a circular shape. The coalescence process was dominated by sharp changes in the geometry of the combined drop. Coalescence speed was influenced by the initial height of the drop. Early growth of the interface scaled with time as $t^{0.5}$. The x-component velocity of the coalesced drop before contact qualitatively represented the spreading velocity; the velocity of the liquid bulk was indicative of the spreading velocity after contact. The slow flux due to spreading in the x-direction played a major role in growing the interface in the early stages because of negligible y-component velocity.

Kapur and Gaskell (2007) studied the morphology and dynamics of coalescence of water drops over a surface while the second drop was brought near the already grown drop (Fig. 3.5). The authors found that the merging and evolution of the combined drop to the final state created a peanut-shaped footprint. The bulge was seen due to inertia along the longer side of the coalesced drop. The authors observed self-excited oscillations and a capillary wave that affected the local pressure field at the wetting line and the speed of interface displacement. The motion of the wetting line was not completely smooth in the later stages of spreading. The half-width of the neck scaled with time as t^{α}; here α ranged from 0.42–0.57.

Liao et al. (2008) experimentally studied the coalescence of a sessile water drop with an adjacent, already placed, drop on an inclined textured surface (Fig. 3.6). The formation of a bridge altered the initial equilibrium state in the early stages of coalescence. Flow inside a coalesced drop was driven by surface tension gradient established between concave interface at the liquid bridge and the convex interface at the top side of the two drops. In the later stages, surface tension gradient was established between highly convex interfaces at the front-end and nearly flat interface at the back-end of a coalesced drop. Fluid flow inside a coalesced drop, movement of the three-phase contact line, and the oscillation process were assisted by an adversely directed gravity. The deformation of the interface led to the release of surface energy. The change of centroidal position of a coalesced drop helped release potential energy that was balanced by viscous dissipation in the bulk and viscous friction associated with the motion of a coalesced drop on the surface. The ratio of energy required to overcome viscous friction to dissipation increased with an increase in the inclination angle and drop diameter. Strong oscillations were seen for the downhill displacement and front-end contact angle whereas oscillations were weak for uphill displacement and the back-end contact angle.

Narhe et al. (2008) experimentally studied flow-induced drying in the early stage of coalescence of drops of diethylene glycol sitting adjacent to each other (Fig. 3.5). Three stages of coalescence were seen. The movement of the three-phase contact line was negligible in the first stage of coalescence. In the second stage, the bridge was seen to grow with time as ~t along the height and as ~$t^{0.5}$ along the length of the bridge. The movement of the three-phase contact line was substantial in the third stage. The bridge relaxed exponentially making the combined drop elliptical. The bridge evolution time was 10 times smaller than the overall relaxation time.

Sellier and Trelluyer (2009) numerically studied coalescence of sessile water drops sitting on a surface and adjacent to each other (Fig. 3.5). The contact angle considered in the study was $68° \pm 2°$. Contact line velocity, and hence, capillary number were larger in the early stages. The early stage of coalescence was not resolved in the model since the lubrication approximation was valid only for small capillary numbers. The half-neck width was correlated in time by

$$R_x(t) = A_1 \exp\left(-\frac{t}{t_c}\right) + A_2 + A_3 t \tag{3.1}$$

where t_c, A_1, A_2, and A_3 are curve-fitting parameters.

Wang et al. (2010) experimentally studied coalescence of sessile water drops kept adjacent to each other (Fig. 3.5) on a wettability gradient surface. The surface energy gradient was created by varying the length and density of the silane molecules. The drop kept on the hydrophobic part of the gradient surface was seen to approach the drop towards the hydrophilic part, followed by coalescence. The authors observed three stages of coalescence. Pinning of the three-phase contact line of coalescing drops changed the shape of the neck region from concave to convex in the initial stage of 0–4 ms. The movement of the back-end of the drop along the direction of energy gradient owing to transition from interfacial energy to surface energy was observed during 4–12 ms. The coalesced droplet moved to the hydrophilic region and transformed into a single near-spherical droplet during 12–27 ms. The peak velocity of the center of the coalescing droplets was found to be 140 mm/s.

Castrejón-Pita et al. (2011) studied the dynamics of coalescence of a sessile drop sitting on a transparent surface with an impacting drop of glycerol (Fig. 3.7). The authors identified three stages of coalescence. The formation of a bridge, its relaxation, and relaxation of the combined drop to the final equilibrium shape were the stages recorded. The authors observed radially inward flow indicating the region of upward fluid motion. The footprint of the coalesced drop was circular for smaller

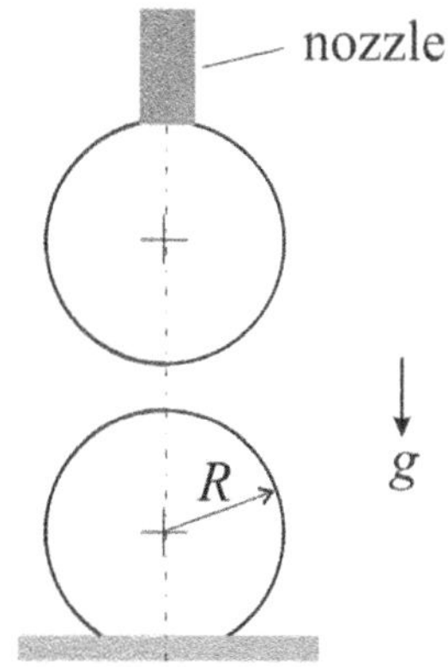

Fig. 3.7 Orientation of vertically aligned coalescing liquid drops, where one has an initial velocity relative to the other; the drops could also be offset to varying degrees

droplet separations. The difference between impact-driven coalescence and coalescence of two static drops was due to growth of the neck caused by larger initial separation between the drops. The formation and evolution of the neck and the final shape of the coalesced drop was strongly influenced by pinning of the contact line. Simulations and experiments showed differences in the length of the coalesced drop for large initial separation.

Nilsson and Rothstein (2011) experimentally studied the effect of contact angle hysteresis on the coalescence of sessile water drops sitting on a surface adjacent to each other (Fig. 3.5). Coalescence was divided into three regimes: one dominated by oscillations, second by rotation and the third, a combination of oscillations and rotations. The regimes were decided by the Weber number, the impact number, and the contact angle hysteresis of the surface. The drop deformation and dynamics with strong oscillations were seen during head-on collisions on a surface with low contact angle hysteresis. Large deformation with high rotation rates were seen during collisions having high impact numbers. Droplet deformation was inversely proportional to the contact angle hysteresis. Angular velocity was directly proportional to the contact angle hysteresis, while growth of the drop decreased with increasing angular velocity.

Sellier et al. (2011) numerically studied coalescence of sessile silicone oil drops sitting on a surface adjacent to each other (Fig. 3.5). The contact angle considered in the study was 40°. The authors observed neck formation initially, growth of the neck in the second stage, and further relaxation of the combined drop to a spherical cap in the third stage of growth. The coalesced droplet was propelled by curvature changes creating a surface tension gradient over the liquid volume. The strength of self-propulsion was controlled by surface tension contrast relative to viscosity.

Wang et al. (2011) theoretically studied the self-propelling behavior of droplets, induced during coalescence of sessile droplets kept adjacent to each other on a superhydrophobic surface. This behavior was observed for droplet sizes smaller than the capillary length. Velocity generated due to coalescence increased with droplet size to a peak and then decreased. Velocities estimated theoretically were 3–4 times

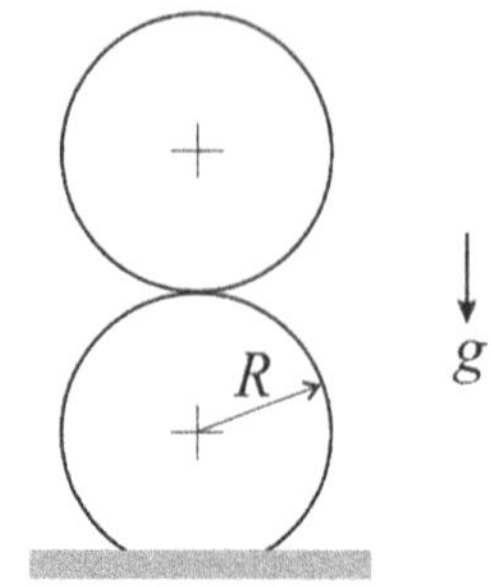

Fig. 3.8 Orientation of vertically aligned coalescing liquid drops with the lower drop placed on a solid surface

larger than the experimental values. The capillary length and timescale considered were $l = \sqrt{\sigma/\rho g}$ and $\tau = \sqrt{\rho R^3/\sigma}$ respectively; here R is the drop radius.

Graham et al. (2012) studied the impact of a falling drop on another, placed on a horizontal surface (Fig. 3.7). Surface texture was varied, ranging from hydrophilic to hydrophobic. The recoil of the coalesced drop was seen when restoring force of surface tension exceeded the inertia force. Deformation and merging of a coalesced drop were observed because of inertia of the impacting droplet. The maximum spreading length increased with hydrophobicity of the surface and the Weber number, for low impact velocities. On the other hand, maximum spreading length decreased with an increase in the offset ratio and hydrophobicity. The maximum spreading length of a coalesced sessile drop was twice that of the equilibrium spreading length. The overall coalescence process, including spreading, revealed the relative importance of surface tension, gravity, inertia, and viscosity.

Yeh et al. (2013) reported experiments on coalescence of small droplets over a textured surface of variable wettability (Fig. 3.5). A drop was introduced adjacent to a previously placed stationary drop. The movement of one towards the other was assisted by the wettability gradient of the substrate. Coalescence initiated strong convection patterns within the combined droplet. A short convection regime was identified during which large fluid velocities were created, followed later by a long diffusion-dominated tail. Mushroom-shaped mixing patterns were seen inside a coalesced drop due to the large surface energy release associated with the drops of large surface tension. Mushroom-shaped mixing patterns were also seen during coalescence of the stationary droplet with a larger surface tension and a moving droplet of smaller surface tension. Mixing patterns of the shape of a round head were seen inside the coalesced droplet during coalescence of a stationary droplet with a smaller surface tension and a moving droplet of larger surface tension showing no enhancement in diffusion-dominated mixing. More than 60% of mixing was observed during the convection-dominated regime. Lower surface tension was seen to reduce the extent of fluid mixing within the drops.

Gunjan et al. (2015) experimentally studied the effect of volume ratio of two drops on coalescence with one drop sitting on a superhydrophobic surface and the other placed above it (Fig. 3.8). The growth of the drop above was seen imparting velocity to the liquid in the upward direction. Growth of the drop below occurred because of the change of direction of fluid velocity in the direction of gravity. Over a shorter timescale, footprint radius became progressively smaller showing recoil of the drop. It was observed from energy analysis that the smaller drop below had larger internal pressure and was sucked into a larger drop above, thus contributing to the mechanism of recoil. The authors imaged recoil for a range of radius ratios around unity. The configuration of a smaller drop below had a greater tendency for recoil. Oscillations in interface velocity diminished subsequently over a longer time period, while revealing multiple timescales.

Nam et al. (2015) numerically studied drop coalescence on a water-repellent surface. Two sessile drops of equal volume were kept adjacent to each other in a computational domain (Fig. 3.5). Contact line relaxation was modeled using damped harmonic oscillations. Droplet relaxation was given by the expression $R_{x(y)} = Ae^{-Bt}(\cos \omega_d t + \phi)$. Here A is amplitude, B is the damping constant, ω_d is natural frequency, and ϕ is the phase angle. The level set method was used to study the evolution of the interface during coalescence. Coalescence showed two regimes, namely bridge formation followed by viscous dissipation. A low-pressure zone was created near the bridge due to negative curvature of the neck of the two drops. Contact line pinning and viscous dissipation were realized. The amount of surface energy released during footprint oscillations increased with a reduction in the base area. The released surface energy triggered jumping in the computational domain for some of the coalescence events.

Coalescing liquid drops experience a variety of forces, each related to surface tension, gravity, viscosity, and inertia, whose magnitudes change with time. Derived quantities such as pressure are decided by the respective values of these forces though it can be estimated by the changes in interface curvature. The ratio of a pair of forces will generate dimensionless parameters and indicate the mechanisms of relevance. Apart from fluid and interfacial properties, the drop size is an important parameter. Coalescence dynamics is influenced indirectly by gravity and has been studied earlier by the authors (Somwanshi et al. 2018).

3.4 Collision of Drops

Drops moving towards each other at a prescribed velocity will generate shapes that are fundamentally different from those seen in static drops just meeting each other at near-zero speed. The outcome of collision can be in the form of rebound, coalescence, or separation depending upon the Weber number and impact number. This section discusses the coalescence regime emerging during collision of liquid drops.

Qian and Law (1997) identified the regimes of collision of equal diameter drops of tetradecane, in an environment of nitrogen at atmospheric pressure (Fig. 3.9). The relevant dimensionless parameters were the Weber number $\mathrm{We} = 2\rho R U_{\mathrm{rel}}^2/\sigma$ and impact parameter $I = b/2R$; here, b is the initial projection of center-to-center distance of droplets on a vertical plane, and U_{rel} is the initial relative velocity between droplets. Five regimes realized were (1) secondary coalescence at

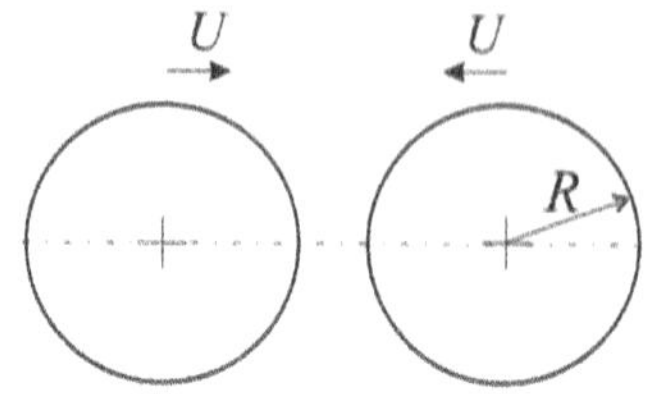

Fig. 3.9 Orientation of horizontally aligned coalescing liquid drops approaching each other with a relative velocity

We = 0.2 and I = 0.2; (2) bouncing at We = 0.5 and I = 0.1 and We = 8.6 and I = 0.08; (3) primary coalescence at We =15.2 and I = 0.08; (4) stretching separation without formation of a satellite drop at We = 19.4 and I = 0.05 and with a satellite drop at We = 32.8 and I = 0.08; and (5) reflexive separation at We = 61.4 and I = 0.06.

Brenn et al. (2001) experimentally observed head-on collision of equal diameter drops of propanol-2 in ambient air (Fig. 3.9). The collision outcome was decided by Weber number defined as We = $8\rho RU_{\mathrm{rel}}^2/\sigma$. Coalescence was observed for We <49 and formation of a satellite drop followed by separation for We >49.

Tang et al. (2012) experimentally observed the regimes of collision of drops of unequal diameter of tetradecane opposed to each other in gaseous nitrogen at atmospheric pressure. The collision outcome was again decided by the Weber number and the radius ratio $\Delta = R_l/R_s$; here, R_s is the radius of the smaller drop and R_l is the radius of the larger drop. Secondary coalescence at We <5; bouncing at We = 7.3, Δ = 1.46; primary coalescence at We = 13.8, Δ = 1.46; and stretching separation without the formation of a satellite drop at We = 52.5, Δ = 1.5, were recorded.

Pan and Suga (2005) studied collision of binary drops of water in ambient air (Fig. 3.9). The outcome of collision showed four regimes, namely bouncing, coalescence, reflexive separation, and stretching separation. The size of the satellite drop that emerged during collision of binary drops was larger in simulations than in experiments. Similarly, the transformation of drop shape was faster, and the neck was thicker in simulations than in experiments. Differences between simulations and experiments in low Weber number regimes were seen due to the modeling steps related to interfacial tension.

Results obtained by various authors show the coalescence process to be configuration specific. Studies of spreading, coalescence, and impact of drops are mostly restricted to sessile drops, while pendant drops, and those on inclined surfaces are under-represented. These configurations are important in many applications, including dropwise condensation. While surface tension and gravity act to stabilize the sessile configuration, gravity opposes surface tension in the case of pendant droplets. Hence, the data of sessile drops cannot be carried forward to the pendant since gravitational forces can upset the relative importance of other forces present.

The following sections describe experiments on coalescence of equal and unequal volume water drops (0.4–3.7 μl) in pendant and sessile configurations when the equilibrium contact angle for each of the surfaces is close to 150°. Measurements pertain to high-speed imaging of the merging drops. Quantities of interest are velocities generated post-coalescence, shear rates, and timescales estimated from experiments, relative to the natural scales inherent to the coalescence of drops of equal and unequal volumes.

3.5 Experimental Arrangement

The experimental arrangement comprises surface texturing, instrumentation, data analysis, and experimental setup. The details can be found in the following sections.

3.5.1 *Surface Texturing and Characterization*

The interaction of forces between the liquid domain and a solid surface in a gaseous medium (or otherwise immiscible fluid environment) determines wettability of the surface with respect to the chosen liquid. Wettability decides spreading or contracting characteristics of the liquid over the chosen substrate. It is strongly influenced by the chemical composition of the liquid and the surface chemistry of the substrate. It is also controlled by physical characteristics of the surface, for example, the micro/nanostructures that may be grown on it by suitable deposition techniques. Wettability ranges from completely wetting surfaces (~0°; superhydrophilic) to non-wetting surfaces (~180°; superhydrophobic) (de Gennes 1985). Quéré (2005) studied various methods of creating hydrophobic surfaces. Wenzel and Cassie-Baxter states were extensively studied. The wetting characteristics are reported in the form of the equilibrium shape of a liquid drop placed on the solid surface and more specifically, the equilibrium contact angle. Drop shape and size during coalescence evolve with time. The mathematical modeling of the evolution of the liquid-air interface can be seen in the work of Ajaev (2012).

The equilibrium shape of the drop placed on a textured surface is governed by the forces acting at the three-phase contact line, where the liquid and solid phases coexist with the environment (air, in the present work). Force balance along the contact line is expressed in the form of the Young's equation (Young 1805)

$$\sigma_{sg} - \sigma_{sl} = \sigma_{lg} \cos \theta \tag{3.2}$$

It can be seen from the Eq. (3.2) that the contact angle vanishes if the surface tension of the solid-gas interface is equal to the sum of the surface tensions of the liquid-solid and liquid-gas interfaces. The surface is wetted easily for vanishing contact angles and is called a high energy surface since it participates strongly in spreading the liquid phase. The maximum contact angle attainable theoretically is ~180° for a low energy surface, which does not have the ability to spread the liquid mass. Surface energies are finite, falling in the range of 0–180°. A surface with a

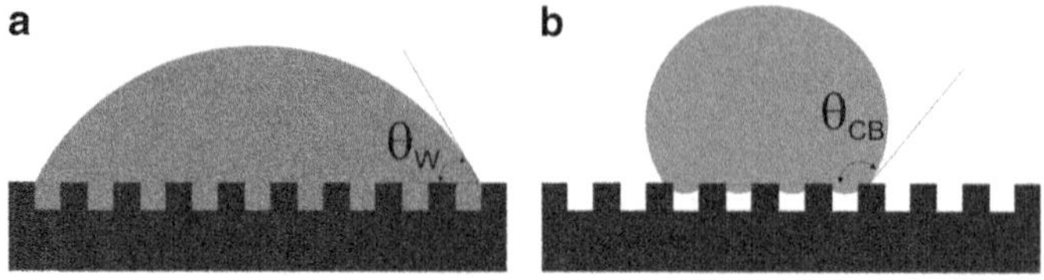

Fig. 3.10 Liquid drop at equilibrium in (**a**) Wenzel and (**b**) Cassie-Baxter state. Liquid spreading in (**a**) leads to a lower (apparent) contact angle

small contact angle is said to be hydrophilic while one with a large contact angle is hydrophobic. Wetting characteristics of a surface (called *hydrophobicity*) can be controlled by adopting a suitable surface engineering approach. These approaches can be physical or chemical. Texturing with micro/nanopillars promises creation of surfaces with an extreme contact angle of ~180° (de Gennes 1985). The method adopted in the present study is chemical texturing as discussed in the following paragraphs.

The wetting characteristics of a physically textured surface are further described by either the Wenzel (Fig. 3.10a) or the Cassie-Baxter (Fig. 3.10b) state. In the Wenzel state, the liquid penetrates the rough surface of pillars at the micro/nano-scale. The increase in interaction between the liquid and the solid surface leads to high wettability. The contact angle of the drop in the Wenzel state is given by (Cassie and Baxter 1944)

$$\cos \theta_{\mathrm{W}} = f \cos \theta \tag{3.3}$$

Here, f is the roughness factor of the surface and θ is the equilibrium contact angle of the original surface. It can be seen from Eq. (3.3) that roughness factor greater than unity ($f > 1$) improves hydrophilicity ($\theta_{\mathrm{W}} < 90°$) and ($f < 1$) improves hydrophobicity ($\theta_{\mathrm{W}} > 90°$), respectively (Cassie and Baxter 1944). In the Cassie-Baxter state (Fig. 3.10b), the liquid in the form of a hydrophobic drop ($\theta > 90°$) sits over the top of the asperities of the solid surface trapping air in-between. It decreases contact between the interacting liquid and the solid surface to maintain low wettability. The contact angle of a liquid drop in the Cassie-Baxter state (Fig. 3.10b) with a fraction of solid surface area ϕ_{s}, and the rest air ($1 - \phi_{\mathrm{s}}$) is given by (Cassie and Baxter 1944)

$$\theta_{\mathrm{CB}} = -1 + \phi_{\mathrm{s}}(\cos \theta + 1) \tag{3.4}$$

It can be seen from Eq. (3.4) that θ_{CB} increases with a reduction in ϕ_{s}, theoretically achieving $\theta_{\mathrm{CB}} = 180°$ for $\phi_{\mathrm{s}} = 0$, i.e., the liquid and the solid surface are physically in contact with trapped air alone. However, the roughness factor decreases with a reduction in ϕ_{s}, promoting the Wenzel state. It is clear from the above discussion (Eq. 3.2) that wettability can be tuned for ideally smooth and homogeneous surfaces by adjusting the three surface tension forces. However, the state of liquid on a solid surface can be Wenzel (Fig. 3.10a) or Cassie-Baxter (Fig. 3.10b) for real surfaces. The critical roughness factor (Eq. 3.5), deduced from Eqs. (3.3) and (3.4), defines the state of the liquid drop to be either Wenzel (Fig. 3.10a) or Cassie-Baxter (Fig. 3.10b). The critical roughness factor that defines this transition is given by (Cassie and Baxter 1944)

$$f_c = \frac{(\phi_s - 1)}{\cos \theta} + \phi_s \tag{3.5}$$

Therefore, wettability can be altered on a surface by tuning surface energy, for example, by using micro/nanotexturing of the surface. The present study uses chemical texturing for this purpose and is discussed in the following sections.

Surface Preparation

An appropriate surface treatment procedure was carried out on a copper substrate to attain superhydrophobicity. This process is described in the following paragraph.

The substrate was first polished using a high-grade grinding wheel. Emery papers of grit size ranging from 600 to 2500 were used sequentially on the already machined substrate. The substrate was cleaned using neutral liquid detergent (Labolene®—Fischer Scientific) and flushed with ethanol and deionized water. It was further cleansed using an ultra-sonication bath for 20–30 min to remove the impurities at the microscale and dried with pure nitrogen. The cleaned substrate was immersed in an aqueous solution of 0.01 mol/l $Ag(NO)_3$ for 20 s, washed with water and dried again with nitrogen. The substrate was subsequently immersed in a solution consisting of 1 mmol/l of 3,3,4,4,5,5,6,6,7,7,8,8,9,9,10,10,10-heptadecafluoro-1-decanethiol ($CF_3(CF_2)_7CH_2CH_2SH$) in CH_2Cl_2 for 5 min and dried. The process, adapted from Larmour et al. (2007), resulted in a robust superhydrophobic surface.

Surface Characterization

Field emission scanning electron microscopy (FESEM; JSM-7100F) was used to image the surface. JEOL® images of the substrate are presented at low (Fig. 3.11a(i)) and high (Fig. 3.11a(ii)) magnifications in the figures below. It is clearly seen that protrusions like micro-flowers of diameter 10–600 nm are uniformly grown because of oxidation of the copper surface. The gap between adjacent micro-flowers is occupied by ambient air. The combination of air gap and the micro-flowers forms a Cassie-Baxter wetting regime. In turn, water drops show superhydrophobic behavior with respect to the substrate. The static contact angle was measured at three different locations over a 100 mm^2 area, while many substrates were tested. Figure 3.12 shows typical set of experimental images wherein a single static drop of volume 1–5 μl was kept on or underside the substrate and imaged. The static (equilibrium) contact angles measured for these pendant and sessile configurations

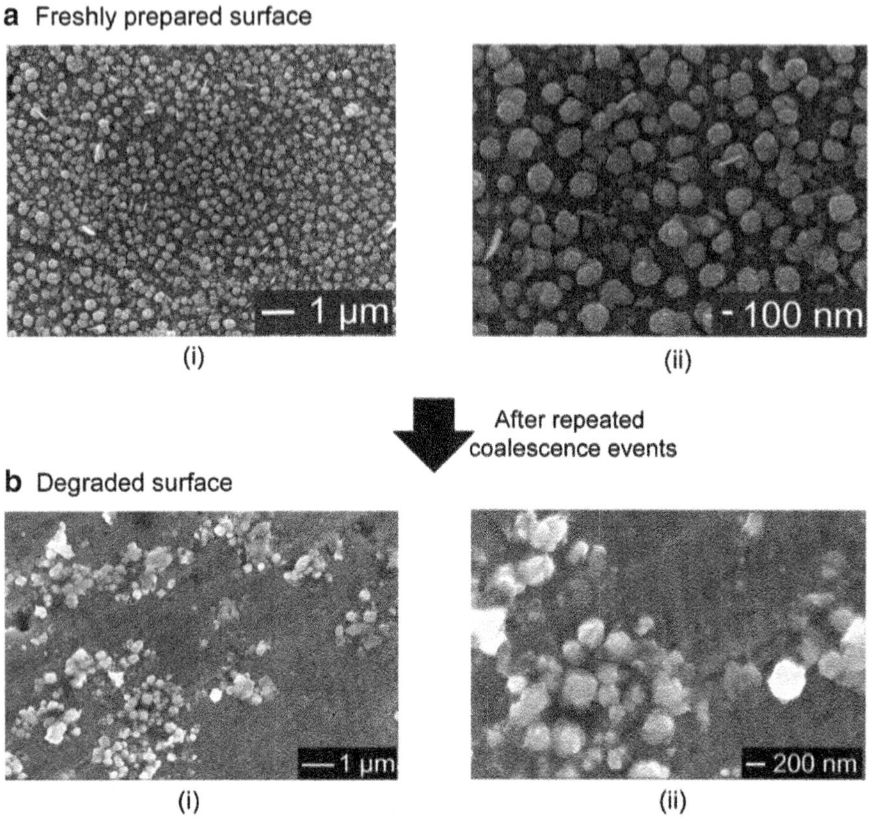

Fig. 3.11 Field emission scanning electron microscopy (FESEM) images of (**a**) the virgin hydrophobic copper substrate as-prepared at (**i**) low (×10,000) and (**ii**) high (×30,000) magnification showing the flowering oxide structures responsible for creating the hydrophobicity; contact angle ~150°. (**b**) The hydrophobic copper substrate is shown after several occurrences of coalescence at (**i**) low (×25,000) and (**ii**) (×75,000) magnification showing leaching of the flowering structures due to repeated shear stress generated on the wall during drop coalescence; contact angle reduces to ~120°

were found to be $150° \pm 4°$ and $145° \pm 3°$, respectively, independent of the volume variation (within Bo $\approx$ 0.01–0.04). On an average, contact angles for pendant drops were slightly greater than the sessile. The uniformity of surface morphology and repeatability of coalescence details were confirmed for the range of experiments reported in the present work.

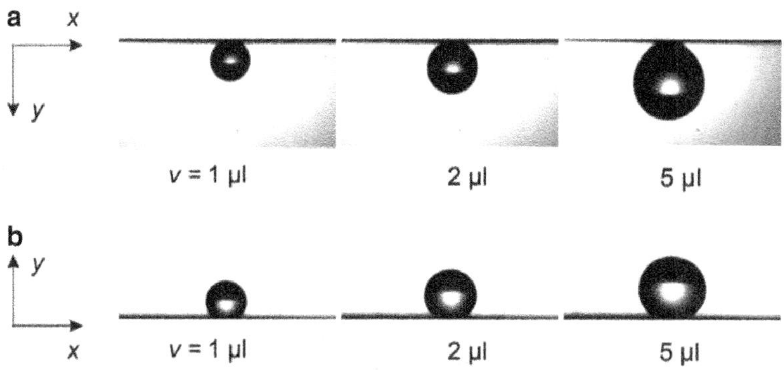

Fig. 3.12 Imaging of a static drop showing nearly constant static contact angle for (**a**) pendant and (**b**) sessile configuration over a range of volumes (1–5 μl)

3.5.2 Instrumentation

Schematic of the experimental setup is shown in Fig. 3.13. Preliminary experiments in droplet coalescence show it to be an intensely dynamic phenomenon, thus requiring a sufficiently high-speed camera. For the camera speed chosen, millisecond-scale resolution was possible, but microsecond-level changes were not detected. Further, oscillations induced during coalescence make the drop movement three dimensional. The primary measurement taken up for discussion here is of the mid-plane of the merging drop. Limited experiments related to imaging the drop footprint were conducted using confocal microscope to explore the third dimension.

High-Speed Camera

A monochrome high-speed camera (Photron® FASTCAM SA-3) selected for the present work has superior performance at a level of megapixel resolution, an ultra-sensitive image sensor which records clear images in low light while the maximum imaging frame rate is 4000 fps. The spatial resolutions per pixel are ~8.75 and ~9.07 μm in the x- and y-directions, respectively.

The camera has an internal memory which enables one to save high-speed images as uncompressed digital data. It has 8-bit digitization with respect to light intensity. The high-speed images are captured blur-free by keeping the exposure shorter than the frame rate. It also has a video output connector to playback the recorded images on a video monitor. The high-speed camera is connected using a Gigabit Ethernet interface to the computer.

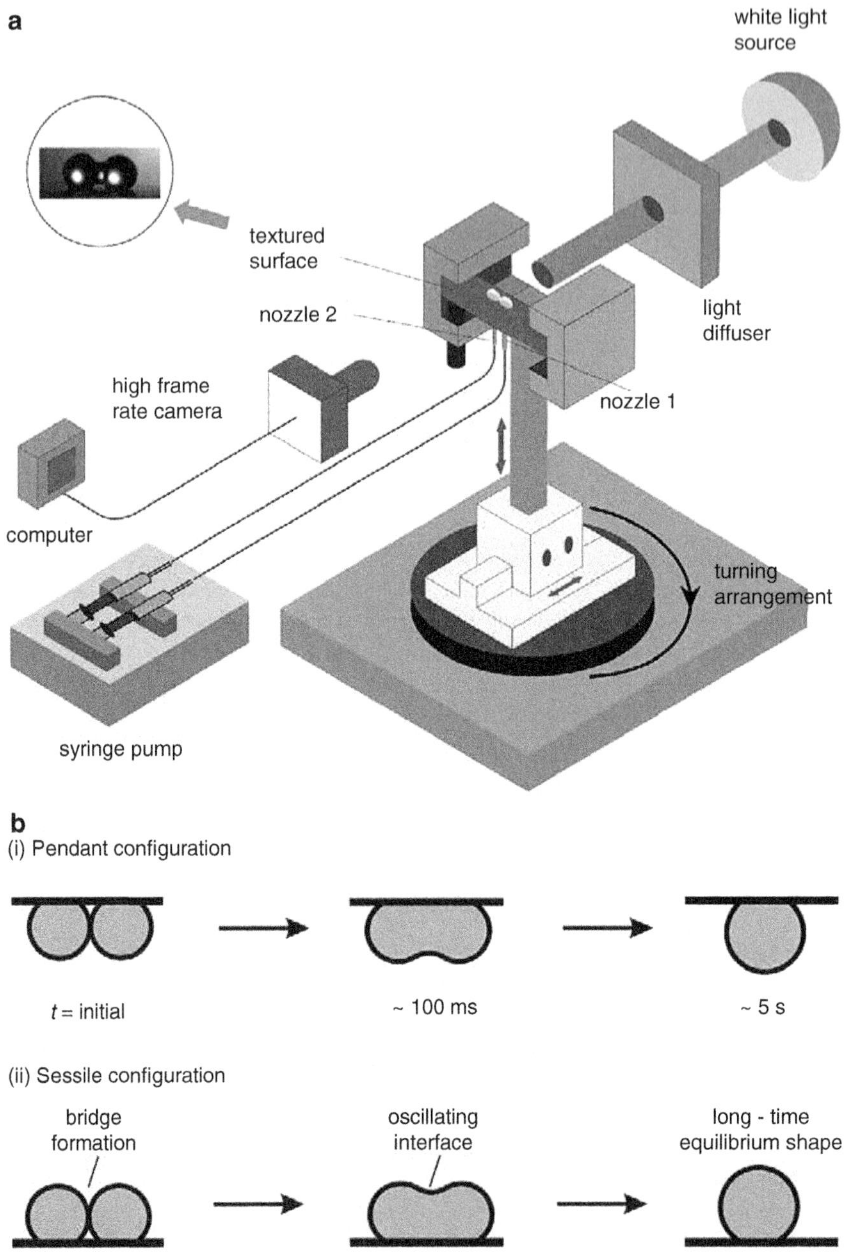

Fig. 3.13 Schematic drawing (**a**) of the experimental setup for studying the coalescence process of two adjacent drops in either pendant or sessile configuration. (**b**) Orientation of (**i**) pendant drops in a range of volumes and (**ii**) sessile drops in a range of volumes and the observed order of magnitude of timescales, immediately after coalescence till equilibrium is attained

Confocal Microscope

A confocal microscope (Leica CTR6500) is used for three-dimensional imaging of a transparent object such as a water drop by collecting stacks of planar (x–y) images. A series of 2D images are taken by stepping either the samples or the objective lens along the z-axis while the whole image is scanned over a horizontal plane. The stacks of planar images may be assembled digitally to create a 3D image of the object. A CCD camera is attached to the confocal microscope to record each of the 2D images. Though the microscope is highly resolved in space, the scanning process over a plane makes the camera slow in time. The framing rate of the microscope used in the present work was 25 fps. It has been used mainly to record the footprint of the merging drops at selected time instants.

3.5.3 Experimental Procedure

Experiments adopt a well-defined protocol to ensure repeatability and accuracy of the measurements. The surface plays an important role in performing high-quality experiments since the three-phase line formed over the surface plays a decisive role in fluid motion. A protocol as discussed in the previous section is used to prepare the surfaces of nearly identical morphology. Drops of specific volume are deposited on or underneath a textured surface using a syringe pump and a micro-liter syringe. A rigid fluid line is used to ensure the conservation of the mass from the syringe pump to the substrate. The remaining steps in imaging measurements proceed as per the following sequence.

Experiments using a High-Speed Camera

The following procedure was adopted for conducting measurements with a high-speed camera.

1. The substrate is prepared using chemical texturing as discussed earlier and the equilibrium contact angle for deionized water on this surface is measured in the pendant and sessile configurations.
2. These substrates are vacuum dried before each experiment to ensure repeatability of the coalescence sequence.
3. Through-holes (size ~0.45 mm) under the substrate with a predetermined center-to-center distance assist in introducing the two drops with two independent micro-nozzles, kept adjacent to each other.
4. A micro-liter syringe with 100 μl capacity and a least count of 0.02 μl is used to deposit sessile/pendant drops, above or on the underside of the substrate, as required.

5. The drops are then grown slowly on or below the substrate. As the drop expands, its footprint covers the nozzles and the three-phase contact line lies entirely on the hydrophobic substrate.
6. The liquid drops are illuminated by a diffuse white light source. The exposure is shorter than the frame rate to record blur-free images.
7. For measurement of distances, the high-speed camera is calibrated in the x- and the y-directions.
8. The merging drops are imaged in a vertical plane using the high-speed camera whose viewing axis is horizontal.
9. The image sequences are recorded in the computer and analyzed as discussed in the following section.

Experiments using a Confocal Microscope

Coalescence of pendant drops in the form of their footprint can be imaged under an upright confocal microscope. Imaging of sessile drops can also be performed with transparent substrates but is not reported. The following procedure was adopted for confocal imaging.

1. Laser and chiller are started before imaging.
2. A micro-liter syringe with 10 μl capacity and a least count of 0.01 μl is used to deposit pendant drops on the underside of the substrate.
3. The drops are then grown slowly on or below the substrate. As the drop expands, its footprint covers the nozzles and the three-phase contact line lies entirely on the hydrophobic substrate.
4. The imaging plane is set at the footprint of the coalescing drops, and the image recording is initiated to capture an image sequence.
5. The merging drops are imaged as a time sequence in a horizontal plane using a confocal microscope whose viewing axis is vertical. Intermediate stages of coalescence, particularly at the initial stages where a bridge is formed, are extracted from the data of contiguous experiments with an offset in the initial time instant. In the present study, confocal data is examined purely from a qualitative perspective.

3.5.4 Data Analysis

Image sequences for equal and unequal volume drops in the sessile and pendant configurations are considered for data analysis. Sample sequences can be seen in Figs. 3.14 and 3.15 for equal volume drops; and Fig. 3.16 for unequal volume drops. In each experiment, imaging commenced a few seconds prior to coalescence and continued for nearly a minute. The significant part of the coalescence event lasted around 300–400 ms, starting with bridge formation leading all the way to equilibrium. The frame prior to the one containing the bridge was taken as the time origin

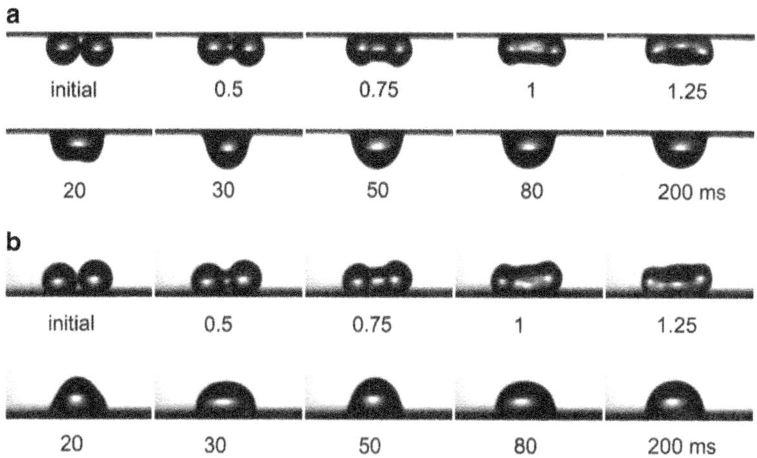

Fig. 3.14 Image sequence of the coalescence process for (**a**) pendant drops (v_1 = 0.6 μl, v_2 = 0.6 μl). (**b**) Sessile drops (v_1 = 0.51 μl, v_2 = 0.50 μl) of nearly equal volume at a lower combined Bond number (Bo = 0.014–0.015)

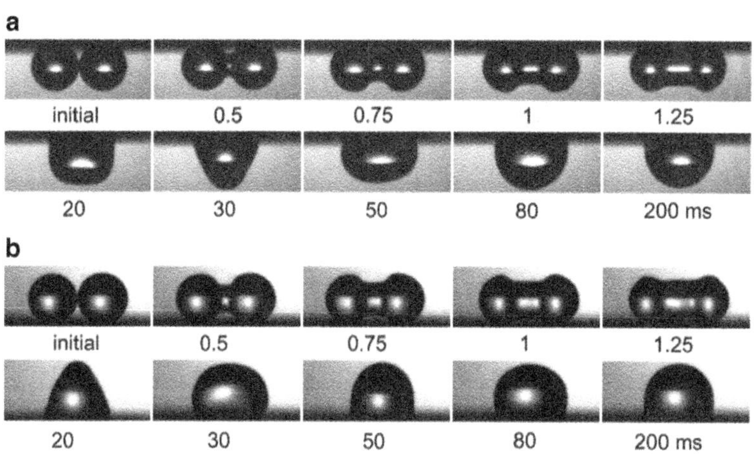

Fig. 3.15 Image sequence of the coalescence process for (**a**) pendant drops (v_1 = 2.0 μl, v_2 = 2.0 μl). (**b**) Sessile drops (v_1 = 1.8 μl, v_2 = 1.8 μl) of nearly equal volume at a higher combined Bond number (Bo = 0.032–0.034)

for image analysis. Using the reduced sequence of images, *x*- and *y*-coordinates of the centroid of the evolving drop shapes have been determined. Here, *x*-coordinate is in the horizontal direction and *y*, the vertical in the upward (sessile) or downward (pendant) direction, as shown in Fig. 3.12.

The recorded images were analyzed to estimate the parameters of interest. Each instantaneous image was represented as an array of 1024 × 1024 pixels. A unique light intensity in the range of 0–255 was assigned at the pixel locations. This data

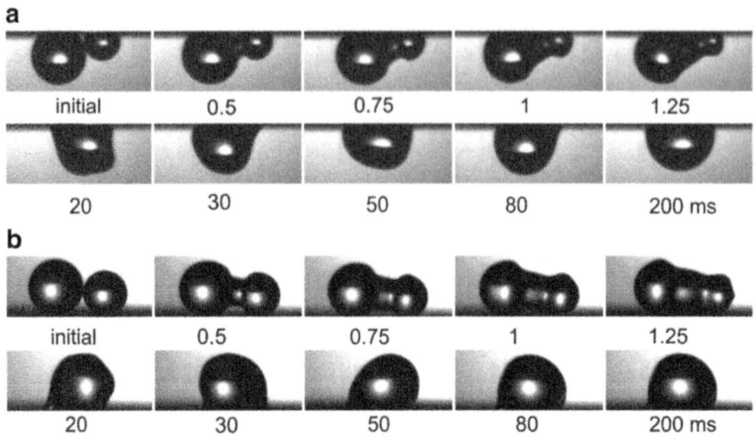

Fig. 3.16 Image sequence of the coalescence process for (**a**) pendant drops ($v_1 = 1.2$ μl, $v_2 = 3.3$ μl). (**b**) Sessile drops ($v_1 = 1.6$ μl, $v_2 = 3.1$ μl) of unequal volume at a combined Bond number of 0.037 and 0.038, respectively

was used in conjunction with MATLAB® for the estimation of the centroid positions of the coalesced drop (Gonzalez et al. 2004; Weeks 2007).

The data of position coordinates as a function of time can be numerically differentiated to yield the two respective velocity components. Drop movement along the *z*-coordinate normal to the plane of the image was found to be small; accordingly, the third component of velocity is neglected in subsequent calculations, specifically for the wall shear stress.

The *x*- and *y*-components of the centroid are calculated from each image of *N* pixels by

$$x_c = \frac{\sum_{i=1}^{N} x_i \cdot w_i}{\sum_{i=1}^{N} w_i}, \quad y_c = \frac{\sum_{i=1}^{N} y_i \cdot w_i}{\sum_{i=1}^{N} w_i} \tag{3.6}$$

Here, the weighting factor is $w_i = 1$ inside the drop and $w_i = 0$ outside. The light intensity contrast within and outside the drop was sufficiently large and Eq. (3.6) could be evaluated without ambiguity. A set of four adjacent coordinates of the combined drop were linearly regressed in the form

$$x_c = a_0 + a_1 t; \quad y_c = b_0 + b_1 t \tag{3.7}$$

Here, time is the independent variable. The coefficients a_1 and b_1 were identified as the *x*- and *y*-components of the respective centroidal velocities. The data set had four adjacent coordinates through which a line was fitted to find the respective centroidal velocities. Following a moving frame approach, the next set of four adjacent values had three values from the previous data set excluding the first. Here, overlap was maintained among the images evaluated to provide continuity in the distribution of velocity with time. The overlap of three, between adjacent sets of four coordinates, was finalized from numerical experimentation. The velocity values converged well with the choice of four frames with a three-frame overlap.

It is appropriate to use the definition of instantaneous shear rate for analyzing coalescence since the shear stress induced on the wall depends on the velocity and timescales of the coalescence event. The x- and the y-components of centroid velocities can be combined with the definition of shear rate to yield

$$\dot{\gamma} = \left(\frac{\partial u_c}{\partial y_c} + \frac{\partial v_c}{\partial x_c}\right) \tag{3.8}$$

The first term of Eq. (3.8) was found to be substantially larger than the second. Accordingly, the instantaneous shear rate at the wall at a time instant t_i was estimated to a first degree of approximation by

$$\dot{\gamma} \approx \frac{(u_c)_i}{(y_c)_i} \tag{3.9}$$

and wall shear stress by $\tau_w(t) = \mu\dot{\gamma}$.

3.5.5 *Experimental Setup*

Two drops are placed on a horizontal surface in a pendant or a sessile configuration (Fig. 3.13). A micro-liter syringe with 100 μl capacity and a least count of 0.02 μl is used to deposit sessile/pendant drops, above or on the underside of the substrate, as required. The substrate is chemically textured to generate micro-flowers (Fig. 3.11a, b). Substrates are vacuum dried before each experiment to ensure repeatability of the coalescence sequence. Figure 3.13b shows the orientations of drops considered in the present study. The moving three-phase contact line is entirely on the hydrophobic substrate for the range of volumes (0.4–3.7 μl) considered. The merging drops are then imaged using a high-speed camera. The camera is set at a frame rate of $4000\ s^{-1}$ with a shutter opening of 2×10^{-2} ms. The recorded images have been analyzed to estimate the parameters of interest.

The experimental procedure involves capturing the image sequence to study the evolution of droplet footprint and its movement along the horizontal plane. The movement of the footprint provides information on the speed with which the three-phase contact line moves over the solid surface; due to low frame rate, the microscope could not resolve the early transients.

Precautions taken during experiments include maintaining surface quality, alignment, and cleanliness so that the initial and final equilibrium contact angles are identical for all drop volumes studied. Injection from the syringe pump is greatly reduced just before coalescence to minimize the influence of initial velocities. Post coalescence, the original contact angle is recovered over a period of several seconds to a minute; this long-time data is not shown in the coalescence images discussed. In the following discussion, the movement of centroid of the merged drops, velocities acquired, and shear rates are presented as a function of time.

3.6 Results and Discussion

In the following discussion, the drops are taken to merge under ambient conditions. Volumes of water drops in experiments range from 0.4 to 3.7 μl, with equal and unequal combinations separately considered. The Bond numbers based on the combined volume are in the range of 0.01–0.04. Sessile and pendant configurations are studied, with the contact angle for the two being 150° ± 4° (pendant) and 145° ± 3° (sessile), as noted earlier. Instantaneous snapshots from the coalescence sequence are shown in Figs. 3.14, 3.15, and 3.16, while the evolution of the footprint is shown in Fig. 3.17. In Figs. 3.14, 3.15, and 3.16, small and long timescales are clearly visible. The initial timescale of ~0–1 ms is for bridge formation during which the individual drops lose their shape identity. The second stage (~20 ms) involves large fluid velocities and oscillations in the drop position. This is followed by the relaxation timescale (>300 ms) when the velocities generated within the coalesced drop are dissipated by viscosity. In similar experiments with low-speed confocal microscopy, the initial timescale was seen to be ~0–25 ms for bridge formation and the relaxation timescale of ~300 ms was longer (Fig. 3.17).

Images recorded from experiments have been analyzed for the positions of the centroid, respective centroidal velocities and shear rates. Velocities and shear rates as a function of time are obtained from this analyses, with their characteristic values summarized for a range of drop volumes and orientation. The data is categorized and discussed in the following sections.

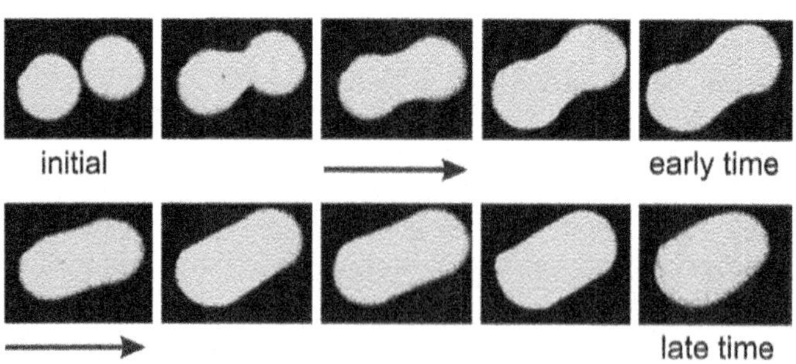

Fig. 3.17 Image sequence of the footprint of a combined drop recorded using a confocal microscope for pendant drops ($v_1 = 0.5$ μl, $v_2 = 0.5$ μl) of equal drop volumes. Early time corresponds to a period less than ~25 ms; late time extends to around ~200 ms. Eventually, the footprint of a combined drop assumes a near circular shape

Results obtained from image analysis are organized in four sections. The first is a discussion on velocity traces generated during coalescence. The second presents scale analysis of instantaneous velocity and timescales connected with the coalescence data for equal and unequal volumes. The third section is concerned with shear rates generated during coalescence of drops of equal volumes in sessile and pendant configurations. The last section includes shear rates generated during coalescence of unequal volume drops. In each instance, the measured velocity and timescales are compared with the natural scales associated with the drop merging process.

3.6.1 Velocity Traces

While the pressure difference between adjacent drops of equal volume will be small, velocities are set up by the pressure difference between the bridge and each drop, where negative curvature is site for pressures below atmospheric. Pressure difference will set the fluid in motion. When the drops merge to form effectively a single drop, internal pressure will scale with the new volume of the combined drop. For drops of unequal volume, pressure difference relates to the difference in volumes themselves, apart from bridge formation, and velocities can be larger than those in coalescing drops of equal volumes. In the pendant mode, gravity and pressure are opposed to each other, pressures are themselves smaller compared to the sessile, and the resulting velocities (and displacements) are accordingly smaller. Interface oscillations are influenced by gravity and will depend on the centroid displacement in the vertical direction, being larger in drops of unequal volumes. The shear rate is proportional to the x-component velocity parallel to the surface and inversely proportional to the normal distance of the centroid from the wall. Hence, sessile drops generate greater shear rate in the longer timescale of > 20 ms relative to the pendant though in the short timescale (< 20 ms) shear rates are uniformly large for pendant and sessile drops.

The peak displacement and velocity in the x-direction are greater initially for unequal drops and soon settle to comparable levels realized in equal pairs. However, the y-component behavior is similar for the coalescence of equal and unequal volume drops. In addition, sustained oscillations can be seen for pendant drops over the time period covered. This trend is a consequence of gravitational oscillations specific to coalescence of drops in the pendant configuration. The vertical oscillations in terms of displacement as well as velocity are practically independent of the resulting drop size. These have a time period of 15–20 ms corresponding to a frequency of 50–66 Hz. The vertical velocity fluctuations lasted for nearly a full second in all the experiments.

The x-component velocity decays rapidly with time while the one in the y-direction oscillates for a longer duration. The initial transients are followed by a longer tail of a few 100 ms, where the velocities initially generated, decay with time owing to viscous dissipation. The overall timescale in the x-component velocity is seen to increase with the combined drop volume.

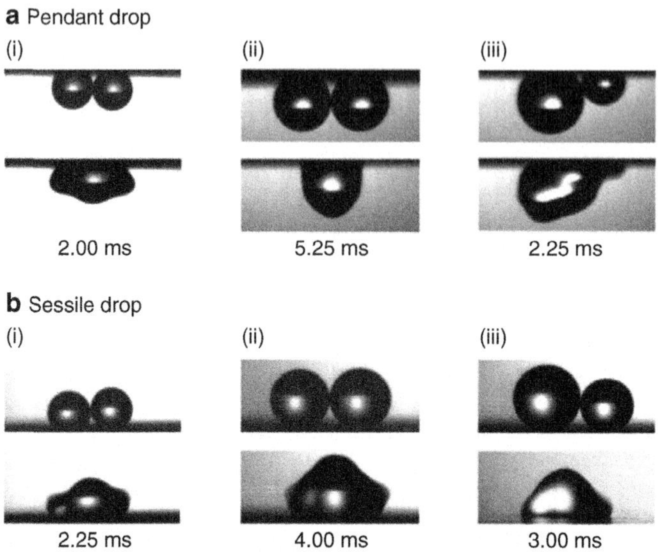

Fig. 3.18 Shape of the combined drop at the instant of peak shear rate for (**a**(**i**)) pendant drops ($v_1 = 0.51$ μl, $v_2 = 0.50$ μl) of equal volumes and (**b**(**i**)) sessile drops ($v_1 = 0.6$ μl, $v_2 = 0.6$ μl) of equal volumes, (**a**(**ii**)) pendant drops ($v_1 = 2.0$ μl, $v_2 = 2.0$ μl) of equal volumes and (**b**(**ii**)) sessile drops ($v_1 = 1.8$ μl, $v_2 = 1.8$ μl) of equal volumes and (**a**(**iii**)) pendant drops ($v_1 = 1.2$ μl, $v_2 = 3.3$ μl) of unequal volumes and (**b**(**iii**)) sessile drops ($v_1 = 1.6$ μl, $v_2 = 3.1$ μl) of unequal volumes

A quantity of interest is the shape of the drop at the instant of highest velocity, soon after bridge formation. The shapes determined in various experiments are summarized in Fig. 3.18; Fig. 3.18a for the pendant and Fig. 3.18b for the sessile. The two individual drops that initiate coalescence are seen to lose their identity and a single fluid mass is set in motion. The motion is oscillatory with opposing influences of the acquired kinetic energy in bulk and departure from equilibrium at the three-phase contact line.

Shear stresses arising in coalescence can be large since velocities are finite and length scales are small. The adhesive force between the hydrophobic coating and the substrate depends substantially on the methodology employed for preparation. Values reported range from few hundreds of nano-Newton to micro-Newton (Azehara et al. 2008; Tambe and Bhushan 2005; Ren et al. 2004). Kim et al. (2006) measured adhesive forces between the substrate and fluorinated sol-gel hybrid materials during UV-based nano-imprint process to be in the range of 15–42 μN. The corresponding stresses were estimated to be in the range 38–1160 mPa. The maximum shear stress during droplet coalescence was seen in the present set of experiments to fall in the range of 80–450 mPa during the early inertial oscillations. These serve as periodic loading over the textured surface. Hence, oscillations generated by coalescence are expected to be large enough to cause leaching and degradation of the hydrophobic coatings over a period. In the present study, the surface was imaged before and after 50–100 sets of coalescence events. While the initial image recorded using FESEM is given in Fig. 3.11a, the final image shown in Fig. 3.11b shows definite leaching of the surface, specifically of the copper oxide flowers that are responsible for hydrophobicity. Substrate damage further reduces the contact angle from ~150° to ~120° (Fig. 3.11b).

3.6.2 Velocity Scales and Timescales

As seen in the image sequences (Figs. 3.14, 3.15, and 3.16), large velocities are generated during early time, followed by slow decay over a longer time span. In addition, there is a distinct dependence of magnitude of velocity on the size of the combined drop. The nature of unsteadiness is characteristic of the chosen velocity component, the vertical velocity providing for long-lived gravity-induced oscillations. Figures 3.14, 3.15, and 3.16 clearly show the appearance of two timescales within the transient coalescence process. The highest Bond number in Table 3.1 is less than unity confirming the importance of surface tension in the analysis. The first is rapid and lasts for around 10–30 ms. The second is longer and persists over a timescale of greater than 200 ms.

Timescales appropriate for the coalescence process can be estimated using dimensional analysis and are discussed in the following paragraphs. Let R be a length scale and U, a velocity scale. In the present discussions, R is taken to be the radius of the footprint of the resulting drop after equilibrium is attained. For volume V and contact angle θ, the radius of the drop footprint R in the sessile configuration is (Khandekar and Muralidhar 2014),

$$R = r \sin\theta \quad \text{where } V = \frac{\pi}{3} r^3 \left(2 - 3\cos\theta + \cos^3\theta\right) \tag{3.10}$$

For the coalescence of two stationary drops, the velocity scale is an internal variable and cannot be independently prescribed. Similarly, the process does not have a single prescribed timescale. These may be estimated by consideration of

Table 3.1 Magnitudes of dimensionless parameters arising during the coalescence experiment of two water drops under ambient conditions; velocity scale U is taken as $(\sigma/\rho R)^{0.5}$, where R is the base radius of the merged drop at equilibrium

Combined volume, μl (U, m/s)	Base radius (mm)	We	Fr	Re	Bo	Oh
0.04 (0.81)	0.107	1	631.46	108	0.002	9.25×10^{-3}
0.4 (0.55)	0.230	1	136.04	159	0.007	6.31×10^{-3}
0.8 (0.49)	0.289	1	85.70	178	0.011	5.61×10^{-3}
1.6 (0.44)	0.364	1	53.99	200	0.018	5.01×10^{-3}
3.2 (0.39)	0.459	1	34.01	224	0.029	4.46×10^{-3}
6.4 (0.35)	0.578	1	21.43	252	0.047	3.97×10^{-3}
12.8 (0.31)	0.729	1	13.49	282	0.074	3.54×10^{-3}

forces relevant to coalescence. For deionized water, the respective fluid properties at 30 °C are $\rho = 9.95 \times 10^2$ kg/m^3; $\mu = 7.97 \times 10^{-4}$ Pa s; $\sigma = 7 \times 10^{-2}$ N/m. These properties have been used in the calculation of the non-dimensional parameters. Forming dimensionless groups of forces, we have the following numbers

$$\text{Inertia-surface tension, Weber number: } \quad \mathrm{We} = \frac{\rho U^2 R}{\sigma} \tag{3.11}$$

$$\text{Inertia-gravity, Froude number: } \quad \mathrm{Fr} = \frac{U^2}{gR} \tag{3.12}$$

$$\text{Inertia-viscosity, Reynolds number: } \quad \mathrm{Re} = \frac{\rho U R}{\mu} \tag{3.13}$$

$$\text{Gravity-surface tension, Bond number: } \quad \mathrm{Bo} = \frac{\rho g R^2}{\sigma} \tag{3.14}$$

$$\text{Viscosity-surface tension, Ohnesorge number: } \quad \mathrm{Oh} = \frac{\mu}{\sqrt{\rho R \sigma}} \tag{3.15}$$

The footprint radius relates to the drop volume via the contact angle.

The dimensionless quantities defined in Eqs. (3.11)–(3.15) are evaluated and presented in Table 3.1. As the range of volumes (0.04–12.8 μl) considered have a length scale of 1–2 mm, surface tension is expected to be uniformly important. The highest Bond number in Table 3.1 is less than unity confirming the importance of surface tension in the analysis.

The characteristic velocity scale based on surface tension and viscosity, surface tension and inertia, and gravity alone are, respectively,

$$U = \frac{\sigma}{\mu}, \quad U = \sqrt{\frac{\sigma}{\rho R}}, \quad U = \sqrt{gR} \tag{3.16}$$

As the velocity scale based only on gravity does not include surface tension, it is not considered further. The characteristic velocity for water using the expression σ/μ is 87.7 m/s. This is quite large and has not been observed in the centroidal velocities of the combined drop. Instead, the velocity based on scaling $(\sigma/\rho R)^{0.5}$ is around 0.2 m/s and can be seen in the experimental observations. The choice of this scale makes Weber number unity and Oh = 1/Re.

With $(\sigma/\rho R)^{0.5}$ as the velocity scale, Table 3.1 shows that drop oscillations associated with coalescence are expected to be driven by inertia and surface tension. Here, Reynolds number is expected to be high, immediately following bridge formation, while viscosity plays a secondary role. Between surface tension and gravity, the former is of greater significance. In agreement with this observation, Froude number is also seen to be on the higher side. Thus, the volumes of coalescing drops are such that the coalescence of these drops is controlled by inertia and surface tension for the observable velocities, viscosity serving the purpose of damping fluid motion over a longer time span.

Table 3.2 Timescales estimated for coalescence of water drops under ambient conditions; IS inertial-surface tension, VS viscous-surface tension, IV inertia-viscous

Combined volume (μl)	Base radius (mm)	t_{IS} (ms)	t_{VS} (ms)	t_{IV} (ms)
0.04	0.107	0.131	1.21×10^{-3}	0.014×10^3
0.4	0.230	0.415	2.62×10^{-3}	0.066×10^3
0.8	0.289	0.586	3.30×10^{-3}	0.104×10^3
1.6	0.364	0.829	4.15×10^{-3}	0.166×10^3
3.2	0.459	1.173	5.23×10^{-3}	0.263×10^3
6.4	0.578	1.659	6.59×10^{-3}	0.417×10^3
12.8	0.729	2.346	8.30×10^{-3}	0.663×10^3

Additional scales of interest are as follows: For small liquid drops in a gaseous environment, the importance of gravity over surface tension is determined by the capillary length $l = \sqrt{\sigma/\rho g}$. For the properties of water at ambient conditions, the capillary length is ~2.6 mm. Since the maximum length scale in Table 3.2 is 0.728 mm, coalescence is dominated by surface tension.

The relevant timescales associated with pairs of forces including inertia-surface tension, viscous-surface tension, and inertia-viscous are

$$t_{IS} = \sqrt{\frac{\rho R^3}{\sigma}}, \quad t_{VS} = \frac{\mu R}{\sigma}, \quad t_{IV} = \frac{R^2}{\nu} \tag{3.17}$$

Based on the properties of water, timescales defined in Eq. (3.17) are summarized in Table 3.2. As shown, the timescale of viscous–surface tension interaction is very small and is not of any specific relevance to the present study. The inertia-surface tension timescale relates to the moment following initial bridge formation when large velocities are generated within the drop, leading to large deformation of the interface. The inertia-viscous timescale is quite large and can be associated to bulk dissipation of kinetic energy of the fluid, leading to long-term relaxation of the coalesced drop towards equilibrium. Additional dissipation occurring at the three-phase contact line cannot be selectively identified since it will be seen jointly with bulk dissipation.

Large shear stresses can be associated with the inertial-surface tension coupling and hence, in balance, the inertia-surface tension-based timescale ($=t_{IS}$) is considered as relevant for inclusion in the dropwise condensation model. The velocity scale $(\sigma/\rho R)^{0.5}$ is considered appropriate in the present context.

The inertia-surface tension timescale relates to the duration beyond the microsecond range, when large but finite velocities are generated within the drop. The instant of attainment of highest velocity in this regime is summarized in Fig. 3.18, along with the corresponding shape of the drop. Magnitudes of the timescale t_{IS} defined in Eq. (3.17) and summarized in Table 3.2 are of the order of 1–2 ms. These numbers agree with the data given in Fig. 3.18.

The inertia-viscous timescale is quite large and can be associated with bulk dissipation of kinetic energy of the fluid, including additional dissipation occurring at the three-phase contact line, leading to long-term relaxation of the drop, finally settling in the equilibrium shape.

Viscous dissipation driven by the movement of the three-phase contact line at the solid surface can be estimated from the time evolution of the drop footprint. Of interest is the time duration over which two drops become a single entity geometrically though velocities within may not have fully reduced to zero. Snapshots of the footprint recorded at select time instants indicate the initial coalescence timescale to arise from surface tension and inertia, while the long-term relation is inertia-viscous driven.

3.6.3 Evaluation of Timescales

The timescales contained in the experimental data have been extracted by the following procedure. The centroid movement in the vertical direction is expressed in terms of two time parameters τ_1 and τ_2,

$$y(t) = a\sin\left(\frac{2\pi t}{\tau_1}\right)\exp\left(-\frac{t}{\tau_2}\right) + b \tag{3.18}$$

where a, b, τ_1, and τ_2 are parameters to be estimated. The oscillatory timescale is indicated by τ_1, while τ_2 is the parameter related to viscous damping.

Parameter estimation has been carried out within MATLAB® using a least squares method. The number of images M to be considered for regression is determined based on sensitivity analysis. Here, a sensitivity function of the following form is defined for the two coordinates

$$E_x(M) = \frac{1}{M}\sum_{i=1}^{M}\left((x_c)_i - \bar{x}_c\right)^2, \quad E_y(M) = \frac{1}{M}\sum_{i=1}^{M}\left((y_c)_i - \bar{y}_c\right)^2 \tag{3.19}$$

For increasing values of M, the two quantities E_x and E_y first show a maximum and then diminish to zero. The choice of M ensures the peaks in E_x and E_y fall in the mid-range, namely around $M/2$. Much larger values of M were found to emphasize the late transient of the coalescence process. The suitability of curve fitting using Eq. (3.18) was confirmed by superimposing the function with the discrete experimental data.

Timescales τ_1 and τ_2 obtained from the regression analysis of centroid position data are summarized in Table 3.3. Oscillations stem from a coupling among inertia effects, surface tension, and the contact line force. In this respect, parameter τ_1 is distinct from t_{IS}, listed in Table 3.2. The latter indicates the first time instant when peak velocities are attained during coalescence and is insensitive to gravity and viscous effects. In contrast, τ_1 in Table 3.3 has an explicit dependence on drop configuration and the component of displacement. Parameter τ_2 arises from viscous

Table 3.3 Timescales estimated from curve fitting of the centroid position data from the experiments

Initial drop volumes (μl)	Drop configuration	τ_1 (ms) (x_c, y_c)	τ_2 (ms) (x_c, y_c)
0.6–0.6	Pendant	11, 7	120, 95
0.5–0.51	Sessile	14, 6	111, 62
2.0–2.0	Pendant	25, 12	76, 198
1.8–1.8	Sessile	7, 7	49, 89
1.2–3.3	Pendant	25, 12	319, 278
1.6–3.1	Sessile	27, 11	234, 197

dissipation of kinetic energy and has a closer bearing with the timescale t_{VS} of Table 3.2. The former contains additional details of drop orientation and initial volumes while the data of Table 3.3 is purely based on dimensional reasoning.

Parameters τ_1 and τ_2 are plotted as a function of initial drop volume, Bond number and the experimental configuration in Fig. 3.19a, b. The viscous timescale is seen to be consistently greater than the oscillatory value for the range of volumes considered. Both timescales are significantly longer for unequal volume drops where the initial pressure difference and hence the initial velocities generated are greater. The decaying timescales τ_2 for the pendant drop in the y-coordinate (open blocks in Fig. 3.19b) are longer compared to the sessile and arise from persistent oscillations in a gravity field. However, the motion in the x- and y-directions remains correlated, first through nearly equal peak velocities developed in the respective directions and later through the mass balance constraint.

In addition, distinct differences between sessile and pendant drops, as well as the timescales of motion in the x- and y-directions, are discussed in the following sections.

For the instantaneous data of Figs. 3.20, 3.21, and 3.22 in dimensional form, the following scales have been adopted for the process of non-dimensionalization

$$x_c^* = \frac{x_c - \overline{x}_c}{R}, \quad y_c^* = \frac{y_c - \overline{y}_c}{R}, \quad u_c^* = \frac{u_c}{U}, \quad v_c^* = \frac{v_c}{U}, \quad \tau = \frac{t}{t_{IV}}, \quad \gamma^* = t_{IV}\dot{\gamma} \tag{3.20}$$

Here, the velocity scale is $U = \sqrt{\sigma/\rho R}$ and the timescale is $t_{IV} = R^2/\nu$. The data of Fig. 3.22 is plotted in dimensionless form in Fig. 3.23. The peak centroid displacement scales to unity and the time-axis is also mapped to unity. The trends seen in the tabular data of Fig. 3.23f are like those in Figs. 3.20, 3.21, and 3.22. There are no new features in the data in dimensionless form and hence, these are not discussed further. The data related to the centroid coordinates, velocity, and shear rates (Figs. 3.20, 3.21, and 3.22) are further discussed below.

It is clear from the data presented that coalescing drops lead to damped harmonic oscillations. With the interplay of surface tension, gravity, viscosity, local interface curvature and hence, the ensuing internal pressure field simultaneously affecting the

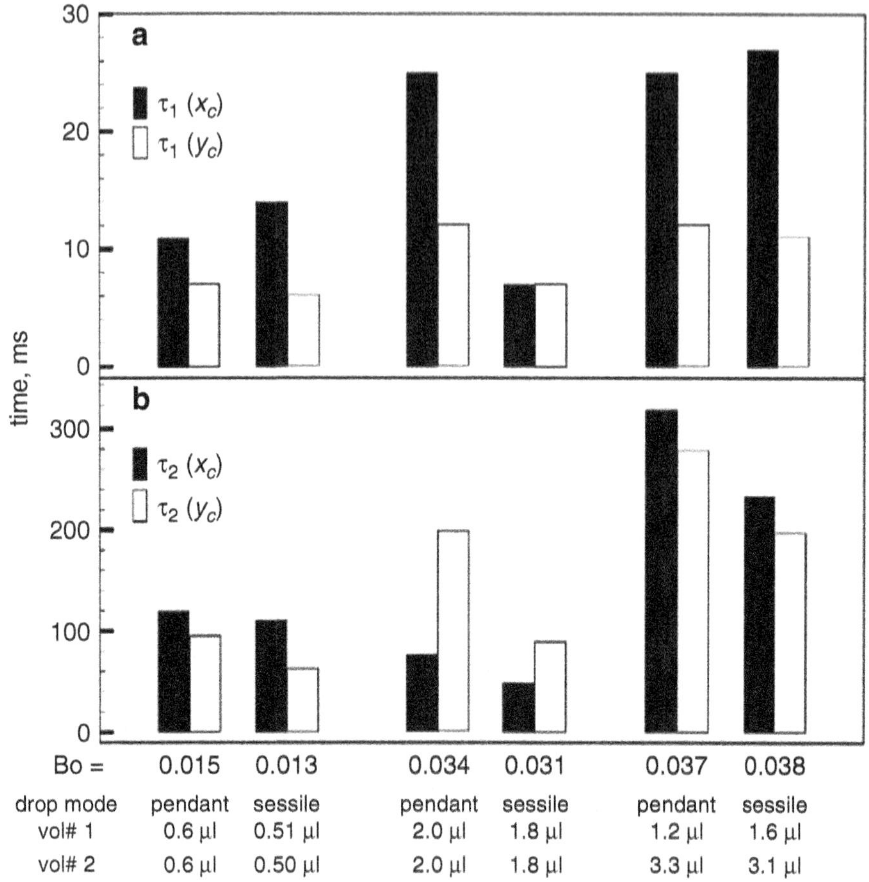

Fig. 3.19 Timescales arising from curve fitting through the data of x_c and y_c of a combined pendant and sessile drop considering. (**a**) The inertia-surface tension forces. (**b**) The inertia-viscous forces. Bond number indicated is based on the combined drop volume

drop dynamics, it is instructive to analyze the nature of these oscillations. The configuration under investigation is schematically shown in Fig. 3.24. Nonlinearity is expected to influence the dependence of effective stiffness of the fluid element on its spatial location within the drop volume and is investigated further.

Following linear analysis of a single degree of freedom system, the equivalent spring stiffness k and damping ratio ζ can be determined in terms of time constants

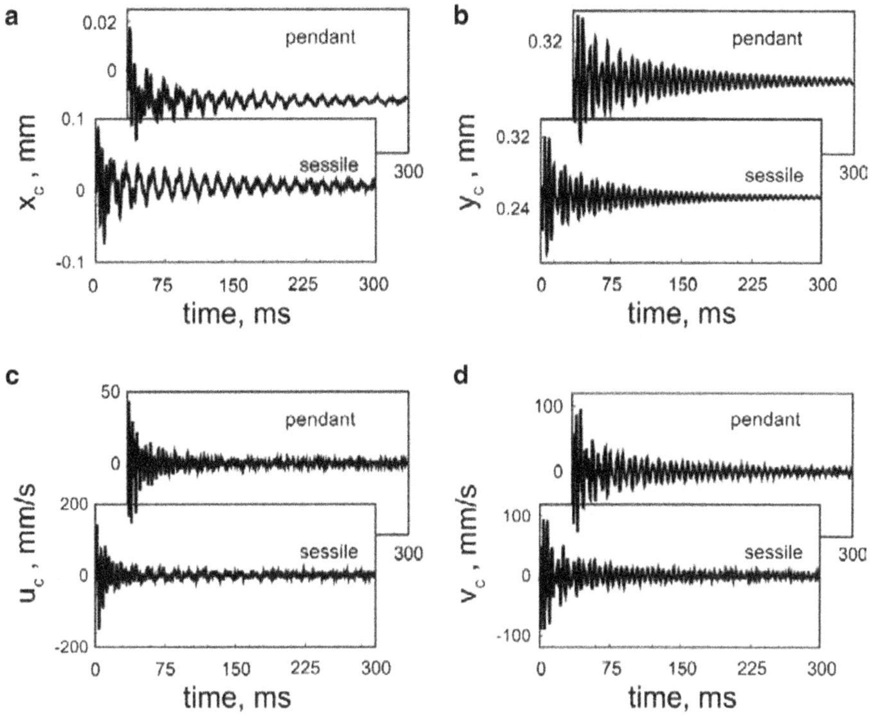

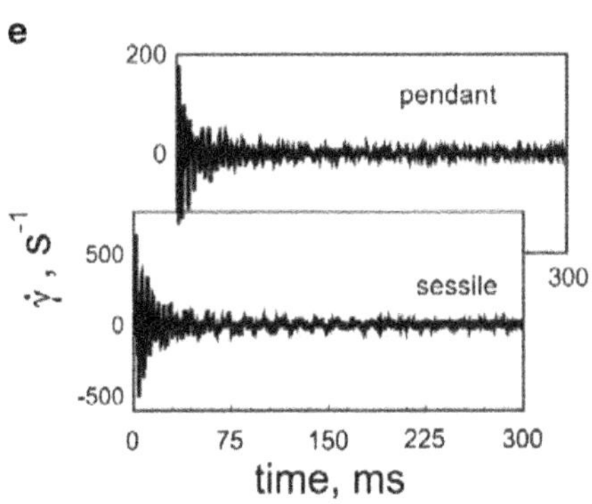

f

	Maximum value		RMS value	
Drop orientation	Pendant	Sessile	Pendant	Sessile
x_c, mm (x_c^*)	0.03 (0.09)	0.08 (0.27)	0.02 (0.05)	0.04 (0.13)
y_c, mm (y_c^*)	0.35 (1.04)	0.32 (1.02)	0.03 (0.09)	0.03 (0.09)
u_c, mm/s (u_c^*)	43 (0.09)	140 (0.31)	22.7 (0.05)	58.2 (0.12)
v_c, mm/s (v_c^*)	93 (0.2)	94 (0.2)	44.39 (0.1)	50.3 (0.1)
$\dot{\gamma}$, s^{-1} (γ^*)	141 (24.5)	640 (77.5)	80.2 (10.97)	236.2 (28.62)

Fig. 3.20 Coalescence of pendant drops ($v_1 = 0.6$ μl, $v_2 = 0.6$ μl) and sessile drops ($v_1 = 0.51$ μl, $v_2 = 0.50$ μl) of equal volumes showing the time-variation of the (**a**) x-coordinate of the centroid, (**b**) y-coordinate of the centroid, (**c**) x-component of centroid velocity, (**d**) y-component of centroid velocity, and (**e**) shear rate. Coordinates x and y are displacements relative to the joint initial centroid position of the merging drops. Maximum and rms values of the respective quantities are tabulated in (**f**)

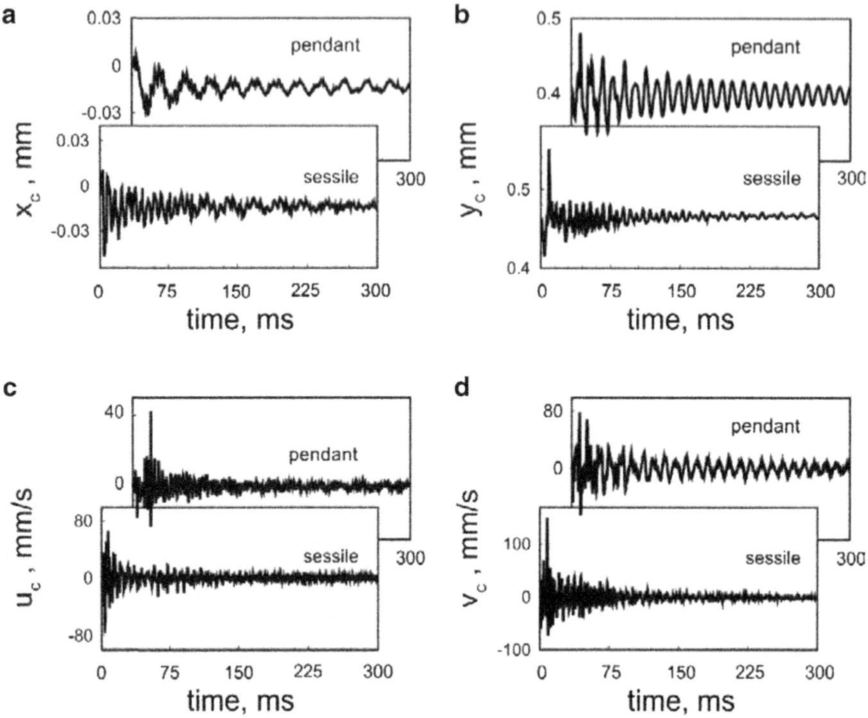

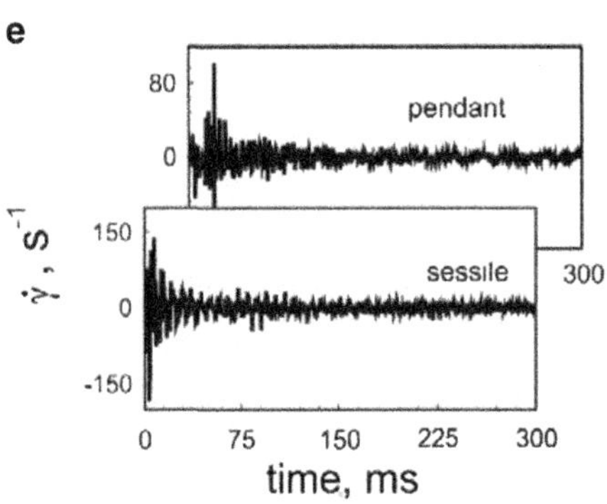

f

	Maximum value		RMS value	
Drop orientation	Pendant	Sessile	Pendant	Sessile
x_c, mm (x_c^*)	0.03 (0.06)	0.05 (0.1)	0.01 (0.02)	0.02 (0.04)
y_c, mm (y_c^*)	0.48 (0.97)	0.55 (1.15)	0.03 (0.03)	0.03 (0.06)
u_c, mm/s (u_c^*)	42 (0.11)	76 (0.19)	8.6 (0.02)	29.1 (0.08)
v_c, mm/s (v_c^*)	77 (0.21)	149 (0.39)	28.5 (0.08)	43.6 (0.08)
$\dot{\gamma}$, s^{-1} (γ^*)	101 (30.98)	183 (52.21)	21.5 (6.58)	66.8 (19.03)

Fig. 3.21 Coalescence of pendant drops ($v_1 = 2.0$ µl, $v_2 = 2.0$ µl) and sessile drops ($v_1 = 1.8$ µl, $v_2 = 1.8$ µl) of equal volumes showing the time-variation of the (**a**) *x*-coordinate of the centroid, (**b**) *y*-coordinate of the centroid, (**c**) *x*-component of centroid velocity, (**d**) *y*-component of centroid velocity, and (**e**) shear rate. Coordinates *x* and *y* are displacements relative to the joint initial centroid position of the merging drops. Maximum and rms values of the respective quantities are tabulated in (**f**)

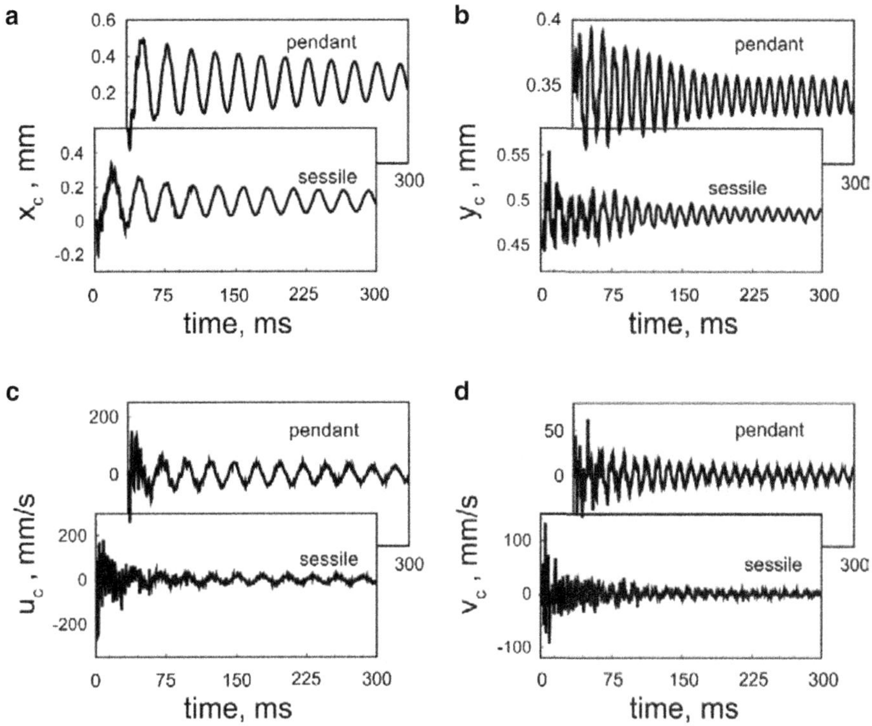

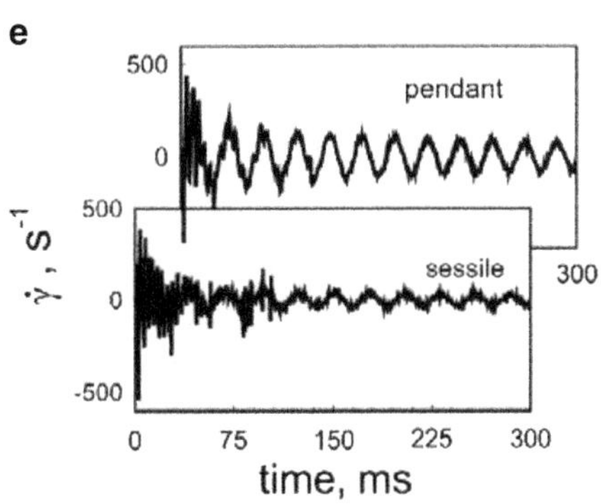

f

	Maximum value		RMS value	
Drop orientation	Pendant	Sessile	Pendant	Sessile
x_c, mm	0.49	0.33	0.19	0.13
y_c, mm	0.39	0.55	0.03	0.03
u_c, mm/s	168	244	68.86	75.52
v_c, mm/s	62	133	21.23	33.67
$\dot{\gamma}$, s^{-1}	468	543	178.4	249.7

Fig. 3.22 Coalescence of pendant drops ($v_1 = 1.2$ μl, $v_2 = 3.3$ μl) and sessile drops ($v_1 = 1.6$ μl, $v_2 = 3.1$ μl) of unequal volumes showing the time-variation of the (**a**) x-coordinate of the centroid, (**b**) y-coordinate of the centroid, (**c**) x-component of centroid velocity, (**d**) y-component of centroid velocity, and (**e**) shear rate. Coordinates x and y are displacements relative to the joint initial centroid positions of the individual drops. Maximum and rms values of the respective quantities are tabulated in (**f**)

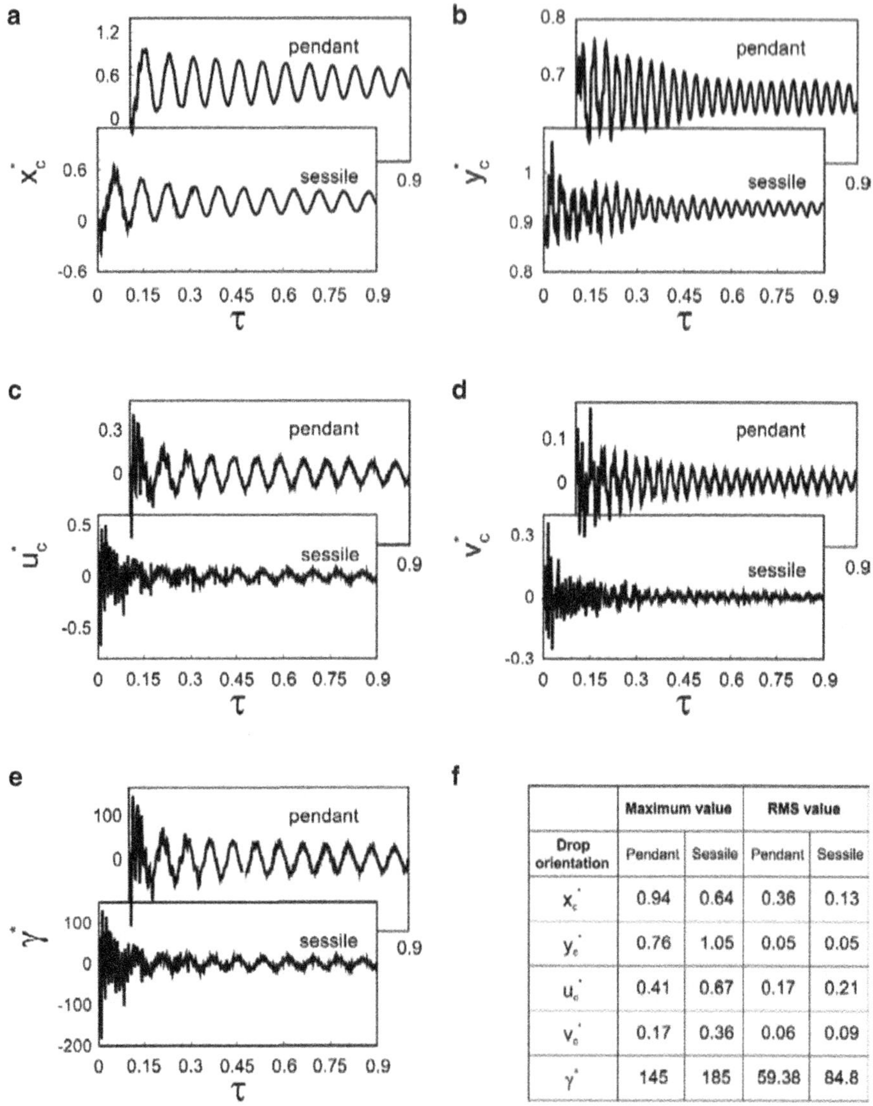

Drop orientation	Maximum value		RMS value	
	Pendant	Sessile	Pendant	Sessile
x_c^*	0.94	0.64	0.36	0.13
y_c^*	0.76	1.05	0.05	0.05
u_c^*	0.41	0.67	0.17	0.21
v_c^*	0.17	0.36	0.06	0.09
γ^*	145	185	59.38	84.8

Fig. 3.23 Coalescence of pendant drops (Re = 237) and sessile drops (Re = 239) of unequal volumes showing the variation of the non-dimensional (**a**) x-coordinate of the centroid, (**b**) y-coordinate of the centroid, (**c**) x-component of centroid velocity, (**d**) y-component of centroid velocity, and (**e**) shear rate with non-dimensional time. Coordinates x and y are displacements relative to the joint initial centroid positions of the individual drops. Maximum and rms values of the respective quantities are tabulated in (**f**)

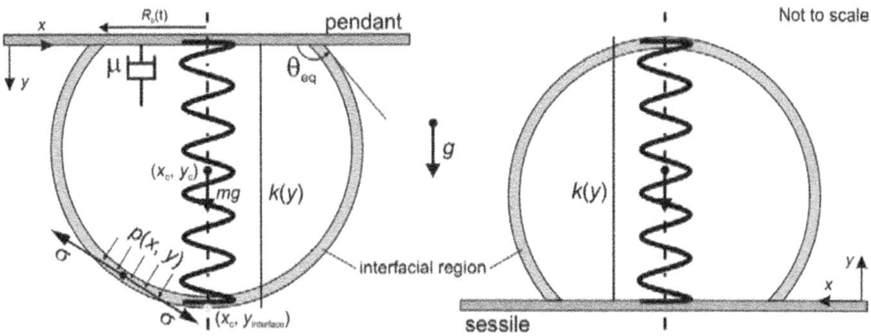

Fig. 3.24 Schematic drawing of an oscillating pendant and sessile drop, after coalescence, showing spring-mass-damper model; the system parameters of this oscillating system depend on the thermophysical properties of the fluid and spatial location within the drop volume

$$
\begin{aligned}
\sqrt{\frac{k}{m}}\sqrt{1-\zeta^2} &= \frac{2\pi}{\tau_1} \\
\zeta\sqrt{\frac{k}{m}} &= \frac{2\pi}{\tau_2}
\end{aligned}
\tag{3.21}
$$

Spring stiffness and damping ratio of the drop oscillations at the centroidal locations (x_c, y_c) are summarized in the left section of Table 3.4, for four different configurations of coalescing drops, with respective individual drop volumes shown. The right side of Table 3.4 shows similar values, for the y-components, for a point $(x_c, y_{\text{interface}})$ located in line with the centroidal point, on the interface Fig. 3.24. The combined volume of coalescing drops of equal volume is smaller than the drops of unequal volumes and hereafter referred as "small drops" and "large drops," respectively.

Comparing the behavior of the centroidal point (x_c, y_c) of pendant and sessile drops, it is clear that the timescale $(\tau_1)_x$ is consistently larger than $(\tau_1)_y$ for these configurations. This is clearly seen from the time series data of Figs. 3.20 and 3.22, where the frequency of y-direction oscillations is higher. The same trend is followed by (τ_2) as well further corroborated by the fact that the respective damping ratios for (x_c, y_c) tend to be greater for the x-direction velocities as compared to the y-counterpart. This trend can be explained by noting that dissipation arises in the bulk as well as at the three-phase contact line. The latter is greater for a larger portion of the decaying timescale. While bulk dissipation affects both components of velocities, the one at the three-phase contact line selectively damps the x-component, the y-component being already zero. However, the image sequence recorded in the present work does not distinguish the two sources of dissipation explicitly, and the numerical data of Table 3.4 are a cumulative effect. Simultaneously, it is interesting to note that the values of k_y is larger than k_x for the centroidal point. Given the fact that the fluid displacement in the y-direction is much larger than the x-direction, is expected to, lead to such a behavior.

Table 3.4 Spring constant and damping ratio determined by fitting a spring-mass-damper model to the velocity fluctuation data for two points (x_c, y_c) and (x_c, $y_{interface}$)

	Values for centroid (x_c, y_c)				Values for interface point (x_c, $y_{interface}$)			
Initial drop volumes of coalescing drops and configuration (μl)	$(\tau_1)_x$, $(\tau_1)_y$ (ms)	$(\tau_2)_x$, $(\tau_2)_y$ (ms)	Damping ratio, ζ_x, ζ_y	Spring constant, k_x, k_y (Nm^{-1})	$(\tau_1)_y$ (ms)	$(\tau_2)_y$ (ms)	Damping ratio, ζ_y	Spring constant, k_y (Nm^{-1})
0.6 and 0.6 (pendant)	11, 7	120, 95	0.092, 0.073	0.391, 0.973	6.9	81.1	0.085	0.988
0.5 and 0.51 (sessile)	14, 6	111, 62	0.125, 0.096	0.224, 1.219	5.8	51.4	0.108	1.407
1.2 and 3.3 (pendant)	25, 12	319, 278	0.078, 0.043	0.282, 1.233	11.9	238.1	0.049	1.306
1.6 and 3.1 (sessile)	27, 11	234, 197	0.115, 0.056	0.252, 1.518	10.9	171.1	0.064	1.547

The timescale $(\tau_1)_y$ estimated for the data of (x_c, y_c) as well as $(x_c, y_{\text{interface}})$ is smaller for a small drop as compared to the larger drop. The frequency of oscillation in small drops is larger than the large drops. Similarly, $(\tau_2)_y$ (estimated from both centroid and interface points) for coalescence of smaller drops is smaller than larger drops. This is expected as larger $(\tau_2)_y$ indicates that the drop oscillations will continue for a larger duration without substantial amplitude decay (Eq. (3.18)).

Both damping ratios, ζ_x and ζ_y of small drops are larger when compared to the large drop for both (x_c, y_c) and $(x_c, y_{\text{interface}})$. Smaller liquid drop stabilizes quickly relative to the large drop. The spring constant is also larger for small drops. The total available surface energy is smaller for small drops. Given the fact that surface tension forces are larger in a small drop resultant displacement will be smaller when small drops are involved. This explains higher stiffness (k_y) values for small drops.

The estimated values of $(\tau_1)_y$ from the oscillation data of the interface point $(x_c, y_{\text{interface}})$ and the centroid (x_c, y_c) are comparable. It signifies that the time period of oscillation is similar along the y-direction. The estimated values of $(\tau_2)_y$ for the interface $(x_c, y_{\text{interface}})$ is smaller ($\approx$13–17%) than the centroid (x_c, y_c). The motion of the interface $(x_c, y_{\text{interface}})$ tends to dampen faster than the centroid (x_c, y_c). Hence, the damping ratio estimated for the interface point $(x_c, y_{\text{interface}})$ is larger ($\approx$12–16%) than the centroid (x_c, y_c). Hence, the interface stabilizes quickly compared to the bulk of the liquid. The spring constant estimated for the oscillations of the interface $(x_c, y_{\text{interface}})$ is larger ($\approx$2–7%) than the centroid (x_c, y_c). It indicates that for coalescing liquid drops, the stiffness (k_y) varies with the location of point inside a drop.

3.6.4 Drops of Equal Volumes

The focus of the present study is on oscillations. Centroid positions are reported after subtracting the long-term displacement of the drops in each direction over the substrate. Since oscillations decay with time, the rms values reported here are those evaluated within the inertia-surface tension timescale τ_1 in the early portion of the data.

Figure 3.20a shows the variation of the x-component of the centroid displacement with time, when the drop volumes are in the range of 0.5–0.6 μl. Figure 3.20b is a similar variation of the y-component with time. Figure 3.21c, d shows the variation of the x- and y-velocity of the centroid of a combined drop with time. Variation of shear rate with time is presented in Fig. 3.20e. Major trends seen in these figures include large oscillations of the interface at early time followed by slow decay over the long run. Velocities generated in the sessile arrangement are larger and consequently, shear rates are also larger compared to the pendant drop. Oscillations of a pendant drop show a greater level of regularity that persists for a longer time. Maximum and rms values of quantities presented in Fig. 3.20a–e are tabulated in Fig. 3.20f. The values given in brackets are in the dimensionless form. Higher shear rates for sessile compared to pendant drops are clearly revealed in the tabulated data.

Velocity and shear rate data for drop volumes in the range 1.7–2.1 μl are shown in Fig. 3.21a–e. Since the initial pressure difference between the larger drops is smaller, velocities acquired during coalescence are smaller, leading to smaller shear rates relative to Fig. 3.21. The corresponding statistical data of displacement, velocity, and shear rate are tabulated in Fig. 3.21f. The difference between sessile and pendant arrangements, in terms of shear rates, is once again revealed, the former being the greater of the two. Shear rates in numerical terms are however smaller, compared to those in Fig. 3.20, where the individual drop volumes are smaller.

The dimensionless timescale of the overall process is close to τ_2, namely the inertia-viscous timescale t_{IV}. It maps to nearly unity for the two experiments of Figs. 3.20 and 3.21. This result is expected, as the range of Reynolds numbers (170–220) generated during fluid motion is quite narrow.

3.6.5 Drops of Unequal Volumes

The drop volumes considered are in the range of 1.6–3.1 μl for the sessile mode and 1.2–3.3 μl in the pendant. The resulting motion is presented in Fig. 3.22. The variation of the x-component of the centroid displacement with time is shown in Fig. 3.22a.

Figure 3.22b presents the variation of the y-component of the centroid movement with time. Time traces of the x- and y-component velocities of the centroid of a combined drop are shown in Fig. 3.22c, d. Shear rate variation with time is given in Fig. 3.22e. The statistical outcome arising from these plots is tabulated in Fig. 3.22f. The corresponding data in dimensionless form are reported in Fig. 3.22.

The defining feature for drops of unequal volumes is the higher initial internal pressure difference, compared to drops of equal volumes. Thus, velocities generated soon after contact are higher in magnitude. The initial velocities created in both sessile and pendant arrangements are comparable showing that viscosity, as well as gravity, play only a secondary role in this regime. For pendant drops, pressure difference is partially compensated by gravity, and the velocities attained subsequently are smaller. During the relaxation phase, velocities and shear rates continue to be higher for the sessile, as compared to the pendant case. Combined with a greater centroid movement in the vertical direction, the shear rates for the pendant arrangement are smaller than for sessile drops. These trends are seen in dimensional (Fig. 3.22) and dimensionless (Fig. 3.23) form and are tabulated in Figs. 3.22f and 3.23f, respectively. These trends in dimensionless form generalize the physical explanation of coalescence presented in the previous sections.

For unequal volume drops, the time period of oscillations in the x-displacement (and velocity) are visibly greater than those in y. Referring to values of τ_1 in Table 3.3, the x-movement shows a range of 7–27 over the range of volumes studied, while it is restricted to 6–11 in the y-direction. This result arises from the x-movement being driven by pressure difference between drops placed adjacent to each other, while the y-movement is affected by gravity.

The Reynolds numbers arising in coalescence experiments with unequal volume drops are 237 (pendant) and 239 (sessile). As in experiments with equal volume drops, the total timescale of drop movement including relaxation is mapped to nearly unity.

3.6.6 Energy Analysis

Coalescence of drops involves the appearance of various forms of energy whose relative magnitudes change sharply with time. In the present discussion, energy rates (units of μW) corresponding to changes in surface energy, gravitational, kinetic energy, as well as viscous dissipation at the contact line and in the bulk, are estimated and compared. While kinetic and gravitational energy are obtained from the centroidal positions available in the image sequences, surface energy requires the determination of the drop curvature. For this purpose, local curvature is estimated at all the pixels of the interface using curve fitting and are used to obtain the average curvature of the coalescing drops. The reciprocal of arithmetic mean of curvatures is taken as an average radius and is used for the estimation of surface energy. Contact line dissipation requires the estimation of the dynamic contact angle for every frame which in turn is obtained by using commercial software (ImageJ®; Stalder et al. 2006).

The definitions of the energy rate components are provided in Eqs. (3.22)–(3.27),

$$\text{Average curvature:} \frac{1}{R_{\text{avg}}} = \frac{1}{N_{\text{p}}} \sum_{i=1}^{N_{\text{p}}} (\text{RC})_i \tag{3.22}$$

$$\text{Surface energy release:} E_{\text{s}} = -2\pi\sigma \left(\frac{1}{R_{\text{avg}}^2} \frac{\mathrm{d}R_{\text{avg}}}{\mathrm{d}t} \right) V_{\text{comb drop}} \tag{3.23}$$

$$\text{Kinetic energy:} E_{\text{k}} = \frac{\mathrm{d}}{\mathrm{d}t}\left(\frac{1}{2} m u_{\text{res}}^2\right) = \left(m u_{\text{res}} \frac{\mathrm{d}u_{\text{res}}}{\mathrm{d}t} \right); \quad u_{\text{res}} = \sqrt{u_{\text{c}}^2 + v_{\text{c}}^2} \tag{3.24}$$

$$\text{Gravitational energy:} E_{\text{g}} = mg \frac{\mathrm{d}y_{\text{c}}}{\mathrm{d}t} = mg v_{\text{c}} \tag{3.25}$$

$$\text{Viscous dissipation:} E_{\text{vis}} = \mu \left(\frac{u_{\text{c}}}{y_{\text{c}}} \right)^2 V_{\text{comb drop}} \tag{3.26}$$

$$\text{Contact line dissipation:} E_{\text{CL}} = \sigma\left(\cos\theta - \cos\theta_{\text{eq}}\right) 2\pi R_{\text{b}} u_{\text{c}} \tag{3.27}$$

Here, θ is the instantaneous dynamic contact angle of the spreading drop.

Instantaneous energy rates of coalescence of equal volume pendant drops ($v_1 = 0.6$ μl, $v_2 = 0.6$ μl) and sessile drops ($v_1 = 0.51$ μl, $v_2 = 0.5$ μl) are shown in Figs. 3.25 and 3.26, respectively. Energy levels in the sessile configuration are

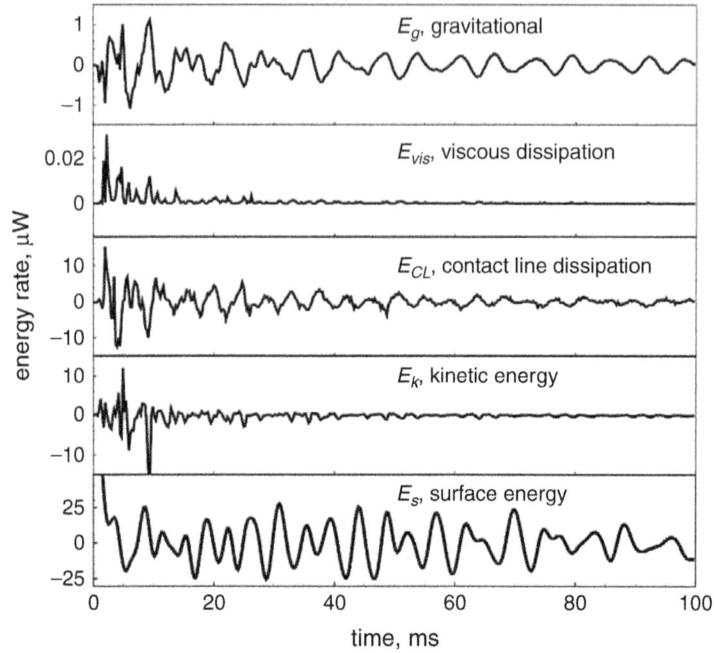

Fig. 3.25 Time-dependent energy rates for coalescence of pendant drops ($v_1 = 0.6$ μl, $v_2 = 0.6$ μl) of equal volumes

greater than the pendant since gravity and pressure oppose each other in the second orientation. Apart from this difference, the trends seen are similar. Changes in surface energy arise from those in the average drop curvature and is a primary source for drop oscillations. Fluctuations in kinetic energy and viscous dissipation in the bulk are short-lived (~40 ms). On a slightly longer timescale, surface energy fluctuations and the dissipation at the three-phase contact line remain finite (~100 ms). For the present set of experiments, viscous dissipation in the bulk is small in comparison to the contact line. The contribution arising from gravitational oscillations are uniformly small for sessile and pendant drops but are long-lived over the 100 ms window presented in Figs. 3.25 and 3.26. Thus, the role of gravity is primarily seen to diminish internal pressure for merging pendant drops relative to the sessile, and hence diminish velocities generated during coalescence.

3.6.7 Vertical Coalescence of Two Liquid Drops

Coalescence details of liquid drops placed adjacent to each other on a hydrophobic surface are discussed in the previous sections. Here, oscillations are set up due to the conversion of an initially high surface energy of the combination of drops to a lower

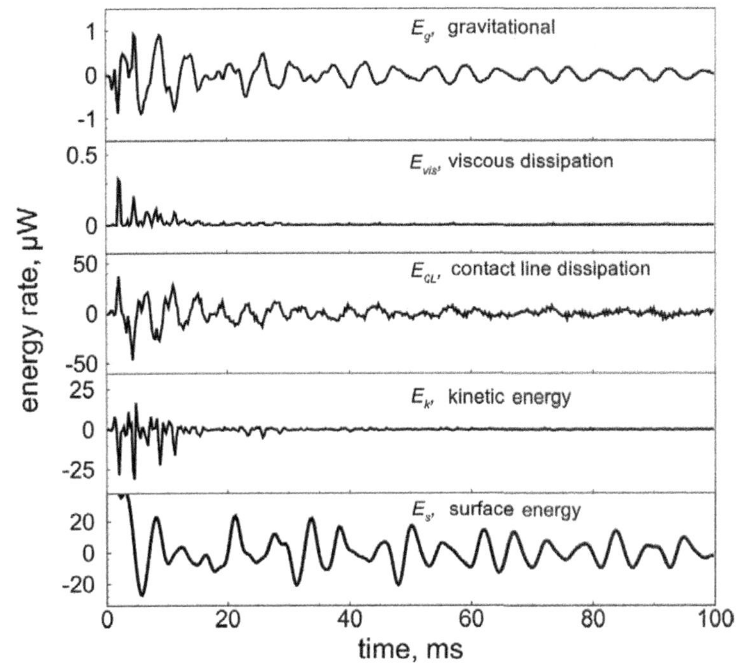

Fig. 3.26 Time-dependent energy rates for coalescence of sessile drops ($v_1 = 0.51$ μl, $v_2 = 0.50$ μl) of equal volumes

value acquired by a single drop. Kinetic energy thus released is dissipated by the mechanism of viscosity in the bulk and at the contact line. Velocity and timescales appearing in the respective processes depend on the hydrodynamic properties of the liquid drops. Two observations emerging from these experiments are as follows:

1. Coalescence starts with the formation of a liquid bridge joining the two drops. The bridge has negative curvature relative to the rest of the liquid body, leading to large internal pressure differences and hence the resulting oscillatory fluid motion.
2. The process is characterized by three distinct timescales related to bridge formation, appearance of fluid oscillations, and long-term decay of velocity.

It is reasonable to examine these trends during coalescence of liquid drops when one is placed above the other in a vertical orientation. Results of coalescence of vertically aligned liquid drops of an equal ($R^* = 1.0$) volumes for a surface with an equilibrium contact angle of 150° are presented here. The combined Bond number is 0.054. The similarity with the coalescence process of drops placed adjacent to each other is considerable with one exception. The vertical orientation leads to recoil of the combined away from the surface, as discussed below.

The extent of recoil can be discussed in terms of energy components. The change in gravitational energy of the system can be estimated from the movement of the centroid of the drops, going from the initial configuration to the final merged drop shape. Similarly, it is possible to determine the reduction in surface energy as two drops combine to form one large entity. Since the surface is strongly hydrophobic, the surface energies can be obtained by approximating the drops as part of a sphere in the calculation of surface curvature and internal pressure. The lowering of gravitational energy and the surface energy will be compensated at intermediate times by the appearance of kinetic energy in the form of a velocity distribution and ultimately by viscous dissipation, when the merged drop eventually comes to rest. On engineered surfaces, inhomogeneities and variation in chemical composition will contribute to pinning, a major source of additional viscous dissipation within the drop.

For the surface energy analysis of static drops, the following equations may be adopted

$$E_1 = \sigma A_1, \quad E_2 = \sigma A_2, \quad E_3 = \sigma A_3 \tag{3.28}$$

Symbols E_1, E_2, and E_3 are the surface energies of the droplet placed below, the one placed above it and the combined value after merger, respectively. Symbols A_1, A_2, and A_3 are the corresponding gas-liquid interfacial surface areas. Symbol σ is the surface tension of the water-air interface. In the present discussion, air is taken to be under ambient conditions of pressure and temperature. Calculation of surface areas of the sessile drop placed below before coalescence and the final drop after coalescence are carried out by considering them as spherical caps and have been formulated as a function of the equilibrium contact angle θ,

$$A_1 = 2\pi R_1^2(1 - \cos\theta), \quad A_2 = 2\pi R_2^2, \quad A_3 = 2\pi R_3^2(1 - \cos\theta) \tag{3.29}$$

Here, drop (1) is placed over the surface and drop (2) above is initially purely spherical. The relative excess surface energy is defined as

$$E^* = \frac{(E_1 + E_2 - E_3)}{E_3} \tag{3.30}$$

The relative excess energy E^* is clearly a function of radius ratio $R^* = R_1/R_2$ and the contact angle. The excess surface energy function plays an important role for the initiation of recoil.

Besides surface tension, the other major forces that control the movement of the bulk of liquid on coalescence are gravity and internal capillary pressure. Though, for low Bond numbers, volume is smaller, and the effect of gravity is diminished, it is still an important factor as far as recoil of drops from the surface is concerned. The role of the initial capillary pressure difference can be studied along the following lines. This pressure difference is given by

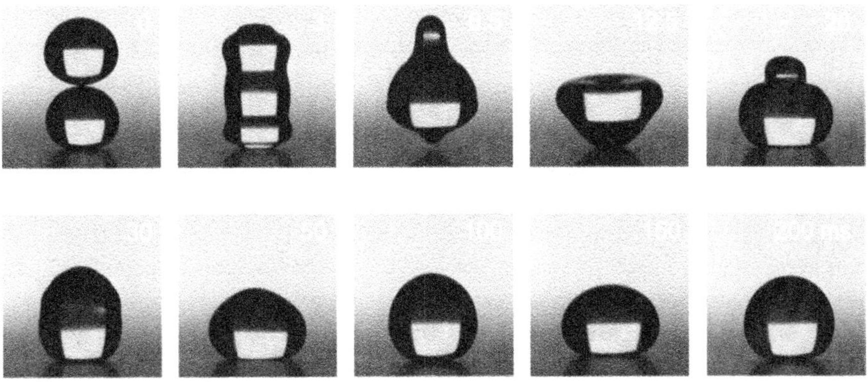

Fig. 3.27 Image sequence of the evolution of the interface shapes observed during coalescence of drops of water drops of equal volumes ($R^* = 1.0$) and a combined Bo = 0.054. The combined drop bounces off the surface at a time of ~6.5 ms in experiments

$$\Delta p = 2\sigma \frac{R_2 - R_1}{R_1 R_2} \tag{3.31}$$

It is zero for equal volume drops. The excess surface energy function is large, but the initial capillary pressure difference is small for radius ratios $R^* = 1.0$. For unequal volume drops, the smaller one has greater internal pressure and is physically displaced towards the larger drop during merger. For equal volume drops, the initial pressure difference being zero, recoil is not related to the bulk movement of one drop into another. Instead, the coalesced drop forms a bridge, large velocities are created, and the instantaneous pressure field is modified, while the footprint shrinks progressively in time. Thus, recoil is delayed but its vertical extent can be greater than at small (or, non-unity) radius ratios, where recoil scales with the smaller drop diameter.

For two drops of equal volumes, Fig. 3.27 shows an experimentally recorded image sequence of intermediate shapes of the coalesced water drop as a function of time. Recoil of the merged drop after coalescence is clearly seen at ~6.5 ms. Other trends are like the coalescence process of two liquid drops that are initially placed adjacent to each other.

3.6.8 *Influence of Finite Time Coalescence on Dropwise Condensation*

This chapter has identified velocity and timescales of coalescence from experiments and dimensional analysis from a viewpoint of improving the mathematical model of the dropwise condensation cycle. Finite time coalescence in terms of the length, the velocity, and the timescales can be incorporated in the mathematical modeling dropwise condensation; this improvement in the mathematical model is described in Chap. 7 for water. The influence of these scales on dropwise condensation in terms of liquid holdup, wall heat flux, wall shear stress, and spatio-temporal drop distribution has been studied in Chap. 10 when bismuth is the condensing fluid. Wall shear is generated for a horizontal surface only during coalescence. For a vertical surface, it is seen during coalescence as well as drop instability. The coalescence-induced wall shear stress is substantial. The instantaneous shear rates during coalescence can be significant, affecting the mechanical life cycle of the hydrophobic surface coating/promoter layers that are used for facilitating dropwise condensation. The surface-averaged wall shear stress is, however, quite small. Heat flux diminishes with time during drop growth, fluctuates with time during the coalescence events as well as drop instability, attaining a peak just afterward. Since coalescence occurs occasionally over a small area, the contribution of the heat flux peaks to the overall substrate-level heat flux is shown to be small.

3.7 Closure

Coalescence process of small water drops of equal and unequal volumes deposited on a chemically textured superhydrophobic surface in pendant (contact angle, $150^\circ + 4^\circ$) and sessile ($145^\circ + 3^\circ$) configurations were imaged with a 4000 fps camera. Imaging experiments were followed by the analysis of centroidal positions of the combined drop. Length, velocity, and timescales contained in the drop position oscillations were estimated using dimensional reasoning. Dimensionless parameters were developed by suitably combining contributions from surface tension, viscosity, gravity, and inertia. The very small timescale corresponding to bridge formation could not be resolved at 4000 fps. The instantaneous wall shear stress appearing in the inertia-surface tension dominated regime was quantified and seen to be in excess of reported strengths of commercially available hydrophobic coatings. Energy analysis revealed oscillations to be driven by the release of surface energy and hence the changes associated with the shape of the drop.

In sessile and pendant arrangements, an initial bridge was formed with large velocities created over a timescale of 1–2 ms. The velocity components generated in the two configurations were oscillatory with a timescale of 20–30 ms. These velocities subsequently decayed over a longer time frame of 225–300 ms due to viscous dissipation. Specific observations arrived at from the experiments are listed below.

1. Following non-dimensionalization, the timescale arising from the pair of inertia forces and surface tension indicates the time instant when velocity is a maximum. The preceding viscous-surface tension timescale is in microseconds and has not been resolved in the imaging experiments. The inertia-viscous timescale relates to the time required by the velocity amplitudes to decay monotonically in the long term by viscous dissipation.

2. Unequal volume drops have a larger internal pressure difference and the horizontal velocity components generated during coalescence are larger by 150–400% compared to coalescing drops of equal volume. Hence, unequal coalescing drops generate larger shear rates (100–400% relative to equal volume drops). In the pendant mode, gravity and pressure are opposed to each other. Hence, the pressure difference between the drops is smaller when compared to the sessile configuration, and the resulting velocities during coalescence are also smaller by 30–60%.
3. Shear rate scales with velocity and inversely with the distance of the centroid from the wall. Velocities are greater and the centroid position in the sessile configuration is closer to the wall. Hence, shear rates in a sessile arrangement are greater than the pendant by 100–350%.
4. Smaller drops stabilize earlier than larger drops after coalescence, indicating a larger damping ratio. In a coalescence event, fluid oscillations diminish faster at the interface region as compared to the fluid at the core. Interfacial oscillations diminish rapidly for sessile drops compared to the pendant mode.
5. Energy analysis shows changes in surface curvature as the primary factor contributing to changes in kinetic and gravitational energy. It is largely dissipated due to motion of the contact line and to a smaller extent in the form of viscous dissipation in the bulk, eventually leading to a temporal equilibrium of the combined drop. The combined dissipative energy is expected to leach the hydrophobic promoter layer and, in an application, prevent a dropwise mode of vapor condensation.

Chapter 4
Introduction to Evaporative Heat Transfer

Manish Bhendura, K. Muralidhar, and Sameer Khandekar

Nomenclature

a	Coefficient of velocity distribution function
c_{pa}	Specific heat capacity of air at constant pressure (J/kg K)
D	Mass diffusivity of water vapor in dry air (m^2/s)
f_M	Maxwellian velocity distribution function
F	Correction factor for total mass flux
G	Correction factor for total y-momentum flux
g	Gravitational acceleration (m/s^2)
h_c	Convective heat transfer coefficient (W/m^2 K)
h_e	Evaporative heat transfer coefficients (W/m^2 K)
h_{fg}	Latent heat of vaporization of water (J/kg)
H	Correction factor for total energy flux
j	Molecular evaporative mass flux in kinetic model (kg/m^2 s)
k_B	Boltzmann's constant (J/K)
k_m	Thermal conductivity of mth fluid (W/m K)
m_e	Evaporative mass flux of water in continuum model (kg/m^2 s)
M_a	Molecular mass of air (kg/kmol)
M_w	Molecular mass of water (kg/kmol)
n	Molecular density (mol/m^3)
p	Saturation pressure of water vapor at a given temperature (kPa)
p'	Partial vapor pressure of water in saline water (kPa)
p_T	Total pressure (kPa)
Q	Binary collision integral
q_e	Evaporative heat flux (W/m^2)
R	Gas constant of water vapor (J/kg K)
S	Dimensionless speed ratio

S. Khandekar, K. Muralidhar, *Drop Dynamics and Dropwise Condensation on Textured Surfaces*, Mechanical Engineering Series,
https://doi.org/10.1007/978-3-030-48461-3_4

t Time (s)
T Temperature (K); suffix c and w for cold and water surface
u Speed of bulk flow (m/s)
u_m Tangential velocity component of mth fluid (m/s)
v_m Velocity vector of mth fluid (m/s)
X_{w} Mole fraction of water in the salt solution
y Directional length normal to the liquid-vapor interface (m); continuum model
Z Pressure ratio

Greek Letters

α Accommodation coefficient of water evaporation
α_{c} Condensation coefficient
α_{e} Evaporation coefficient
α_m Thermal diffusivity of mth fluid (m^2/s)
β_m Thermal expansion coefficient of mth fluid (K^{-1})
γ Unit vector directed vertically upward
ξ_y y-component (normal) of molecular velocity vector (m/s)
θ Temperature ratio
μ_m Dynamic viscosity of mth fluid (Pa s)
ν_m Kinetic viscosity of mth fluid (m^2/s)
ξ Molecular velocity vector (m/s)
ρ_m Density of mth fluid (kg/m^3)
ψ Arbitrary function for total fluxes of mass, y-momentum, and energy
ω Absolute humidity of air

Subscripts and Superscripts

∞ Vapor side in kinetic model
0 Liquid side in kinetic model
e Emitted
r Reflected
+ Outward flux
− Inward flux

4.1 Introduction

Evaporation is a spontaneous liquid-to-vapor phase change process, seen in a variety of contexts including the natural hydrological cycle of earth. It serves as a natural thermal control mechanism of living and breathing animals. In the process industry, evaporation is an important intermediate step in drying operations and in water purification via distillation. Low temperature applications of evaporation can be seen in devices providing indoor thermal comfort while evaporation of thin films has been proposed as an effective technique for thermal management of intense heat generating devices such as those in miniature electronics.

Evaporation is a heterogeneous phase change phenomenon, in which a surface molecule of matter in the low energy liquid state jumps to a high energy vapor state. The energy transition is accompanied by the transfer of latent heat of vaporization from adjacent molecules, causing a thermal perturbation at the liquid-vapor interface. The transfer of latent heat from the molecule in the liquid phase to the vapor is isothermal but will tend to cool the liquid body. The molecule that has passed into the vapor phase has definite kinetic energy and will diffuse into its surroundings, thus initiating vapor phase mass transport. Evaporation of a liquid body into a gas phase (air, in the present discussion) is spontaneous if the gas phase above is dry, relative to saturation conditions, defined by the prevailing pressure and temperature. Therefore, evaporation involves heat transfer in the liquid phase and simultaneous heat and mass transport in the vapor phase, with the liquid-vapor interface serving as a thermally active boundary.

In the simplest form, evaporation is considered a multiscale, multiphase, and multiphysics phenomenon. The scales arise from the gas and liquid phases, apart from molecular scales of diffusion and larger scales of fluid convection. In terms of distinct physical processes, we have heat transfer in the individual phases coupled at the interface, mass transfer of moisture across the interface and within the gaseous region, possible transport of solutes within the liquid body as well as buoyant- and Marangoni-driven flow in the liquid body. These processes are intrinsically coupled and require an elaborate mathematical model for analysis. In device modeling, for example, solar stills, evaporation is represented empirically in terms of an evaporative mass flux stated in terms of a sink temperature and the temperature of the evaporating liquid surface. Such models need to be validated against a full experiment backed by detailed numerical simulation.

4.2 Evaporation Models

These are broadly divided into two categories depending on the treatment of the air-water interface, as shown in Fig. 4.1. In a continuum model, interface temperature is continuous across the liquid and gas phases and is referred to as quasi-equilibrium (or, a local equilibrium) model, while variations in temperature, pressure, velocity, and other properties are permitted elsewhere in the fluid domain. Consideration of temperature jump at the interface originating from microscale processes constitutes the non-equilibrium model. Models arising from these two criteria are discussed here.

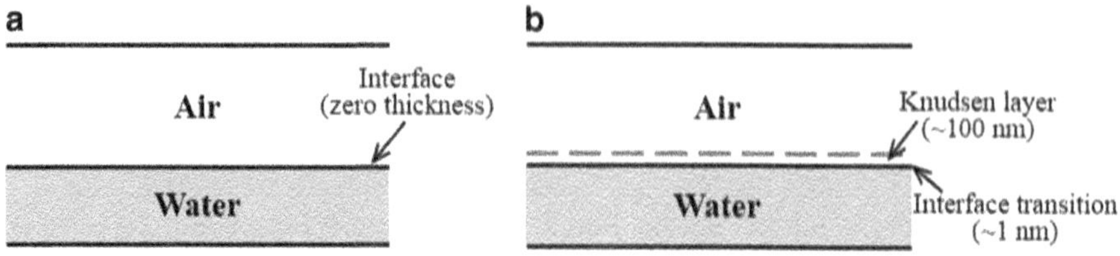

Fig. 4.1 (**a**) Air-water interface assumed to be in quasi-equilibrium with each of the phases; (**b**) non-equilibrium conditions revealed the appearance of a Knudsen layer

4.2.1 Quasi-Equilibrium Model

In a quasi-equilibrium model, temperature varies continuously from the liquid side to the gas side, while the jump in heat flux at the liquid-gas interface is fully accounted for in terms of the latent heat absorbed. Specifically, temperature jump is zero at the interface. Model statement developed under this framework in the form of differential equations is presented below.

Following evaporation, water vapor diffuses in the surrounding air and towards a region of lower humidity. Water vapor is transported in air and sets up a moisture concentration field. The gradient of moisture concentration at the air-water interface controls the evaporation rate. One can now state a species transport equation for moisture in the gas phase. In applications where the air-water interface is practically stationary, transport of water vapor by evaporation may be taken as diffusion dominated and the mass flux of water vapor (kg/m^2-s) calculated using the Fick's law

$$\dot{m}_e = -\rho_a D \frac{\partial \omega}{\partial y} \tag{4.1}$$

Here, ω is the absolute humidity of air (kg vapor/kg of dry air), ρ_a is the density of dry air (kg/m^3), and D is mass diffusivity of water vapor in dry air (m^2/s). The corresponding evaporative heat flux (W/m^2) at the interface is the product of evaporation rate and the latent heat of vaporization (J/kg), namely

$$\dot{q}_e = \dot{m}_e h_{fg} \tag{4.2}$$

To obtain the evaporation rate using Eq. (4.1), the moisture field must be known in the surrounding air. This distribution can be determined by solving an advection-diffusion equation for humidity along with suitable initial and boundary conditions.

In a simplified quasi-equilibrium model (Tiwari and Sahota 2017), an evaporative heat transfer coefficient is empirically specified to estimate the evaporative heat flux, and in turn, the evaporation rate. For a body of warm water evaporation into a cold surface, this prescription is

$$\dot{q}_{\mathrm{e}} = h_{\mathrm{e}}(T_{\mathrm{w}} - T_{\mathrm{c}}) \tag{4.3}$$

Here, q_{e} and h_{e} are the evaporative heat flux and evaporative heat transfer coefficient, respectively, and T_{w} and T_{c} are the temperature of water surface and the average cold surface temperature, respectively. The evaporative heat transfer coefficient can be connected to the convective heat transfer coefficient (h_{c}) using the definitions of partial pressures ($p_{\mathrm{w}}, p_{\mathrm{c}}$) and humidity ratio

$$\frac{h_{\mathrm{e}}}{h_{\mathrm{c}}} = \left(\frac{h_{\mathrm{fg}}}{c_{\mathrm{pa}}}\right)\left(\frac{M_{\mathrm{w}}}{M_{\mathrm{a}}}\right)\left(\frac{p_{\mathrm{w}} - p_{\mathrm{c}}}{T_{\mathrm{w}} - T_{\mathrm{c}}}\right)\left(\frac{p_{\mathrm{T}}}{(p_{\mathrm{T}} - p_{\mathrm{w}})(p_{\mathrm{T}} - p_{\mathrm{c}})}\right) \tag{4.4}$$

Here, h_{fg} is the latent heat of vaporization of water (J/kg), c_{pa} is the specific heat of air (J/kg K), M_{w} and M_{a} are the molecular masses of water and air, respectively, and p_{T} is the total pressure. Equation (4.4) is quite useful for the estimation of the evaporative heat transfer coefficient since correlations are available in the literature for the convective heat transfer coefficient. Dunkle's correlation is one of these and is derived for non-isothermal convection in an air-water system inside an enclosure. It is expressed as

$$h_{\mathrm{c}} = 0.844\left[(T_{\mathrm{w}} - T_{\mathrm{c}}) + \frac{(p_{\mathrm{w}} - p_{\mathrm{c}})T_{\mathrm{w}}}{(268.9 - p_{\mathrm{w}})}\right]^{\frac{1}{3}} \tag{4.5}$$

Here, T_{w} and T_{c} are the bulk temperature of water and average temperature of the cold surface, respectively. Pressures are expressed in units of kPa and temperature in Kelvin.

The quasi-equilibrium model with a correlation for the interfacial heat transfer coefficient is extensively used in the analysis of solar distillation systems. Here, the liquid body comprises saline water and the condensate is potable water. Salinity affects the evaporation rate of water since it regulates the saturation pressure for a given temperature. This change in saturation pressure is accounted for by Raoult's law, which states that the vapor pressure of a solvent (water) in a solution (saline water) is

$$p'_{\mathrm{w}} = X_{\mathrm{w}} p_{\mathrm{w}} \tag{4.6}$$

Here, X_{w} is the mole fraction of water in the salt solution and is less than unity. Thus, the effect of salt in water is to diminish the saturation pressure and thus lower evaporation rates. Vapor pressure depression can be explained in terms of the reduction of chemical potential of water due to an increase in the salt ion activity. The influence is more pronounced at higher temperatures since the ionic activity is also elevated (Kokya and Kokya 2008).

4.2.2 Transport Equations in a Two-Layer Air-Water System

In water purification applications, water is heated by an external agency, while the surrounding air is cooled externally using an active cold surface to increase the evaporation rate and maximize condensation of pure water. A negative temperature gradient is set up in both air and water, which results in two-layer natural convection. The gradient in humidity is opposed to gravity and can stabilize convection in air. Heat transfer rates in air and water influence interfacial temperature, which, in turn affects the evaporation rate. Hence, the device performance depends on transport phenomena at all scales of transport prevailing in the cavity.

The complete transport model of evaporation with associated initial and boundary conditions are laid out in this section for a rectangular cavity that is partially filled with water, the rest being air. The water body is taken to be initially warm; the top surface is cold, other surfaces are insulated, and air is initially dry. Evaporation from the air-water interface consumes the latent heat of evaporation from the water body that progressively cools with time. The temperature differential in water is gravitationally unstable and generates buoyancy-driven convection. Surface tension gradients over the gas-liquid interface may generate additional convection as well. From thermal considerations alone, density gradient in air is unstable and heat transfer would be controlled by buoyant convection. The interface region contains practically moisture-saturated air while it is relatively dry at the cooler surface where water has condensed. Hence, a gravitationally stable density gradient may form in air and the resulting strength of convection currents will respond to the difference between the body forces related to the heat and mass transfer mechanisms.

A rectangular enclosure partially filled with initially hot water is considered for the formulation of transport models in air and water, Fig. 4.2. Governing equations applicable for a two-dimensional geometry are written out in the following discussion. Flow is taken to be unsteady but laminar; buoyancy is included but surface tension gradients are neglected. The two phases are clearly separated across an interface. The air-water interface is always flat and horizontal; change in water level with time is neglected though it can be easily accounted for. Water condensing at the top cooled wall is assumed to be drained away immediately so that a single-

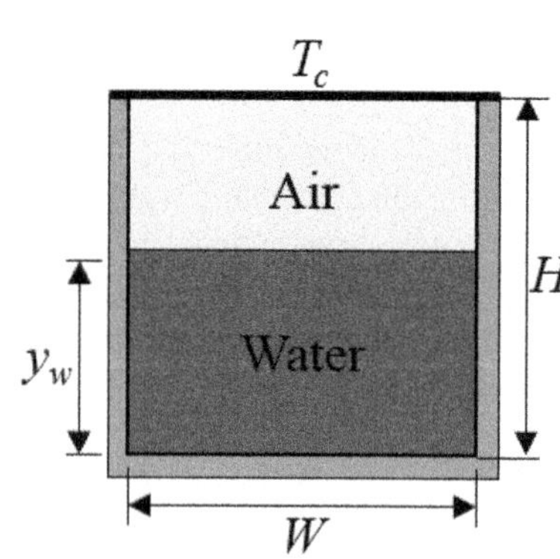

Fig. 4.2 Schematic of a rectangular enclosure, thermally insulated from all sides except the top and partially filled with initially hot water

phase model of transport in air is applicable here. The gradient in temperature within the cavity will be responsible for Rayleigh-Benard convection that is intrinsically three dimensional. The two-dimensional treatment may correspond to an average temperature field determined with respect to the dimension along the length of the cavity.

The mass, momentum, and energy balance equations for air and water are as follows (Nepomnyashchy and Simanovskii 2004):

$$\nabla \cdot \vec{v}_m = 0 \tag{4.7}$$

$$\frac{\partial \vec{v}_m}{\partial t} + \left(\vec{v}_m \cdot \nabla\right)\vec{v}_m = -\frac{1}{\rho_m}\nabla p_m + \upsilon_m \nabla^2 \vec{v}_m - g[1 - \beta_m(T_m - T_c)]\hat{\gamma} \tag{4.8}$$

$$\frac{\partial T_m}{\partial t} + \vec{v}_m \cdot \nabla T_m = \alpha_m \nabla^2 T_m \tag{4.9}$$

Here, index m is used to denote the two fluids; $m = 1$ is water, while 2 is air. Quantities $\vec{v}_m, p_m,$ and T_m are the velocity vector, pressure, and temperature fields of the mth fluid. Additionally, ρ_m, υ_m, β_m, and α_m are the density, kinetic viscosity, thermal expansion coefficient, and thermal diffusivity of the respective fluid, while $\hat{\gamma}$ is a unit vector directed vertically upward.

Initially ($t = 0$), water inside the enclosure is hot ($T_1 = T_h$) and filled up to a depth of y_w, while air above is at the cold temperature ($T_2 = T_c$). In most applications, the cold temperature is the ambient value. The fluid phases are both initially stagnant $\left(\vec{v}_m\,(t=0) = 0\right)$.

The two vertical side walls and the lower surface are insulated. Hence, $\hat{n} \cdot \nabla T_m = 0$, whereas the top surface is maintained at the cold temperature. All four walls enforce the no-slip boundary condition, $\vec{v}_m = 0$. At the air-water interface, tangential and normal velocity components are continuous, shear stresses are continuous, and phase heat fluxes are separated by the latent heat release. In addition, the flatness of the interface condition requires the normal velocity component to be zero. Symbolically, these interface conditions are expressed as

$$\mu_1 \frac{\partial u_1}{\partial y} = \mu_2 \frac{\partial u_2}{\partial y} \quad \text{and} \quad u_1 = u_2; \quad v_1 = v_2 = 0 \tag{4.10}$$

$$-k_1 \frac{\partial T_1}{\partial y} = -k_2 \frac{\partial T_2}{\partial y} + q_e \quad \text{and} \quad T_1 = T_2 \tag{4.11}$$

where μ_m and k_m are the dynamic viscosity and thermal conductivity of the mth fluid. The evaporative heat flux q_e is estimated by utilizing one of the evaporation models described earlier, via a correlation (via Eq. (4.3) with the correlation Eq. (4.4)) or via the gradient of humidity (specifically, the evaporation rate) multiplied by the latent heat of vaporization (Eqs. 4.1 and 4.2).

Computation of the humidity gradient requires the solution of the moisture transport equation in air. The governing equation for moisture transport in air is of the advection-diffusion type

$$\frac{\partial \omega}{\partial t} + \vec{v}_2 \cdot \nabla \omega = D \nabla^2 \omega \tag{4.12}$$

where ω is the absolute humidity of air, and D is the mass diffusivity of water vapor into air.

The binary mass diffusivity of air and water vapor, D is the strong function of temperature and pressure, which can be estimated using the following expression (Bird et al. 2002):

$$D = a \frac{T^b}{p} \tag{4.13}$$

Here, D is in m^2/s, p in atm, and T in K and the constants are $a = 0.434 \times 10^{-10}$ and $b = 2.334$. A typical value of the binary mass diffusivity is 2.5×10^{-5} m^2/s.

Air is initially assumed to be dry, i.e., $\omega = 0$ (or partially saturated), whereas at the interface and at the top surface, air is saturated with water vapor at the local temperature; in other words, the value of ω is specified at these boundaries. The moisture content under saturation conditions is calculated using

$$\omega_s = \frac{M_a}{M_w} \times \frac{p_s}{p_T - p_s} \tag{4.14}$$

where M_a and M_w are the molecular masses of the air and water, respectively, and p_s and p_T denote the saturation pressure and total pressure (in Pa), respectively. The saturation pressure of water can be calculated using empirical relations proposed by numerous researchers. The correlation used in the design of solar distillation systems is

$$p_s(T) = \exp\left(25.317 - \frac{5144}{T}\right) \tag{4.15}$$

where T is in units of Kelvin and pressure is recovered in Pa.

The governing equations of phase velocity, pressure, temperature, and humidity are clearly nonlinear and coupled. These can only be solved numerically using sophisticated computational tools.

4.2.3 Non-Equilibrium Model

We present here a simplified form of the non-equilibrium model by neglecting the presence of air. Air is treated as a non-condensable gas and significantly alters the evaporation rate. Specifically, the evaporation rate in humid air is smaller than in pure vapor for a given temperature difference. For equal evaporation rates, humid air would require a higher temperature difference when compared to pure vapor. In addition, evaporation in humid air will give rise to a diffusion layer that is rich in vapor. This contrasts with condensation of vapor from moist air, where the diffusion layer is rich in air. These corrections can be determined from kinetic theory of gases but is beyond the scope of the present discussion.

In the non-equilibrium model of a liquid-vapor system, two microscopic layers appear between the bulk phases of water and vapor—the interface transition layer and the Knudsen layer (Fig. 4.3). The interface transition thickness is of few molecular diameters (~0.25–1 nm), across which the liquid density transitions to vapor density. Evaporation causes a small but a nearly discontinuous temperature jump in this layer. The thickness of this layer may be approximated as zero and considered to be a part of the liquid surface.

The Knudsen layer is located between the interface transition layer and the bulk vapor phase. Its thickness is of the order of a few free molecular mean-free paths (~10–150 nm). It involves interaction among molecules arising from three sources. These are molecules emerging from the liquid surface, molecules impinging on the liquid surface from the vapor side and molecules of the vapor phase reflected from the liquid surface. The molecules are indistinguishable and do not represent any chemical reaction. A net positive molecular flux at the outer boundary of the Knudsen layer results in evaporation. A negative flux of molecules ensures condensation. In case of zero net flux, the liquid-vapor system around the interface is taken to be in thermal, mechanical, and chemical equilibrium.

Molecular interaction in thin regions such as the Knudsen layer can result in large temperature changes and hence alter heat and flow characteristics of bulk phases involved in phase change. However, first principles modeling of transport in the Knudsen layer is difficult since continuum principles of density, pressure, and

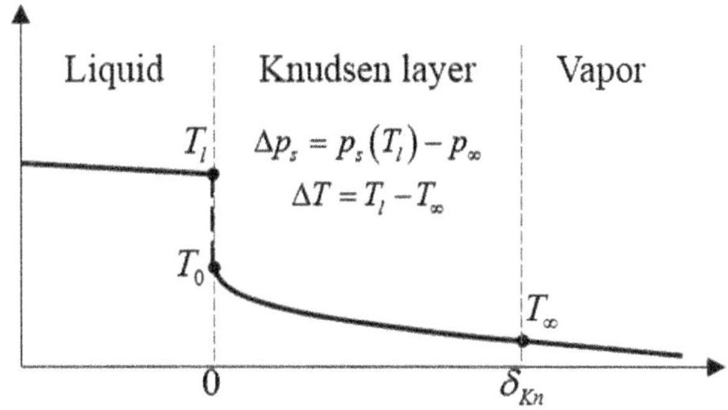

Fig. 4.3 Schematic representation of temperature drop along the interface transition layer ($T_l - T_0$) and the Knudsen layer ($T_0 - T_\infty$)

temperature are not applicable (Gerasimov and Yurin 2018). An alternative is a molecular approach, for example, the molecular kinetic theory (MKT) that can determine changes in the macroscopic quantities due to molecular interaction in a gap of a few molecular mean-free paths.

4.2.4 Kinetic Theory of Gases

The molecular kinetic theory involves the velocity distribution function for molecules in a probabilistic context. In a collection of particles moving in time and space with a wide spread of velocities, the distribution function represents the fraction of molecules in a specified velocity interval. The commonly used Maxwell's distribution function is an example of such a variation but holds only under conditions of thermodynamic equilibrium. Vapor molecules leaving the liquid surface into the Knudsen layer will generally show departure from equilibrium and the distribution function will deviate from the Maxwellian. Differences are expected to be large at temperatures closer to the boiling point of the liquid. The molecular velocity distribution function, once found by other methods, can be utilized to estimate macroscopic parameters such as number density, bulk velocity, and temperature, including the temperature jump. The molecular interactions inside the Knudsen layer can be analyzed using the transport equation of velocity distribution function and is known as the Boltzmann's kinetic equation (BKE).

For closure of BKE, boundary conditions for the Knudsen layer at the inlet (towards the transition layer) and the outlet (towards the vapor bulk phase) are required. The inlet velocity distribution function depends on the molecule emission rate from the transition layer. It can be predicted reasonably well by molecular dynamic simulations (MDS). The outlet boundary condition (velocity distribution) is defined at the thermodynamic state of the bulk vapor phase. Hence, the solution of MDS is the boundary condition for BKE, whereas the solution of BKE is utilized as the boundary condition for the continuum analysis of the bulk phases (Shishkova et al. 2017).

The involvement of two statistical approaches (MDS and MKT) results in a comprehensive but difficult and computationally expensive methodology. The complexity can be reduced by avoiding MDS and using an approximate velocity distribution function at the inlet boundary of the Knudsen layer (Frezzotti et al. 2018). The half-space Maxwellian distribution is one such distribution well-known in the kinetic theory of evaporation and has been utilized in the following discussion. Under this simplification, evaporation mass flux can be estimated though tempertaure jump and the thickness of the Knudsen layer are no longer be resolved.

The evaporation models based on the molecular kinetic theory are further discussed below (Aursand and Ytrehus 2019).

Using the kinetic theory of gases approach, the velocities of molecules are denoted probabilistically by the Maxwell velocity distribution. The probability density function of velocity represents the fraction of the total number of molecules at a given point (*x*, *t*) having velocities in a specific range is

$$f_{\mathrm{M}}(\xi; T, n, u) = \frac{n}{(2\pi RT)^{3/2}} \exp\left(-\frac{(\xi - u\hat{y})^2}{2RT}\right) \tag{4.16}$$

Here, x is position vector, t is time, ξ is a molecular velocity vector, n is number density of elementary particles, and u is the speed of bulk flow along the y-axis. This distribution is further utilized to compute the fluxes of mass, y-momentum, and kinetic energy of molecules evaporating from the interface into the Knudsen layer. The outgoing particles ($\xi > 0$) from the liquid surface are postulated to be half of the Maxwellian distribution corresponding to the reference state. Total fluxes of mass, y-component momentum, and kinetic energy of a molecule are expressed in terms of the Maxwellian distribution. A function ψ_i ($i = 1$, 2, and 3) is defined to represent these fluxes

$$\psi_i = \left(m, m\xi_y, \frac{1}{2}m\xi^2\right) \tag{4.17}$$

Hence, the net outgoing fluxes ($y > 0$) are

$$\int_{\xi>0} \xi_y \psi_i f_{\mathrm{M}} \mathrm{d}\xi = \begin{cases} \rho\sqrt{\dfrac{RT}{2\pi}}F^{+}(S), & i = 1 \\ \rho\dfrac{RT}{2}G^{+}(S), & i = 2 \\ 2\rho RT\sqrt{\dfrac{RT}{2\pi}}H^{+}(S), & i = 3 \end{cases} \tag{4.18}$$

The net inward fluxes ($y < 0$) are

$$\int_{\xi<0} \xi_y \psi_i f_{\mathrm{M}} \mathrm{d}\xi = \begin{cases} -\rho\sqrt{\dfrac{RT}{2\pi}}F^{-}(S), & i = 1 \\ \rho\dfrac{RT}{2}G^{-}(S), & i = 2 \\ -2\rho RT\sqrt{\dfrac{RT}{2\pi}}H^{-}(S), & i = 3 \end{cases} \tag{4.19}$$

where $F^{\pm}(S)$, $G^{\pm}(S)$, and $H^{\pm}(S)$ are correction factors to include the effect of bulk molecular flow; these functions are defined

$$F^{\pm}(S) = \sqrt{\pi}S(\mathrm{erf}(S) \pm 1) + \mathrm{e}^{-S^2} \tag{4.20}$$

$$G^{\pm}(S) = (2S^2 + 1)(\mathrm{erf}(S) \pm 1) \pm \frac{2}{\sqrt{\pi}} \mathrm{e}^{-S^2} \tag{4.21}$$

$$H^{\pm}(S) = \frac{\sqrt{\pi}S}{2}\left(S^2 + \frac{5}{2}\right)(\mathrm{erf}(S) \pm 1) + \frac{1}{2}(S^2 + 2)\mathrm{e}^{-S^2} \tag{4.22}$$

The dimensionless speed ratio (S) is

$$S(u, T) = \frac{u}{\sqrt{2RT}} \tag{4.23}$$

The three correction factors approach unity, when S is small.

In the calculation of the evaporative mass flux, a collision-free transport of molecules through the Knudsen layer under a high-vacuum condition is assumed for each of its boundaries. The net mass flux at a boundary is the difference between the incoming molecules and outgoing molecules. The two boundaries of the Knudsen layer are at temperatures T_{l} (liquid side) and T_{∞} (vapor side). Bulk flow of the vapor molecules is not considered in the present formulation ($u = 0$). Therefore, the distribution of molecules emerging from the liquid surface is

$$f_{\mathrm{e}} = f_{\mathrm{M}}(\xi; T_{\mathrm{l}}, n_{\mathrm{e}}, u = 0) \tag{4.24}$$

The distribution of molecules coming out of the vapor side towards the liquid boundary is

$$f_{\infty} = f_{\mathrm{M}}(\xi; T_{\infty}, n_{\infty}, u = 0) \tag{4.25}$$

The resulting evaporation flux is derived as

$$j = \alpha[\int\limits_{\xi>0} \xi_y \psi_1 f_{\mathrm{e}} \mathrm{d}\xi + \int\limits_{\xi<0} \xi_y \psi_1 f_{\infty} \mathrm{d}\xi] \tag{4.26}$$

$$= \alpha\left[\frac{p_{\mathrm{s}}(T_{\mathrm{l}})}{\sqrt{2\pi R T_{\mathrm{l}}}} - \frac{p_{\infty}}{\sqrt{2\pi R T_{\infty}}}\right] \tag{4.27}$$

where α is a pre-factor, known as the accommodation coefficient of evaporation, falling in the range of 0 to 1. It is connected to the extent to which molecules originating from the vapor side are reflected at the vapor-liquid interface.

In Eq. (4.27), temperature T_{∞} (the vapor side temperature) is unknown and must be the outcome of the evaporation model. In order to improve the utility of Eq. (4.27), an ad hoc assumption is often implemented. The assumption states that the vapor at the outer boundary of the Knudsen layer is fully saturated, namely

$$T_{\infty} = T_{\mathrm{s}}(p_{\infty}) \tag{4.28}$$

For weak evaporation, the temperature difference across the Knudsen layer is small and the evaporation flux is further reduced to the Hertz-Knudsen (HK) formula

$$j = \alpha\left[1 - \frac{1}{2}\frac{RT_\mathrm{l}}{h_\mathrm{fg}}\right]\frac{\Delta p_\mathrm{s}}{\sqrt{2\pi RT_\mathrm{l}}} \tag{4.29}$$

where, as before, h_{fg} is the latent heat of vaporization, and $\Delta p_\mathrm{s} = p_\mathrm{s}(T_\mathrm{l}) - p_\infty(T_\mathrm{s})$.

An improvement over the Hertz-Knudsen formula was made by including the effect of bulk vapor molecular flow in the formulation. The outgoing Maxwellian distribution for the vapor side boundary is written

$$f_\infty = f_\mathrm{M}(\xi; T_\infty, n_\infty, u_\infty) \tag{4.30}$$

The resulting evaporation flux, known as the Schrage-Mills (SM) formula is

$$j = \alpha\left[\frac{p_\mathrm{s}(T_\mathrm{l})}{\sqrt{2\pi RT_\mathrm{l}}} - \frac{p_\infty}{\sqrt{2\pi RT_\infty}}F^-(S_\infty)\right] \tag{4.31}$$

The bulk velocity correction factor $F^-(S_\infty)$ is calculated using Eq. (4.20) with dimensionless speed ratio evaluated at temperature T_∞. The temperature T_∞ is again an unknown and the difficulty is accounted for using the assumption of Eq. (4.28).

Equation (4.31) reduces for weak evaporation ($S_\infty \ll 1$) to

$$j = \frac{\alpha}{1 - 0.5\alpha}\left[1 - \frac{1}{2}\frac{RT_\mathrm{l}}{h_\mathrm{fg}}\right]\frac{\Delta p_\mathrm{s}}{\sqrt{2\pi RT_\mathrm{l}}} \tag{4.32}$$

In the Boltzmann Equation Moment Method (BEMM) model, the y-momentum and energy conservation equations across the Knudsen layer are included, contrary to the earlier formulations. Accordingly, the model predicts the vapor side boundary temperature (T_∞) as well. Three conservation equations are solved for the Knudsen layer using the one-dimensional, steady-state version of the Boltzmann equation

$$\xi_y \frac{\partial f}{\partial y} = Q(ff_1) \tag{4.33}$$

Equation (4.33) is an integro-differential equation for the y-velocity component with the binary collision integral (Q) on the right-hand side. It is the rate of change of the distribution function f during collision with another particle with a partner distribution function f_1 (Ytrehus and Østmo 1996).

The Knudsen layer consists of three probabilistic velocity distributions at its two boundaries: the outgoing distribution ($f^+(0, \xi)$) at the liquid side boundary ($y = 0$); the incoming distribution ($f^-(\infty, \xi)$) and the outgoing ($f^+(\infty, \xi)$) distribution on the vapor side boundary ($y \rightarrow \infty$). The following forms of the three distributions as boundary conditions for Eq. (4.33) facilitate a closed-form solution:

$$f(y,\xi) = a_0^+(y)f^+(0,\xi) + a_\infty^+(y)f^+(\infty,\xi) + a_\infty^-(y)f^-(\infty,\xi) \tag{4.34}$$

The y-dependent coefficient a's have boundary conditions

$$\begin{array}{llll} \text{At } y=0 & a_0^+=1 & a_\infty^+=0 & a_\infty^-=\beta \\ \text{At } y=\infty & a_0^+=0 & a_\infty^+=1 & a_\infty^-=1 \end{array}$$

Using these conditions with Eq. (4.34), the probability distribution for velocity on the vapor side boundary can be written as

$$f(\infty,\xi) = f^+(\infty,\xi) + f^-(\infty,\xi) \tag{4.35}$$

This distribution may be taken as prevailing in the bulk vapor state. Similarly, the outgoing velocity distribution for the liquid surface can be written as

$$f(0,\xi) = f^+(0,\xi) + \beta f^-(\infty,\xi) \tag{4.36}$$

where the symbol β is an unknown parameter related to the incoming velocity distribution at the outer boundary of the Knudsen layer ($y \to \infty$). The outgoing distribution at the liquid surface can further be divided into an emission part and a reflection part

$$f^+(0,\xi) = \alpha_e f_e^+(0,\xi) + (1-\alpha_c)f_r^+(0,\xi) \tag{4.37}$$

where α_e and α_c are the evaporation-coefficient and condensation-coefficient, respectively. Functions $f_e^+(0,\xi)$ and $f_r^+(0,\xi)$ are the emission and reflection distributions, respectively. These distributions are assumed to be Maxwellian (Eq. 4.16) at the liquid temperature, but with a number densities n_e and n_r, respectively.

The number density n_r in the reflection distribution is estimated by equating the incoming flux on the vapor side boundary to the reflected flux on the liquid side boundary

$$\frac{n_r}{n_\infty} = \beta\sqrt{\frac{T_\infty}{T_l}}F^-(S_\infty) \tag{4.38}$$

The boundary condition at $y=0$, can be written as

$$f(0,\xi) = \begin{cases} \left(\alpha_e + (1-\alpha_c)\dfrac{n_r}{n_e}\right)f_e^+(0,\infty) & \xi_y > 0 \\ \beta f_\infty^- & \xi_y < 0 \end{cases} \tag{4.39}$$

Equation (4.33) of the kinetic theory model is solved next with the boundary conditions (Eqs. 4.35 and 4.39). The input variables are the properties of the bulk liquid surface (T_l) and corresponding number density $n_e(=\rho_l/(k_B T_l))$ and the outputs of the model are the properties of the bulk vapor phase, T_∞, u_∞, and β.

These equations are solved using the moment method, in which the mass, y-momentum, and energy are stated to be collision-invariants. This statement is expected to be valid for the kinetic theory model when it is solved for the outer boundary parameters (T_∞ and u_∞) of the Knudsen layer, instead of a full solution in the entire vapor phase. The approximation eliminates the need of a collision model, namely function $Q(ff_1)$. Thereafter, the moment equations are generated using the function ψ_i defined in Eq. (4.17). The basic form of moment equations is expressed as

$$\frac{\partial}{\partial y}\int \psi_i(\xi)\xi_y f(y,\xi)\,\mathrm{d}\xi = 0 \quad (i = 1, 2, 3) \tag{4.40}$$

These integrals can be split into an outgoing part ($y > 0$) and incoming part ($y < 0$), which are further evaluated for the half Maxwellian distribution (Eq. 4.16) using Eqs. (4.18) and (4.19). This formulation leads to three conservation equation for mass, momentum, and energy in non-dimensional form, with four dependent variables ($S_\infty, \mathbb{Z}, \theta$, and β). Among these, parameters S_∞, $\mathbb{Z}$, and θ are the dimensionless speed ratio $\left(u_\infty/\sqrt{2RT_\infty}\right)$, pressure ratio ($p_s(T_l)/p_\infty$), and temperature ratio (T_∞/T_l), respectively. These equations are solved by specifying one of the unknown variables and solving for the other three parameters. Based on these values, the evaporative mass flux can finally be calculated

$$j = \frac{p_\infty}{RT_\infty}u_\infty; q = j \times h_{fg} \tag{4.41}$$

In case a known pressure (p_∞ or $\mathbb{Z}$) boundary condition is imposed outside the Knudsen layer, Eq. (4.41) can be reduced to

$$j = \frac{p_\infty}{\sqrt{RT_l}}\sqrt{\frac{2}{\theta}}S_\infty \tag{4.42}$$

The kinetic theory models are likely to be accurate since they account for the temperature jump across the air-water interface. Among the kinetic theory models, the BEMM model is preferable under strong evaporation conditions, while SM should be comparable to BEMM for weak evaporation (Aursand and Ytrehus 2019).

Polikarpov et al. (2019) compared the experimental data of Badam et al. (2007) and Kazemi et al. (2017) on interfacial evaporation rate and temperature jump to the available SM and BKE-based non-equilibrium models. The SM formula overestimated the evaporation rate whereas the BKE-based kinetic model predicted the evaporation rate reasonably well. The interfacial temperature jump from BKE did not show good agreement with experiments.

The evaporative mass flux obtained from the molecular kinetic theory can also be used as an interfacial boundary condition in the quasi-equilibrium model (Qin and Grigoriev 2015). In this approach, the transport equations (Eqs. 4.7–4.11) are solved with evaporative heat flux calculated from Eq. (4.2). The evaporation rate in this formulation is approximated by using the SM formula (Eq. 4.32), instead of Eq. (4.1).

4.2.5 Accommodation Coefficient

The accommodation coefficient is a macroscopic measure of the actual (experimental) evaporation/condensation flux when compared with the estimated (theoretical) flux. In molecular terms, the accommodation coefficient of evaporation is the ratio of the molecular flux emitted from liquid surface and the molecular flux transferred to the vapor phase. Similarly, the accommodation coefficient of condensation is defined as the ratio of molecular flux absorbed by the liquid surface to the molecular flux impinging on it, the difference being reflected from the interface.

In the literature, these evaporation and condensation coefficients are assumed to be equal. A value of unity will indicate an equilibrium condition though evaporation and condensation are strongly non-equilibrium phenomena. Hence, a unit accommodation coefficient may be applicable only for weak evaporation (Kryokov and Levashov 2011). These coefficients, though less than unity, are treated to be constant, but are expected to show dependence on temperature, pressure, and contaminants spread over the interface (Marek and Straub 2001). Accommodation coefficient may also be estimated by comparing kinetic theory predictions of macroscopic quantities such as heat fluxes and water production rate with a fully continuum-scale model. In principle, the quantification of accommodation coefficients requires molecular dynamic simulation and is a topic of research.

4.3 Closure

The quasi-equilibrium model of evaporation utilizes continuum tools and is conceptually simple. However, the resulting mathematical problem is coupled and nonlinear and can only be solved using approximate computational tools. The non-equilibrium model based on the kinetic theory is rich in physics and postulates the appearance of a Knudsen layer at the air-water interface. The extent of jump across the Knudsen layer in continuum-scale properties such as temperature, pressure, and velocity are expected to be important for microscale device-level applications of evaporation and condensation. These calculations are formulated using the framework of kinetic theory of gases. The significance of such Knudsen layer at the liquid-vapor interface in engineering applications remains to be conclusively established.

Part II
Modeling Dropwise Condensation

Chapter 5
Introduction to Condensation

Sameer Khandekar and K. Muralidhar

Nomenclature

A_1, A_2	Area fractions occupied by patches 1 and 2
f	Equivalent roughness of the substrate
F_g	Force due to gravity (N)
F_σ	Force due to liquid-vapor interfacial tension (N)
g	Acceleration due to gravity (m/s^2)
K	Shape constant in droplet shape calculations
$l_c = r_c$	Capillary length (m)
r	Characteristic drop radius (m)
R	Length-scale, typically the footprint radius (m)
m	Mass of the drop (kg)

Greek Symbols

α	Angle of inclination of the substrate (°)
ρ	Liquid density (kg/m^3)
$\widehat{\sigma}$	Accommodation coefficient
σ_{lv}	Liquid-vapor interfacial tension (N/m)
θ	Equilibrium contact angle of the chemically textured surface (°)
θ_a	Equilibrium contact angle of the pillared surface with a layer of air (°)
θ_c	Equilibrium contact angle in the Cassie state (°)
θ_w	Equilibrium contact angle in the Wenzel state (°)
θ_{adv}	Advancing contact angle, namely the limiting value of θ_{max} (°)
θ_{max}	Maximum contact angle when a drop rests on an inclined surface (°)
θ_{min}	Minimum contact angle when a drop rests on an inclined surface (°)
θ_{rcd}	Receding contact angle, namely the limiting value of θ_{min} (°)

S. Khandekar, K. Muralidhar, *Drop Dynamics and Dropwise Condensation on Textured Surfaces*, Mechanical Engineering Series,
https://doi.org/10.1007/978-3-030-48461-3_5

5.1 Classification

Condensation involves change of phase from the vapor state to the liquid state. It is associated with mass transfer, during which vapor migrates towards a liquid-vapor interface or a solid substrate and is converted into liquid. Transport is driven by a pressure reduction that occurs at the phase boundary. Condensation process is initiated by a temperature difference, called subcooling, between the bulk vapor and the solid surface. Subsequently, energy in the form of the latent heat must be removed from the interfacial region either by conduction or convection. Apart from natural phenomena, condensation is an essential part of many energy conversion, water harvesting, and thermal management systems. Improvement in heat and mass transfer rates during the phase change process, therefore, can have beneficial effects.

Classification of the condensation process and the corresponding pictorial view of possible condensation patterns are depicted in Figs. 5.1 and 5.2, respectively. Homogeneous condensation occurs in free space in the absence of any foreign material. It takes place stochastically as a result of fluctuations in the vapor molecules. Such a process is only occasionally seen and barely plays an important role in heat transfer devices. As an example, Fig. 5.2a shows saturated steam flowing in a pipe, where sudden expansion creates a thermodynamically favorable condition for the condensation of vapor. Heterogeneous condensation occurs when vapor condenses on or underneath the surface of any other material (either liquid or solid) or on spatially distributed nuclei, Fig. 5.2b–e. According to the type of condensing surface, heterogeneous condensation is divided into volume condensation and surface condensation. An example of volume condensation is formation of clouds, mist, or fog. Surface condensation takes place on or underneath a subcooled surface that is exposed to vapor. The resulting heat transfer coefficient is orders of magnitude greater than the single-phase convective paradigm. Hence, it plays an important role in many heat transfer devices and systems.

The phase change process may result in either (1) the formation of a continuous film of liquid on the cold substrate (filmwise condensation) or (2) the formation of an ensemble of droplet (dropwise condensation). There can be a mixed mode as well,

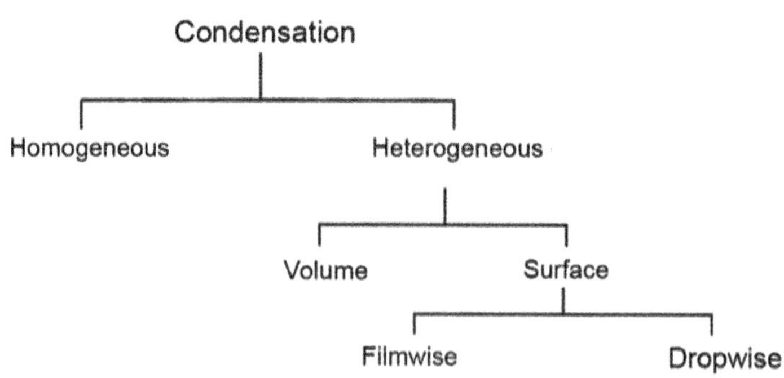

Fig. 5.1 Classification of vapor condensation phenomena

having fuzzy overlapping characteristics of drops and a liquid film. The condensation form that is realized depends on the wettability of the surface and surrounding media, related to the free energy of the condensing wall and the surface tension of the condensate. Filmwise condensation occurs when the liquid wets the substrate while dropwise condensation takes place when the liquid does not have high affinity for the substrate. These processes are of interest from an engineering point-of-view, as condensation occurs quite often in industrial equipment.

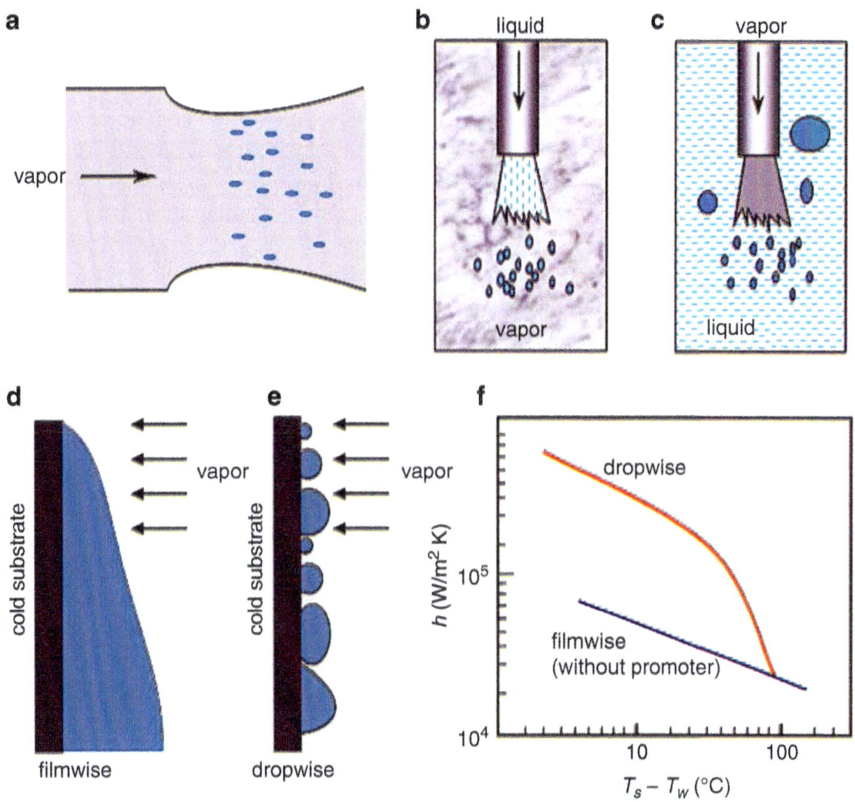

Fig. 5.2 Pictorial views of various types of condensation; (**a**) homogeneous condensation of steam due to pressure drop in a steam nozzle, (**b**) heterogeneous volume condensation, (**c**) heterogeneous condensation on a liquid surface, (**d**) Filmwise condensation, and (**e**) dropwise condensation on a vertical cold substrate. (**f**) Comparison of dropwise and filmwise condensation heat transfer trend for steam at atmospheric pressure

5.2 Filmwise Condensation

Filmwise condensation is preferred when the liquid wets the condenser surface, resulting in the complete coverage of the surface by a liquid film. It is commonly observed in various phase change heat transfer devices. The film is removed from the surface under the action of the gravity, acceleration, or any other interacting body forces and shear stresses due to the surrounding vapor flow. The film usually renders a high thermal resistance to heat transfer and therefore, a relatively large temperature gradient may prevail across it, depending on the thermophysical properties of the condensing liquid.

5.3 Dropwise Condensation

Vapor-to-liquid phase change process in the form of discrete drops on or underneath a cold substrate is called dropwise condensation. It is realized when the condensate does not wet the substrate except at locations where well-wetted contaminant nuclei exist, Fig. 5.2e. The heat transfer coefficient during this process is usually an order-of-magnitude larger than for filmwise condensation, Fig. 5.2f under comparable imposed boundary conditions. This makes dropwise condensation a very attractive mechanism for industrial applications.

It is usually believed that dropwise condensation begins with drop formation at preferred nucleation sites at the atomic scale. These droplets grow by direct condensation, up to a size of the order of the distance between neighboring nucleation sites. Beyond this point, coalescence among neighboring drops takes place and subsequent growth of drops occurs by the combination of direct condensation and coalescence. When a drop reaches a size, at which the body forces exceed surface tension holding it to the solid surface, the drop departs and sweeps the surface clear, permitting new nucleation sites to become available. Hence, coalescence and sliding droplets re-expose substrate area to provide a continuous source of nucleation sites. Overall, dropwise condensation is a quasi-cyclic process, as represented in Fig. 5.3. Several sub-processes of distinct length and timescales interact in space and time to form a closed cycle of events. Various researchers (Mikic 1969; Griffith 1985; Tanasawa 1991; Rose 2002; Carey 2008) have confirmed that dropwise condensation is a complex phenomenon involving an interplay of several factors—from molecular level forces at the three-phase contact line of the droplets to the body forces acting on the condensing liquid droplets.

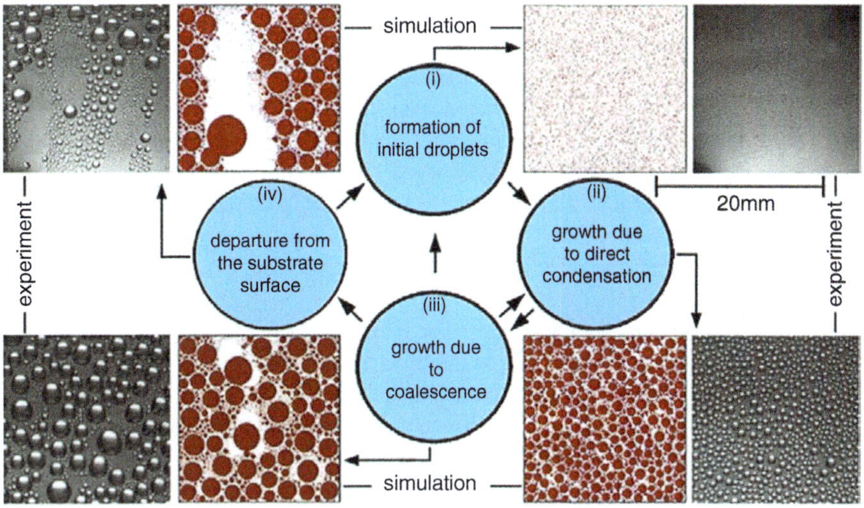

Fig. 5.3 Cycle of processes in dropwise condensation of vapor over an inclined cold substrate

5.4 Understanding Dropwise Condensation

Industrial applications of dropwise condensation have not been very successful. This is because of the intricacies faced in controllability and long-term sustainability of the process on textured substrates. Its dependence on many parameters, such as nucleation site density, hydrophobicity, thickness of any promoter layer if present, substrate orientation, degree of subcooling, and environmental conditions, adds to the difficulty in modeling. In addition, many issues are unresolved. These include drop formation and its dynamics at the atomic scale, hierarchical phenomena, growth and coalescence mechanisms, contact angle hysteresis, dynamics of the three-phase contact line, instability of drops, and leaching.

Small changes in the surface morphology on a micro/nanoscale lead to changes in the droplet distribution, thereby affecting drop mobility. The overall temperature difference is small making measurement of heat transfer coefficient quite difficult. In addition, the statistical nature of droplet distribution in the ensemble contributes to the intricacy of analysis and interpretation. Reported heat transfer data of dropwise condensation invariably shows large scatter.

Dropwise condensation can be sustained only if the chemical and/or physical morphology of the condensing surface is maintained over long periods of time. In practice, condensing surface features are altered during growth, coalescence, and slide-off of the drops. Sustaining dropwise condensation for long time periods on engineered surfaces is a major challenge. Issues that require attention can be broadly summarized as follows:

Longtime sustainability. The foremost is devising reliable means of promoting a cyclic dropwise condensation process. The ideal substrate has low thermal resistance, high durability, and low surface energy. With the advent of nano-technology, breakthroughs in thin film coating, physical and chemical texturing, and availability of superior experimental techniques, definite possibilities arise for sustaining dropwise condensation over longer durations of time.

Heat transfer measurement. Experimental heat transfer investigations report widely scattered data because of inconsistency and difficulty in measurement of very low temperature differences that are usually applicable in condensation processes.

Substrate thermal conductivity. The effect of thermal diffusivity of the substrate on heat transfer, and the role it plays during dropwise condensation is rather controversial.

Mechanism of dropwise condensation. The interrelationship among the hierarchy of processes from the atomic scale to the drop is not addressed. A few fundamental questions remain unanswered. What part of the surface forms nucleation sites? What is the relation between the surface morphology and nucleation site density? How are drops distributed on or underneath the substrate? Is the equation determining minimum drop radius correct? How are critical sizes of drop at slide-off and fall-off calculated? Details of drop coalescence during condensation process are unaddressed. Role of pressure and temperature fluctuations on leaching needs to be understood. Effects of substrate orientation, hydrophobicity, and surface energy gradient on heat transfer rates are of importance.

Controllability. Dropwise condensation depends on thermo-physical properties of the condensing fluid (both liquid and vapor state), physico-chemical properties of the cold substrate, its orientation, surface energy profile, subcooling, and saturation pressure. Thus, close control of dropwise condensation is difficult, and one needs clarity in understanding the importance of parameters on various length and timescales involved in the process.

Multiscale phenomenon. The overall mechanism of formation of a droplet on a textured surface involves varied length scales, from atomistic orders at early stage of nucleation to scales affected by the body forces, while surface tension, viscosity, and gravity are important at intermediate scales. Thus, nuances of dropwise condensation can be properly understood only when a multiscale modeling approach is adopted.

5.5 Intermediate Steps in Dropwise Condensation

Dropwise condensation begins at an atomistic level in which vapor atoms impinge on the cold substrate. These individual atoms will form stable clusters which eventually lead to microscopic droplets at a specific location on a surface. These grow by direct condensation of the vapor and by coalescence between droplets, until a certain critical size is reached. Drops then leave the surface by the action of body forces and vapor shear and re-expose the substrate area to bulk vapor. Fresh nucleation occurs at the re-exposed area and the complete condensation cycle begins at the atomic level once again. The atomistic model captures the initial stage of

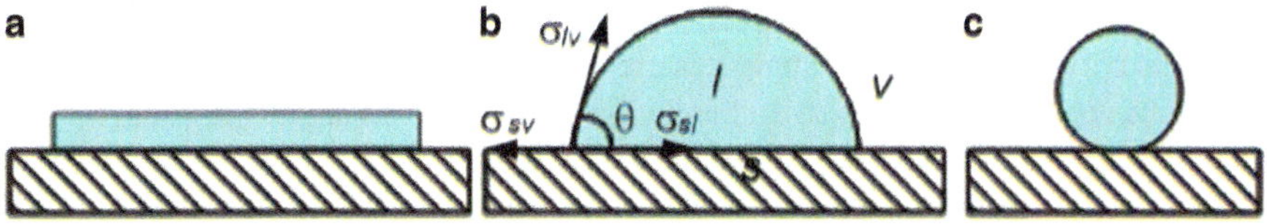

Fig. 5.4 Schematic representation of surface-drop interaction on a planar substrate at the continuum scale. (**a**) Complete wetting ($\theta = 0$), (**b**) partial wetting ($0^\circ < \theta < 180^\circ$), and (**c**) non-wetting ($\theta = 180^\circ$)

condensation, which leads to a stable cluster. Once a cluster is formed, bulk thermophysical properties of the liquid and physico-chemical properties of substrate become relevant and start influencing growth. Dropwise condensation can be sustained if the condensate does not wet the cold surface. Figure 5.4 shows the schematic diagram of various types of solid-liquid drop interaction on a planar surface. Wetting characteristics can be established by the measurement of the apparent contact angle θ, specific to the choice of the liquid and the surface material (de Gennes 1985). It is defined as the angle between the tangents drawn at the liquid/vapor interface and the liquid/solid interface. The wettability of a surface by a liquid is a consequence of a combination of complex processes. Some of these originate at the microscale and can be understood in terms of surface chemistry and van der Waals forces. Certain factors are purely statistical and may vary from sample to sample. These include wetting transitions and the pinning of the contact line. Fluid motion inside the droplet commences when it starts moving due to a force imbalance. The shape of the droplet will then depend on the principles of fluid dynamics as well. As a first step, the solid-liquid interaction in a drop may be characterized uniquely by the apparent contact angle θ and determined by measurements when the drop is in mechanical equilibrium. Liquid is said to wet a solid surface completely if it spreads over a considerable distance with a limiting value of $\theta = 0^\circ$, Fig. 5.4a. If it retains a full spherical drop on contact with a solid surface, it is said to be fully non-wetting with contact angle $\theta = 180^\circ$, Fig. 5.4c. In between the wetting and non-wetting regimes, there can be a situation when a liquid has a contact angle $0 < \theta < 180^\circ$. This situation is known as partial wetting and the liquid has a finite liquid/solid interface, as shown in Fig. 5.4b. In the real engineering context, most systems involving solids and interacting liquids are invariably between the fully wetting and the non-wetting limits.

The contact angle contains details of the interactions at various interfaces including solid/liquid, liquid/gas, solid/gas, and solid/liquid/gas. The adoption of the apparent contact angle simplifies analysis and helps understand the behavior of drops from a mechanics perspective. For a given liquid or gas/vapor system, a wide variety of solid substrates, natural and engineered, will produce a range of contact angles. These are classified as hydrophilic ($0^\circ < \theta \leq 90^\circ$), hydrophobic ($90^\circ < \theta \leq 140^\circ$), and superhydrophobic ($140^\circ < \theta \leq 180^\circ$) (Berthier 2008).

Several important phenomena in condensation rely on partial wetting of the solid substrate by the condensing liquid. Surface heterogeneities on the condensing substrate—chemical and topographical, have profound effect on the apparent contact angle and give rise to contact angle hysteresis and probability of local pinning of the three-phase contact line. Based on the knowledge of contact angle, the behavior of a liquid drop on, or underneath a solid surface, is obtained. As the contact angle influences the equilibrium shape of the drop and hence its curvature, it can be related to interfacial tension and the surface energy distribution of the solid substrate.

5.5.1 Measurement of Apparent Contact Angle

Three-phase contact lines are formed when materials in different phases, e.g., solid, liquid, and gas (or vapor) intersect. Common examples are a liquid drop spreading on a solid surface or a liquid meniscus in a capillary tube. In the presence of the third phase (gas or vapor), a liquid spreading on a solid surface can reach two distinct equilibrium states. These are: (a) partial wetting and (b) complete wetting. The condition for static equilibrium of a triple contact line involving an ideal solid (perfectly smooth and chemically homogeneous), liquid, and a gas/vapor surrounding is stated in the form of the classical Young's equation

$$\sigma_{\mathrm{lv}} \cos\theta = \sigma_{\mathrm{sv}} - \sigma_{\mathrm{sl}} \tag{5.1}$$

Here, the symbol σ_{ab} is the surface tension between phases a and b. Symbols s, l, and v in Eq. (5.1) stand for solid, liquid, and the gaseous phases, respectively. The symbol θ is the apparent contact angle at each point of the solid-liquid boundary. Equation (5.1) holds for an ideally smooth solid surface with no chemical heterogeneities. Real solid surfaces depart from an ideal behavior as they are not perfectly smooth. In addition, their composition may also vary slightly with location. Molecules, atoms, or ions of other chemical species may be adsorbed on the surface. Effectively, the static contact angle turns out to be non-unique on real surfaces and can only be experimentally determined.

The experimentally observed contact angle depends on the way the surface is prepared, Fig. 5.5. One of the first attempts on understanding the influence of surface roughness on wetting is due to Wenzel (1936) who proposed the following relationship for the apparent contact angle

$$\cos\theta^{*} = f \cdot \cos\theta \tag{5.2}$$

Here, θ^{*} is the apparent contact angle, f is the degree of roughness (with $f = 1$ for a smooth surface, $f > 1$ for a rough surface, and θ the local apparent contact angle). Equation (5.2) embodies two types of behavior for rough surfaces. For hydrophilic behavior, we have $\theta^{*} < \theta$ since $f > 1$, as shown in Fig. 5.5a(i). Likewise, for hydrophobic, we have $\theta^{*} > \theta$, as depicted in Fig. 5.5a(ii). Many researchers (Huh and Mason 1977; Leger and Joany 1977; De Gennes 1985) have shown that a wetting experiment is extremely sensitive to heterogeneities of the solid surface.

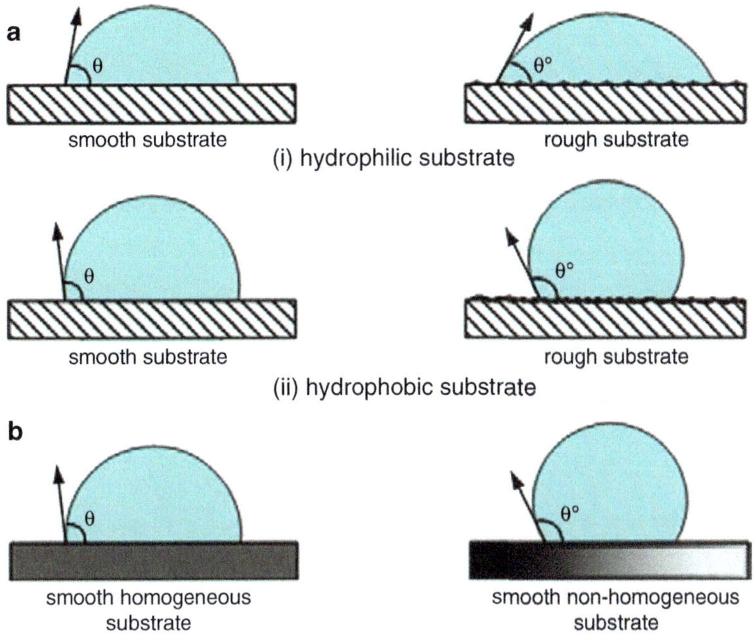

Fig. 5.5 Measurement of apparent contact angle on a textured substrate. (**a**) Effect of roughness on contact angle (Wenzel state) for a hydrophilic substrate and hydrophobic substrate. (**b**) Effect of chemical non-homogeneity on local contact angle (Cassie-Baxter state)

Shibuichi et al. (1996) have shown that contact angle can be tuned by varying solid roughness in the hydrophilic region ($\theta < 90°$). Other groups (Lenz and Lipowsky 1998; Li and Amirfazli 2007; Chen et al. 2007; Berthier 2008; Hsieh et al. 2008) have shown that the substrate roughness amplifies the hydrophilic or hydrophobic character.

Similar reasoning can be applied to a surface that is planar but chemically heterogeneous. The contact angle on chemically homogeneous and non-homogeneous surfaces is shown in Fig. 5.5b. Viewing a chemically heterogeneous surface, as composed of distinct patches (e.g., A_1 and A_2) of various species, the apparent contact angle is as follows:

$$\cos\theta^* = A_1 \cdot \cos\theta_1 + A_2 \cdot \cos\theta_2 \tag{5.3}$$

Equation (5.3) is called the Cassie-Baxter relation; θ^* is the apparent contact angle, θ_1 and θ_2 are the local contact angles for surface patches 1 and 2, respectively, A_1 and A_2 are the fractional areas occupied by surface patches 1 and 2, respectively. Therefore, the apparent angle θ^* (restricted to the interval (θ_1, θ_2)) is given by an average involving the cosines of the angles characteristic of each constituent specie.

This discussion clarifies why the three-phase contact line of a liquid drop resting on a surface gets locally deformed: chemical and topographical heterogeneities play

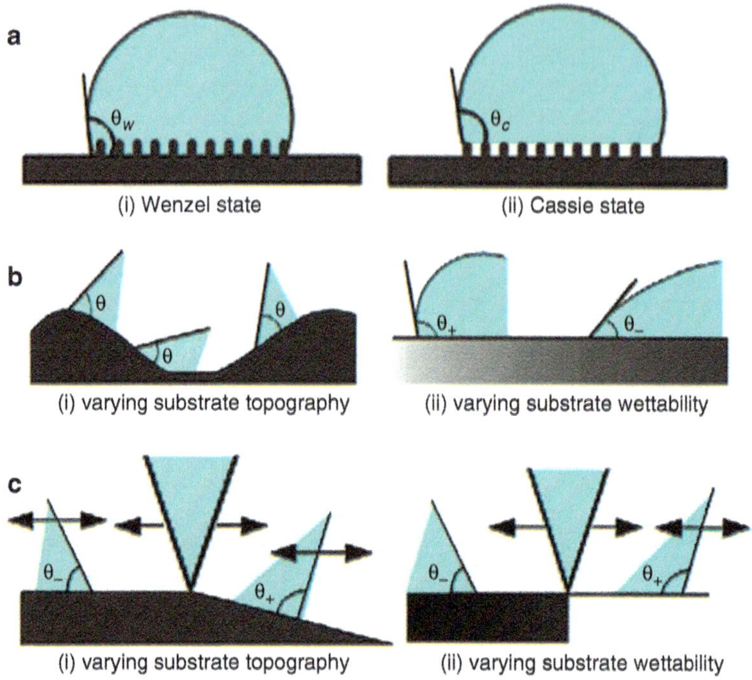

Fig. 5.6 (**a**) Two possible states of a drop sitting on a physically textured surface. (**b**) Effect of physical roughness or variation in topography and continuously varying wettability of substrate on contact angle. (**c**) Pinning due to sudden physical and chemical discontinuity

an important role. Certain surfaces have roughness in the form of micro-pillars, creating a superhydrophobic substrate. It has been observed that the drop can sit on a textured (rough) surface in two distinct configurations, Fig. 5.6a(i). For the Wenzel state, drop penetrates the pillars. For Cassie state, it does not contact the actual surface and indeed, may stay on the top of the pillars, Fig. 5.6a(ii). In such cases, the contact angle is obtained as follows. For Wenzel state, Eq. (5.2) (Wenzel's law) applies,

$$\cos\theta_{\mathrm{w}} = f \cdot \cos\theta \tag{5.4}$$

Here, θ_{w} is the apparent contact angle of the Wenzel state and f is the equivalent roughness of the substrate. For the Cassie state, one can write Cassie's law (Eq. 5.3),

$$\cos\theta_{\mathrm{c}} = f \cdot \cos\theta + (1 - f) \cdot \cos\theta_0 \tag{5.5}$$

where θ_{c} is the apparent contact angle of Cassie state of drop and θ_0 is the contact angle with the layer of air, and f is the ratio of the contact surface (top of the pillars) to the total horizontal surface. If the pillars are not too far from each other, the value of θ_0 approaches π.

5.5.2 Pinning of the Contact Line

In Fig. 5.6c, a sketch of the wetting behavior of a drop of liquid on a substrate, with a continuously varying topography and continuously varying wettability, is depicted. Ondarçuhu (1995) and Lenz and Lipowsky (1998) showed the three-phase contact line to be pinned at a surface defect and a sharp transition of wettability on the substrate. Local chemical and geometrical defects locally modify the contact angle. If the drop size is smaller than the length scale of the topography, Fig. 5.6c(i) shows that the drop shape is not affected by the topography. If the drop is larger than the topographical features, the global shape of the drop will be affected by the deformation of the three-phase contact line. Similarly for a substrate with a gradient in wettability (chemically non-homogeneous surface, Fig. 5.6c(ii)), the drop is deformed due to the peripheral changes in the contact angle of the three-phase contact line.

If the substrate has a sharp topography or wettability pattern, the situation is different (Fig. 5.6c(i)). At the discontinuities, Young's Eq. (5.1) becomes ill-defined. As a result, the three-phase contact line becomes locally immobilized. This effect is known as the pinning of the contact line. The pinning of an advancing contact line towards a convex edge over the substrate with homogeneous wettability is illustrated in Fig. 5.6c(ii). The apparent contact angle at the boundary can have any value in between the smaller angle θ^- on the hydrophilic part and the larger value θ^+ on the hydrophobic part. As a consequence, the position of the contact line is fixed to the line of discontinuity as long as the contact angle falls in the range of θ^- to θ^+. The contact angle now depends on the local wettability of the substrate and the global shape of the liquid-vapor interface at equilibrium. Contact angles will change further under dynamic conditions when, owing to fluid motion, a non-uniform pressure field is created within the drop. Hence, the contact angle is not only governed by the local wettability of the substrate but also depends on global shapes of the liquid-vapor interface. The wettability pattern on or underneath the substrate may act as an anchoring point for the contact line of a wetting liquid.

5.5.3 Capillary Length Scale

Surface tension, a negligible weak force in the macroscopic world, is dominant at smaller scales. This is because the force due to surface tension decreases linearly with size whereas weight scales down as the third power (Trimmer 1989). The cross-over occurs at around the capillary length. Well below this cross-over, the force due to surface tension is dominant and well above, the force of gravity is important.

To determine the capillary length, consider a liquid droplet underneath a substrate. It hangs under the substrate due to surface tension. It is stable until it grows large enough to be separated by the force due to gravity, i.e., its own weight. The force due to surface tension is approximated as

$$F_\sigma \approx r\sigma \tag{5.6}$$

The force due to the gravity is approximated as

$$F_g \approx r^3 \rho g \tag{5.7}$$

The two forces are equal when the drop separates from the substrate. The critical radius of drop, when it separates, is obtained by balancing Eqs. (5.6) and (5.7) and it equals the capillary length

$$l_c = r_c = \sqrt{\sigma/\rho g} \tag{5.8}$$

Capillary length defines the length scale below which surface tension dominates gravity. It is in the millimeter range for water at usual conditions. It depends on the thermo-physical properties of the liquid-solid combination. Leach et al. (2006) and Leipertz (2010) reported that small drops are locations of high heat transfer rates. Heat transfer diminishes with increasing drop radius. The largest drop diameter depends the interfacial forces at phase boundaries and the body force, and hence, capillary length. Therefore, heat transfer coefficient in dropwise condensation crucially depends on capillary length or capillary radius.

5.5.4 *Contact Angle Hysteresis*

Partially immersing a thin solid sheet in a liquid and moving it slowly, Furmidge (1962) reported the appearance of two distinct contact angles. These angles are known as the advancing angle θ_{adv} and receding angle θ_{rcd}, depending on the direction of motion of the plate, Fig. 5.7a. Arising from this experiment, the difference between the advancing and receding contact angles is known as contact angle hysteresis. For an idealized solid surface that is perfectly smooth, clean, and homogeneous in composition, there would appear to be no reason for θ_{adv} and θ_{rcd} to be different. However, such an idealized surface does not exist. Real condensing surfaces are typically metallic and are never perfectly smooth, composition may vary slightly with location, and molecules, atoms, or ions of other substances may be adsorbed on the surface. Contact angle hysteresis is acknowledged to be a consequence of three factors: (1) surface inhomogeneity, (2) surface roughness, and (3) impurities present on the surface.

If a drop of liquid is placed underneath a horizontal surface, it achieves an equilibrium shape and leaves the droplet with almost constant angle all around its perimeter (ideal contact angle hysteresis is zero). If the surface is then turned through angle α, the drop will deform to balance the gravity force, parallel to plate inclination and perpendicular to the substrate, Fig. 5.7b. The contact angle reaches its maximum value for an advancing liquid edge of the drop and the minimum value for a receding

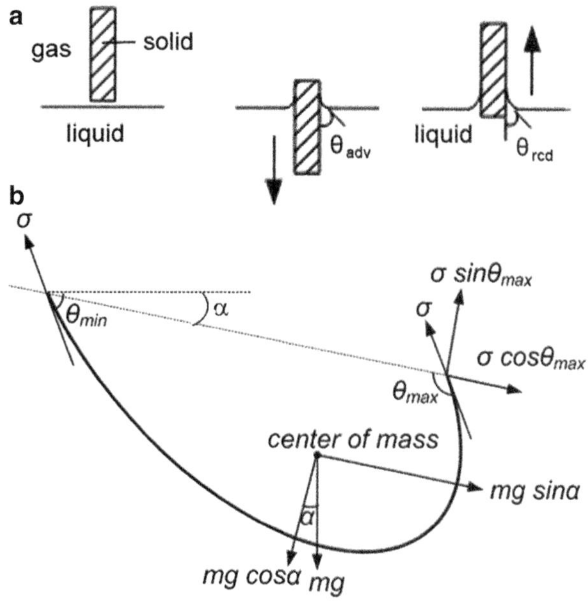

Fig. 5.7 (**a**) Basic definition of advancing (θ_{adv}) and receding (θ_{rcd}) angles during immersion and removal of a solid plane surface in a liquid medium. (**b**) Droplet angle on an inclined plane with leading side angle (θ_{max}) and rear side angle (θ_{min})

liquid edge of the drop. Many researchers (Brown et al. 1980; Lawal and Brown 1982; Extrand and Kumagai 1995; Elsherbini and Jacobi 2004a, b; Dimitrakopoulos and Higdon 1999) argued that the maximum and minimum contact angles, on or underneath an inclined substrate are equivalent to advancing and receding contact angles, respectively. The static advancing and receding contact angles can be easily observed from a tilted pendant droplet as shown in Fig. 5.7b. In a dynamic context, the contact angle changes from static values so that the advancing angle increases and receding angle decreases as a function of the speed of the three-phase contact line. The literature on the interrelation between the static and dynamic contact on or underneath an inclined substrate is rather scarce.

Contact angle hysteresis plays an important role in the stability of a drop on or underneath an inclined substrate. The difference between the top and bottom side contact angle makes it possible for the droplet to adopt a shape that may support the weight of the liquid against gravity. Hence, contact angle hysteresis offers resistance against the motion of drop. Note in Fig. 5.7b that the interface radius of curvature is smaller over the upper portion of the droplet and greatest near the base. The resulting variation of the surface forces over the interface may serve to balance the hydrostatic pressure difference across the interface, allowing the droplet to hold its position on the vertical or inclined wall against gravity.

Various investigators (Dussan 1979, 1985; Dussan and Chow 1983; Briscoe and Galvin 1991b; Extrand and Kumagai 1995; Miwa et al. 2000; Elsherbini and Jacobi 2006) have derived an expression relating the criticality of drop sliding on or underneath an inclined plane in dropwise condensation, as a function of the contact angle hysteresis ($\theta_{adv} - \theta_{rcd}$),

$$\sin\alpha = \sigma_{\mathrm{lv}} \cdot (R \cdot K/m \cdot g)(\cos\theta_{\mathrm{rcd}} - \cos\theta_{\mathrm{adv}}) \tag{5.9}$$

In Eq. (5.9), α is the critical sliding angle, σ_{lv} the surface tension, m the mass of the drop, and R and K are an appropriate length scale and shape constant for the contour of the drop, respectively.

Öner and McCarthy (2000) made it clear that contact angle hysteresis can be a qualitative indication of drop mobility. Yet, Krasovitski and Marmur (2005) and Pierce et al. (2008) argued that advancing and receding contact angles are measured on a level surface and should theoretically not be used in numerical predictions of the sliding angles. Instead, they define the maximum and minimum contact angles ($\theta_{\max}$ and $\theta_{\min}$), which are those that occur at the leading and trailing edges of a drop profile on a surface inclined at the sliding angle, Fig. 5.7b. The modified form of Eq. (5.9) is

$$\sin\alpha = \sigma_{\mathrm{lv}}(RK/mg)(\cos\theta_{\max} - \cos\theta_{\min}) \tag{5.10}$$

Theoretical and experimental evidence suggest that the relationship between $\theta_{\max}$ and θ_{adv} as well as $\theta_{\min}$ and θ_{rcd} respectively, varies with the surface-liquid combination. Elsherbini and Jacobi (2004a, b) reported empirical data that exhibits $\theta_{\max}$ and $\theta_{\min}$ approximately equal to θ_{adv} and θ_{rcd} for all surface-water combinations. Krasovitski and Marmur (2005) reported that the upper side contact angle ($\theta_{\min}$) tends to be approximately equal to the receding contact angle, while the lower side contact angle ($\theta_{\max}$) may be much lower than the advancing contact angle. Hence, there is some controversy on the value of the leading angle and trailing angle of a deformed drop on an inclined substrate at criticality. This information is quite important from the viewpoint of dropwise condensation and has attracted attention. In the present work, the leading side angle ($\theta_{\max}$) is assumed equal to the advancing angle (θ_{adv}) of drop and trailing side angle ($\theta_{\min}$) equal to the receding angle (θ_{rcd}) at criticality for determining the size of drop at criticality.

Various researchers (Leach et al. 2006; Ma et al. 2008; Kim and Kim 2011; Rykaczewski 2012) have reported that heat transfer rate increases with diminishing contact angle hysteresis, as the criticality of drop slide-off /fall-off is inversely proportional to it. Large hysteresis will provide adequate forces along and normal to the wall and improve the stability of the drop. Conversely, the drop slide-off or fall-off will occur early on a surface that has small hysteresis. The repeated sweep and removal of drops from a surface results in fresh condensation and an overall improvement in the heat transfer rate.

On a roughened hydrophobic substrate, a liquid drop can exhibit either the Cassie state, where the drop sits on the air/vapor-filled textures or the Wenzel state where the drop wets cavities of the textures (Fig. 5.6). The apparent contact angle of a roughened hydrophobic surface is enhanced in both the Cassie and Wenzel states; however, the Cassie state is the preferred superhydrophobic state in which a drop has a much smaller contact angle hysteresis and therefore a higher mobility. Till date, none of the reported condensation studies on engineered superhydrophobic surfaces has exhibited a sustained Cassie state; instead, the condensate drops partially or fully penetrates into the cavities over the course of long time condensation (Ma et al. 2012).

5.5.5 *Vapor Accommodation Coefficient*

When condensation occurs, kinetic theory of gases suggests that the flux of vapor molecules going into the liquid-phase must exceed the flux of molecules escaping the liquid phase. Accommodation coefficient denoted as $\hat{\sigma}$ defines the fraction of the striking vapor molecules that get condensed on the vapor-liquid interface. The remaining fraction $(1 - \hat{\sigma})$ is due to reflection of vapor molecules that strike the interface but do not condense. The accommodation coefficient indirectly measures the interfacial resistance of the liquid-vapor interface to condensation. Higher the accommodation coefficient, lower the interfacial resistance of the liquid-vapor interface of the condensed drop. Quoted values of $\hat{\sigma}$ in literature widely vary. Mills and Seban (1967) reported that the accommodation coefficient is less than unity only when the interface is impure. For pure liquid-vapor interface, the value reported in the literature is unity. Because extreme purity is unlikely in most engineering systems, a value of less than unity can be expected. Sukhatme and Rohsenow (1966) reported its values ranging from 0.37 to 0.61 for condensation of metallic vapors. For liquid ethanol, methanol, alcohol, and water, the reported values of accommodation coefficient range from 0.02 to 0.04 (Carey 2008). On the other hand, for benzene and carbon tetrachloride, reported values are closer to unity. Marek and Straub (2001) reported that accommodation coefficient decreases with increasing temperature. The interfacial resistance may be particularly important in the condensation of liquid metals since their bulk thermal conductivity is large.

The variation of interfacial heat transfer resistance per unit area (see Chap. 2), at experimental conditions of 30 °C saturation temperature and 0.015 bar saturation pressure for water vapor, are presented in Fig. 5.8. There is considerable variation of the interface heat transfer coefficient of over an order of magnitude for small values of the accommodation coefficient, $0.01 < \hat{\sigma} < 0.1$. Beyond $\hat{\sigma} > 0.1$ the interface heat transfer coefficient does not tend to change appreciably.

5.6 Closure

This chapter introduces briefly the classification and significance of various physical processes in dropwise condensation. The importance of wettability and contact angle on the formation of drops is highlighted. The shape of the drop plays a central role in determining its conduction resistance to heat flow through it, the onset of instability with respect to static equilibrium, as well as its motion over the substrate. Once large drops move out of the surface, fresh nucleation ensures that the condensation process

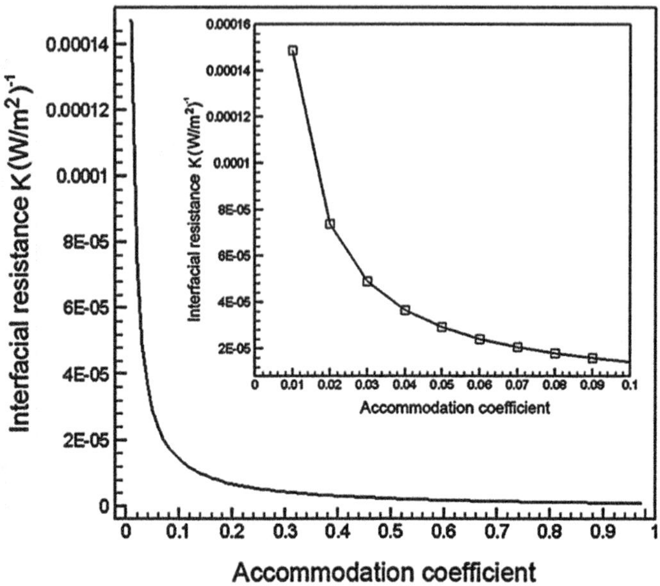

Fig. 5.8 Variation of interfacial heat transfer coefficient with respect to the accommodation coefficient for water vapor

is cyclic, with a characteristic timescale, area coverage, and drop size distribution. Mathematical modeling of the dropwise condensation process forms the topic of the following chapters.

Chapter 6
Modeling Dropwise Condensation: From Atomic Scale to Drop Instability

Sumeet Kumar, Smita Agrawal, Basant Singh Sikarwar, N. K. Battoo, K. Muralidhar, and Sameer Khandekar

Nomenclature

A	Surface area, m^2; suffixes sl for solid-liquid, lv for liquid vapor, cd for substrate average
C_p	Specific heat at constant pressure, W/kg K
d_b	Base diameter of drop, m
D	Diffusion coefficient of adatoms
$f_i(s/S)$	Scaling function for the island size distribution, –
f	Degree of roughness of substrate, –
F	Force, N; suffix σ for surface tension and g for gravity; also, mass flux of monomers in the atomistic model
g	Acceleration due to gravity, m/s^2
h_{lv}	Latent heat of vaporization, J/kg
h	Heat transfer coefficient, W/m^2 K
$\hat{i}, \hat{j}, \hat{k}$	Unit vectors in x, y, and z directions
j_n	Average velocity of vapor molecules, m/s
k	Thermal conductivity of condensate, W/m K
l_{ij}	Distance between two nucleation sites i and j, m
$\overline{M}$	Molecular weight of the condensing liquid, kg/kmol
M	Maximum size of unstable clusters, –
m	Mass of droplet, kg; also, mass of a vapor phase particle; Δm is the incremental mass condensed over a time step Δt
m''	Mass flux, kg/m^2 s
n_s	Number of clusters of size s at coverage ϑ
n_d	Number of atoms/molecules in a drop of minimum radius
n_1	Number density of monomers, monomers/m^2
n_j	Number density of clusters containing j atoms/molecules, m^{-2}

S. Khandekar, K. Muralidhar, *Drop Dynamics and Dropwise Condensation on Textured Surfaces*, Mechanical Engineering Series,
https://doi.org/10.1007/978-3-030-48461-3_6

n_i	Number density of critical clusters i refers to that cluster size which does not decay but may change due to growth by the addition of clusters, m^{-2}
n_x	Number density of stable clusters $n_x = \sum_j n_j$ for all $j > i$, m^{-2}
N	Number of nucleation sites, cm^{-2}; N_f for a textured surface
N_A	Avogadro number
p	Pressure, $\mathrm{N/m^2}$; v for vapor and l for liquid; sat for saturation
q_d	Surface heat transfer rate, W
q''	Average heat flux, $\mathrm{W/m^2}$
R	Specific gas constant J/kmol K; $\overline{R}$ is the universal gas coefficient
r	Radius of drop, m; suffix b is for base radius
r_min	Radius of thermodynamically smallest drop, m
r_cap	Capillary length, $\sqrt{\sigma/g(\rho_\mathrm{l} - \rho_\mathrm{v})}$, m
r_max	Size of the drop at instability due to fall-off, m
r_crit	Size of the drop at instability due to slide-off, m
$S = \frac{\sum s \cdot n_s}{\sum n_s}$	Average island size; $s = 1, 2, 3, \ldots$
t, Δt	Time, time step, s
T	Temperature, K; l, v, and w are for liquid, vapor, and wall; sat for saturation
ΔT, $(T_\mathrm{sat} - T_\mathrm{w})$	Temperature difference between the saturated vapor and condensing wall, K
u, v, w	Velocity component in x, y, and z directions, m/s
U	Relative velocity between the wall and the drop, m/s; also, terminal velocity
V	Volume of the drop, $\mathrm{m^3}$; c, i, and j are for the centroid and locations i and j
x, y, z	Cartesian coordinates
X	Characteristic distance for a graded surface
Z	Fraction of surface covered by a stable cluster

Non-dimensional Parameters

C_f	Local skin friction coefficient, $2\tau_\mathrm{w}/\rho U^2$
$\overline{C_\mathrm{f}}$	Area-averaged skin friction coefficient, $2\overline{\tau_\mathrm{w}}/\rho U^2$
Ja	Jakob number, $(C_\mathrm{p}/h_\mathrm{lv})_\mathrm{ref}(T_\mathrm{sat} - T_\mathrm{w})$
Nu	Nusselt number, hr_cap/k
$(\mathrm{Nu})_\mathrm{sd}$	Local Nusselt number, $h_\mathrm{sd}d_\mathrm{b}/k$
$(\overline{\mathrm{Nu}})_\mathrm{sd}$	Average Nusselt number, $\overline{h_\mathrm{sd}}d_\mathrm{b}/k$
Pr	Prandtl number, $\mu C_\mathrm{p}/k$
Re	Reynolds number, $\rho U d_\mathrm{b}/\mu$

Dimensionless Quantities

u/U	Dimensionless velocity in x direction
v/U	Dimensionless velocity in y direction
w/U	Dimensionless velocity in z direction
$p/(\frac{1}{2}\rho U^2)$	Dimensionless pressure
$(T - T_w)/\Delta T$	Dimensionless temperature

Greek Symbols

α	Inclination angle of the substrate from horizontal, radians
α_l	Thermal diffusivity, m^2/s
β	Volumetric expansion coefficient, K^{-1}
Γ	Progress velocity in the vapor phase
δ	Thickness of promoter layer, mm
δ_j	Decay constant of a cluster with j particles, s^{-1}
μ	Dynamic viscosity, Pa s
ξ	Azimuthal angle, °
ρ	Density, kg/m^3; suffix l for liquid and v for vapor
σ	Surface tension, N/m; suffix lv, sv, and ls are the solid (s), liquid (l), and vapor (v) interfaces
$\widehat{\sigma}$	Accommodation coefficient, –
σ_1	Capture rate of monomers by formation of dimers
σ_j	Capture number of clusters containing j atoms
$\tau_w, \overline{\tau}_w$	Local and average wall shear stress, N/m^2
θ	Contact angle, radians, or degrees; * for a rough surface, suffixes w and c for Wenzel and Cassie states; adv and rcd are advancing and receding angles

6.1 Mechanisms of Dropwise Condensation

The large body of literature available on the subject suggests the following three independent mechanisms of dropwise condensation (Leipertz 2010):

Mechanism 1: The vapor condenses primarily between the droplets, i.e., the droplet-free area. This condensate layer gets transported to the droplets in their vicinity by surface diffusion. According to this model, the thin film between the droplets and the free surface of the droplets contribute to overall heat transfer.

Mechanism 2: While vapor condensation begins in a filmwise mode (filmwise condensation), the film reaches a critical thickness and ruptures due to surface tension-driven instability forming droplets. It is postulated that major part of the heat transfer takes place at this very thin condensate film, while the droplets mainly act as liquid collectors. This model of the dropwise condensation process was proposed by Jakob (1936). Song et al. (1991) have put forward a droplet and condensate film mechanism for the formation of droplets during dropwise condensation. These authors observed that a thin film of condensate exists on open areas in-between the droplets and a film of condensate remains at the spots from where large droplets have departed.

Mechanism 3: Droplets are only formed at individual nucleation sites, while the area between the droplets is regarded to be inactive with respect to condensation. In this model, heat transfer occurs only through the droplets and is primarily limited by their heat conduction resistance. This model was first proposed by Eucken (1937). Majority of the studies support this mechanism, in which the condensate is in the form of discrete drops located at the nucleation sites on or underneath a lyophobic substrate. Further studies that support this model are discussed below.

McCormick and Baer (1963) described the heat transfer process in mechanism 3 of dropwise condensation. The analysis indicated that heat is transferred through active areas on the condenser surfaces which are continually produced by numerous drop coalescence. These areas remain active for a short portion of the cycle time. During this time, numerous submicroscopic drops grow from randomly distributed sites. McCormick and Westwater (1965) studied nucleation of water drops during dropwise condensation on a horizontal surface of copper coated with a monolayer of benzyl mercaptan. Their photographic evidence showed no visible condensate liquid film among the droplets. Drops nucleated at natural cavities on the condenser surface. Some cavities were nucleation sites because they contained trapped, liquid water.

Umur and Griffith (1965) found that, at least for low temperature difference, the area between growing droplets on the surface was, in fact dry. Their results indicate that no film greater than monolayer thickness exists between the droplets, and no condensation can take place in these areas. Further evidence of nonexistence of a condensate film between droplets was furnished by Ivanovskii et al. (1967), using a different fluid. By measuring the electrical resistance between the two electrodes embedded in a glass surface on which dropwise condensation of mercury was taking place, they concluded that no thin condensate layer existed between the droplets.

Photographs taken through a microscopic with magnification of up to 400 showed nucleating and growing droplets which eventually coalesce with neighboring droplets (McCormick and Westwater 1965). New drops form on the sites vacated by the coalescing droplets.

Based on mechanism 3 of dropwise condensation, the first dropwise condensation model was proposed by Le Fevre and Rose (1966). In this model, a calculation for heat transfer through a single drop was combined with that of the drop size distribution to obtain the average heat flux passing through the condensing surface. For deriving heat transfer rates through a single drop, the following three thermal resistances are considered: (a) conduction resistance, (b) vapor-liquid interfacial resistance, and (c) surface curvature resistance. The thickness of the promoter layer coated on the surface was neglected.

Gose et al. (1967) developed a model for heat transfer during dropwise condensation on randomly distributed nucleation sites. Simulation was performed on a 100×100 grid with 200 randomly distributed nucleation sites. The model accounted for growth, coalescence, vacating active sites beneath the smaller of the coalescing drops, renucleation on the newly exposed sites, and drop removal. For steady-state condensation, the theory showed that small drops grow by vapor condensation, and that larger drops grow predominantly by coalescence. The authors observed that higher nucleation sites and drop removal from the substrate were factors for a large heat transfer coefficient.

Glicksman and Hunt (1972) simulated the condensation cycle in several stages, covering the equilibrium drop size to the departing drop size, with a large nucleation site density. The initial nucleation site density considered was 10^5 cm^{-2} with 1000 sites on a surface of size 33×33 μm. The area of the second stage was increased ten times and the droplets from the first stage were redistributed on this surface. In this way, the simulation was repeated until the departure droplet size was reached.

Rose and Glicksman (1973) proposed a universal form of the distribution function for large drops which grow primarily by coalescence with smaller drops though smaller drops themselves grow by direct condensation.

Tanaka (1975a, b) used a precise expression for the calculation of drop size distribution. The author considered the transient change of local drop size distribution, considering the processes of growth and coalescence of drops. From this point of view, the author put forward a theory of dropwise condensation. The theory is based on the following assumptions: (1) primary droplets nucleate at discrete sites distributed randomly on the condensing surface; (2) drops are hemispherical; (3) the governing heat transfer resistance through a single drop is heat conduction; and (4) temperature of the condensing surface is uniform. Basic integro-differential equations describing the transient process of dropwise condensation on a newly swept region were derived. By introducing a model for the cycle of drop departure, a general expression for the average heat transfer coefficient was obtained.

Wu and Maa (1976) used the population balance method to find the drop size distribution of small drops which grow mainly by direct condensation. They estimated the heat transfer coefficient by considering the conduction resistance through the drop. Maa (1978) later utilized the population balance equation derived for dropwise condensation, considering both drop growth due to direct condensation and coalescence of drops, to obtain the resulting drop size distributions. Results confirmed that the drop size distribution and heat flux passing during dropwise condensation depend strongly on the concentration of active nucleation sites on the substrate.

Meakin (1992) described the following stages of the dropwise condensation: (a) nucleation and growth, (b) growth and coalescence, (c) growth and coalescence, with renucleation in exposed regions and (d) growth, coalescence and renucleation, with removal of larger droplets. All the four stages were simulated, and their results were described in terms of simple scaling theories.

Abu-Orabi (1998) used the population balance approach to predict the drop size distribution for small drops that grow by direct condensation. Resistances to heat

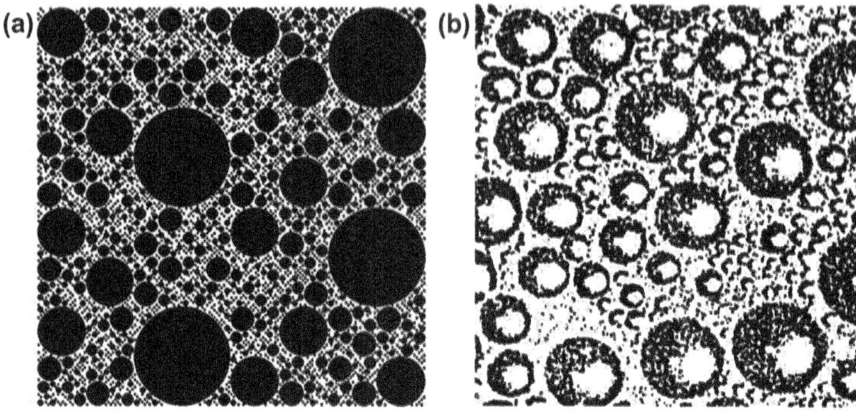

Fig. 6.1 Comparison of drop distribution between (**a**) random fractal model and (**b**) direct photography (Wu et al. 2001)

transfer due to the drop (conduction through the drop, vapor-liquid interfacial resistance, and drop curvature), the promoter layer and the sweeping effect of the drops were incorporated into the model. The total heat flux was calculated from the drop size distributions and the heat transfer rate through a single drop. Drop size distribution for large drops that grow by coalescence was obtained from the work of Rose and Glicksman (1973).

Burnside and Hadi (1999) simulated dropwise condensation of steam from an equilibrium droplet to a detectable size on 240 × 240 μm surface with 10^8 cm^{-2} randomly spaced nucleation sites, stopping when the maximum drop radius was about 4 μm. The authors observed a maximum drop radius of 3.9 mm, 0.21 ms after the start of condensation and peak heat transfer coefficient immediately after the condensing surface gets wiped up by the drop.

Wu et al. (2001) presented a fractal model to simulate drop size and its spatial distribution in dropwise condensation. The boundary conditions of heat conduction through the condensing surface were established using the heat transfer model through a single drop proposed by Rose (1981). Photographs of dropwise condensation at various instants were like experiments, as seen in Fig. 6.1.

Vemuri and Kim (2006) modeled dropwise condensation for hemispherical drops which grow by direct condensation, using the population balance method. The primary resistances to heat transfer, such as conduction through the drop and vapor-liquid interface were considered. The derivation of steady-state distribution for small drops within the size range of negligible coalescence was based on the conservation of the number of drops with no accumulation. Contact angles other than 90° were not considered in this model.

On growth kinetics, Leach et al. (2006) reported from experiments that the smallest drops grow principally by the accretion of liquid molecules diffusing along the substrate surface, while drops larger than about 50 μm in diameter grow by the deposition of condensing vapor directly onto the drop surface. The effects of contact angle, degree of imposed subcooling, and inclination of substrate for a hydrophobic polymer film and silanized glass surface for sessile droplets were reported.

Liu et al. (2007) experimentally proved that the state of initial condensate formed on the surfaces is not in the form of a thin film but as nuclei. These results demonstrate that the mechanism of formation of initial condensate drops for dropwise condensation accords with the hypothesis of individual and distinct nucleation sites. Consequently, recent analytical models have assumed that droplets form on nucleation sites, neglecting any heat transfer taking place from surfaces between the drops. It is also assumed that condensation occurs only on the free surface of droplets, and that the latent heat is transferred through the droplets to the solid surface.

Similarly, experimental and theoretical work of Carey (2008) casts serious doubt on the existence of films and supports the earlier view of McCormick and Baer (1963) that nucleation is an essential feature of dropwise condensation.

Kim and Kim (2011) modeled dropwise condensation over a superhydrophobic surface. The overall methodology, similar to those described earlier, has the following differences: (a) Heat transfer through a single droplet is analyzed as a combination of the vapor-liquid interfacial resistance, the resistance due to the conduction through the drop itself, the resistance from the coating layer, and the resistance due to curvature, (b) Population balance model is adapted to develop a drop distribution function for the small drops that grow by direct condensation, (c) Drop size distribution for large drops that grow mainly by coalescence is obtained from the empirical equation of Tanaka (1975b). Results showed that the single droplet heat transfer and drop distribution are significantly affected by the apparent contact angle of the condensing medium.

A complete simulation of dropwise condensation, from drop formation at the atomic scale to the departing droplet size, accounting for the effect of saturation temperature, contact angle and contact angle hysteresis, wettability gradient on the condensing substrate and the inclination of the substrate along with its experimental validation was reported by the authors in their previous publications. The model presented here is based on these studies and relies on the postulation that drop embryos form and grow at nucleation sites, while the portion of the surface between the growing drops remains dry. The vapor condenses on the free surface of drops at each of the nucleation sites. Latent heat released during condensation is transferred through the liquid drop to the cold wall. Thus, heat transfer in dropwise condensation is primarily limited by the thermal resistance of the liquid drop and the available nucleation site density. Dropwise condensation is a combination of various processes occurring over a wide spectrum of length and timescales. A comprehenisve mathematical model of various subprocesses arising in dropwise condensation underneath an inclined textured substrate is reported in the present chapter.

A framework that explains hierarchical modeling of dropwise condensation in terms of the processes involved is depicted in Fig. 6.2. The atomistic model, which relies on population balance, is the starting point for determining the size of the smallest stable drop. The nucleation sites are randomly distributed on the substrate, and all the sites are initially occupied by the drops of smallest possible radius—namely, the maximum size of a stable cluster in the atomistic model. The subsequent

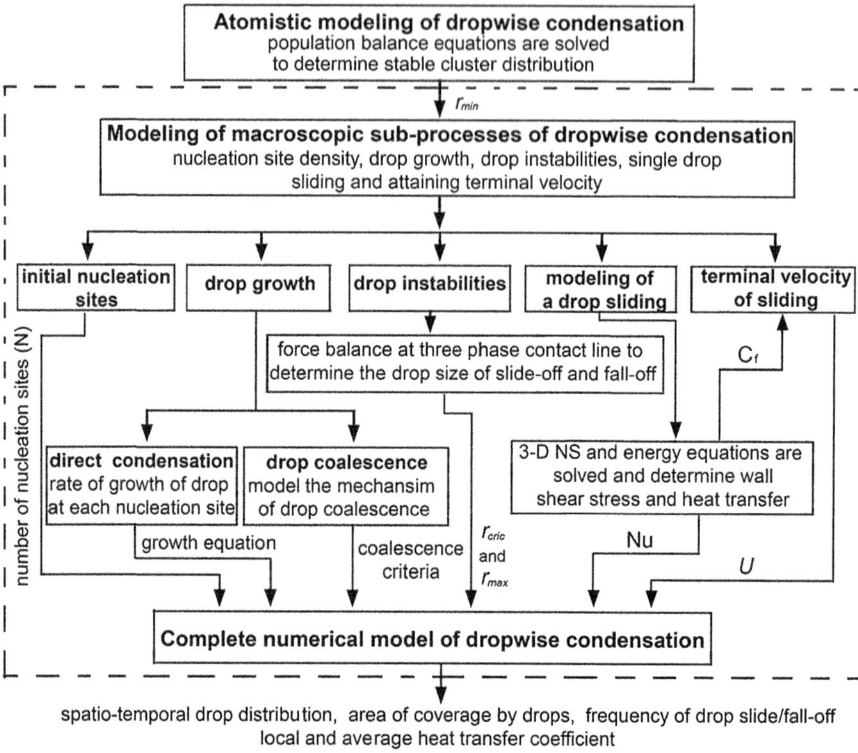

Fig. 6.2 Schematic diagram of hierarchical modeling in dropwise condensation

steps that follow are growth by direct condensation, coalescence, instability, drop motion, and computation of transport coefficients for sliding drops. The model yields the instantaneous drop size distribution, instantaneous rate of growth of drops, area of coverage by drops, frequency of drop slide/fall-off, and local and average heat transfer coefficient over inclined surfaces. The simulation discussed here is confined to condensation underneath cold inclined substrates forming pendant drops.

6.2 Drop Formation at Atomistic Scale

Condensation in the form of discrete drops can be homogeneous, namely distributed in the vapor phase, or heterogeneous, as in the presence of a cooler solid substrate. It is now accepted that phase change, whether homogeneous or heterogeneous, is induced by nucleation, triggered by molecular clustering. In view of experimental limitations, the physical picture, right at nucleation, is not very clear. From a heat transfer viewpoint, an important question is—how do drops form, grow, and get mobilized over a textured solid surface. At the atomic level, vapor atoms may

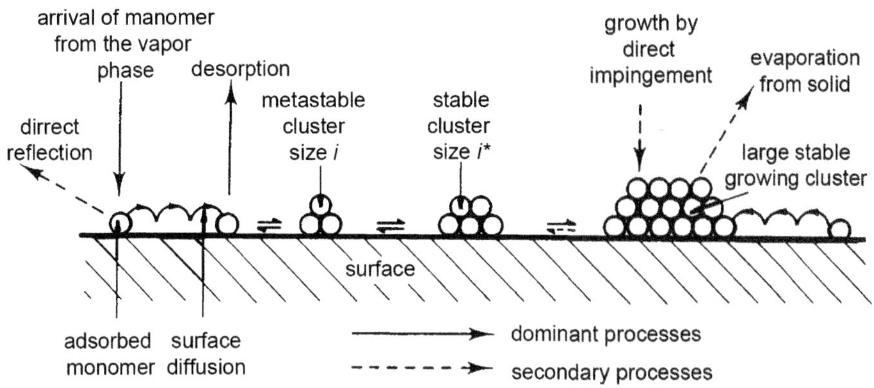

Fig. 6.3 Processes involved in deposition of condensate atoms on a cold substrate (Bentley and Hands 1978)

impinge on the surface with a direct velocity, or alternatively, the vapor may be quiescent. Individual quiescent vapor atoms may form stable clusters by combining with neighbors and grow on the substrate with time by losing their latent heat.

Bentley and Hands (1978) reported various processes occurring at the atomic scale (Fig. 6.3), from arrival of monomers to formation of stable clusters on the cold substrate. The surface adatoms undergo a sequence of processes such as adsorption, diffusion, reflection, agglomeration, transfer of energy, and formation of stable clusters, eventually manifesting as a distribution of condensed liquid nuclei. Atoms/molecules bound to the surface form an adatom and a group of adatoms leads to a cluster. Although it is possible to form clusters in the vapor phase before they get deposited on the surface, with large substrate subcooling, one can expect all the condensation to occur at the surface leading to heterogeneous condensation.

Lee and Maa (1991) observed the mechanism of vapor deposition by using an electron microscope. The process was composed of adsorption of the vapor molecules on the substrate, surface diffusion, growth, and coalescence of the deposited clusters.

Hashimoto and Kotake (1995) reported that molecules approaching the cooled wall have higher energies than the departing molecules that have transferred their energy to the wall. The energy exchange between the incoming and departing molecules at distinct temperatures provides a sufficient condition for molecular clustering (Fig. 6.4). The cluster size formed increases near the wall, and the thickness of the cluster zone depends on the thermal condition of the molecular system and the processes of energy transfer.

Kotake (1998) reported that the existence of clusters depends on the condition of energy transfer between molecules of the vapor to the cold substrate. The rate of condensation depends on the relative strengths of three intermolecular attractions: (a) the energy of attraction between two adsorbed molecules, (b) the adsorption energy on the substrate, and (c) the adsorption energy of a vapor molecule on an adsorbed layer of its own species.

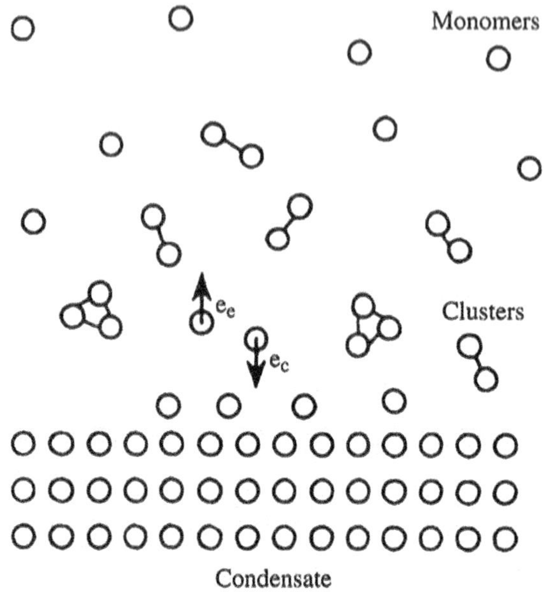

Fig. 6.4 Cluster formation in the vicinity of the condensate or a cooled wall (Kotake 1998): evaporating molecules have lower energy (ee) than condensing molecules (ec); ee < ec is a favorable condition for cluster formation

Peng et al. (2000) reported that clustering of molecules on a cold substrate is like reaction kinetics with appropriate reaction rates. The authors reported that the driving force for the coalescence of clusters results from the tendency of minimization of surface energy. The transport of mass occurs via the routes of evaporation–condensation and surface diffusion.

McCoy (2000) presented a theory based on cluster distribution kinetics for single monomer addition and dissociation. Population balance equation was used to describe the dynamics of cluster mass distribution during homogeneous and heterogeneous nucleation in unsteady closed and steady flow systems. The distribution kinetics approach was based on the recognition that nucleation and growth from vapor led to droplets larger than the nuclei and distributed in mass. Cluster growth by addition-dissociation was found to be like polymerization-depolymerization reaction. It was shown that heterogeneous nucleation preserves the number of clusters, equal to the nucleation sites.

Wang et al. (2003) proposed an idea of critical aggregation concentration of active molecules to describe the moment just before nucleus formation. Tian et al. (2004) studied the aggregation of active molecules inside a metastable bulk-phase using thermodynamics. The authors derived an expression for the critical aggregation concentration, energy distribution of active molecules inside the bulk-phase at superheated and supercooled limits and used the molecular aggregation theory to describe the gas-liquid phase-transition process.

Song et al. (2009) suggested that steam molecules become clusters prior to condensation on the cooled surface (Fig. 6.5). The authors argued that clustering begins in the vapor phase itself close to the cold wall. Clusters formed closer to the wall are larger than those formed in the bulk vapor phase.

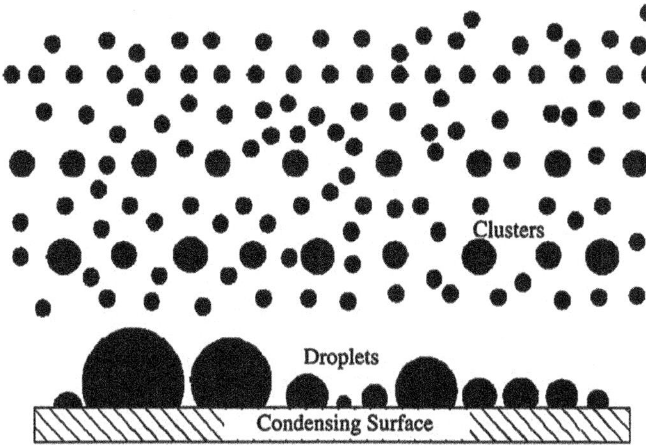

Fig. 6.5 Physical model of vapor condensation proposed by Song et al. (2009)

Based on the available literature, it can be concluded that drop formation during condensation commences with the impingement of vapor atoms on or underneath a cold substrate. Alternatively, vapor may be quiescent. The individual quiescent atoms may form stable clusters by combining with the neighbors and grow on the surface with time (Fig. 6.6a). An atom/molecule bound to the surface is an adatom and a group of adatoms leads to a cluster (Fig. 6.6b). It is also possible to form clusters in the vapor phase before they get deposited on the surface. With large substrate subcooling, one can expect all condensation to occur at the surface level. The stability of the cluster depends on mutual energy interactions between the cluster, the atoms of the surrounding vapor and the cold wall. Molecules/atoms approaching the cold wall have a higher temperature than departing molecules/atoms that have transferred their energy to the wall. This energy difference determines whether a given cluster clinging to the surface will be stable, grow with time, or diminish in size. Many stable clusters growing together may form an atomic/molecular monolayer of condensate on the substrate (Fig. 6.6b, c).

There are at least two possibilities of drop formation (dewetting, Fig. 6.6c, d). In the first model, it is postulated that the condensation initially occurs in a filmwise manner, forming an extremely thin film on the solid surface (Fig. 6.6c(i)). As condensation continues, this film ruptures due to intrinsic interfacial instabilities and distinct drops are formed (Fig. 6.6d(i)). The second theory is based on the premise that drop formation is a heterogeneous nucleation process (Fig. 6.6c(ii)). Here, stable clusters get located at specific nucleation sites over the substrate, such as pits and grooves, grow in the continuum domain, while the portion of the surface between growing drops essentially remains dry. Small droplets are formed by direct condensation via nucleation at locations with local minima of the free surface energy (Fig. 6.6d(ii)). Hence, the processes such as molecular potential, adatoms dynamics, cluster dynamics, surface diffusion, stable cluster size, nucleation density, film stability and rupture, topography interaction, and stable cluster formation appear

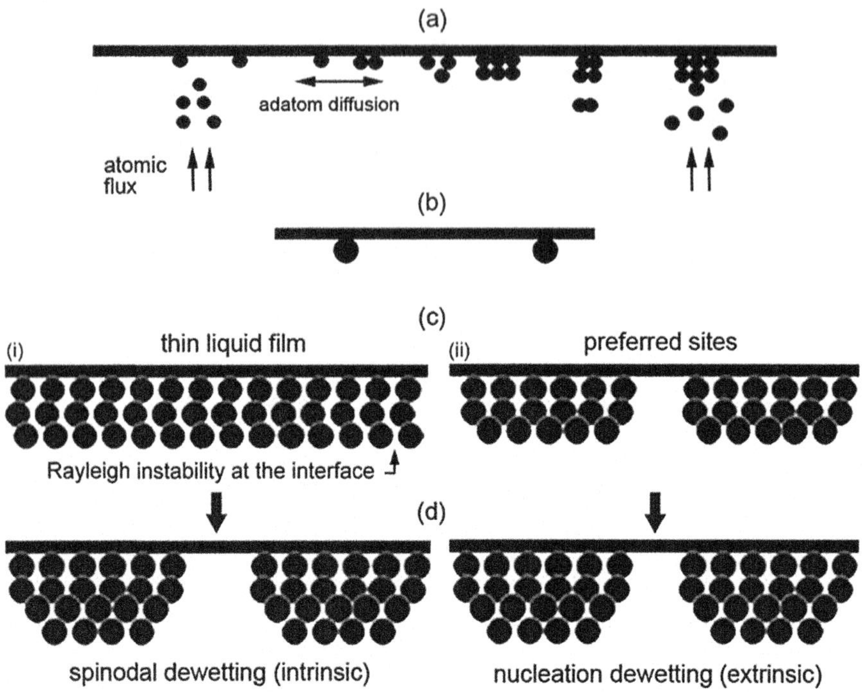

Fig. 6.6 Mechanism of liquid drop formation underneath a cold substrate. From angstroms to nanometers, individual vapor molecules come closer, a system of adatoms form and a group of adatoms leads to a cluster. Many growing clusters together may form a molecular monolayer of liquid. At this stage, there are at least two possibilities: droplet formation (dewetting) and film formation. The liquid film ruptures and forms droplets. (**a**) Atoms, adatoms, and unstable cluster, (**b**) stable cluster, (**c**) stable cluster growth with two possibilities - thin liquid film and droplet formation at preferred sites, (**d**) drop formation

as condensation proceeds from the atomistic scale to the microscale. From the heat transfer point-of-view, control of these processes is of critical concern in atomistic modeling of dropwise condensation.

6.3 Atomistic Modeling of Dropwise Condensation

The basic aim of atomistic modeling of dropwise condensation is to determine the size of the stable cluster at nucleation and connect phenomena occurring at the atomic scale to the macroscale. Formation of drops during condensation commences with the impingement of vapor atoms on a cold substrate kept at a temperature below saturation. Atoms approaching the cold wall have higher energies than departing atoms and hence transfer energy to the wall. The energy exchange between incoming and departing atoms of different energies provides a sufficient condition for molecular clustering.

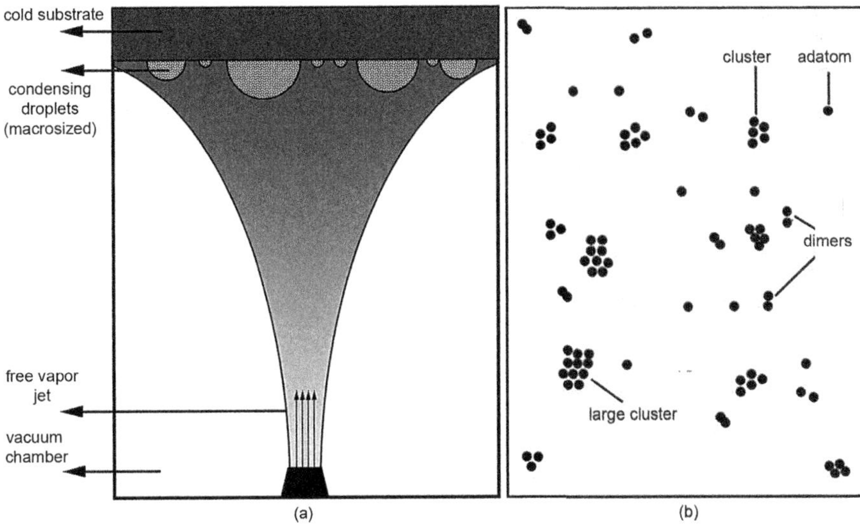

Fig. 6.7 Physical modeling of droplet formation underneath a substrate. (**a**) Schematic representation of the vapor flux impinging vertically on the underside of a horizontal substrate. (**b**) Schematic drawing of the distribution of clusters on the substrate

The vapor mass flux F is obtained in the form of an over-expanded jet from a nozzle discharging into an evacuated chamber (Fig. 6.7a). When the vapor is stationary, the mass flux is set to zero. The substrate on which all the condensation takes place is initially clean and free of any condensate. Atoms are deposited on the substrate at a constant rate. An adsorbed layer of atoms, called adatoms, is first formed prior to nucleation (Fig. 6.7b). These adatoms can diffuse on the surface with a characteristic time period that is the mean resident time (τ_{ads}) and then re-evaporate back to the vapor phase. They may collide with other adatoms or clusters during their migration, thus causing nucleation to be initiated. The adatom population on the substrate changes with time due to desorption, capture, or release of an adatom by a cluster. The population may redistribute itself over the surface as a result of diffusion at a speed determined by the diffusion coefficient. If two adatoms occupy neighboring sites, they will stick to form a cluster. More adatoms may be captured by a cluster or two clusters may combine to form large clusters. The population of cluster of a certain size will thus change due to adatom capture or release, coalescence with other clusters, or breakage into smaller clusters and desorption. In the growth stages, the condensate clusters grow, not only by capturing adatoms on the surface, but also by direct capture of impinging vapor molecules/atoms. The residence time is taken to be large enough so that enough time is available for all the adatoms existing in vapor phase to lose their latent heat and get condensed.

When the temperature of the substrate is significantly lower than the saturation temperature, condensation will be complete, in the sense that all the atoms contained in the vapor phase stick to the substrate. Under these conditions, the following additional assumptions facilitate the computation of cluster densities:

Adatoms alone diffuse while dimers and larger clusters are stable, namely they do not disintegrate or diffuse within the substrate.

Direct impingement of free atoms on adatoms and clusters, and the coalescence of clusters can be neglected. Thus, the atoms and clusters diffusing within the substrate arise exclusively from the condensate and do not have contributions to their population from the vapor phase.

6.3.1 Mathematical Model

With the approximation discussed above, the rate equations of the atomistic model (Brune 1998; Amar et al. 1999; Venables 2000; Oura et al. 2003) reduce to

$$\frac{dn_1}{dt} = F - \frac{n_1}{\tau_{\text{ads}}} + \left(-2\sigma_1 \cdot D \cdot n_1^2 - n_1 \cdot \sigma_x \cdot D \cdot n_x\right) \tag{6.1}$$

$$\frac{dn_j}{dt} = n_1 \cdot \sigma_{j-1} \cdot D \cdot n_{j-1} - n_1 \cdot \sigma_j \cdot D \cdot n_j \quad \text{for } j = 2 \text{ to } 1000 \tag{6.2}$$

$$\frac{dn_x}{dt} = n_1 \sigma_i \cdot D \cdot n_i \tag{6.3}$$

Complete information on the local distribution of clusters is contained in the capture and decay rates, σ_j and δ_j respectively. In the present study, these quantities are given parameters. The capture coefficients are nearly constant with $\sigma_1 = 3$ and $\sigma_x = 7$. A first principles calculation of these parameters involves solving a Helmholtz type diffusion equation for clusters in two dimensions in the presence of a certain density of stable islands. The analytical expressions obtained with this approach are (Brune 1998; Venables 2000)

$$\sigma_x = \frac{4\pi(1-z)}{\ln(1/Z) - (3-Z)(1-Z)/2} \tag{6.4}$$

$$\sigma_1 = 4\pi(1-n_1)\frac{n_x}{n_1}\frac{1}{\ln(1/Z) - (3-Z)(1-Z)/2} \tag{6.5}$$

Here, $Z = \vartheta - \sum_{j=1}^{i} n_j$ is the fraction of the surface covered by the stable clusters, and ϑ is the total coverage area. Using constant values of σ_1 and σ_x one can obtain the island size distribution for a specified value of i. For the present discussion, it is assumed that dimers as well as clusters with three or more atoms are stable; consequently, the decay constants $\delta_j (j \geq 2)$ are effectively zero. The assumption is equivalent to stating that clusters that are held together by the long-range van der Waal forces do not have any intrinsic break-up mechanism. The long-range forces appear over length scales of a few nanometers while repulsive forces become significant over considerably shorter length scales of a few angstroms. Thus, number densities of clusters change purely because of the addition of monomers.

The condition that complete condensation of the impinging vapor takes place is equivalent to the inequality $\sigma_x \cdot n_x \cdot D \cdot \tau_{\text{ads}} \gg 1$. It neglects the effect of re-evaporation (Venables 2000). For the complete condensation regime modeled here, the mean residence time τ_{ads} is high. It was found that the model predictions reported in the present study were not sensitive to changes in this quantity for $\tau_{\text{ads}} \geq 0.1$ s.

6.3.2 Numerical Methodology

Numerical simulation of Eqs. (6.1)–(6.3) was run for a large set of cluster sizes varying from adatoms (cluster containing one atom/molecule) to clusters containing 1000 atoms/molecules. The largest cluster with a non-zero number density was found from simulation to have 100–200 atoms/molecules. Hence, the choice of a cluster with 1000 atoms as an upper limit was considered adequate.

The initial conditions were specified during simulation as $n_j(t = 0) = 0$ (for $j = 1$ to 1000) and $n_x(t = 0) = 0$. The model parameters were taken as $\sigma_1 = 3$ (for $j = 1$), $\sigma_j = 7$ (for $j = 2$ to 1000) and $\sigma_x = 7$.

Brune (1998) has showed that the values of the capture coefficients specified above give meaningful results; the corresponding computational effort is also lower since they need not be repeatedly calculated from Eqs. (6.4) and (6.5). A vapor flux of $F = 0.005\ \text{s}^{-1}$ has been adopted for the study. The diffusion constant D was calculated with the ratio (D/F) taking on the values of 10^5, 10^6, 10^7. The residence time of $\tau_{\text{ads}} = 2.3$ s was chosen from numerical experiments to model the complete nucleation regime.

Equation (6.1)–(6.3) constitute a system of 1001 coupled ODEs. The fourth-order Runge-Kutta method was implemented in a C++ language program to solve the system of simultaneous differential equations. The model and the computer program were validated against the benchmark results and are presented next.

6.3.3 Validation

The validation of cluster growth simulation is discussed here. The number of islands/clusters of size s can be expressed in terms of the scaling function (Bartelt et al. 1993; Ratsch and Zangwill 1994; Stroscio and Pierce 1994; Brune et al. 1999; Ratsch and Venables 2003; Shi et al. 2005).

$$n_s(\vartheta) = \frac{\vartheta \cdot f_i(s/S)}{S^2} \tag{6.6}$$

where the symbol n_s is the number of islands of size s at coverage, ϑ,

$$\vartheta = \sum_{s \geq 1} s \cdot n_s \tag{6.7}$$

The average island size is

$$S = \frac{\sum s \cdot n_s}{\sum n_s} \tag{6.8}$$

The quantity $f_i(s/S)$ is the scaling function for the island size distribution corresponding when the critical sized island is equal to i.

The variation of the scaled island size distribution with the scaled island size is reported by Shi et al. (2005). A comparison of the data generated in the present work against Shi et al. (2005) is shown in Fig. 6.8. A close match between the two is obtained. The variation of monomer density and saturation island density with coverage in Fig. 6.9 also show a good match.

6.3.4 Parametric Study with Atomistic Model

After validation, a parametric study has been carried out for studying the variation in monomer density, saturation island density, and density of stable clusters with respect to parameters D, F, and τ_{ads}. Simulation is conducted for the limiting case of zero flux deposition rate ($F = 0$). The results plotted in Fig. 6.10 show that the initial spike in the number density distribution vanishes when the deposition rate is zero. The number density distribution of the condensing clusters on the substrate as a function of the model parameters D, F, and τ_{ads} is shown in Fig. 6.11a, b. The first peak at the origin of the coordinate system corresponds to single adatoms originating from the impingement of the vapor flux. The second peak indicates the most probable cluster size of the condensate. The tail of the distribution shows that sizes beyond a certain value do not appear on the substrate. The size distribution determined from Eqs. (6.1)–(6.3) is purely from microscopic considerations and does not include macroscopic influences such as surface tension and gravity. Hence, the largest cluster, corresponding to the smallest number density in Fig. 6.11a, b can be interpreted as the smallest drop that would appear on a macroscopic viewpoint. Beyond this size, factors such as gravity, surface tension, and mutual coalescence would be operative in determining the increase in the drop diameter.

The preceding expectation has been examined with reference to the thermodynamic estimate (Eq. 6.24) as follows. At atmospheric pressure and a surface

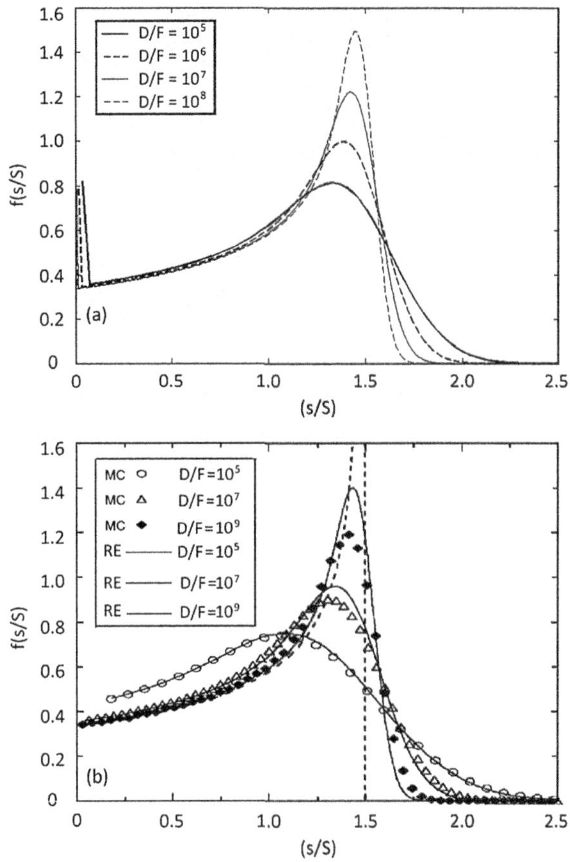

Fig. 6.8 (**a**) Numerical simulation of the rate equations governing the nucleation process. (**b**) The results given by Shi et al. (2005) by using both the rate equations (RE) approach (solid lines) and Monte Carlo (MC) simulation (symbols) are presented here

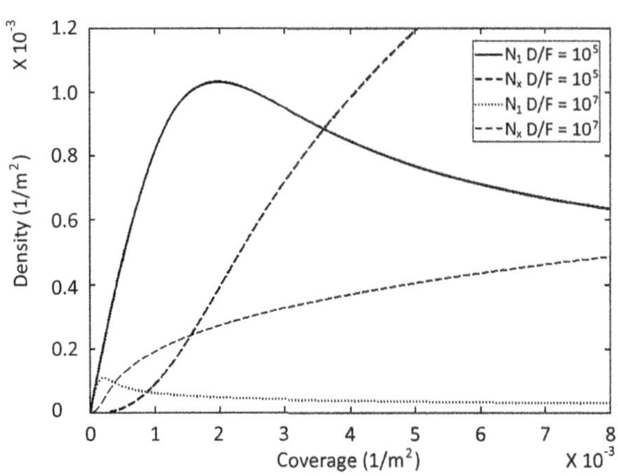

Fig. 6.9 Variation of monomer density (n_1) and saturation island density (n_x) with coverage at various values of D/F

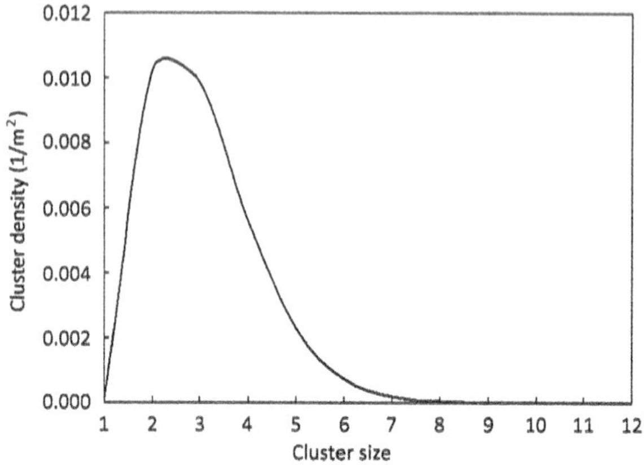

Fig. 6.10 Variation of the number density of clusters with their size

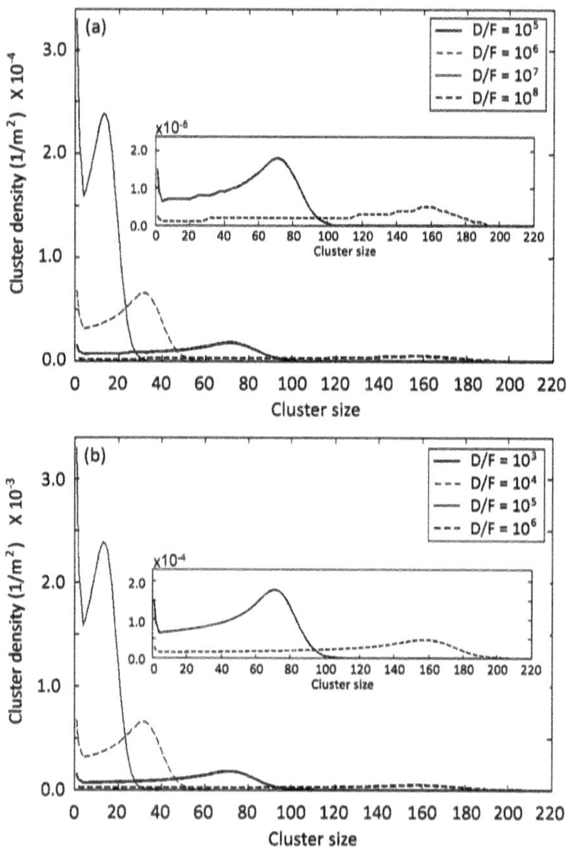

Fig. 6.11 (**a**) Variation of cluster density with cluster size at $F = 0.005\ \mathrm{s}^{-1}$. The cluster size where the number density becomes zero yields the maximum cluster size. Inset shows the details of the island density profiles for $D/F = 10^7$ and 10^8. (**b**) Variation of the cluster density with cluster size at $F = 0.05\ \mathrm{s}^{-1}$. Inset shows the details of the island density profiles for $D/F = 10^5$ and 10^6

maintained at 80 °C, one can calculate $r_{\min} = 9.617 \times 10^{-10}$ m for water. The number of molecules in the drop can be found from

$$n_{\mathrm{d}} = \frac{N_{\mathrm{A}}\pi \cdot r^{3}}{3M \cdot v_{\mathrm{l}}}\left(2 - 3\cos\theta + \cos^{3}\theta\right) \tag{6.9}$$

Using properties of water, namely molecular weight $\overline{M}$ of 18 g/mol, N_{A}, the Avogadro number and $\theta = 90°$

$$n_{\mathrm{d}} = \left(2\pi \cdot r^{3} N_{\mathrm{A}}\right) / \left(3\overline{M} \cdot v_{\mathrm{l}}\right) \tag{6.10}$$

The volume referred to in Eqs. (6.9) and (6.10) is that of the spherical cap of a droplet whose radius is $r_{\min}$ (Eq. 6.24) and contact angle is θ. The number of molecules corresponding to the minimum radius of 9.617×10^{-10} m can now be estimated as $n_{\mathrm{d}} = 60$. In the cluster model, the results were

$$D = 5000 \text{ and } F = 0.005, \quad n_{\mathrm{d}} = 53$$
$$D = 500 \text{ and } F = 0.05, \quad n_{\mathrm{d}} = 58$$
$$D = 50 \text{ and } F = 0.5, \quad n_{\mathrm{d}} = 62$$

The number of molecules thus calculated in the smallest drop corresponds quite well to the data of Fig. 6.8.

The sensitivity of the drop size to the diffusion parameter D and the impinging flux F are shown in Fig. 6.12a, b. The minimum drop size is seen to increase with D as well as F though the change is not substantial. For an increase of four orders of magnitude in the diffusion coefficient, the minimum drop radius increases by a factor of about two. For an increase of one order of magnitude in the vapor flux, the minimum drop radius increases by about 30%. These changes are related to the slight broadening of the cluster density, and hence there is an increase in the size of the largest possible cluster. A higher mass flux increases the number density of adatoms over the substrate and consequently diminishes the extent of diffusion away from the clusters. A higher diffusion constant encourages the association process of monomers and permits the clusters of larger sizes to form. Both the factors lead to an increase in the number of molecules in the largest cluster and hence, in the minimum drop radius.

Apart from the material properties of the condensing medium, the diffusion coefficient is a function of the surface properties and temperature of the substrate. The vapor flux is a process parameter and can be independently controlled. The cluster model given by Eqs. (6.1)–(6.3) predicts that by varying D, in effect varying the surface properties, the minimum drop radius is altered. One method available for altering the surface characteristics is physical texturing. As discussed by Chen et al. (1996),

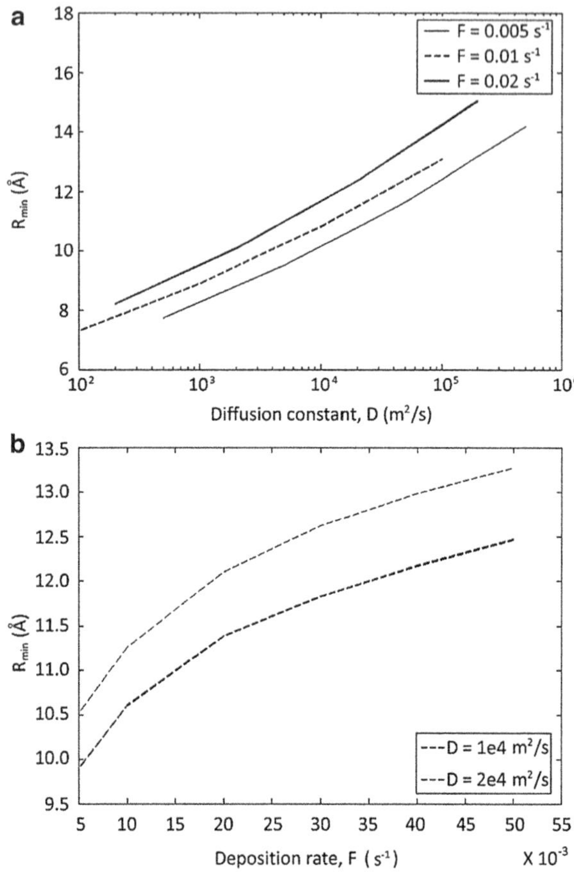

Fig. 6.12 (**a**) Variation of the minimum drop radius with diffusion constant D at three different deposition rate F. (**b**) Variation of the minimum drop radius with deposition rate F at two different values of diffusion constant D

$$D \propto 1/\overline{\eta} \tag{6.11}$$

The symbol $\overline{\eta}$ represents friction coefficient of the surface. When $\overline{\eta}$ is small,

$$D \propto 1/\overline{\eta}^{0.5} \tag{6.12}$$

While texturing decreases the friction coefficient, the diffusion coefficient increases, with a corresponding increase in the minimum drop diameter (Chen et al. 1996, 1999). The increase is, however, marginal as shown in Fig. 6.11a. For chemical texturing of a surface, first principles calculations can be used to predict the diffusion constant (Bloch et al. 1993; Ratsch et al. 1997).

The sensitivity of the drop size distribution on the macroscale to the initial minimum drop minimum radius is examined in Fig. 6.13. Here, the question of special interest is whether the drop size distribution can be influenced by controlling the minimum drop radius. To answer this question, two different $r_{\min}$ values were

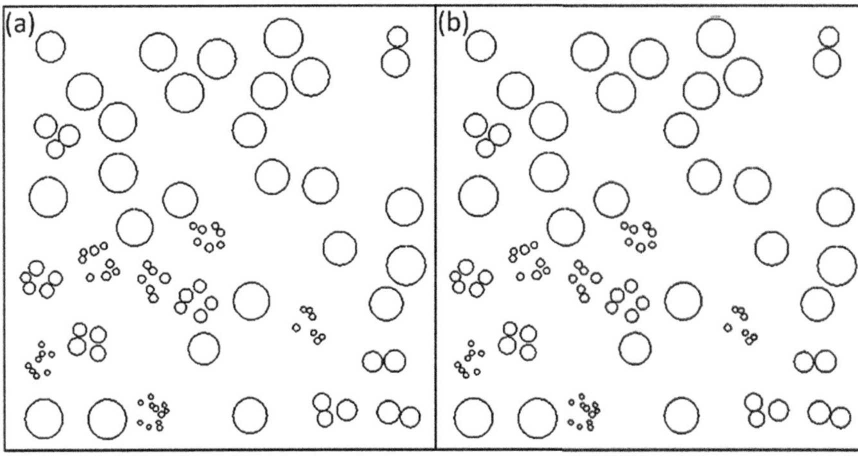

Fig. 6.13 Drop size distribution on a surface for (**a**) $r_{\mathrm{min}} = 10$ Å and $r_{\mathrm{max}} = 5$ mm, (**b**) $r_{\mathrm{min}} = 100$ Å, and $r_{\mathrm{max}} = 5$ mm during condensation of water vapor

started with, and droplet growth simulation was carried out till drops were large enough for fall-off. The two distribution patterns which emerge are practically identical, suggesting that the macroscale drop distribution is determined by coalescence dynamics, rather than the minimum drop radius.

6.4 Macroscopic Modeling of Dropwise Condensation

Dropwise condensation at the macroscale is a consequence of the time-dependent sub-processes associated with the formation of drops at nucleation site, growth by direct condensation and coalescence, sliding motion, fall-off, and then by renucleation on or underneath the substrate. It is a complex intricately linked phenomenon. A mathematical model of these sub-processes is required to describe the entire dropwise condensation process.

Atomistic modeling of drop formation reveals that the maximum stable cluster obtained by atomistic nucleation process is equal to the size of the minimum stable radius obtained from thermodynamics consideration. Simulations show that condensation patterns at longer timescales are not sensitive to the atomic level processes that fix the minimum drop radius. Therefore, atomic level modeling of condensation is dispensed with and drops formed at the initial nucleation sites are directly assigned the minimum possible stable radius from thermodynamic considerations. The expression for the minimum radius is derived first.

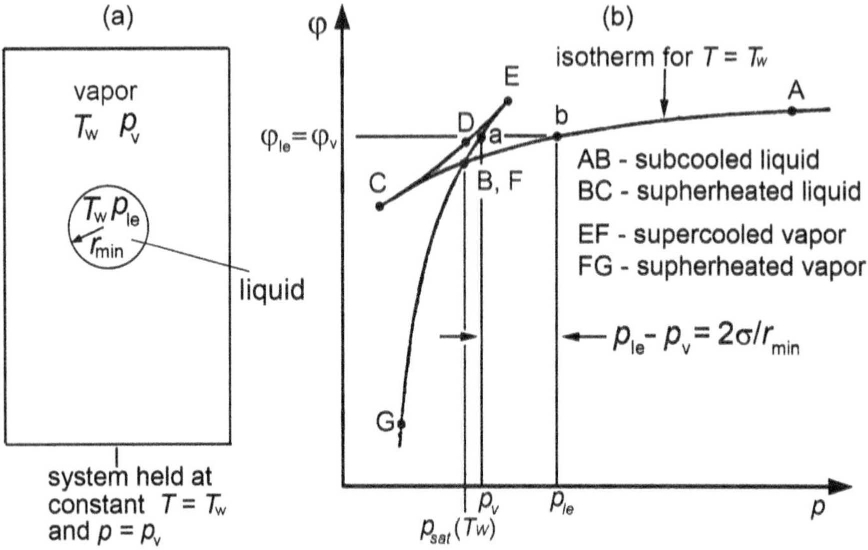

Fig. 6.14 (**a**) System considered in the analysis of the smallest possible stable droplet. (**b**) Liquid is in equilibrium with the surrounding vapor

6.4.1 Determination of Minimum Droplet Radius

Consider a system (Fig. 6.14a), containing a liquid droplet of radius $r_{\min}$ in equilibrium with supersaturated vapor held at constant temperature (T_w) and pressure (p_v). Vapor is supercooled or in a supersaturated state without a phase transformation since condensation transfers heat to the adjacent wall. The vapor temperature is equal to the condensing wall temperature and the saturation temperature (T_s) corresponding to vapor pressure (p_v) is higher than T_w. The liquid and the vapor state for a liquid droplet in equilibrium with surrounding vapor in a phase diagram. "EF" is supercooled, and "FG" is superheated state of vapor at a given wall temperature (Fig. 6.14b). Similarly, "AB" is supercooled, and "BC" is the superheated state of liquid. At equilibrium, temperature and chemical potential (φ) in the vapor and droplet must be the equal

$$\varphi_{ve} = \varphi_{le} \tag{6.13}$$

The pressure in the two phases (liquid and vapor) are related through the Young-Laplace equation

$$p_{\mathrm{le}} = p_{\mathrm{v}} + \frac{2\sigma}{r_{\min}} \tag{6.14}$$

The chemical potential of vapor and liquid phases at equilibrium is evaluated by using the integrated form of the Gibbs-Duhem equation for a constant temperature process

$$\varphi - \varphi_{\mathrm{sat}} = \int_{P_{\mathrm{sat}}}^{p} v\,\mathrm{d}p \tag{6.15}$$

We evaluate the integral on the right side (Eq. 6.15) using the ideal gas law ($v = RT_{\mathrm{w}}/p$) for vapor. Therefore, the chemical potential of the vapor phase is

$$\varphi_{\mathrm{ve}} = \varphi_{\mathrm{sat,v}} + RT_{\mathrm{w}} \ln\left[\frac{p_{\mathrm{v}}}{p_{\mathrm{sat}}(T_{\mathrm{w}})}\right] \tag{6.16}$$

For the liquid phase inside the droplet, the chemical potential can again be evaluated using Eq. (6.15). The liquid is taken to be incompressible, with v equal to the value for saturated liquid at T_{w}. With this assumption, the chemical potential of liquid phase is

$$\varphi_{\mathrm{le}} = \varphi_{\mathrm{sat,l}} + v_{\mathrm{l}}[p_{\mathrm{le}} - p_{\mathrm{sat}}(T_{\mathrm{w}})] \tag{6.17}$$

Equating the values of φ_{ve} and φ_{le} given by Eqs. (6.16) and (6.17) to satisfy Eq. (6.13) and using the fact $\varphi_{\mathrm{sat,\ v}} = \varphi_{\mathrm{sat,l}}$

$$p_{\mathrm{v}} = P_{\mathrm{sat}}(T_{\mathrm{w}}) \exp\left\{\frac{v_{\mathrm{l}}[p_{\mathrm{le}} - P_{\mathrm{sat}}(T_{\mathrm{w}})]}{RT_{\mathrm{w}}}\right\} \tag{6.18}$$

As seen in Fig. 6.14b, if the vapor state point is on the metastable supercooled vapor curve at point a, the liquid state corresponding to equal φ must lie on the subcooled liquid line at point b. For the liquid droplet with finite radius, equilibrium can be achieved only if the liquid is subcooled and the vapor is supersaturated relative to its normal saturation state for a flat interface. Equation (6.18) indicates that if p_{v} is greater than $p_{\mathrm{sat}}(T_{\mathrm{w}})$, then p_{le} must also be greater than $p_{\mathrm{sat}}(T_{\mathrm{w}})$, consistent with the state points in Fig. 6.14b. Substituting Eq. (6.14) to eliminate p_{le}, Eq. (6.18) becomes

$$p_{\mathrm{v}} = P_{\mathrm{sat}}(T_{\mathrm{w}}) \exp\left\{\frac{v_{\mathrm{l}}[p_{\mathrm{v}} - P_{\mathrm{sat}}(T_{\mathrm{w}}) + 2\sigma/r_{\min}]}{RT_{\mathrm{w}}}\right\} \tag{6.19}$$

In most instances, the steep slope of the subcooled vapor line in Fig. 6.14b results in values of p_{v} that are much closer to $p_{\mathrm{sat}}(T_{\mathrm{w}})$ than p_{le}. Therefore, $p_{\mathrm{v}} - p_{\mathrm{sat}}(T_{\mathrm{w}}) \ll 2\sigma/r_{\min}$ and Eq. (6.19) is well approximated by

$$r_{\min} = \frac{2\sigma}{(RT_{\mathrm{w}}/v_{\mathrm{l}}) \ln\left[\frac{p}{p_{\mathrm{sat}}(T_{\mathrm{w}})}\right]} \tag{6.20}$$

The Clapeyron equation is combined with the ideal gas law of vapor to obtain

$$\frac{\mathrm{d}p}{\mathrm{d}T} = \frac{ph_{\mathrm{lv}}}{RT_{\mathrm{w}}^2} \tag{6.21}$$

Integrating Eq. (6.21) between the p_{v} and p_{sat} and rearranging

$$\int_{P_{\mathrm{sat}}(T_{\mathrm{w}})}^{p_{\mathrm{v}}} \frac{\mathrm{d}p}{p} = \frac{h_{\mathrm{lv}}}{RT_{\mathrm{w}}^2} \int_{T_{\mathrm{w}}}^{T_{\mathrm{sat}}} \mathrm{d}T \tag{6.22}$$

$$\ln\left[\frac{p_{\mathrm{v}}}{P_{\mathrm{sat}}(T_{\mathrm{w}})}\right] = \frac{h_{\mathrm{lv}}}{RT_{\mathrm{w}}^2}(T_{\mathrm{sat}} - T_{\mathrm{w}}) \tag{6.23}$$

Substituting Eq. (6.23) in Eq. (6.20) yields

$$r_{\min} = \frac{2\sigma v_{\mathrm{l}} T_{\mathrm{w}}}{h_{\mathrm{lv}}(T_{\mathrm{sat}} - T_{\mathrm{w}})} \tag{6.24}$$

This is the smallest droplet possible corresponding the equilibrium conditions for a specified subcooling of ($T_{\mathrm{sat}} - T_{\mathrm{w}}$).

6.4.2 Nucleation Site Density

The initial, thermodynamically determined drops have a diameter of the order of a few nanometers for fluids encountered in heat transfer applications. Therefore, from an engineering standpoint, it is difficult to experimentally capture the initial nucleation phenomenon on a surface freshly exposed to vapor. Nucleation site density is itself influenced by the thermophysical properties of the condensing fluid, physico-chemical properties of the substrate, degree of subcooling, and the substrate morphology. Thus, it is also difficult to determine the nucleation site density on a substrate, either from theory or from experiments. Leach et al. (2006) reported initial site densities close to 10^6 cm^{-2} for temperature differences in the range of 50–100 °C. For condensation of water at 30 °C, the authors suggested that the initial nucleation site density is in the range of 10^4–10^5 cm^{-2}, and gradually increases to 10^6 cm^{-2} before the first coalescence. Earlier a theoretical expression for nucleation site density (in units of cm^{-2}) over an untreated surface was given by Rose (1976)

$$N = \frac{0.037}{r_{\min}^2} \tag{6.25}$$

where N is the number of sites on the substrate per unit area where the initial drops, identifiable as liquid, are formed. Zhao and Beysens (1996) observed no significant connection between the initial nucleation site density and the wettability of the condensing fluid. Rose (2002) indicated that the parameter N in the range of 10^5–10^6 is close to the experimental data of dropwise condensation underneath a chemically textured substrate. Mu et al. (2008) found that the nucleation density varies with surface topography, the rougher substrate resulting in a higher nucleation density. Based on the work of Rose (1976, 2002) and Mu et al. (2008), one can conclude that the nucleation density is influenced not only by the degree of surface topography but also by the extent of subcooling. Nucleation density might be influenced by these two factors, i.e., changes in surface energy induced by a chemical species (chemical texturing) and varying roughness morphology of the substrate (physical texturing). The modified expression for the nucleation density of a textured substrate can be expressed

$$N_f = fN \tag{6.26}$$

Here, N_f is the nucleation site density of the textured substrate, f is the degree of roughness and N is initial nucleation density of a smooth surface, as calculated by Eq. (6.25), Rose (1976). For a general textured substrate—physical or chemical, factor f needs to be established and is a topic of research.

6.4.3 Nucleation Site Distribution

Heterogeneous nucleation is an important process for phase transitions, including the initial droplet formation of dropwise condensation. The initial droplets form only at the natural nucleation sites on the condenser surfaces, and the number of nucleation sites influence the dropwise condensation heat transfer significantly. On the other hand, the number of nucleation sites is directly related to the surface properties. Thus, it is important to study the relationship between surface topography and nucleation number.

Many researchers have investigated the problem of nucleation site density of dropwise condensation. Glicksman and Hunt (1972) numerically simulated nucleation, growth, coalescence, and renucleation of drops ranging in size from the smallest nucleating drops to the departing drops. Their simulated results agreed well with the data of Krischer and Grigull (1971). Wu and Maa (1976) used the population balance model to derive the size distributions for small pre-coalescence drops. Their calculations showed that the nucleation site density was around $N = 2 \times 10^7\ \text{cm}^{-2}$. Graham (1969) and Graham and Griffith (1973) studied the nucleation site density with optical microscope photographs. Their results indicated that the site density was 2×10^8 cm^{-2}. Tanasawa et al. (1974) investigated the nucleation site density with electron microscope photographs. The density exceeded 10^{10} cm^{-2} in these measurements. Rose (1976) computed nucleation densities as 5.9×10^9 and 2.9×10^{11} cm^{-2}, respectively, for minimum nucleation radii r_{min} being 0.07 and 0.01 μm. Leach et al. (2006) reported initial drop densities close to 10^6 cm^{-2} for temperature differences in the range of 50–100 °C. For condensation of water at 30 °C, the initial nucleation site density is in the range of 10^4–10^5 cm^{-2} and gradually increases to 10^6 cm^{-2} before the first coalescence. These numbers from various researchers show that the nucleation site density may not be determined by r_{min} alone.

The differences in the nucleation sites densities may also result from the methods used to evaluate the parameters from the images. There are quite few studies related to the nucleation step of dropwise condensation getting influenced by surface characteristics. McCormick and Westwater (1965) applied an optical microscope and showed that the drops nucleated not only at natural cavities on the condenser surface but also at those produced by needles and by erosion and scratches on the surface. Therefore, surface properties of the material affect nucleation site density. Fractal dimension can be used to describe the irregularity and complexity of a rough surface. Yang et al. (1998) and Wu et al. (2001) showed that the droplet distribution had self-similarity, an important feature of fractal behavior. However, the authors studied only the fractal character of droplet pattern without considering the fractal behavior of the condensation surface. In the present work, nucleation site density has been parametrically varied to gage its sensitivity on the resulting heat transfer rate.

From the viewpoint of the current model, the nucleation sites are randomly distributed over the substrate area by using a random seed generator function in C ++. The function returns a matrix containing pseudo random numbers with a uniform probability density function in the range (0, 1). The distribution of sites over the area proceeds column-wise till all the sites are occupied. Once this distribution is carried out, it remains fixed for a given simulation.

Parameters, including the average contact angle, contact angle hysteresis and the nucleation site density of chemically textured surfaces can be quite different from the physically textured counterparts. These parameters are an input to the condensation model reported in the present study.

Physically textured surfaces are unique in many ways for the following reasons. For a single drop of liquid sitting over a physically patterned surface, multiple droplet configurations are possible, making the determination of the apparent contact angle a challenge. For example, the static drop could exhibit wetting transitions between Cassie state or the Wenzel state (Berthier 2008; Miljkovic et al. 2012); such configurations of droplets over physically textured surfaces can be seen, for example, in Berthier (2008), Ma et al. (2012) and Rykaczewski (2012). However, for a continuous cyclic process of dropwise condensation on randomized hydrophobic textured surfaces encountering an ensemble of drops of various sizes, it may be argued that the bulk behavior of physically and chemically textured surfaces could be comparable, except for some differences in apparent dynamic contact angles and the mobility of the three-phase contact line. These differences should be small when the drop size is large in comparison to the characteristic scale of the surface roughness. As the drop size at criticality is of the order of a few mm, the drop is much larger than the surface features and therefore, the proposed condensation model is expected to be uniformly valid for physically as well as chemically textured surfaces.

6.4.4 Growth by Direct Condensation

Geometric parameters of a drop located underneath textured surfaces of various orientations are estimated. A drop with radius r underneath a horizontal substrate is considered as a part of sphere of contact angle θ (Fig. 6.15a). The contact angle θ on the coated surface is assumed to be constant regardless of the drop size r and the vapor and surface temperature. Therefore, the average contact angle is $\theta_{\text{avg}} = \theta$.

For a horizontal substrate with wettability gradient and inclined substrate without wettability gradient, the drop gets deformed and is not a part of a spherical frustum (Fig. 6.15b, c). In these cases, the geometric parameters of a deformed drop are calculated using the spherical cap approximation. It is assumed that volume and areas of the deformed drop are equivalent to the part of sphere of contact angle θ_{avg}, as shown in Fig. 6.15b, c. There is some ambiguity in the calculation of volume of deformed drop and its experimental validation for sessile drops on an inclined surface.

Dussan and Chow (1983) and Elsherbini and Jacobi (2004a, b) suggested that approximating the drop shape as spherical cap can lead to 10–25% error in volume. Based on experimental evidence, others (Extrand and Kumagai 1995; Dimitrakopoulos and Higdon 1999) believe that such approximation is quite valid for a small drop. As pendant drops tend to be small, the spherical cap approximation is used in the present work.

The drops deform according to the applicable value of wettability gradient underneath the horizontal substrate. Therefore, the θ_{avg} of a given ith drop is

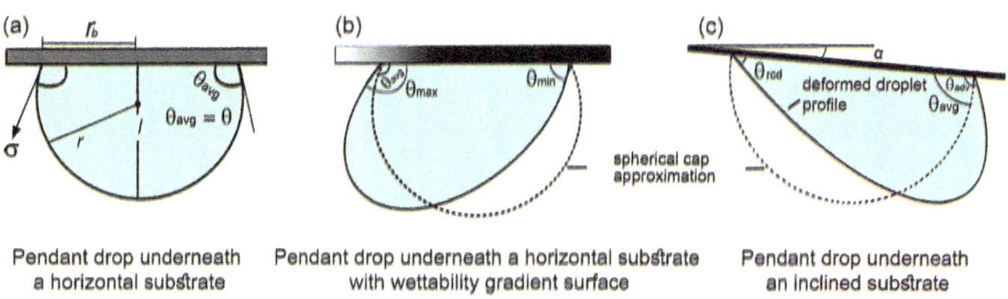

Fig. 6.15 Drop shape underneath (**a**) horizontal substrate, (**b**) horizontal substrate with unidirectional wettability gradient, and (**c**) inclined substrate

$$\left(\theta_{avg}\right)_i = \frac{\left[(\theta_{max})_i + (\theta_{min})_i\right]}{2} \tag{6.27}$$

where the $(\theta_{max})_i$ and $(\theta_{min})_i$ are contact angles at the two sides of the drop (Fig. 6.15b). For a horizontal substrate with wettability gradient, $(\theta_{max})_i$ and $(\theta_{min})_i$ vary according to the drop position.

For an inclined substrate, drops deform according to the advancing angle (θ_{adv}) and receding angle (θ_{rcd}) of the liquid-substrate combination. Hence, the average contact (θ_{avg}) angle is

$$\theta_{avg} = \frac{(\theta_{rcd} + \theta_{adv})}{2} \left(\geq \frac{\pi}{2}\right) \tag{6.28}$$

Here, θ_{rcd} and θ_{adv} are assumed to be constant regardless of the position of drop on the substrate. The drop volume V, area of liquid-vapor interface A_{lv}, maximum drop height from the free surface to wall l, base radius r_b and area of the solid-liquid interface A_{sl} are

$$V = \frac{\pi r^3}{3}\left(2 - 3\cos\theta_{avg} + \cos^3\theta_{avg}\right) \tag{6.29}$$

$$A_{lv} = 2\pi r^2\left(1 - \cos\theta_{avg}\right) \tag{6.30}$$

$$A_{sl} = \pi r^2\left(1 - \cos^2\theta_{avg}\right) \tag{6.31}$$

$$l = r\left(1 - \cos\theta_{avg}\right) \tag{6.32}$$

$$r_b = r\sin\theta_{avg} \tag{6.33}$$

6.4.5 *Temperature Drop due to Various Thermal Resistances*

In the proposed model, condensation occurs only over the free surface of the drops. The latent heat release at the free surface is transferred through the volume of liquid to the cold substrate. The substrate area between drops is inactive with respect to heat transfer.

A drop of contact angle θ_{avg} with radius r underneath a textured substrate, which is coated with a promoter layer of thickness δ, shows various thermal resistances in the path of heat transfer (Fig. 6.16). The rate of condensation on the free surface depends on its ability to transfer latent heat released to the cooler substrate. The following thermal resistances are considered in the model

1. Interfacial resistance (R_{int}) associated with liquid-vapor interface.
2. Capillary resistance (R_{cap}) indicating a loss of driving temperature potential due to droplet interface curvature.

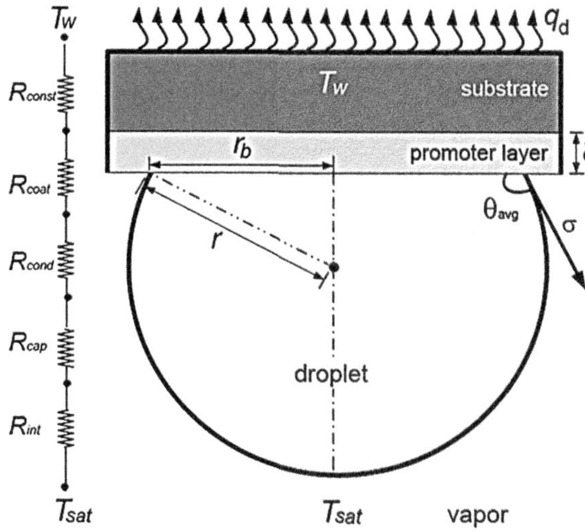

Fig. 6.16 Schematic diagram of a pendant drop with thermal resistances in the droplet growth equation. The promoter layer has a thickness δ, T_{sat} is the vapor saturation temperature, and T_w, the wall temperature

3. Conduction resistance (R_{cond}) associated with the conduction of heat through the droplet.
4. Drop promoter layer resistance (R_{coat}) associated with the thickness of the promoter layer.
5. Constriction resistance (R_{const}) associated with the thermal conductivity of the substrate and non-uniform temperature distribution on the condensing wall due to variable size of drops.

The total temperature difference between the vapor and the substrate ($T_{sat} - T_w$) is the sum of temperature drops due to the individual resistances

$$(T_{sat} - T_w) = \Delta T_{cond} + \Delta T_{int} + \Delta T_{cap} + \Delta T_{coat} + \Delta T_{const} \tag{6.34}$$

The component temperature drops are determined as follows.

Temperature Drop due to Interfacial Resistance

During dropwise condensation, various researchers have observed that there is transport of molecules crossing the liquid-vapor interface in both directions (Carey 2008). When condensation occurs, the flux of vapor molecules joining the liquid must exceed the flux of liquid molecules escaping into the vapor phase. If the temperature of the interface is equal to the saturation temperature, no net condensation must take place. For net condensation to occur, there should be a finite difference between the saturation temperature of vapor, T_s and temperature of liquid-vapor interface, T_{int}. The temperature difference ($\Delta T_{int} = (T_s - T_{int})$) due to film resistance at the vapor-liquid interface is

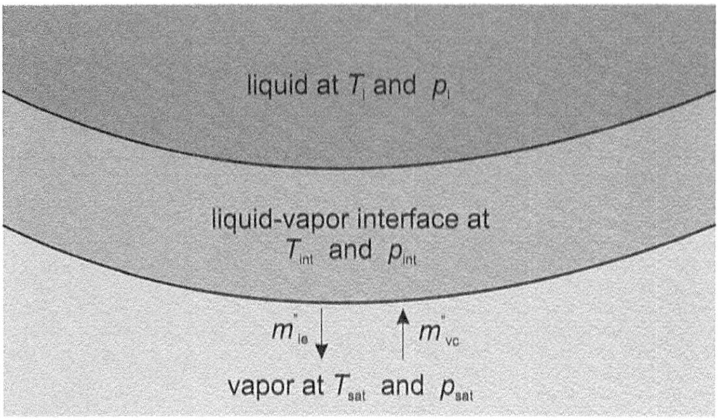

Fig. 6.17 Liquid-vapor interface and mass fluxes at the liquid-vapor interface

$$\Delta T_{\text{int}} = \frac{q_{\text{d}}}{A_{\text{lv}} h_{\text{int}}} \tag{6.35}$$

Therefore,

$$\Delta T_{\text{int}} = \frac{q_{\text{d}}}{2\pi r^2 \left(1 - \cos\theta_{\text{avg}}\right) h_{\text{int}}} \tag{6.36}$$

where h_{int} is the interfacial heat transfer coefficient, which is usually large and strongly depends on the vapor pressure. It is thus possible to transfer thermal energy at high heat flux levels with relatively low driving temperature difference in a phase change process.

To determine h_{int}, we consider the liquid-vapor interface at the molecular level as shown in Fig. 6.17. The motion of vapor molecules in the vicinity of the interface plays a central role in heat flux limitation during the condensation process. According to the kinetic theory of gases, the statistical behavior of vapor at a certain temperature is described by the Maxwellian velocity distribution

$$\frac{\text{d}n_{uvw}}{n} = \left(\frac{m}{2\pi k_{\text{b}} T}\right)^{3/2} \text{e}^{-m\left(u^2+v^2+w^2\right)/2k_{\text{b}}T} \text{d}u\text{d}v\text{d}w \tag{6.37}$$

If the velocity of vapor molecules obeys the Maxwell distribution, the total rate at which molecules passes through the surface A_{lv} per unit mass and per unit area is

$$j_{\text{n}} = \left(\frac{\overline{M}}{2\pi\overline{R}}\right)^{1/2} \frac{p}{mT^{1/2}} \tag{6.38}$$

The mass flux of vapor molecules from the vapor phase that impinge on the surface is

$$m''_{\mathrm{vc}} = m\hat{\sigma}\,\Gamma\, j_{\mathrm{n}} \tag{6.39}$$

$$m''_{\mathrm{vc}} = \hat{\sigma}\Gamma\left(\frac{\overline{M}}{2\pi\overline{R}}\right)^{1/2}\frac{p_{\mathrm{sat}}}{T_{\mathrm{sat}}^{1/2}} \tag{6.40}$$

The term Γ corresponds to the fact that the vapor progresses towards the substrate if net condensation takes place. This progress velocity should be superimposed on the Maxwell velocity distribution.

The net mass flux of vapor molecules in the direction opposite to the substrate is

$$m''_{\mathrm{le}} = m\hat{\sigma}\, j_{\mathrm{n}} \tag{6.41}$$

$$m''_{\mathrm{le}} = \hat{\sigma}\left(\frac{\overline{M}}{2\pi\overline{R}}\right)^{1/2}\frac{p_{\mathrm{int}}}{T_{\mathrm{int}}^{1/2}} \tag{6.42}$$

The net mass flux per unit area m''_{int} condensing at the liquid vapor interface is equal to the difference between m''_{vc} and m''_{le}

$$m''_{\mathrm{int}} = \left(\frac{\overline{M}}{2\pi\overline{R}}\right)^{1/2}\left(\hat{\sigma}\Gamma\frac{p_{\mathrm{sat}}}{T_{\mathrm{sat}}^{1/2}} - \hat{\sigma}\frac{p_{\mathrm{int}}}{T_{\mathrm{int}}^{1/2}}\right) \tag{6.43}$$

where

$$\Gamma = 1 + \frac{m''_{\mathrm{int}}}{p_{\mathrm{v}}\sqrt{2\overline{M}/\pi\overline{R}T_{\mathrm{sat}}}} \tag{6.44}$$

Combining Eqs. (6.43) and (6.44),

$$m''_{\mathrm{int}} = \left(\frac{2\hat{\sigma}}{2-\hat{\sigma}}\right)\left(\frac{\overline{M}}{2\pi\overline{R}}\right)^{1/2}\left(\frac{p_{\mathrm{sat}}}{T_{\mathrm{sat}}^{1/2}} - \frac{p_{\mathrm{int}}}{T_{\mathrm{int}}^{1/2}}\right) \tag{6.45}$$

Equation (6.45) can be put into the following form if $((T_{\mathrm{sat}} - T_{\mathrm{int}})/T_{\mathrm{sat}}) \ll 1$:

$$m''_{\mathrm{int}} = \left(\frac{2\hat{\sigma}}{2-\hat{\sigma}}\right)\left(\frac{\overline{M}}{2\pi\overline{R}T_{\mathrm{sat}}}\right)^{1/2} p_{\mathrm{sat}}\left(\frac{p_{\mathrm{sat}} - p_{\mathrm{int}}}{p_{\mathrm{sat}}} - \frac{T_{\mathrm{sat}} - T_{\mathrm{int}}}{2T_{\mathrm{sat}}}\right) \tag{6.46}$$

When the two terms in the parentheses on the right-hand side of Eq. (6.46) are compared, the first term, $(p_{\mathrm{sat}} - p_{\mathrm{int}})/p_{\mathrm{sat}}$, is usually much larger than the second, $(T_{\mathrm{sat}} - T_{\mathrm{int}})/2T_{\mathrm{sat}}$. Thus Eq. (6.46) is written as

$$m''_{\text{int}} = \left(\frac{2\widehat{\sigma}}{2-\widehat{\sigma}}\right)\left(\frac{\overline{M}}{2\pi\overline{R}T_{\text{sat}}}\right)^{1/2}(p_{\text{sat}} - p_{\text{int}}) \tag{6.47}$$

Further, from the Clausius-Clapeyron relation

$$\frac{p_{\text{sat}} - p_{\text{int}}}{T_{\text{sat}} - T_{\text{int}}} = \frac{\rho_{\text{lv}}h_{\text{lv}}}{T_{\text{sat}}} \tag{6.48}$$

Hence, Eq. (6.47) becomes

$$m''_{\text{int}} = \left(\frac{2\widehat{\sigma}}{2-\widehat{\sigma}}\right)\left(\frac{\overline{M}}{2\pi\overline{R}T_{\text{sat}}}\right)^{1/2}\frac{\rho_{\text{lv}}h_{\text{lv}}(T_{\text{sat}} - T_{\text{int}})}{T_{\text{sat}}} \tag{6.49}$$

Therefore,

$$h_{\text{int}} = \frac{m''_{\text{int}}h_{\text{lv}}}{(T_{\text{sat}} - T_{\text{int}})} \tag{6.50}$$

$$h_{\text{int}} = \left(\frac{2\widehat{\sigma}}{2-\widehat{\sigma}}\right)\frac{h_{\text{lv}}^2}{T_{\text{sat}}v_{\text{lv}}}\left(\frac{\overline{M}}{2\pi\overline{R}T_{\text{sat}}}\right)^{1/2} \tag{6.51}$$

Here, the accommodation coefficient $\widehat{\sigma}$ defines the fraction of the striking vapor molecules that gets condensed on the vapor-liquid interface. The remaining fraction $(1-\widehat{\sigma})$ is the reflection of vapor molecules that strike the interface but do not condense. The accommodation coefficient indirectly measures the interfacial resistance of the liquid-vapor interface to condensation. Higher the accommodation coefficient, lower the interfacial resistance of the liquid-vapor interface of the condensed drop. For liquid ethanol, methanol, alcohol, and water, the reported values of the accommodation coefficient range from 0.02 to 0.04. On the other hand, reported values for benzene and carbon tetrachloride are closer to unity. It has values ranging from 0.37 to 0.61 for condensation of metallic vapor.

Temperature Drop due to Capillary Resistance

As discussed earlier, a pressure difference occurs at the liquid-vapor interface. Therefore, interface temperature is below the saturation temperature of vapor. The depression of the equilibrium interface temperature below the normal saturation temperature for the droplet of radius r can be estimated by replacing $(T_{\text{sat}} - T_{\text{w}})$ by ΔT_{cap} and r_{min} by the radius r in Eq. (6.24). The resulting relation is

$$\Delta T_{\text{cap}} = \left(\frac{2\sigma}{r}\right)\left(\sigma\frac{v_{\text{l}}T_{\text{w}}}{h_{\text{lv}}}\right) = \frac{(T_{\text{sat}} - T_{\text{w}})r_{\text{min}}}{r} \tag{6.52}$$

Temperature Drop due to Conduction Resistance

The drop itself acts as resistance to heat conduction. Accordingly, the conduction resistance through a liquid drop from the wall to liquid-vapor interface is such that the effective temperature drop associated with this resistance is

$$\Delta T_{\text{cond}} = \frac{q_{\text{d}}(l/2)}{A_{\text{lv}}k} \tag{6.53}$$

Substituting Eqs. (6.31) and (6.32) into Eq. (6.53) yields the following relation for the temperature drop due to conduction,

$$\Delta T_{\text{cond}} = \frac{q_{\text{d}}r(1 - \cos\theta_{\text{avg}})}{4\pi r^2 k(1 - \cos\theta_{\text{avg}})} \tag{6.54}$$

Temperature Drop due to Promoter Layer

The temperature drop due to the resistance offered by the coating material on the substrate is

$$\Delta T_{\text{coat}} = \frac{q_{\text{d}}\delta}{k_{\text{coat}}A_{\text{sl}}} \tag{6.55}$$

Substituting Eq. (6.30) into Eq. (6.55) yields the following for the temperature drop:

$$\Delta T_{\text{coat}} = \frac{q_{\text{d}}\delta}{k_{\text{coat}}\pi r^2(1 - \cos^2\theta)} \tag{6.56}$$

Temperature Drop due to Constriction Resistance

It measures the effect of substrate thermal conductivity on dropwise condensation. There has been continuing controversy about whether the thermal conductivity of the condensing surface plays a significant role in determining effective of heat transfer in dropwise condensation. Results of several investigators (Rose 2002;

Bansal et al. 2009) have been interpreted as indicating a strong effect of the thermal conductivity of the substrate on dropwise condensation. But others (Rose 1978a, b; Wilmshurst 1979) show negligible effect of thermal conductivity of the condensing wall. Tsuruta (1993) has indicated an additional thermal resistance due to the non-uniform heat flux distribution over the condensing surface. In the present simulation, the substrate temperature is assumed uniform. Therefore, the temperature drop due to constriction resistance is absent ($\Delta T_{\text{const}} = 0$).

The total temperature drop will balance the total available subcooling and thus

$$\Delta T_{\text{t}} = \Delta T_{\text{cond}} + \Delta T_{\text{int}} + \Delta T_{\text{cap}} + \Delta T_{\text{coat}} = (T_{\text{sat}} - T_{\text{w}}) \tag{6.57}$$

Therefore, the heat transfer rate through a drop of radius r is

$$q_{\text{d}} = (T_{\text{sat}} - T_{\text{w}})\left(1 - \frac{r_{\min}}{r}\right) \times \left(\frac{1}{2\pi r^2 h_{\text{int}}(1 - \cos\theta_{\text{avg}})} + \frac{r(1 - \cos\theta_{\text{avg}})}{4k\pi r^2(1 - \cos\theta_{\text{avg}})} + \frac{\delta}{k_{\text{coat}}\pi r^2(1 - \cos^2\theta_{\text{avg}})}\right)^{-1} \tag{6.58}$$

The heat transfer rate through a drop of radius r equals the product of the rate of mass condensate formation at the free surface and the latent heat of vaporization.

$$q_{\text{d}} = (\rho_{\text{l}} h_{\text{lv}})\frac{\text{d}v}{\text{d}t} \tag{6.59}$$

$$\frac{\text{d}V}{\text{d}t} = \frac{\text{d}V}{\text{d}r} \times \frac{\text{d}r}{\text{d}t} = \pi r^2\left(2 - 3\cos\theta_{\text{avg}} + \cos^3\theta_{\text{avg}}\right)\frac{\text{d}r}{\text{d}t} \tag{6.60}$$

$$q_{\text{d}} = \left(\pi r^2 \rho_{\text{l}} h_{\text{lv}}\right) \times \left(2 - 3\cos\theta_{\text{avg}} + \cos^3\theta_{\text{avg}}\right) \times \left(\frac{\text{d}r}{\text{d}t}\right) \tag{6.61}$$

From the above set of equations, one can show that the rate of growth of individual drops follows the equation

$$\frac{\text{d}r}{\text{d}t} = \left(\frac{4(T_{\text{sat}} - T_{\text{w}})}{\rho_{\text{l}} h_{\text{lv}}}\right)\left(1 - \frac{r_{\min}}{r}\right)\left[\left(\frac{2}{h_{\text{int}}} + \frac{r(1 - \cos\theta_{\text{avg}})}{k} + \frac{4\delta}{k_{\text{coat}}(1 + \cos\theta_{\text{avg}})}\right)\right]^{-1} \times \left[\frac{(1 - \cos\theta_{\text{avg}})}{(2 - 3\cos\theta_{\text{avg}} + \cos^3\theta_{\text{avg}})}\right] \tag{6.62}$$

Equation (6.62) is valid for horizontal and inclined surfaces without wettability gradient as well as the ith drop (average contact angle $= \theta_{\text{avg}}$) underneath a horizontal substrate with wettability gradient.

Fig. 6.18 Observing drop coalescence during a dropwise condensation experiment (Leach et al. 2006). (**a**) Drops prior to coalescence; the drop coalescence is marked with ×. The arrow identifies a drop-free region where drops will be observed later. (**b**) The same region during coalescence. (**c**) The next images, where drops coalesce are marked with ×. (**d**) The same region during coalescence and (**e**) after coalescence. (**f**) The same region seconds later, after newly nucleated drops become visible following coalescence. Some of these drops appear in the region marked in (**a**), which was drop-free prior to the coalescence events

6.4.6 Growth by Coalescence

In dropwise condensation, two or more drops on or underneath a cold substrate grows large enough to touch one another, coalesce, and form a single larger drop (Fig. 6.18). Leach et al. (2006) studied dropwise condensation of water vapor coming from a hot water reservoir onto a naturally cooled hydrophobic polymer film and a silanized glass slide. The authors observed that the coalesced drop is at the center of mass of the original drops. The smallest detectable droplets were seen to grow and eventually fall off, after repeated cycles of nucleation to coalescence. The spatio-temporal coalescence scales were also reported. Images acquired before and after coalescence events confirmed that drop coalescence re-exposed the substrate area for nucleation of new liquid drops.

Many researchers (Vemuri and Kim 2006; Leipertz 2010; Dietz et al. 2010; Miljkovic et al. 2012) have experimentally recorded the smallest detectable droplets that grow and eventually fall off after repeated cycles of coalescence. The growth rate of drops depends on their respective size: small drops grow by direct condensation as well as occasional coalescence, but large drops grow mainly by coalescence. The growth rate of small drops is related to heat transfer. Smaller drops offer less thermal resistance, thus permitting rapid condensation. Larger drops offer a higher thermal resistance and grow primarily by coalescence. Hence, coalescence plays a primary role in determining the drop size distribution on the macroscale while direct condensation is of secondary importance. Coalescence also plays a direct role in the frequency of attainment of drop criticality, either for sliding motion or fall-off. Subsequently, nucleation occurs over the re-exposed area of the substrate. Nucleation, slide-off or fall-off and droplets coalescence are the fundamental processes that enhance heat transfer coefficient at later stages of growth in dropwise condensation. Since the associated heat transfer rates are high, one can imagine coalescence dynamics as one of the important factors contributing to the enhanced heat transfer during dropwise condensation.

Coalescence-induced instability in the pendant mode is an effective means of passively enhancing heat transfer coefficient during dropwise condensation. Inclined substrates have natural advantage in terms of sweeping of drops from the substrate, thereby, exposing fresh sites for nucleation. As compared to coalescence of sessile droplets, flow instabilities are induced faster in pendant drops, enhancing the associated heat transport characteristics.

Although coalescence of pendant drops underneath an inclined hydrophobic surface is an efficient process in dropwise condensation, discussion on the subject is scarce in the literature. Much of the research available is on the formation of a liquid bridge and the relaxation time coalescence in sessile drops.

Eggers et al. (1999) focused on early-time behavior of the radius of the small bridge between two drops. When two liquid droplets touch each other, a liquid bridge is formed between them. A negative curvature or negative pressure is created at the point of joining. This bridge quickly expands under the influence of interfacial stresses and the resultant fluid motion pulls the two drops together, forming a large drop with a smaller surface area. This motion is viscously dominated in the initial stages. Based on the above concept the authors proposed a scaling law for a variation of the liquid bridge radius with time.

For two drops merging together, Andrieu et al. (2002) experimentally recorded and theoretically described the kinetics of coalescence of two water drops on a plane solid surface. Immediately after coalescence, an ellipsoidal shape results, eventually relaxing into a hemispherical shape, in a few milliseconds. The characteristic relaxation time is proportional to the drop radius R at final equilibrium. This relaxation time is nearly 10^7 times larger than the bulk capillary relaxation time $t_\eta = R\mu/\sigma$, where σ is the vapor-liquid surface tension and μ is the liquid shear viscosity.

Duchemin et al. (2003) studied coalescence of two liquid drops driven by surface tension. The fluid was considered to be ideal and velocity of approach, zero. Using the boundary integral method, the walls of the thin retracting sheet of air between the drops were seen to reconnect in finite time to form a toroidal enclosure. After initial reconnection, retraction starts again, leading to a rapid sequence of enclosures. Averaging over the discrete events, the minimum radius of the liquid bridge connecting the two drops were scaled as r_b proportional to $t^{0.5}$.

Using high-speed imaging, Wu et al. (2004) studied early-time evolution of the liquid bridge that is formed upon the initial contact of two liquid drops in air. Experimental results confirmed the scaling law that was proposed by Eggers et al. (1999). Further, their experimental study demonstrated that the liquid bridge radius

(r_b) follows the scaling law $r_b \propto (t)^{0.5}$ in the inertial region. The pre-factor of the scaling law, $r_b/(t)^{0.5}$, is shown to be proportional to $R^{1/4}$, where R is the inverse of the drop curvature at the point of contact. The dimensionless pre-factor is measured to be in the range of 1.03–1.29, which is lower than 1.62, a pre-factor predicted by the numerical simulation of Duchemin et al. (2003) for inviscid drop coalescence.

Narhe et al. (2004) investigated the dynamics of coalescence of two sessile water drops and compared them with the spreading dynamics of a single drop in the partially wetting regime. The composite drop formed due to coalescence relaxed exponentially towards equilibrium with a typical relaxation time that decreases with contact angle. The relaxation dynamics is larger by 5–6 orders of magnitude than the bulk hydrodynamics which is of the order of a few milliseconds, due to the high dissipation in the contact line vicinity. Narhe et al. (2005) studied the dynamics of drop coalescence in the sessile mode of dropwise condensation of water vapor onto a naturally cooled hydrophobic polymer film and silanized glass slide. The authors reported that coalescence is affected by surface orientation and composition, vapor and surface temperatures, humidity and vapor flow rate.

Aarts et al. (2005) studied droplet coalescence in a molecular system with variable viscosity and a colloid-polymer mixture with an ultra-low surface tension. When either the viscosity is large or the surface tension is small enough, the liquid bridge opening initially proceeds with capillary velocity. Inertial effects are dominant at a Reynolds number of about 1.5 ± 0.5 and the neck then grows as the square root of time. In a second study, decreasing the surface tension by a factor of 10^5 opened the way to a more complete understanding of the hydrodynamics involved.

Thoroddsen et al. (2005) studied pendant as well sessile drop coalescence. The authors used an ultra-high-speed video camera to study coalescence, over a range of drop sizes and liquid viscosities. For low viscosity, the outward motion of the liquid contact region is successfully described by a dynamic capillary-inertial model based on the local vertical spacing between two drop surfaces. This model can also be applied to drops of unequal radii. Increasing viscosity slows down coalescence. For the largest viscosity, the neck region initially grows at a constant velocity. The authors compared their results with the previously predicted power law, finding slight but significant deviation from the predicted exponents.

Ristenpart et al. (2006) investigated experimentally and theoretically the coalescence dynamics of two spreading drops on a highly wettable substrate. They found that the width of the growing meniscus bridge between the two droplets exhibits power-law behavior, growing at early times as $(t)^{0.5}$. Moreover, the growth rate is highly sensitive to both the radii and heights of the drops at contact, scaling as $h^{3/2}/R_o$. This size dependence differs significantly from the behavior of freely suspended drops, in which the coalescence growth rate depends only weakly on the drop size.

Kapur and Gaskell (2007) experimentally investigated coalescence of a pair of drops on a surface with high quality images from flow visualization revealing the morphology of the process. The drops merge and evolve to a final state with a footprint that is peanut-like in shape, with bulges along the longer sides resulting from the effects of inertia during spreading. The associated dynamics involve a subtle interplay between (a) the motion of the wetting process due to relaxation of the contact angle and (b) a rapid rise in free surface height above the point where coalescence begins due to negative pressure generated by curvature. During the early stages of motion, a traveling wave propagates from the point of initial contact up the side of each drop as liquid is drawn into the neck region, and only when it reaches the apex of each do their heights start decrease. A further feature of the rapid rise in height of the neck region is that the free surface overshoots significantly from its final equilibrium position; it reaches a height greater than that of the starting drops, producing a self-excited oscillation that persists long after the system reaches its final morphological state in relation to its footprint.

Thoroddsen et al. (2007) studied drop coalescence of two different miscible liquids and found that the coalescence speed is governed by the liquid having weaker surface tension. Marangoni waves propagate along the drop with stronger surface tension. Surface profiles and propagation speeds of these waves were reported from experiments with a pendant water drop coalescing with a flat ethanol surface or with a sessile drop of ethanol. In the former, capillary-Marangoni waves along the water drop showed self-similar character in terms of arc length along the original surface.

Liao et al. (2008) performed an experimental investigation on coalescence of two equal-sized water drops on inclined surfaces. The effects of inclination angle and the drop size were studied with respect to the liquid bridge, fore/back contact angle, and the evolving three-phase contact line.

Sellier and Trelluyer (2009) proposed a power-law growth of the bridge between the drops describe the coalescence of sessile drops. The exponent of the power-law depends on the driving mechanism for the spreading of each drop. The authors validated the experiment against numerical simulations.

Boreyko and Chen (2009) linked the coalescence with heat transfer rate in dropwise condensation. The authors experimentally showed the drop shifting on a substrate and releasing interfacial energy during coalescence. Energy released is higher for higher contact angle and is responsible for the drop movement and enhancement of heat transfer during coalescence.

Wang et al. (2010) conducted an experiment to study the behavior of liquid drop coalescence on a surface with gradient in surface energy. The microscopic contour of the gradient energy surface was fabricated on the base of a silicon chip by diffusion controlled silanization of alkyl-trichloro-silanes and characterized by an atomic force microscope. The effect on the three-phase contact line and contact angle was obtained. The process of drop coalescence was seen to accelerate the drop speed on the gradient surface.

Sellier et al. (2011) studied coalescence of sessile drops of distinct liquids, arising from Marangoni stresses due to surface tension gradient. The analysis revealed two dimensionless numbers that govern flow characteristics. One is related to the strength of surface tension gradient and the other to the diffusion timescale. Numerical results confirmed the occurrence of the self-propulsion behavior.

Paulsen et al. (2011) used an electrical method and high-speed imaging to investigate drop coalescence down to 10 ns after the drops touch. Viscosity was varied over two decades. At a sufficiently low approach velocity where deformation is not present, the drops coalesced with an unexpectedly late cross-over time between a regime dominated by viscous and one by inertial effects.

Much of the research on the topic covers the formation of a liquid bridge and the relaxation time of sessile drops during coalescence. Coalescence of pendant drops and its role of coalescence in heat transfer enhancement are not readily available.

6.4.7 Modeling Growth by Coalescence

A simple model of drop coalescence is adopted in the present work based on experimental observations reported in the literature. Consider two drops of radius r_i and r_j at nucleation sites, i and j, on the substrate (Fig. 6.19).

The distance between the two nucleation sites i and j is

$$l_{ij} = \sqrt{\left(x_i - x_j\right)^2 + \left(y_i + y_j\right)^2 + \left(z_i - z_j\right)^2} \tag{6.63}$$

The coalescence criterion can be stated as

$$l_{ij} - \left[r_i + r_j\right] < 10^{-6} \tag{6.64}$$

If the coalescence criterion is met, a drop of equivalent volume on the mass averaged center of the original coalescing droplets is introduced. The time for coalescence is taken to be much smaller than the other timescales of the condensation process. Hence, as soon as the two droplets contact each other (or three droplets, or, very rarely, four contact each other simultaneously), they are substituted with an equivalent single drop with equal total volume, located at the weighted center of mass of the individual coalescing drops. Two drops i and j in a coalescence process forming drop c, is shown in Fig. 6.19. The volume, position, and radius of drop c are

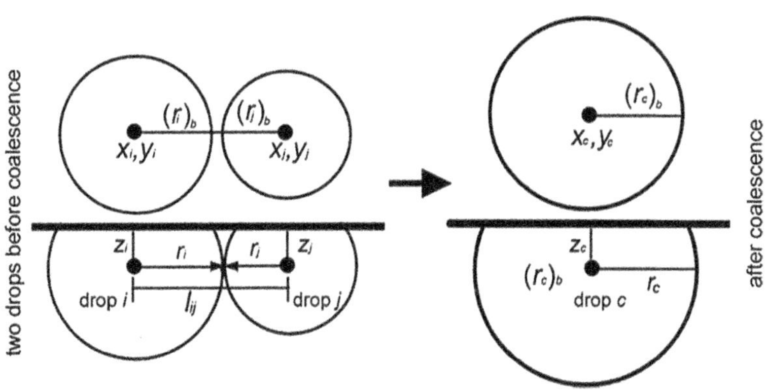

Fig. 6.19 Schematic showing coalescence of two drops and criteria of coalescence

$$V_c = \left(V_i + V_j\right) \tag{6.65}$$

$$x_c = \frac{\left(V_i x_j + V_j x_i\right)}{V_c} \tag{6.66}$$

$$y_c = \frac{\left(V_i y_j + V_j y_i\right)}{V_c} \tag{6.67}$$

$$r_c = \left[\frac{3\left(V_i + V_j\right)}{\pi\left(2 - 3\cos\theta_{\text{avg}} + \cos^3\theta_{\text{avg}}\right)}\right]^{1/3} \tag{6.68}$$

The base radius of the drop formed after coalescence is

$$\left(r_c\right)_{\text{b}} = r_c \sin\theta_{\text{avg}} \tag{6.69}$$

Equations (6.63)–(6.69) are valid for horizontal and inclined surfaces with and without wettability gradient. More than two drops (i, j and k) are coalesced by

$$l_{ij} = \sqrt{\left(x_i - x_j\right)^2 + \left(y_i - y_j\right)^2 + \left(z_i - z_j\right)^2} \tag{6.70}$$

$$l_{ik} = \sqrt{\left(x_i - x_k\right)^2 + \left(y_i - y_k\right)^2 + \left(z_i - z_k\right)^2} \tag{6.71}$$

$$l_{jk} = \sqrt{\left(x_j - x_k\right)^2 + \left(y_j - y_k\right)^2 + \left(z_j - z_k\right)^2} \tag{6.72}$$

The coalescence criterion can be stated as

$$l_{ij} - \left[r_i + r_j\right] < 10^{-6}, \quad l_{ik} - \left[r_i + r_k\right] < 10^{-6} \quad \text{and } l_{jk} - \left[r_j + r_k\right] < 10^{-6} \tag{6.73}$$

The volume, position, and radius of the drop (drop c) formed by coalescing more than two drops is

$$V_c = \left(V_i + V_j + V_k\right) \tag{6.74}$$

$$x_{c,i} = \frac{\left(V_i x_j + V_j x_i\right)}{V_i + V_j} \tag{6.75}$$

$$y_{c,i} = \frac{\left(V_i y_j + V_j y_i\right)}{V_i + V_j} \tag{6.76}$$

$$x_c = \frac{((V_i + V_j)x_k + V_k x_{c,i})}{V_i + V_j + V_k} \tag{6.77}$$

$$y_c = \frac{((V_i + V_j)y_k + V_k y_{c,i})}{V_i + V_j + V_k} \tag{6.78}$$

$$r_c = \left[\frac{3(V_i + V_j + V_k)}{\pi(2 - 3\cos\theta_{\text{avg}} + \cos^3\theta_{\text{avg}})}\right] \tag{6.79}$$

The main assumption in the approach adopted for coalescence is that its timescale (in ms) is small in comparison with the cycle time of dropwise condensation (usually in excess of a second). The coalesced drops relax over a longer time period, but this process can be neglected because most often it would become gravitationally unstable, leading to fall-off or slide-off from the substrate. Hence, the assumption of instantaneous coalescence is expected to be reasonable in dropwise condensation.

6.4.8 Drop Instability

When a certain size is reached, several authors (Citakoglu and Rose 1968a, b; Meakin 1992; Leipertz and Fröba 2006, 2008) have shown that the gravitational force on the droplet exceeds the adhesive force between the liquid and condensing substrate, and the droplet begins to move (Fig. 6.20). Drop motion plays an important role in the enhancement of heat transfer. The sliding drop wipes other droplets off, resulting in re-exposed surface area. New drops are formed again in the re-exposed area of the substrate. The diffusional resistance of the liquid contained in these drops forms the primary thermal resistance of the energy released from the free surface to the condensing wall.

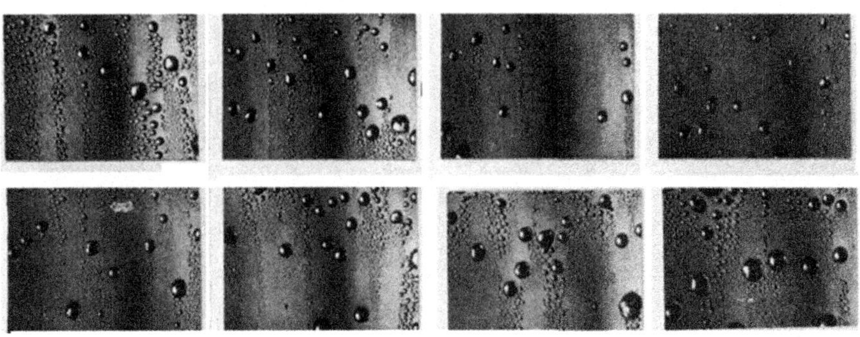

Fig. 6.20 Image of drops siding in dropwise condensation on copper substrate at different inclination with the horizontal (Citakoglu and Rose 1968b)

Enhancement of heat transfer necessitates that these drops be swept away from the substrate as soon as possible to reduce the most prominent thermal resistance in the passages of heat, from the vapor to the substrate. Sliding may be achieved either by (1) inclining the substrate, or alternatively, (2) by creating as additional force imbalance at the three-phase contact line. The later strategy is most suitable for induced motion on horizontal surfaces. Contemporary manufacturing/coating techniques can provide such a wettability gradient by physico-chemical action, leading to additional surface forces required for inducing droplet motion.

6.4.9 Drop Sliding on an Inclined Substrate

Literature on drop sliding on or underneath a textured inclined surface is limited. Most of the existing work (Brown et al. 1980; Dussan 1985; Briscoe and Galvin 1991b; Elsherbini and Jacobi 2006) have considered the critical state of static sessile drop on an inclined surface and focused on the apparent contact angle hysteresis, drop shape, and drop retention with tiltable surfaces for various combinations of hydrophobic surfaces and liquids.

Though a large volume of work exists on predicting the drop shape under static condition, only a few researchers have reported the sliding behavior of the drop on an inclined surface as well as horizontal wettability gradient substrate.

Kim et al. (2002) reported that a liquid drop which partially wets a solid surface will slide down the plane when it is tilted beyond a critical inclination. Experiments for measuring the steady sliding velocity of different liquid drops were performed on an inclined surface leading to a scaling law to relate velocity with wetting characteristics.

Grand et al. (2005) reported experiments on the shape and motion of millimeter-sized drops sliding down a plane in a situation of partial wetting. An unexpected shape change was seen when the velocity of drop is increased. In theoretical analysis, the viscous force was scaled as $\mu U V^{1/3}$ and the drop sliding velocity was found to be a linear function of the Bond number. Rio et al. (2005) examined the microscopic force balance close to a moving contact line to investigate boundary conditions around viscous drops sliding down an inclined plane.

Gao and McCarthy (2006) postulated two mechanisms for a drop moving down the plane. Drops move by sliding, when the particles near the solid-liquid interface exchange their position with those at the gas-liquid interface, while the bulk of the fluid remains unaffected. On the other hand, there could be rolling motion where the entire fluid mass undergoes a circulatory movement. Sakai et al. (2006) used particle image velocimetry (PIV) to observe the internal fluidity of water droplets during slide on various chemically textured surfaces. On normal hydrophobic surface with contact angle of around 100°, both slipping and rolling controlled velocity during slide. On the superhydrophobic surface, however, with a contact angle of 150°, the droplet fell at high velocity by slipping. Yoshida et al. (2006) did not consider the viscous force in their study of the sliding behavior of water drops on a flat polymer

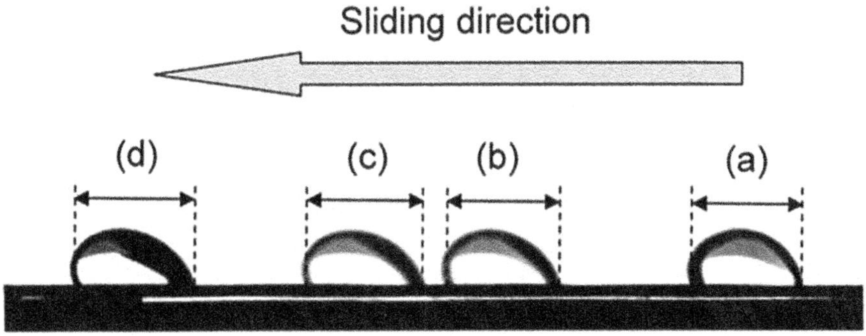

Fig. 6.21 Photographs of the sliding of a 45 mg water droplet on the sample coated with FAS tilted at 35°. Each sliding distance at (**a**), (**b**), (**c**), and (**d**) is 0.000, 0.010, 0.015, and 0.030, respectively. The droplet length at corresponding distance is 5.28, 5.34, 5.58, and 5.78, respectively (Suzuki et al. 2006)

surface. The authors reported that sliding motion changed from constant velocity to one of constant acceleration with an increase in the contact angle. Suzuki et al. (2006) reported a photograph of a sliding 45 mg water droplet on the surface coated with fluoroalkylsilane and tilted at 35° (Fig. 6.21). The authors reported that apparent length of water droplets increases when the sliding velocity increases.

Sakai and Hashimoto (2007) experimentally determined the velocity vector distribution inside a sliding sessile drop using PIV. The authors reported that the velocity gradient near the liquid-solid interface is higher than locations elsewhere inside a drop.

Hao et al. (2010) investigated the internal flow pattern in a water droplet sliding on the superhydrophobic surface by employing PIV. Both rolling and slipping motion were seen inside the drop during sliding though rolling occurred only at the edge of the water droplet.

6.4.10 Drop Sliding over a Horizontal Surface with a Wettability Gradient

The possibility of drop movement resulting from a wettability gradient was noted by Greenspan (1978) and experimentally demonstrated by Chaudhury and Whitesides (1992). Daniel et al. (2001) performed the experiment of condensation on a wettability gradient substrate. The authors observed more rapid motion (1.5 m/s) when condensation occurred over a horizontal wettability gradient surface. Drops moved hundreds to thousands of times faster than the speeds of typical Marangoni flows. Moumen et al. (2006) measured the velocity of a drop along a wettability gradient surface. At steady-state, the driving force for drop movement comes from the gradient of free energy of adhesion of the drop with the substrate and balanced by viscous drag generated within the liquid drop.

6.4.11 Modeling Drop Instability

During condensation, drops grow first by direct deposition of vapor and then by coalescence. Continuously, the weight of the drop increases and can be a destabilizing influence. Force imbalance at the three-phase contact line leads to instability. For definiteness, a free body diagram of a pendant drop underneath a flat horizontal substrate and an inclined substrate with corresponding forces acting at the three-phase contact line (Fig. 6.22) is considered. For determining the onset of instability, pressure, surface tension, and gravity are taken as dominant forces. The size of the drop at slide-off and fall-off are estimated in the following sections.

Liquid pressure within the drop will be in excess of the surrounding vapor pressure. The pressure difference is larger in the smaller drop. As the drop size increases, pressure difference decreases. It is a minimum at the onset of instability, for which the drop diameter has increased to its largest possible size. When the drop is about to slide, pressure acts normal to the surface and does not contribute to the force calculation. For fall-off, the excess pressure has a component in the vertical direction. Since it is small for the large drops, excess pressure has been neglected in

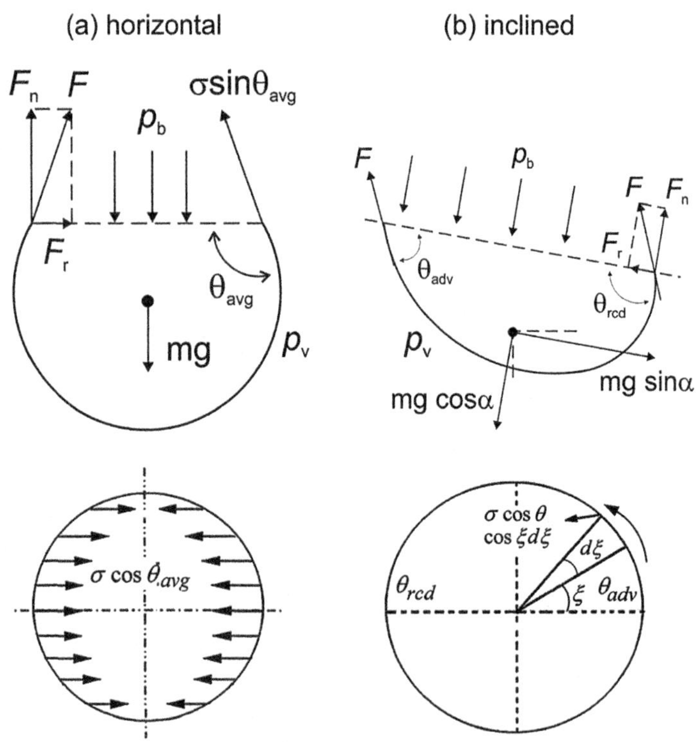

Fig. 6.22 Direction of the retention force shown on the footprint and free body diagram of drop underneath (**a**) horizontal substrate and (**b**) inclined substrate

the fall-off instability calculation. Accordingly, the critical drop diameter is expected to be slightly over-predicted.

6.4.12 *Horizontal Substrate*

A pendant drop underneath a horizontal substrate is a part of a sphere of radius r with a contact angle $\theta = \theta_{avg}$. For a horizontal hydrophobic substrate, surface tension and gravity are in competition (Fig. 6.22a). As droplets grow, body forces (gravity) eventually surpass the limiting surface force (surface tension) at the three-phase contact line. As discussed above, the contribution of excess pressure in the determination of the critical drop diameter is negligible.

The component of surface tension force normal to the substrate is

$$F_n = 2\pi r_b \sigma \sin\theta_{avg} = 2\pi r \sigma \sin^2\theta_{avg} \tag{6.80}$$

The weight of the drop is

$$F_{g\perp} = (\rho_l - \rho_v)\frac{\pi r^3}{3}(2 - 3\cos\theta_{avg} + \cos^3\theta_{avg})g \tag{6.81}$$

Equating Eqs. (6.80) and (6.81), the maximum radius (the size of droplet fall-off) is

$$r_{max} = \sqrt{\left(\frac{6\sin^2\theta_{avg}}{2 - 3\cos\theta_{avg} + \cos^3\theta_{avg}}\right)\left(\frac{\sigma}{(\rho_l - \rho_v)g}\right)} \tag{6.82}$$

Uniform distribution of the contact angle at the three-phase contact line makes the net retention force (F_r) underneath a horizontal plane zero. The arguments leading to Eq. (6.82) do not include adhesion of the liquid with the substrate at the base since higher order effects on the meso-scale and microscales are neglected. Various authors (Li and Amirfazli 2007; Miljkovic et al. 2012; Rykaczewski 2012) showed that such effects are not important on engineering scale calculations.

6.4.13 *Inclined Substrate*

Inclining the substrate causes an imbalance in the forces and results in drop deformation to achieve necessary static balance. A deformed drop underneath an inclined substrate at incipient sliding is shown in Fig. 6.22b. The leading side contact angle is equal to the advancing angle and trailing side contact angle is the receding angle of the liquid substrate combination. The figure highlights the relevant forces at contact line which is taken to be circular. The force balance on a drop underneath an inclined substrate at incipient sliding is also shown. The component of body force (gravity) parallel to the substrate tries to slide the drop and surface tension provides the retention force for stability. Similarly, body force component normal to the substrate leads to fall off while the normal component of surface tension provides stability to hold the drop. Hence, the critical size at which slide/fall-off commences depends not only on the thermophysical properties of the liquid but also on physico-chemical properties of the substrate.

Under dynamic conditions, the applicability of static force balance is questionable due to the presence of capillary waves, distortion in local equilibrium droplet shapes, droplet pinning, variation in dynamic contact angle due to inertia effects, sudden acceleration, and three-dimensional flow structures inside the droplets. Therefore, there is considerable debate in the literature on the applicability of static conditions on the real-time condensation process (Fang et al. 2008; Annapragada et al. 2010). The bulk composite effect of these real-time dynamic situations and local contact line perturbations is manifested in the form of hysteresis of advancing and receding angles. The static force balance conditions should be representative of the dynamic situation since absolute contact angles and hysteresis are accounted for. Expressions for the maximum base radius of a drop that will first slide (r_{crit}) or fall-off (r_{max}) underneath an inclined substrate are derived in the following discussion.

6.4.14 *Critical Radius of Slide-off underneath an Inclined Substrate*

The critical radius of the droplet at slide-off underneath an inclined substrate is obtained by force balance parallel to the substrate. Accordingly, the retention force arising from contact angle hysteresis, namely the difference in the advancing angle and receding angle is equal to the component of weight parallel to the substrate.

The component of retention force acting on the drop in the direction of substrate inclination is found by integrating the differential force over the base of the drop.

$$F_{r\parallel} = 2\int_0^{\pi} r_b \sigma \cos\theta \cos\xi \mathrm{d}\xi \tag{6.83}$$

The normal component of surface tension at the base of the drop is

$$F_{r\perp} = 2\int_0^{\pi} r_b \sigma \cos\theta \sin\xi \mathrm{d}\xi \tag{6.84}$$

Due to symmetry at the base, $F_{r\perp} = 0$; the resulting retention force F_r due to surface tension acts in the direction of substrate inclination $F_{r||}$.

The contact angle hysteresis, namely the variation in the advancing to receding contact angle, is taken to vary linearly along the contact line with respect to azimuthal angle. The base of the droplet is taken to be circular as discussed earlier. The variation of contact angle, with respect to azimuthal angle along the drop contact line is

$$\cos\theta = \cos\theta_{adv} + \left(\frac{\cos\theta_{rcd} - \cos\theta_{adv}}{\pi}\right)\xi \tag{6.85}$$

Substituting the Eq. (6.85) into Eq. (6.83) and integrating,

$$F_r = -(4/\pi)\sigma r_b(\cos\theta_{rcd} - \cos\theta_{adv}) \tag{6.86}$$

The minus sign indicates the direction of force is opposite to the direction of inclination. The drop volume V, area of liquid-vapor interface A_{lv}, and area of solid-liquid interface A_{sl} of deformed drop underneath an inclined substrate is calculated using the spherical cap approximation. Accordingly, the volume of the deformed drop is

$$V = \frac{\pi r_b^3 (2-3)\cos\theta_{avg} + \cos^3\theta_{avg}}{3\sin^3\theta_{avg}} \tag{6.87}$$

The force component due to gravity that is parallel to the substrate is

$$F_{g||} = \frac{\pi r_b^3 (2 - 3\cos\theta_{avg} + \cos^3\theta_{avg})}{3\sin^3\theta_{avg}}(\rho_l - \rho_v)g\sin\alpha \tag{6.88}$$

where r_b is the base radius of drop and is related to the drop radius

$$r_b = r\sin\theta_{avg} \tag{6.89}$$

A balance of forces acting direction parallel to substrate inclination yields

$$F_{g||} + F_{r||} = 0 \tag{6.90}$$

Hence, the critical radius of the droplet at slide-off on the inclined substrate is

$$r_{\mathrm{crit}} = \sqrt{\left(\frac{1.215\sin\theta_{\mathrm{avg}}}{(2 - 3\cos\theta_{\mathrm{avg}} + \cos^3\theta)}\right)(\cos\theta_{\mathrm{rcd}} - \cos\theta_{\mathrm{adv}})\left(\frac{\sigma}{(\rho_1 - \rho_{\mathrm{v}})g\sin\alpha}\right)} \tag{6.91}$$

For $r_{\mathrm{b}} > r_{\mathrm{crit}}$, the drop becomes unstable and slides over the surface.

Estimation of Critical Radius of Fall-Off Underneath an Inclined Substrate

Surface tension perpendicular to the inclined substrate is

$$F_{\mathrm{n}} = 2\int_0^{\pi} r_{\mathrm{b}}\sigma\sin\theta\,\mathrm{d}\xi \tag{6.92}$$

The variation of contact angle, with respect to azimuthal angle ξ along the contact line is

$$\sin\theta = \sin\theta_{\mathrm{adv}} + \left(\frac{\sin\theta_{\mathrm{rcd}} - \sin\theta_{\mathrm{adv}}}{\pi}\right)\xi \tag{6.93}$$

Substituting Eq. (6.93) into Eq. (6.92) and integrating, the surface tension component perpendicular to the substrate is

$$F_{\mathrm{n}} = \pi\sigma r_{\mathrm{b}}(\sin\theta_{\mathrm{adv}} + \sin\theta_{\mathrm{rcd}}) \tag{6.94}$$

The gravity force component perpendicular to the substrate is

$$F_{\mathrm{g}\perp} = \frac{\pi r_{\mathrm{b}}^3(2 - 3\cos\theta_{\mathrm{avg}} + \cos^3\theta_{\mathrm{avg}})}{3\sin^3\theta_{\mathrm{avg}}}(\rho_1 - \rho_{\mathrm{v}})g\cos\alpha \tag{6.95}$$

The maximum radius r_{max} of the drop that will initiate fall-off is obtained by balancing the forces perpendicular to substrate

$$F_{\mathrm{g}\perp} + F_{\mathrm{n}} = 0 \tag{6.96}$$

Hence, the critical radius of fall-off (r_{max}) is

$$r_{\max} = \sqrt{\left(\frac{3(\sin\theta_{\text{avg}})(\sin\theta_{\text{rcd}} + \sin\theta_{\text{adv}})}{(2 - 3\cos\theta_{\text{avg}} + \cos^3\theta_{\text{avg}})}\right)\left(\frac{\sigma}{(g\cos\alpha)(\rho_l - \rho_v)}\right)} \tag{6.97}$$

For $r_b > r_{\max}$, the drop becomes unstable and falls off.

6.4.15 Horizontal Substrate Having Unidirectional Wettability Gradient

A horizontal substrate with wettability gradient is shown in Fig. 6.23a. The contact angle at the lower wettability side is θ_1, while that at the higher wettability side is θ_2. The contact angle varies linearly in one direction from $x = 0$ to X. Here, X is the substrate length in x-direction (Fig. 6.23b).

Consider the footprint of the ith drop of radius r at nucleation site (x_i, y_i) underneath a substrate having unidirectional wettability gradient. The side view of ith drop is shown in Fig. 6.23b. For intermediate calculations, the drop shape is taken to be a spherical cap in the sense that drop volume V_i, area of liquid-vapor interface $(A_{\text{lv}})_i$, and area of solid-liquid interface $(A_{\text{sl}})_i$ are calculated using the average contact angle

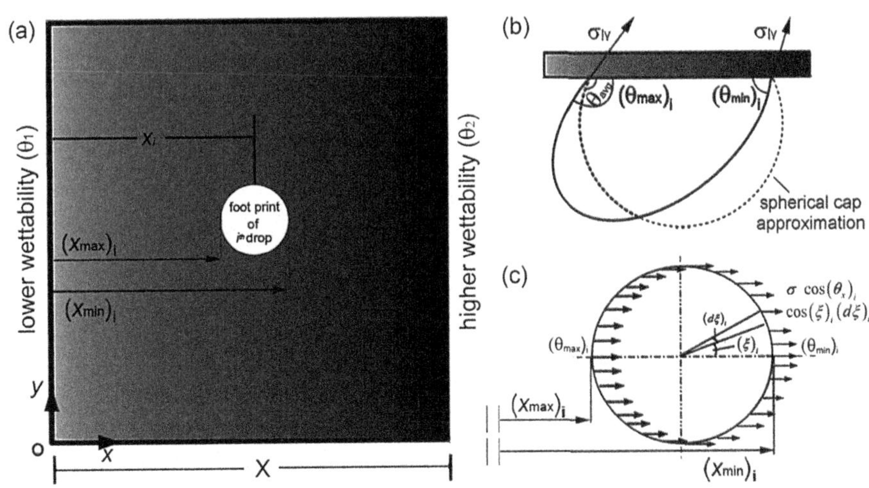

Fig. 6.23 Condensation over a substrate with wettability gradient. (**a**) The footprint of the ith drop is assumed to be circular. (**b**) Side view of each drop is determined by the two-circle approximation. (**c**) Direction of forces acting over the three-phase contact line—the substrate-vapor-liquid boundary

$$\left(\theta_{\mathrm{avg}}\right)_i = \frac{\left[(\theta_{\mathrm{max}})_i + (\theta_{\mathrm{min}})_i\right]}{2} \tag{6.98}$$

Here, the contact angle of the drops depends on their position on the substrate. Consequently, force imbalance is generated primarily because of variation of contact angle from $(\theta_{\mathrm{max}})_i$ to $(\theta_{\mathrm{min}})_i$ at the three-phase contact, arising from the substrate wettability gradient (Fig. 6.23b). Drop motion can be expected even before the shapes are greatly altered by gravity or flow-related pressure non-uniformity. Gravity and pressure will not have component parallel to the horizontal substrate. The unbalanced surface tension will then mobilize the drop along the substrate.

The footprint of the spherical cap shape corresponding to the ith drop is shown in Fig. 6.23c as a circle. The net force at the three-phase contact line of a deformed drop acts in the x-direction towards the higher wettability side. It can be calculated as follows:

The base radius of the ith drop is

$$(r_{\mathrm{b}})_i = (r_i)\sin\left(\theta_{\mathrm{avg}}\right)_i \tag{6.99}$$

Quantities x_{min} and x_{max} for the ith drop are

$$(x_{\mathrm{max}})_i = x_i - (r_{\mathrm{b}})_i \quad \text{and} \quad (x_{\mathrm{min}})_i = x_i - (r_{\mathrm{b}})_i \tag{6.100}$$

Angles $(\theta_{\mathrm{max}})_i$ and $(\theta_{\mathrm{min}})_i$ are

$$(\theta_{\mathrm{max}})_i = \theta_1 + \left(\frac{\theta_2 - \theta_1}{X}\right)(x_{\mathrm{max}})_i \tag{6.101}$$

$$(\theta_{\mathrm{min}})_i = \theta_1 + \left(\frac{\theta_2 - \theta_1}{X}\right)(x_{\mathrm{min}})_i \tag{6.102}$$

The net force acting at the footprint of ith the drop (Fig. 6.24a) towards higher wettability side is

$$(F_{\mathrm{r}})_i = 2\sigma \int_0^{\pi} \cos(\theta_x)_i \cos(\xi)_i (r_{\mathrm{b}})_i (\mathrm{d}\xi)_i \tag{6.103}$$

The value of $\cos(\theta_x)_i$ is linearly interpolated

$$\cos(\theta_x)_i = \cos(\theta_{\mathrm{min}})_i + \left(\frac{\cos(\theta_{\mathrm{max}})_i - \cos(\theta_{\mathrm{min}})_i}{(x_{\mathrm{min}})_i - (x_{\mathrm{max}})_i}\right)(x)_i \tag{6.104}$$

Substituting Eq. (6.104) into Eq. (6.103) and integrating, the retention force parallel to substrate is

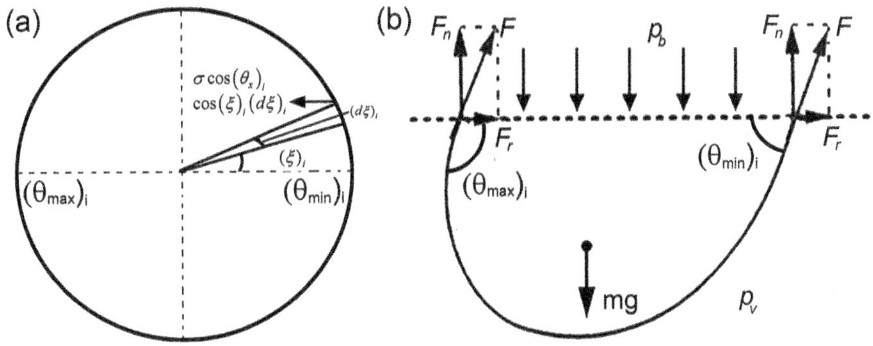

Fig. 6.24 (**a**) Variation of contact angle with respect to the azimuthal angle at base of the drop is assumed to be a circle. (**b**) Free body diagram of *i*th drop underneath wettability gradient horizontal substrate

$$(F_\mathrm{r})_i = (4/\pi)\sigma(r_\mathrm{b})_i\left[\cos(\theta_\mathrm{min})_i - \cos(\theta_\mathrm{max})_i\right] \tag{6.105}$$

The retention force parallel to the substrate towards the higher wettability side is balanced by wall shear between the drop and the substrate. This balance requires that the drop will slide at a constant speed. The estimation of terminal velocity of a drop is discussed in Sect. 6.4.16–6.4.18. Hence on the wettability gradient substrate, every drop becomes unstable. The sliding of the drop wipes off other drops that lie in its path, and its mass and volume change during the motion. If the weight of the drop is higher than the net retention force evaluated at the three-phase contact line normal to the surface, the drop will fall off. The critical radius of the drop at fall-off is estimated as follows:

The surface tension component normal to substrate is

$$(F_\mathrm{n})_i = 2\int_0^\pi \sigma\sin(\theta_x)_i(r_\mathrm{b})_i \mathrm{d}\xi \tag{6.106}$$

The variation of the contact angle, with respect to azimuthal angle along the contact line is linearly interpolated as

$$\sin(\theta_x)_i = \sin(\theta_\mathrm{min})_i + \left(\frac{\sin(\theta_\mathrm{max})_i - \sin(\theta_\mathrm{min})_i}{(x_\mathrm{min})_i - (x_\mathrm{max})_i}\right)(x)_i \tag{6.107}$$

Substituting Eq. (6.107) into Eq. (6.106) and integrating, the surface tension component perpendicular to the substrate is

$$(F_\mathrm{n})_i = \pi\sigma(r_\mathrm{b})_i\left[\sin(\theta_\mathrm{min})_i + \sin(\theta_\mathrm{max})_i\right] \tag{6.108}$$

The gravity force component perpendicular to the substrate is

$$(F_{g\perp})_i = \frac{\pi (r_b^3)_i \left[2 - 3\cos(\theta_{avg})_i + \cos^3(\theta_{avg})_i\right]}{3\sin^3(\theta_{avg})_i} (\rho_l - \rho_v)_i g \tag{6.109}$$

If the weight of the drop is higher than the net retention force evaluated at the three-phase contact line in a direction normal to the surface, the drop will fall off. The corresponding critical radius is

$$(r_{max})_i = \sqrt{\frac{3\sigma \sin(\theta_{avg})_i [\sin(\theta_{max})_i + (\sin\theta_{min})_i]}{(\rho_l - \rho_v) g \left[2 - 3\cos(\theta_{avg})_i + \cos^3(\theta_{avg})_i\right]}} \tag{6.110}$$

6.4.16 Modeling Terminal Velocity

In dropwise condensation, droplets undergo instability and start sliding over the substrate that is either inclined or is horizontal with a wettability gradient. The speed increases with time till the unbalanced force is matched by wall shear, resulting in a constant terminal velocity. In the present model, it is assumed that drops attain terminal velocity immediately after instability.

6.4.17 Inclined Substrate

For an inclined substrate, the drop achieves terminal velocity when the component of weight parallel to the surface, retention force of the deformed drop at the three-phase contact line owing to surface tension, and wall friction are in balance (Fig. 6.25). Hence,

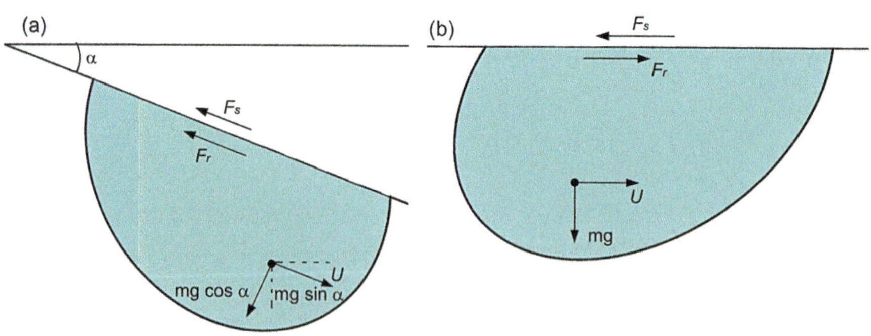

Fig. 6.25 Representation of various forces on sliding drop underneath substrate with terminal velocity (**a**) inclined substrate and (**b**) horizontal substrate with wettability gradient

$$F_{g\|} + F_{r\|} + F_s = 0 \tag{6.111}$$

where $F_{g\|}$ is the component of weight parallel to the inclined substrate, $F_{r\|}$ is retention force opposing drop motion, and F_s is wall shear associated with the relative velocity between the fluid and the substrate. The viscous force acting between the wall and fluid is

$$F_s = \frac{1}{2} C_f A_{sl} \rho_l U^2 \tag{6.112}$$

The average skin friction coefficient (C_f) is determined at the scale of the individual drop by a CFD model described by the authors in their earlier study. The terminal velocity of the drop over an inclined surface that makes an angle α with the horizontal is

$$U = \sqrt{\frac{2\left(F_{g\|} - F_r\right)}{C_f \rho_l A_{sl}}} \tag{6.113}$$

6.4.18 *Horizontal Substrate with a Wettability Gradient*

For an inclined substrate, when the component of surface tension parallel to it balances wall shear stress, the drop will slide with constant speed. The hydrodynamic force that resists motion of the *i*th drop is

$$\left(F_{hyd}\right)_i = C_f\left(0.5\rho U_i^2\right)\left(A_{sl}\right)_i \tag{6.114}$$

The skin coefficient of friction C_f is obtained from the correlations derived from the CFD model described by the authors elsewhere (Sikarwar et al., 2013b). Equating the expressions of forces given by Eqs. (6.105) and (6.114)

$$\left(F_r\right)_i + \left(F_{hyd}\right)_i = 0 \tag{6.115}$$

The terminal velocity is

$$U_i = \left[\frac{0.022\left(F_r\right)_i\left(\theta_{avg}\right)_i^{1.58}}{\rho^{0.03}\mu^{0.97}\left(r_b^{1.03}\right)_i}\right]^{1/1.03} \tag{6.116}$$

The velocity thus obtained is a function of the drop size and its position for a wettability gradient surface.

6.4.19 Wall Heat Transfer

Heat transfer during dropwise condensation can be calculated from the rate of condensation at the free surface of the drop at each nucleation site of the substrate. The gaps between drops are assumed to be inactive for heat transfer. The heat transfer q is a function of the nucleation site density and the rate of growth of drop radius at each nucleation site. The latter is estimated by using a quasi-one-dimensional approximation for thermal resistances, including the interfacial and capillary resistance at the vapor-liquid boundary and conduction resistance through drop, as discussed earlier and given by Eq. (6.62). The rate of condensation of vapor at each nucleation site can now be determined.

Estimate the number of available nucleation sites N. The mass of condensate accumulated at the ith nucleation site over a time interval Δt is

$$\Delta m_i = \rho \frac{\pi}{3}\left(2 - 3\cos\theta_{\mathrm{avg}} + \cos^3\theta_{\mathrm{avg}}\right)\left(r_{\mathrm{new}}^3 - r_{\mathrm{old}}^3\right)_i \tag{6.117}$$

With N, the number of active nucleation sites, the total quantity of condensate at a given time step (Δt) is

$$\Delta m = \sum_{i=1}^{i=N} (\Delta m)_i \tag{6.118}$$

Therefore, the average rate of condensation underneath a substrate is

$$m_{\mathrm{avg}} = \left(\sum_{j=1}^{j=K} \Delta m_j\right)/t \quad \text{where,} \quad t = \sum_{j}^{j=K} \Delta t_j \tag{6.119}$$

where t is the time period of condensation and $K = t/\Delta t$ is the number of time steps. The heat transfer rate is simply the latent heat released during the condensation divided by the time elapsed. The sensible cooling of the liquid is neglected. The average heat transfer coefficient over an area (A) of the substrate during dropwise condensation is

$$h = \frac{m_{\mathrm{avg}} h_{\mathrm{lv}}}{A(T_{\mathrm{sat}} - T_{\mathrm{w}})} \tag{6.120}$$

6.4.20 Area of Coverage

At each time step, the number of available nucleation sites and size of the drop at these locations is obtained from simulation. The area of substrate covered by the condensate is determined as follows:

N is the available nucleation sites at a given time t. The area of coverage at the ith nucleation site is

$$(A_{\mathrm{sl}})_i = \pi r_i^2 \left(1 - \cos^2\theta_{\mathrm{avg}}\right) \tag{6.121}$$

With N and A, the number of active nucleation sites and the total substrate area, the area covered by drops at a given time is

$$A_{\mathrm{cd}} = \frac{\sum_{i=1}^{i=N} (A_{\mathrm{sl}})_i}{A} \tag{6.122}$$

where A_{cd} is the fraction of area of substrate covered by drops at a given time instant. With K, the number of time instants within a cycle of time period t, the average percentage of area covered for a cycle is given by the expression

$$\frac{\left(\sum_i^K A_{\mathrm{cd}}/t\right)}{A} \times 100 \tag{6.123}$$

6.4.21 Available Liquid-Vapor Interface Area

Condensation takes place only over the liquid-vapor interface of the drop. The area of liquid-vapor interface at the ith nucleation site drop at a given time is

$$(A_{\mathrm{lv}})_i = 2\pi r_i^2 \left(1 - \cos\theta_{\mathrm{avg}}\right) \tag{6.124}$$

With N, the number of active nucleation sites, the fraction of liquid-vapor interface area available for condensation on a substrate of area A at a given time instant is

$$A_{\mathrm{alv}} = \frac{\sum_{i=1}^{i=N} (A_{\mathrm{lv}})_i}{A} \tag{6.125}$$

6.5 Numerical Algorithm for Modeling Dropwise Condensation

The modeling of dropwise condensation considers the details of each sub-process of the condensation cycle and interrelates them in such way as to form a full cycle. The important steps in the numerical algorithm can now be stated as follows:

i. Initialize all the variables such as thermophysical properties, physico-chemical properties of the substrate, type of substrate (with or without wettability gradient), orientation of substrate, nucleation site density, time step, and total time of simulation.
ii. Distribute the nucleation sites on the substrate using a random number generator and place the drop of minimum radius at all the nucleation sites.
iii. Solve Eq. (6.62) by a fourth-order Runge-Kutta method over a time step and find the new radius.
iv. Calculate the distance between nucleation sites.
v. Check for the coalescence.
vi. Identify the nucleation sites covered by the resulting coalesced drops and keep them deactivated till the drop covers them.
vii. Simultaneously, search for newly exposed sites created due to drop coalescence and provide a minimum radius drop on such newly exposed sites.
viii. For all the drops, check for the critical radius of slide-off and the sliding velocity.
ix. Re-activate the exposed sites created due to drop slide-off and provide a minimum radius drop on newly exposed sites.
x. Check for drop fall-off.
xi. Re-activate the exposed sites created due to drop fall-off and provide a minimum radius drop on the newly exposed sites.
xii. Repeat (iii)–(ix) till a dynamic steady-state is reached.

A computer program in C++ is written to carry out simulation as per the proposed algorithm. It is run on a high-performance computing machine. The flow chart of the mathematical model of dropwise condensation underneath the inclined substrates is depicted in Fig. 6.26, while the subroutines are detailed in Fig. 6.27. Whenever a drop is removed or shifted from its location due to sliding and coalescence all the hidden nucleation sites underneath the drop become active and are immediately supplied with thermodynamically stable droplets of the minimum radius. It is to be noted that the simulation needs to track multiple generations of the droplets—nucleating, growing by direct condensation, by coalescence and some slide/falling-off, when the virgin surface thus exposed, is renucleated. The computations are, hence, quite intensive.

While the mathematical model developed is quite general, simulations have been carried out under the following assumptions:

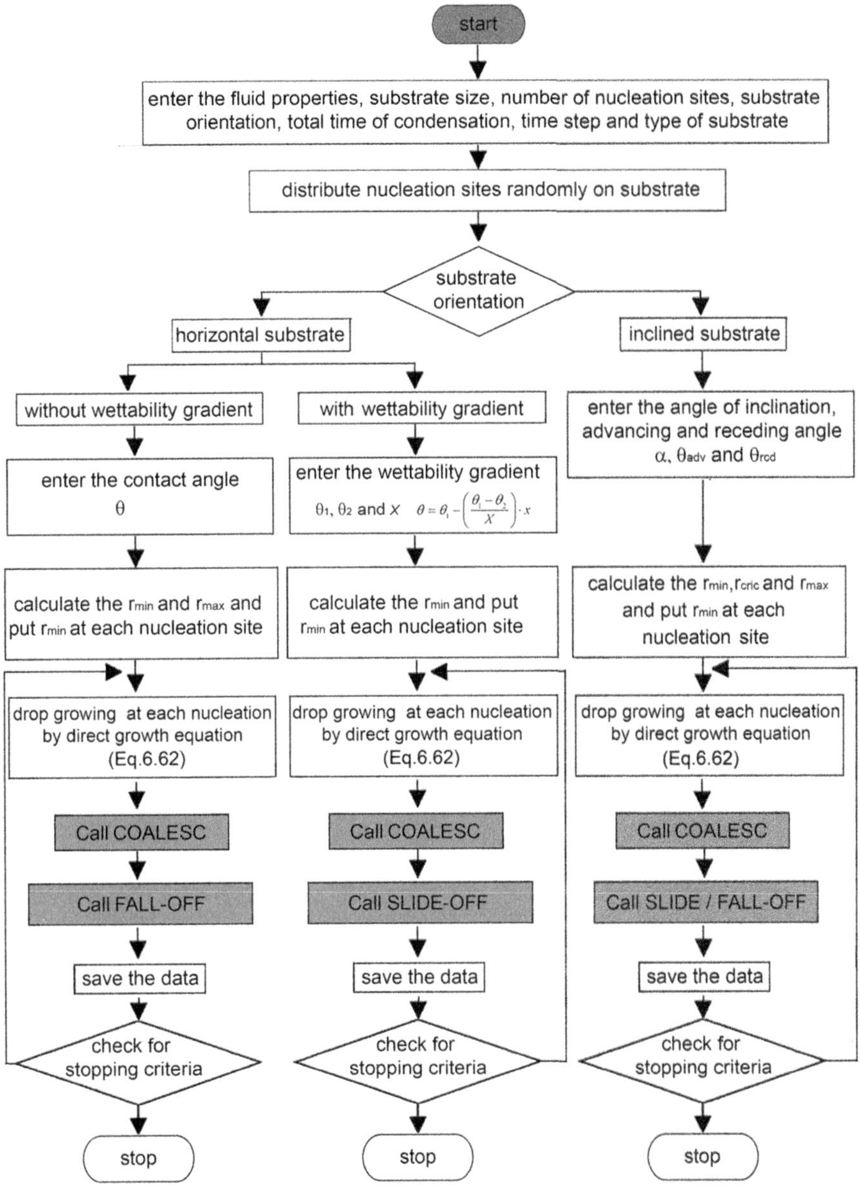

Fig. 6.26 Flow diagram of the dropwise condensation model

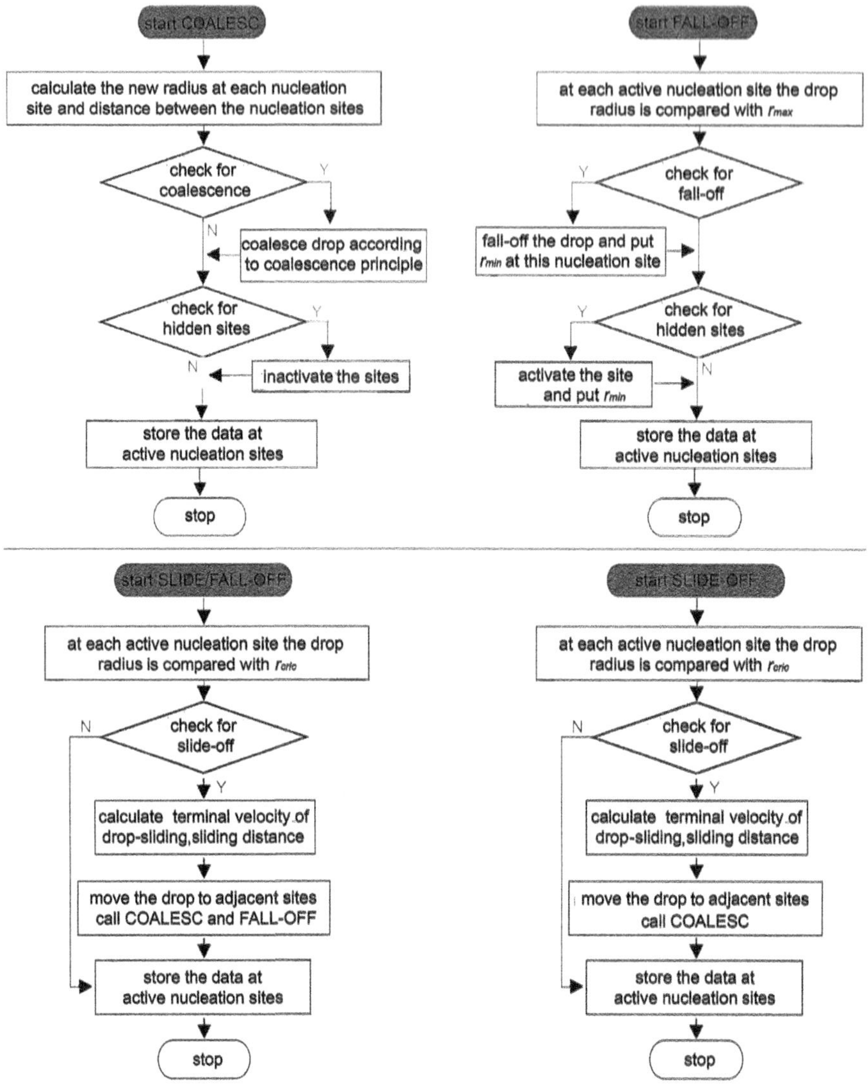

Fig. 6.27 Various subroutines of the dropwise condensation model

i. Nucleation sites are randomly distributed on the surface. Unless stated otherwise, all computations have been performed with an initial nucleation site density of 10^{-6} cm^{-2}.
ii. Thermodynamically constrained smallest radius is taken as the minimum radius in the simulation. Initially, the substrate is dry, and all the nucleation sites are instantaneously occupied by the droplet of minimum radius.

iii. Heat transfer resistance arises due to the liquid-vapor interface, curvature, and conduction, driven by imposed subcooling of the substrate. Convective transport of heat is neglected for static drops but is included for a moving drop. Constriction resistance is neglected.
iv. The accommodation coefficient is taken to be 0.035 for water, 0.45 for mercury, and 0.21 for sodium and potassium (Carey 2008).
v. Droplet coalescence is assumed to be instantaneous and resulting droplet attains instantaneous mechanical stability; interface oscillations are neglected. Also, change in the shape of the drop due to acceleration is neglected.
vi. An equivalent spherical cap approximation has been incorporated to model the droplet shapes. For drops on inclined surfaces, the two-circle approximation is used (ElSherbini and Jacobi 2004a, b).
vii. Though contact angles are obtained (from theory or experiments) under static conditions, these values have been used under dynamic conditions as well.
viii. Partial fall-off of the drops is neglected in the sense that instability results in the complete volume of the drop being removed.
ix. The entire substrate is assumed to be at a constant temperature; drop motion leads to changes in the wall heat flux; local wall temperature fluctuations observed by Bansal et al. (2009) have been neglected.
x. Thermophysical properties of the vapor and liquid phases are taken to be independent of the temperature; the vapor is saturated; all the properties are evaluated at the average of the substrate and saturation temperatures.

6.6 Substrate Leaching

A consequence of the time-dependent processes in dropwise condensation associated with the movement of the drop, first by coalescence and then by sliding motion, is to reduce sustainability on or underneath an inclined chemically textured substrate. Hence, the life of a condensing surface depends on the wall shear interaction of sliding droplets with the drop promoter layer. The phenomenon of removal of the promoter layer over the substrate is called surface leaching. It arises primarily from viscous forces at the contact surface and chemical reactions between the condensing liquid and the promoter. Heat transfer and temperature fluctuations affect these interactions. Accordingly, the long-term sustainability of the process is greatly reduced. Hence, coalescence and sliding of drop in dropwise condensation are significant for improving heat transfer coefficient but reduce the substrate life. Even if there is no chemical reaction between the promoter and condensing liquid, the wall shear stress becomes the primary quantity that controls leaching. A prediction of shear stress requires a complete knowledge of the flow field inside the droplets during coalescence and sliding. Given a shear stress distribution for an individual drop, the net effect due to a drop ensemble can be determined from the time-averaged drop size distribution.

Literature on surface leaching due to drop motion is limited. Therefore, a detailed simulation of flow and heat transfer in a liquid drop sliding underneath a hydrophobic surface and determination of local distribution wall shear stress and wall heat transfer of individual drop form one of the motivations of the present study.

6.7 Closure

A comprehensive mathematical model of dropwise condensation underneath an inclined substrate with and without wettability gradient is presented. The dropwise condensation process is hierarchical because it starts from the atomic scale and progresses on to the engineering scale. The mathematical models of various sub-processes in dropwise condensation have been reported and these are interrelated according to the experimental observations. The overall flow chart of simulation is shown in Fig. 6.26, while Fig. 6.27 shows the various subroutines. A C++ program is written to carry out model simulations.

Chapter 7
Finite Time Coalescence in Dropwise Condensation

Praveen Somwanshi, K. Muralidhar, and Sameer Khandekar

Nomenclature

A	Surface area, m^2
C_p	Specific heat of condensate at constant pressure, J/kg K
F	Force acting on drop, N
h_{int}	Interfacial heat transfer coefficient, $q''/(T_{sat} - T_w)$, W/m^2 K
h_{fg}	Latent heat of vaporization, J/kg
k, k_{coat}	Thermal conductivity of condensate; "coat" for coating, W/m K
L	Distance between two nucleation sites, m
m_{avg}	Space and time-averaged mass of condensate, kg
M	Molecular mass, kg/kmol
N	Nucleation site density, cm^{-2}
p	Pressure, N/m^2
q	Surface heat transfer, W
q''	Surface heat flux, W/m^2
$(q'')_{coal}$	Surface heat flux arising from coalescence, W/m^2
r_b	Base radius of drop (diameter d_b), m
r_{cap}	Capillary length, m
r_{crit}	Radius of drop at instability due to slide-off, m
r_{max}	Radius of drop at instability due to fall-off, m
r_{old}, r_{new}	Radius of the coalesced drop before and after coalescence, m
$r_{b,new}$	Base radius of the coalesced drop after coalescence, m
r_{min}	Minimum radius of the thermodynamically stable drop, m
R, R_{avg}	Base radius of the coalesced drop at equilibrium; "avg" is average over space and time, m
t	Time, s; suffix IS is inertia-surface tension; IV is inertia-viscous; VS is viscous-surface tension

S. Khandekar, K. Muralidhar, *Drop Dynamics and Dropwise Condensation on Textured Surfaces*, Mechanical Engineering Series,
https://doi.org/10.1007/978-3-030-48461-3_7

t_{coal}	Coalescence timescale, s
T_{avg}	Average temperature of condensation = $(T_{sat} + T_w)/2$
T_{sat}	Saturation temperature in vapor phase, K
T_w	Substrate temperature, K
ΔT	Degree of subcooling $(T_{sat} - T_w)$, K
Δt	Time step, s
U, U_{coal}	Relative velocity between substrate and drop; also, characteristic velocity of drop coalescence, m/s
v_l, v_v	Specific volume of liquid and vapor, m^3
V	Volume of liquid drop, m^3

Dimensionless Quantities

Bo	Bond number, $\frac{\rho g R^2}{\sigma}$
$\overline{\overline{C}}_f$	Local and surface-averaged skin friction coefficient, $\tau_w/(1/2)\rho U^2$
Fr	Froude number, $\frac{U^2}{gR}$
$(\mathrm{Nu})_{sd}$	Local Nusselt number of single sliding drop, hr_b/k
Oh	Ohnesorge number, $\frac{\mu}{\sqrt{\rho R \sigma}}$
Pr	Prandtl number, $\mu C_p/k$
Re	Reynolds number, $\rho U d_b/\mu$
We	Weber number, $\frac{\rho U^2 R}{\sigma}$

Greek Symbols

α	Inclination angle, °
μ	Dynamic viscosity of liquid phase, Pa s
ρ	Fluid density ("l" for liquid and "v" for vapor), kg/m^3
σ	Surface tension coefficient at liquid-vapor interface, N/m
$\tau_w, \overline{\tau}_w$	Local and average wall shear stresses, N/m^2
$\tau_{w,\ coal}$	Local wall shear stresses induced during coalescence, N/m^2
θ	Contact angle, °
θ_{adv}	Advancing contact angle, °
θ_{rcd}	Receding contact angle, °
$\Delta\theta$	Contact angle hysteresis, $(\theta_{adv} - \theta_{rcd})$, °
θ_{avg}	Average contact angle, °

7.1 Introduction

Model development of dropwise condensation from saturated vapor over a subcooled surface proceeds along the following lines. Drop embryos form instantaneously at the nucleation sites while the portion of the surface between the growing drops remains dry. Drops subsequently grow by vapor condensation over their surfaces exposed to vapor. Energy release in the form of latent heat is transferred to the cold substrate through the liquid drop. Thus, the limiting factor for heat transfer in dropwise condensation is by way of conduction. The thermal resistance offered by the liquid drop and the available nucleation site density over the substrate play a prominent role in fixing the local wall heat flux.

Condensation at the drop scale additionally involves multiple processes occurring over a wide range of length and timescales. These include coalescence of adjacent drops and gravitational instability of large ones, followed by fresh nucleation at the exposed nucleation sites. Hence, dropwise condensation needs to be modeled by considering applicable processes involving distinct length and timescales. A mathematical model of dropwise condensation underneath a textured substrate involving various sub-processes is outlined in this chapter.

The model described in this chapter is an extension of Chap. 6 and the work of Sikarwar et al. (2011, 2012, 2013a, b). The improvement is mainly in terms of representing finite time coalescence events. Fundamentals of coalescence are discussed in Chap. 3.

7.2 Modeling Dropwise Condensation

Dropwise condensation is a complex phenomenon involving multiple parameters internally connected with each other. It is a consequence of the time-dependent sub-processes associated with the formation of nanometer-sized drops at nucleation sites, growth by direct condensation and coalescence, sliding instability and/or fall-off followed by renucleation on or underneath the substrate. The overall model is thus a collection of the intermediate processes that occur at the scale of individual droplets (Fig. 7.1).

The minimum radius of the thermodynamically stable drop is in the range of a few nanometers for fluids encountered in heat transfer applications. The formation of nanometer-sized droplets at the nucleation sites requires an atomic-scale treatment of vapor molecules interacting with the substrate. It is evidently difficult to capture the initial nucleation phenomenon of these tiny drops on a surface. Sikarwar et al. (2011) showed that the maximum stable cluster obtained by atomistic simulation is equal to the size of minimum stable radius obtained from thermodynamics consideration. This approach is adopted in the present study. In addition, it may be pointed out that the condensation patterns at longer timescales, particularly close to and after instability are found to be insensitive to the atomic level processes. Therefore, the condensation cycle commences with drops of thermodynamically minimum radius and atomic level modeling of smaller liquid drops is not pursued in the present study.

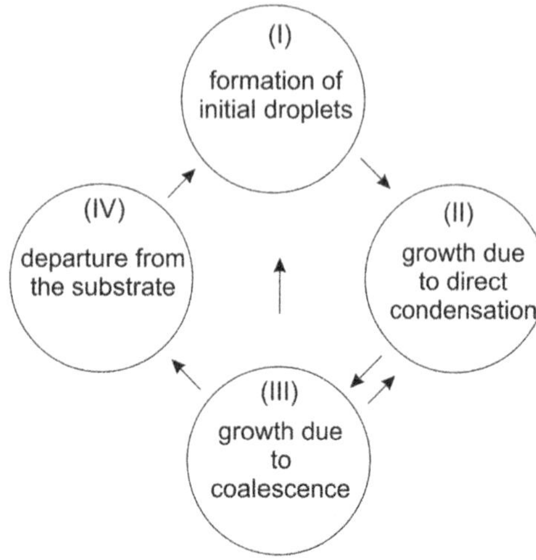

Fig. 7.1 Schematic diagram showing stages of dropwise condensation cycle

7.2.1 *Nucleation Site Density*

Experimental as well as theoretical estimation of the nucleation site density is a formidable task. From a modeling perspective, nucleation site density is dependent on various parameters such as the thermophysical properties of the condensing fluid, physico-chemical properties of the substrate, degree of subcooling, and the substrate morphology. It is a model parameter in the present study.

Nucleation site density has been inferred from indirect measurements of water vapor condensation in the literature. The size of a thermodynamically stable drop at nucleation is of the order of a few nanometers at preferred nucleation sites and suggests a nucleation site density of around 10^6 cm^{-2}. Leach (2006) reported nucleation site density in the range of 10^4–10^6 cm^{-2}. Rose (2002) quoted 10^6 cm^{-2} as a possible nucleation site density and connected this quantity with the radius of the thermodynamically stable drop. With this choice of nucleation site density, simulations of mercury vapor condensation were found to be in good agreement with experiments for wall heat flux (Khandekar and Muralidhar 2014). However, careful determination of nucleation site density for metal vapor condensation has not been considered in the literature.

Liu and Cheng (2015) and Niu et al. (2017) reported in their work a correlation for the available nucleation site density during the condensation process. Liu and Cheng (2015) studied theoretically the effect of subcooling, contact angle, thickness, and thermal conductivity of the coating layer on nucleation site density and condensation heat flux during dropwise condensation. The nucleation site density predicted by Rose (1976) for water vapor based on minimum radius was seen to be an overprediction. The nucleation site density predicted by Liu and Cheng (2015) matched with the experiments when the minimum radius was replaced with critical radius derived by minimizing the free energy. Niu et al. (2017) studied the effect of liquid-solid interfacial thermal resistance on dropwise condensation through modeling and experiments. The nucleation-free energy derived by Liu and Cheng (2015) was minimized to find the critical radius by including the effect of the temperature distribution in the bulk droplet on the liquid-solid interfacial thermal resistance. Incorporation of liquid-solid interfacial thermal resistance increased the critical nucleation radius and reduced the nucleation density. The literature quoted above is for water and a study on nucleation site density including experimental validation for liquid metals has not been reported. It is expected that the nucleation site density will vary substantially with the choice of the characteristics of the surface. Hence, it is chosen as a parameter for numerical simulation that can take on a range of values.

7.2.2 Nucleation Site Distribution

Nucleation sites are taken as randomly distributed over the substrate area. Site locations are generated using the random number generator function available in the standard library of C++. The function returns random numbers in the range of (0, 1) following a Gaussian distribution. The distribution of nucleation sites is assigned column-wise over the substrates until all the sites are occupied. The assigned nucleation sites remain fixed for the entire simulation.

Parameters such as average contact angle, contact angle hysteresis, and nucleation site density are unique for a given substrate-fluid combination which are provided to the model in the form of input data.

Physically textured and chemically textured surfaces are different from each other in terms of the ambiguity involved in the measurement of the apparent contact angle. The apparent dynamic contact angles and the movement of three-phase contact line also show differences for physically and chemically textured surfaces. The differences are small for a large drop close to instability. The drop size is of the order of a few millimeters for the fluids of interest. This is much larger than the surface features such as pillars for a hydrophobic surface. The macroscopic behavior of physically and chemically textured surfaces is expected to be comparable because dropwise condensation is a cyclic process and many drops of varying sizes are spread randomly over the substrate. The proposed model is thus expected to be valid for both physically and chemically textured surfaces.

7.2.3 Determination of Minimum Radius of the Drop

To start the simulation, nucleating liquid drops are distributed over randomly located nucleation sites of the substrate. As discussed in Chap. 6, the radius of the thermodynamically stable smallest drop is

$$r_{\min} = \frac{2\sigma v_{\mathrm{l}} T_{\mathrm{w}}}{h_{\mathrm{lv}}(T_{\mathrm{w}} - T_{\mathrm{sat}})} \tag{7.1}$$

7.2.4 Drop Growth by Direct Condensation

Initially, drops grow by direct condensation of vapor over their exposed surfaces. The growth by direct condensation in not influenced in the presence of finite time coalescence. The growth rate of a drop in terms of its radius, derived in Chap. 6 is

$$\frac{\mathrm{d}r}{\mathrm{d}t} = \left(\frac{4(T_{\mathrm{sat}} - T_{\mathrm{w}})}{\rho_{\mathrm{l}} h_{\mathrm{lv}}}\right)\left(1 - \frac{r_{\min}}{r}\right) \times \left[\frac{2}{h_{\mathrm{int}}} + \frac{r(1 - \cos\theta_{\mathrm{avg}})}{k} + \frac{4\delta}{k_{\mathrm{coat}}(1 + \cos\theta_{\mathrm{avg}})}\right]^{-1} \left[\frac{1 - \cos\theta_{\mathrm{avg}}}{2 - 3\cos\theta_{\mathrm{avg}} + \cos^3\theta_{\mathrm{avg}}}\right] \tag{7.2}$$

7.2.5 Drop Growth by Coalescence

Following growth from condensing vapor, adjacent drops may touch each other and coalesce. From a modeling approach, coalescence is a short duration event and be considered in two ways. First, it may be treated as instantaneous, and the second, it may be taken to prevail over a finite time duration with a characteristic velocity associated with it.

Drop Growth by Instantaneous Coalescence

Sikarwar et al. (2011, 2012, 2013a, b) assumed coalescence to be instantaneous in the sense that the two drops touching each other are immediately replaced by another of the combined volume. The dropwise condensation model with this approach to coalescence is described in Chap. 6.

Finite Time Drop Coalescence

Coalescence of drops of equal and unequal volumes have been experimentally investigated in Chap. 3. An individual coalescence event lasts for a few milliseconds which is very small compared to the cycle time of dropwise condensation of a few minutes. The detailed observations and discussion in Chap. 3 show that treatment of coalescence is necessary despite its short-lived nature.

As seen in the image sequences (Figs. 3.13–3.16, and 3.27), large velocities are generated during early time, followed by slow decay over a longer time span. In addition, there is a distinct dependence of magnitude of velocity on the size of the combined drop. The nature of unsteadiness is characteristic of the chosen velocity component, the vertical velocity providing for long-lived gravity-induced oscillations. Figures 3.13–3.16, and 3.27 clearly show the appearance of two timescales within the transient coalescence process. The first is rapid and lasts for around 10–30 ms. The second is longer and persists over a timescale of greater than 200 ms. These experimental trends can be supported by scale analysis.

Timescales appropriate for the coalescence process can be estimated using dimensional analysis and are discussed in the following paragraphs. Let R be a length scale and U, a velocity scale. In the present discussions, R is taken to be the radius of the footprint of the resulting drop after equilibrium is attained. For the coalescence of two stationary drops, the velocity scale is an internal variable and cannot be independently prescribed. Similarly, the process does not have a single prescribed timescale. These may be estimated by consideration of forces relevant to coalescence. Forming dimensionless groups of forces, we have the following numbers:

$$\text{Inertia-surface tension, Weber number}: \quad \mathrm{We} = \frac{\rho U^2 R}{\sigma} \tag{7.3}$$

$$\text{Inertia-gravity, Froude number}: \quad \mathrm{Fr} = \frac{U^2}{gR} \tag{7.4}$$

$$\text{Inertia-viscosity, Reynolds number}: \quad \mathrm{Re} = \frac{\rho U R}{\mu} \tag{7.5}$$

$$\text{Gravity-surface tension, Bond number}: \quad \mathrm{Bo} = \frac{\rho g R^2}{\sigma} \tag{7.6}$$

$$\text{Viscosity-surface tension, Ohnesorge number}: \quad \mathrm{Oh} = \frac{\mu}{\sqrt{\rho R \sigma}} \tag{7.7}$$

The footprint radius relates to the drop volume via the contact angle (Eqs. 6.29 and 6.33).

The dimensionless quantities defined in Eqs. (7.3–7.7) are evaluated and presented in Table 7.1. As the range of volumes (0.05–7.5 μl) considered have a length scale of 1–2 mm, surface tension is expected to be uniformly important. The highest Bond number in Table 7.1 is less than unity confirming the importance of surface tension in the analysis.

The characteristic velocity scale based on surface tension and viscosity, surface tension and inertia, and gravity alone, are respectively

Table 7.1 Magnitudes of dimensionless parameters estimated for water under ambient conditions. Velocity scale U is taken as $(\sigma/\rho R)^{0.5}$, where R is the base radius of the coalesced drop at equilibrium

Volume, μl (U, m/s)	Base radius (mm)	Bo	We	Fr	Re	Oh
0.05 (0.54)	0.237	0.008	1	127.882	161.022	0.62×10^{-3}
0.5 (0.37)	0.510	0.037	1	27.551	236.348	4.23×10^{-3}
1.5 (0.31)	0.736	0.075	1	13.245	283.839	3.52×10^{-3}
2.5 (0.28)	0.872	0.106	1	9.422	309.063	3.24×10^{-3}
7.5 (0.24)	1.258	0.221	1	4.530	371.165	2.69×10^{-3}

$$U = \frac{\sigma}{\mu}, \quad U = \sqrt{\frac{\sigma}{\rho R}}, \quad U = \sqrt{gR} \tag{7.8}$$

As the velocity scale based only on gravity does not include surface tension, it is not considered further. The characteristic velocity for water using the expression σ/μ is 87.7 m/s. This is quite large and has not been observed in the centroidal velocities of the combined drop (Chap. 3). Instead, the velocity based on scaling $(\sigma/\rho R)^{0.5}$ is around 0.2 m/s and can be seen in the experimental observations of Chap. 3. The choice of this scale makes Weber number unity and Oh = 1/Re.

With $(\sigma/\rho R)^{0.5}$ as the velocity scale, Table 7.1 shows that drop oscillations associated with coalescence are expected to be driven by inertia and surface tension. Here, Reynolds number is expected to be high, immediately following bridge formation, while viscosity plays a secondary role. Between surface tension and gravity, the former is of greater significance. In agreement with this observation, Froude number is also seen to be on the higher side. Thus, the volumes of coalescing drops are such that the coalescence of these drops is controlled by inertia and surface tension for the observable velocities, viscosity serving the purpose of damping fluid motion over a longer time span.

Additional scales of interest are the following: For small liquid drops in a gaseous environment, the importance of gravity over surface tension is determined by the capillary length $l = \sqrt{\sigma/\rho g}$. For the properties of water (Table 7.2), the capillary length is ~2.6 mm. Since the maximum length scale in Table 7.3 is 1.258 mm, coalescence is dominated by surface tension.

The relevant timescales associated with pairs of forces including inertia-surface tension, viscous-surface tension, and inertia-viscous are

$$t_{\mathrm{IS}} = \sqrt{\frac{\rho R^3}{\sigma}}, \quad t_{\mathrm{VS}} = \frac{\mu R}{\sigma}, \quad t_{\mathrm{IV}} = \frac{R^2}{\nu} \tag{7.9}$$

Based on the properties of water, timescales defined in Eq. (7.9) are summarized in Table 7.3. As shown, the timescale of viscous–surface tension interaction is very small and is not of any specific relevance to the present study. The inertia-surface tension timescale relates to the moment following initial bridge formation when large velocities are generated within the drop, leading to large deformation of the

Table 7.2 Thermophysical properties of water

Property	Water
Density, ρ (kg/m^{-3})	995
Latent heat of vaporization, h_{lv} (kJ/kg)	2426
Specific volume, υ (m^3/kg)	0.001 (l), 29.74 (v)
Surface tension, σ (N/m)	0.0709
Thermal conductivity, k (W/mK)	0.62
Molecular weight, M (kg/kmol)	18
Dynamic viscosity, μ (Pa s)	0.769×10^{-3}
Advancing contact angle, θ_{adv} (°)	118.5
Receding contact angle, θ_{rcd} (°)	101.5
Saturation temperature, T_{sat} (K)	303
Wall temperature, T_w (K)	298
Prandtl number (−)	5.182

Table 7.3 Timescales estimated for coalescence of water drops under ambient conditions

Volume (μl)	Base radius (mm)	t_{IS} (ms)	t_{VS} (ms)	t_{IV} (ms)
0.05	0.237	0.434	2.70×10^{-3}	0.070×10^3
0.5	0.510	1.374	5.81×10^{-3}	0.325×10^3
1.5	0.736	2.379	8.38×10^{-3}	0.675×10^3
2.5	0.872	3.072	9.94×10^{-3}	0.949×10^3
7.5	1.258	5.32	1.43×10^{-3}	1.975×10^3

IS inertial-surface tension, *VS* viscous-surface tension, *IV* inertia-viscous

interface. The inertia-viscous timescale is quite large and can be associated to bulk dissipation of kinetic energy of the fluid, leading to long-term relaxation of the coalesced drop towards equilibrium. Additional dissipation occurring at the three-phase contact line cannot be selectively identified since it will be seen jointly with bulk dissipation.

Large shear stresses can be associated with the inertial-surface tension coupling and hence, in balance, the inertia-surface tension-based timescale ($=t_{IS}$) is considered as relevant for inclusion in the dropwise condensation model. The velocity scale $(\sigma/\rho R)^{0.5}$ is considered appropriate in the present context.

7.3 Drop Instability

The growth of a drop during dropwise condensation is initially by direct condensation and subsequently by coalescence. The weight of the drop continuously increases until it reaches a critical value. The imbalance of forces at the three-phase contact line causes instability in the form of drop movement. The drop falls off from a horizontal surface and slides away from vertical and inclined surfaces. Fall-off of a sliding drop is also a possibility.

7.3.1 *Horizontal Substrate*

The critical drop size at fall-off is (Chap. 6)

$$r_{\max} = \left[\frac{6 \sin^2 \theta_{\text{avg}}}{\left(2 - 3 \cos \theta_{\text{avg}} + \cos^3 \theta_{\text{avg}}\right)} \frac{\sigma}{(\rho_{\text{l}} - \rho_{\text{v}}) g} \right]^{\frac{1}{2}} \tag{7.10}$$

7.3.2 *Inclined Substrate*

On an inclined surface, the drop will deform giving rise to contact angle hysteresis. Sliding motion is initiated when the drop radius just exceeds a critical radius r_{crit}. During the sliding process, the drop coalesces with others along its path, may become larger than an appropriate critical radius $r_{\max}$ and fall-off. These possibilities have been included in the computer program of the dropwise condensation model.

Estimation of Critical Radius (r_{crit}) Underneath an Inclined Substrate

As derived in Chap. 6, the maximum size of the drop at which it becomes unstable and starts to slide-off is

$$r_{\text{crit}} = \left[\frac{1.215 \sin \theta_{\text{avg}} \left(\cos \theta_{\text{rcd}} - \cos \theta_{\text{adv}} \right)}{\left(2 - 3 \cos \theta_{\text{avg}} + \cos^3 \theta_{\text{avg}}\right) (\rho_{\text{l}} - \rho_{\text{v}})} \frac{\sigma}{g \sin \alpha} \right]^{\frac{1}{2}} \tag{7.11}$$

Estimation of Maximum Radius ($r_{\max}$) Underneath an Inclined Substrate

As described in Chap. 6, the maximum size of the drop at which it falls off is

$$r_{\max} = \left[\frac{3 \sin \theta_{\text{avg}} \left(\sin \theta_{\text{rcd}} + \sin \theta_{\text{adv}} \right)}{\left(2 - 3 \cos \theta_{\text{avg}} + \cos^3 \theta_{\text{avg}}\right) (\rho_{\text{l}} - \rho_{\text{v}})} \frac{\sigma}{g \cos \alpha} \right]^{\frac{1}{2}} \tag{7.12}$$

7.3.3 *Modeling Terminal Velocity*

In dropwise condensation, drop starts sliding motion post instability. The speed increases with time until the unbalanced force matches wall shear. In the present work, it is assumed that drops attain terminal velocity immediately after attaining instability. The calculation of terminal velocity is as discussed below.

For an Inclined Substrate

In terms of a skin friction coefficient, terminal velocity over an inclined surface is

$$U = \sqrt{\frac{2(F_{\mathrm{g}} - F_{\mathrm{r}})}{C_{\mathrm{f}} A_{\mathrm{sl}} \rho_{\mathrm{l}}}} \tag{7.13}$$

The derivation of an expression for terminal velocity is given in Chap. 6.

For a Horizontal Substrate

A heavy drop will drop-off from a horizontal surface and does not have a terminal velocity.

7.4 Liquid Hold-Up

Condensation model described in Figs. 7.1 and 7.2 can be used to estimate the parameters such as hold-up, wall heat transfer, wall heat flux, and spatio-temporal drop distribution.

Liquid hold-up is the total mass of liquid retained at the substrate in the form of drops at various nucleation sites. In a distillation process, the condensate is the most important quantity. Hence, the liquid hold-up is required to be a minimum while the drained quantity should be a maximum. For calculations, the number of available nucleation sites and size of the drop at these locations is obtained from simulation at every time step. The mass held-up at the ith nucleation site is

$$(\text{hold-up})_i = \rho \frac{\pi}{3} r_i^3 \left(2 - 3\cos\theta_{\mathrm{avg}} + \cos^3\theta_{\mathrm{avg}}\right) \tag{7.14}$$

The total mass held-up on the substrate accumulated at all the available nucleation sites (except those submerged under the individual drops) is

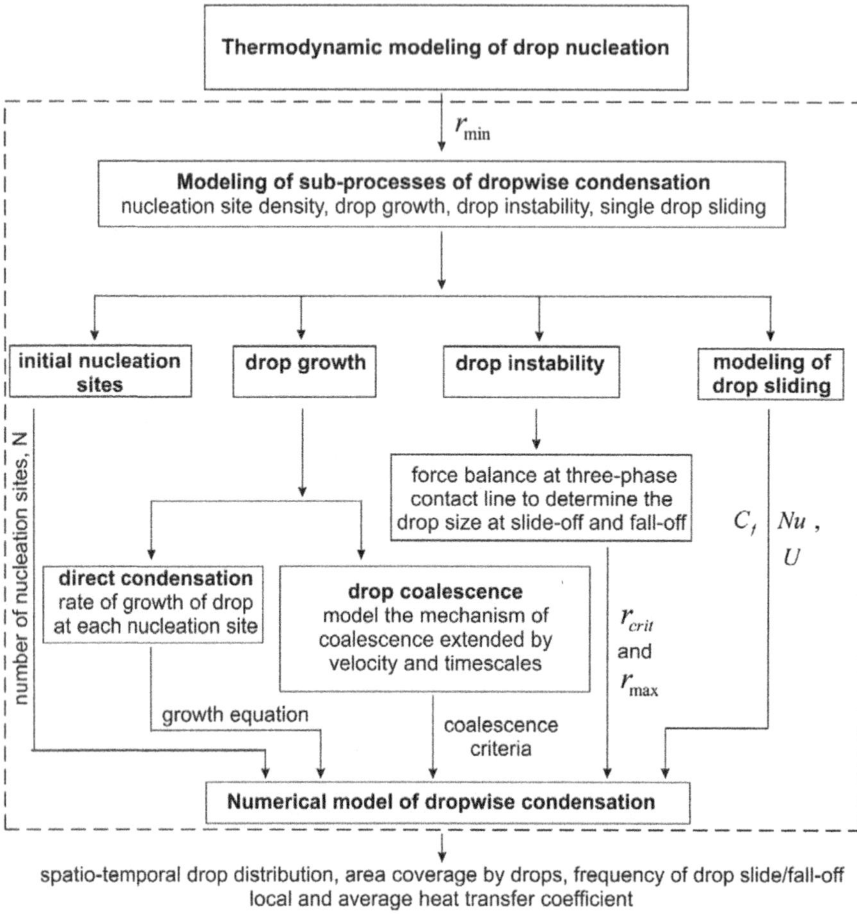

Fig. 7.2 Flow chart presenting a multiscale model of dropwise condensation with the incorporation of finite time coalescence

$$\text{hold-up} = \frac{\sum_{i=1}^{N} (\text{hold-up})_i}{A} \tag{7.15}$$

7.5 Wall Heat Transfer

The release of energy in the form of latent heat to the substrate is through the drop and hence the open areas on the substrate are inactive for heat transfer. The heat transfer rate depends on nucleation site density and the growth rate of the drop at each nucleation site. The growth rate of the drop is calculated from Eq. (7.2). Heat transfer rate during dropwise condensation is then calculated at free surface of the drop located at each nucleation sites of the substrate.

7.5.1 During Direct Condensation

Heat transfer during dropwise condensation is calculated based on direct condensation of vapor on drops at each nucleation site as follows. This approach carries over to a model wherein coalescence is taken to be instantaneous.

The available nucleation sites N are first estimated. The accumulated mass of the condensate for ith nucleation site over a time step is

$$(\Delta m)_i = \rho \frac{\pi}{3}\left(2 - 3\cos\theta_{\text{avg}} + \cos^3\theta_{\text{avg}}\right)\left(r_{\text{new}}^3 - r_{\text{old}}^3\right)_i \tag{7.16}$$

The total quantity of mass of condensate over the substrate for a given time step is

$$\Delta m = \sum_{i=1}^{N} (\Delta m)_i \tag{7.17}$$

The total number of time steps $K = t/\Delta t$ is the ratio of time period of condensation t to the time interval Δt. The average rate of mass of condensate is

$$m_{\text{avg}} = \frac{\sum_{j=1}^{K} (\Delta m)_j}{t} \tag{7.18}$$

The time and space-averaged heat transfer coefficient is

$$h = \frac{m_{\text{avg}} h_{\text{lv}}}{A(T_{\text{sat}} - T_{\text{w}})} \tag{7.19}$$

7.5.2 During Coalescence of Drops

Coalescence is incorporated in the dropwise condensation model as discussed below.

The volume of the drop located at ith and jth nucleation sites is

$$V_i = \frac{\pi}{3} r_i^3 \left(2 - 3\cos\theta_{\text{avg}} + \cos^3\theta_{\text{avg}}\right) \tag{7.20}$$

$$V_j = \frac{\pi}{3} r_j^3 \left(2 - 3\cos\theta_{\text{avg}} + \cos^3\theta_{\text{avg}}\right) \tag{7.21}$$

Post coalescence, the radius of the merged drop is estimated by

$$r_{\text{old},i} = \left[\frac{3\left(V_i + V_j\right)}{\pi\left(2 - 3\cos\theta_{\text{avg}} + \cos^3\theta_{\text{avg}}\right)}\right]^{1/3} \tag{7.22}$$

The relevant timescale of coalescence as discussed in Chap. 3 are

$$t_{\text{coal},i} = \left[\frac{\rho r_{\text{old},i}^3}{\sigma}\right]^{1/2} \tag{7.23}$$

The radius of the drop after coalescence following growth over the relevant timescale is

$$r_{\text{new}} = r_{\text{old},i} + \left(\frac{\mathrm{d}r}{\mathrm{d}t}\right)_{\text{coal}} \frac{t_{\text{coal},i}}{2} \tag{7.24}$$

where

$$\left(\frac{\mathrm{d}r}{\mathrm{d}t}\right)_{\text{coal}} = \left(\frac{4\left(T_{\text{sat}} - T_{\text{w}}\right)}{\rho_{\text{l}} h_{\text{lv}}}\right) \left[\frac{1 - \frac{r_{\min}}{r_{\text{old},i}}}{\frac{2}{h_{\text{int}}} + \frac{r_{\text{old},i}\left(1 - \cos\theta_{\text{avg}}\right)}{k} + \frac{4\delta}{k_{\text{coat}}\left(1 + \cos\theta_{\text{avg}}\right)}}\right] \times \left[\frac{1 - \cos\theta_{\text{avg}}}{2 - 3\cos\theta_{\text{avg}} + \cos^3\theta_{\text{avg}}}\right] \tag{7.25}$$

The updated base radius is

$$r_{\text{b,new}} = r_{\text{new},i}\sin\theta_{\text{avg}} \tag{7.26}$$

The relevant velocity scale as discussed in Chap. 3 is

$$U_{\text{coal},i} = \left[\frac{\sigma}{\rho r_{\text{b,new}}}\right]^{1/2} \tag{7.27}$$

The condensed mass at the ith coalescing event is

$$(\Delta m)_{\mathrm{coal},i} = \frac{\rho \frac{\pi}{3}\left(2 - 3\cos\theta_{\mathrm{avg}} + \cos^3\theta_{\mathrm{avg}}\right)\left(r_{\mathrm{new},i}^3 - r_{\mathrm{old},i}^3\right)}{t_{\mathrm{coal}}\pi r_{\mathrm{b,new}}^2} \tag{7.28}$$

The total mass of n coalescing events is obtained by summation by

$$(\Delta m)_{\mathrm{coal}} = \sum_{i=1}^{n} (\Delta m)_{\mathrm{coal},i} \tag{7.29}$$

The heat flux transferred to the substrate during coalescence is obtained by summing over all the coalescing events as

$$(q'')_{\mathrm{coal}} = \frac{(\Delta m)_{\mathrm{coal}} h_{\mathrm{lv}}}{A} \tag{7.30}$$

7.6 Wall Shear Stress

Wall shear stress is generated in dropwise condensation during drop sliding on or underneath substrate, as well as during coalescence of drops.

7.6.1 Drop Sliding

Instability in the form of slide-off (and possibly later fall off) is observed during dropwise condensation on or underneath an inclined substrate. The velocity associated with the drop causes shear strain and hence wall shear stress. Local wall shear stress is obtained from the skin friction coefficient as (Chap. 6)

$$\overline{\tau}_{\mathrm{w,slide}} = \frac{\overline{F}_{\mathrm{s}}}{A} = \frac{1}{A}\left[\frac{\sum_{i=1}^{N} [F_{\mathrm{s}}]_i^j}{t}\right] \tag{7.31}$$

The space and time-averaged skin friction coefficient on the substrate is

$$\overline{\overline{C}}_{\mathrm{f}} = \frac{\overline{\tau}_{\mathrm{w,slide}}}{\frac{1}{2}\rho U_{\mathrm{rsp}}^2} \quad \text{where} \quad U_{\mathrm{rsp}}^2 = \sqrt{g \cdot r_{\mathrm{crit}}} \tag{7.32}$$

7.6.2 Coalescence of Drops

Coalescence events will occur on or underneath surfaces of all orientations. Large oscillations pertaining to the bulk mass of the coalesced drop cause shear strain and hence shear stress on the wall. Starting from an estimated velocity, shear stress induced during coalescence of drops is

$$\tau_{\mathrm{w,coal}} = \mu\left[\frac{U_{\mathrm{coal},i}}{(r_{\mathrm{b,new}}/2)}\right] \tag{7.33}$$

7.7 Numerical Simulation of Dropwise Condensation of Water Vapor on Horizontal and Vertical Surfaces

The choice of water as the working fluid is motivated by the range of applications, where water vapor is encountered. The advancing and receding contact angles considered are 118.5° and 101.5°, respectively. At high levels of subcooling, drops are expected to form a liquid film on the substrate, and hence the degree of subcooling between the saturated vapor and the substrate is taken to be low, as 1 °C. Since the temperature interval is small, a constant thermophysical property approximation at the average temperature is utilized. Fluid motion in dropwise condensation is seen during (a) the instability of the liquid drop, either during its sliding motion or fall-off and (b) coalescence. Both events occur for less than 0.1% of the condensation cycle time. The coalescence characteristics are determined by a variety of properties such as surface tension, contact angle, density, and volume.

7.7.1 Mathematical Modeling

Models of the dropwise condensation process have been reported in the literature with varying levels of approximations (Vemuri and Kim 2006; Kim and Kim 2011). The model described by Khandekar and Muralidhar (2014) (also see Chap. 6) is the starting point for the present work. The model proceeds from nucleation of drops to their growth and instability but treats coalescence to be an instantaneous process. The model, comprising these four steps, is schematically shown in Fig. 7.3. The model input parameters include nucleation site density, contact angle and hysteresis, interfacial properties, thermodynamics of phase change, and the thermophysical properties of the condensing phase. Thermophysical properties of the vapor and liquid phases are evaluated at the average of the substrate and saturation temperatures. The model predicts the instantaneous condensation pattern, the mass of condensate leaving the surface, and wall heat flux. Additional quantities of interest are the cycle time of instability, liquid hold-up, instantaneous and time-averaged heat fluxes and wall shear stress. Validation studies against experiments of Rose (2002) have been discussed by the authors in their monograph.

The instantaneous space-averaged heat transfer coefficient over an area A of the substrate during dropwise condensation is given in terms of the mass of the vapor condensing over the surface per unit time m_{avg} and the degree of applied subcooling ΔT is estimated using Eq. (7.19).

Shear forces are generated by each drop when it begins to slide over the substrate. Prior to such instability, shear stress is also generated when droplets coalesce. The mass of the coalesced drop will usually increase during its travel on the substrate. These are discrete events and are cumulatively determined over the cycle time of

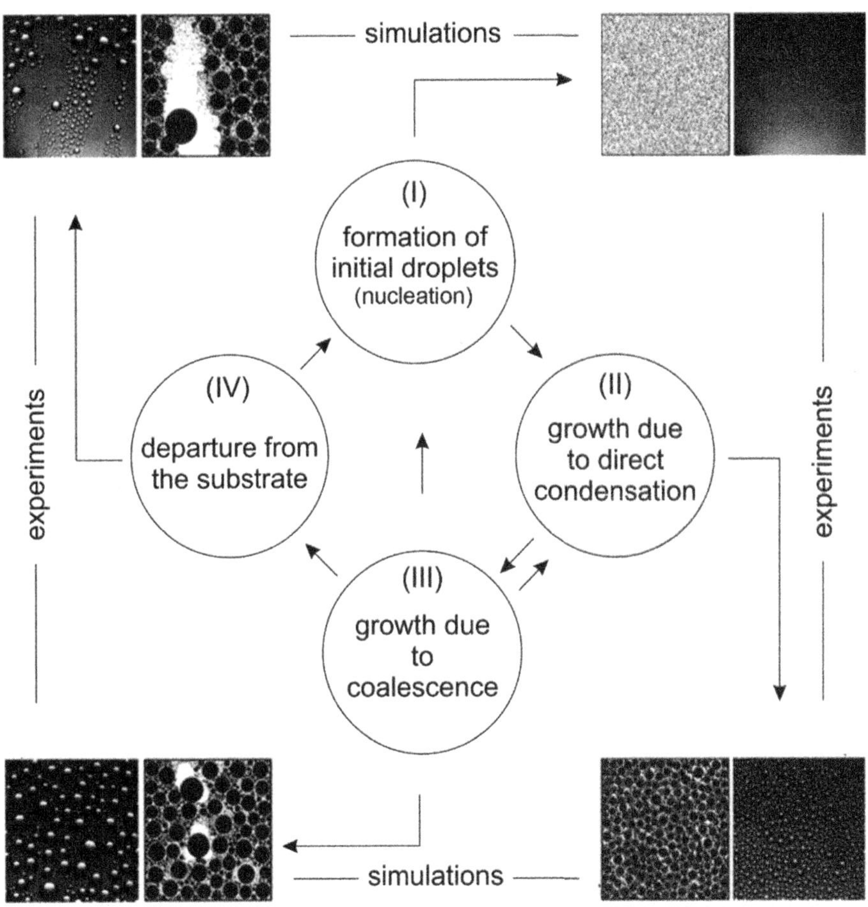

Fig. 7.3 Schematic drawing of a mathematical model representing the dropwise condensation cycle. The model represents the condensation process in four steps, namely nucleation, growth by direct condensation of vapor, growth by coalescence, and drop instability. Experimentally recorded images in water vapor (left) are shown jointly with those obtained from simulation

condensation. The space and time-averaged shear stress on the substrate arising from drop movement is calculated using Eq. (7.31).

The space and time-averaged skin friction coefficient on the substrate can be calculated using Eq. (7.32).

The condensation model of the present study extends the past work of Sikarwar et al. (2013a) in two significant respects. First, coalescence is not treated as instantaneous. Velocity scale and timescale are estimated from non-dimensional analysis and in turn used to estimate additional heat flux as well as the wall shear stress. Secondly, simulations are adapted to a high-performance computing system using domain decomposition that permits the use of MPI to run the computer code in parallel. The second step permits simulations over surfaces as large as 50 × 50 mm in a reasonable amount of time.

7.7.2 Condensation Model with Finite Time Coalescence

In the condensation context, coalescence takes place over a horizontal surface in the pendant configuration. The drops are composed of the same liquid while coalescence takes place in a vapor environment. The drops are not pinned at the three-phase contact line and the footprint evolves with time. Coalescence is taken to introduce additional velocity, length, and timescales that are suitably incorporated in the condensation cycle model.

Timescales related to coalescence in water drops are summarized in Table 7.3. The timescale of viscous–surface tension interaction is very small and is not of any specific relevance to the present study. The inertia-surface tension timescale relates to the moment following bridge formation when large velocities are generated within the drop, leading to large deformation of the interface. The inertia-viscous timescale is quite large and can be associated to bulk dissipation of kinetic energy of the fluid, leading to long-term relaxation of the drop towards equilibrium. Additional dissipation occurring at the three-phase contact line cannot be selectively identified since it will be seen jointly with bulk dissipation. Large shear stresses can be associated with the inertial-surface tension coupling and hence, in balance, the inertia-surface tension-based timescale ($=t_{IS}$) is considered as relevant for inclusion in the dropwise condensation model.

Timescales and velocity scales emerging from drop coalescence have been addressed by the authors in experiments (Somwanshi et al. 2018). The experimental apparatus is shown in Fig. 3.1. A sequence of images showing the coalescence process at selected time instants is also shown in Figs. 3.14–3.17. The match between timescales recorded in images and the inertia-surface tension timescale of Table 7.3 show good agreement. Velocities acquired by the centroid of the merging drops were estimated from the image sequence and were found to be of the order of $(\sigma/\rho R)^{0.5}$, confirming inertia and surface tension forces as most representative of the coalescence process.

The shear stress generated during dropwise condensation due to coalescence is estimated from the scales applicable for inertia–surface tension interaction. The length scale, namely the base radius of the coalesced droplet is determined from the total volume of the coalescing drops (Eq. 7.22). The characteristic velocity scale arising at this instant is given by $(\sigma/\rho R)^{0.5}$, and the strain rate is the ratio of velocity and length scales. Hence, the instantaneous shear stress is determined using Eq. (7.33).

The additional heat flux associated with coalescence is obtained as follows. The average mass of liquid condensed is estimated using the initial and final base radius over the coalescence timescale t_{IS}. The additional heat flux over the substrate is determined as the product of the average mass of vapor condensed during coalescence and the latent heat release during phase change divided by the base area and the timescale.

7.7.3 Simulation of Dropwise Condensation of Water Vapor

Instantaneous condensation patterns of water on vertical and horizontal surfaces are discussed in the present work. The effect of the size of the substrate and the condensation patterns at the core of the surface at the instant of instability are considered. Condensation parameters of water are compared. A quantity of importance is liquid hold-up at the substrate. A larger average hold-up indicates a greater heat transfer resistance and is undesirable in many applications. It is to be expected that the hold-up would be greater for a horizontal surface compared to a vertical, being related to the cycle time and the drop size at instability. The degree of subcooling employed in simulations for water is 1 K at a saturation temperature of water vapor of 303 K.

For horizontal and near-horizontal surfaces with up to an inclination of 10° from the horizontal, instability was seen to be mainly due to fall-off and the condensation patterns thus obtained were indistinguishable. The results discussed in the following paragraphs use a 10° orientation for the near-horizontal surface but the surface, for simplicity, is addressed as "horizontal."

Heat transfer rates during dropwise condensation are controlled by the sum of the conduction resistance that scales with the average radius of drop and convection resistance, namely the reciprocal of the vapor-liquid heat transfer coefficient. In previous studies related to water (Sikarwar et al. 2012), conduction through the drops was found to be the predominant mechanism of heat transfer. With an increase in the nucleation site density, frequent coalescence events lead to early instability and fresh nucleation, thus diminishing the size of the average drop radius. Hence, in water, average heat transfer coefficient increased with the nucleation site density. For the discussion in the following sections, nucleation site density of 10^6 cm^{-2} has been selected.

Condensation Patterns of Water Vapor on Horizontal and Vertical Surfaces

Spatio-temporal drop distributions, from initial nucleation to drop instability are shown in Fig. 7.4 for condensation of water vapor. The underside of a horizontal surface and a vertical surface are individually studied. Surface dimensions of 20 × 20 mm are specified. Sizes of equipment of interest are larger than the dimensions studied but a truly multiscale simulation starting from nucleation is computationally intractable on a device scale. The approach adopted in the present work is to examine surfaces of increasing dimensions and the influence of confinement on the condensation dynamics.

The simulation begins with drops of minimum radius randomly placed at the nucleation sites. Subsequently, they are grown over a time step by direct condensation of the surrounding vapor. Partially grown drops are coalesced with their neighbors if they touch each other. Otherwise, they continue to grow by direct condensation. Fall-off on a near-horizontal and slide-off on a vertical surface are observed if the gravitationally instability criteria are fulfilled. During sliding motion, the size of the sliding drop increases by coalescence with those which come in its path. Instability exposes the surface to fresh nucleation and the entire cycle of nucleation to instability is resumed. When large drops merge or slide/fall-off, fresh sites are exposed, and small drops are placed at these locations. As a result, the relative population of small drops increases. Just ahead of criticality, the average

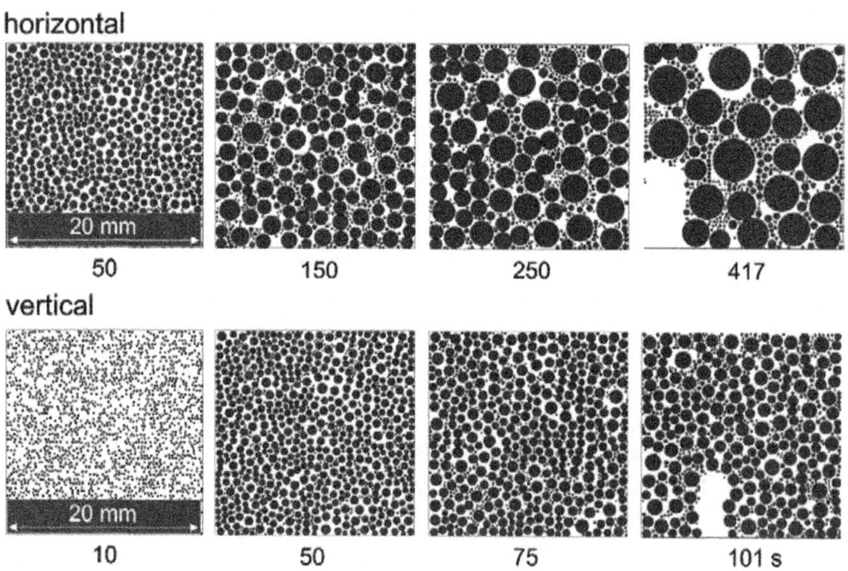

Fig. 7.4 Comparison of full surface condensation patterns in water vapor over 20 × 20 mm surfaces for both, near-horizontal and vertical (gravity pointing downward) surfaces at four time instants (time in seconds)

drop diameter over the surface is the greatest. The combined effect of direct condensation, coalescence, and instability creates drops of varying sizes over the surface, as seen in Fig. 7.4.

The size of a typical drop increases with time till criticality. Jointly, thermal resistance of this drop increases with time. For a given level of subcooling, namely the vapor-substrate temperature difference, heat flux will be large for small drops and small for large drops. Thus, heat fluxes are large at fresh nucleation and will attain a minimum ahead of instability. Coalescence of drops on horizontal and vertical surfaces will further increase the local heat flux. For a vertical surface, shear stresses are generated during the sliding motion of the drop. For a strictly horizontal surface, fall-off is taken to be instantaneous and shear stresses are absent. Coalescence-induced velocity generates shear stresses for both surfaces and could be comparable or even greater than that arising from sliding motion.

Referring to Fig. 7.4 and Table 7.4, the size of the drop at criticality is larger for horizontal surfaces than the vertical. This is because gravity must overcome the contact angle for a horizontal surface while it must overcome the advancing angle to receding angle difference for a vertical surface. Hence, the retention of mass is larger for a horizontal surface when compared to the vertical. It may be realized that drop growth rates are limited by the sum of conduction and vapor-liquid interfacial resistance and diminish for larger drops. In addition, since the drop grows to a larger size over a horizontal surface, the condensation cycle time is clearly greater here when compared to the vertical.

Resistance to heat transfer arises from conduction through the drops and the interfacial heat transfer coefficient. For smaller drops, conduction resistance is small and is a maximum at instability. A smaller cycle time will experience frequent nucleation and show a preference for smaller drops, and hence a smaller conduction resistance. Table 7.4 shows the interfacial resistance in water to be much smaller than conductive resistance. Hence, in water, conduction, namely drop size controls thermal resistance and for a given level of subcooling, determines the overall heat transfer coefficient. Accordingly, in water, one can see a strong correlation among small drop sizes, small cycle time, and large heat transfer coefficient. Specifically, heat transfer coefficient is higher for a vertical surface when compared to the horizontal while the average drop size is smaller.

Liquid Hold-Up

From an application viewpoint, liquid hold-up over the condensing surface is a quantity of great importance. It is the difference between the amount of vapor condensed and the amount drained away. Since the condensate represents the productive aspect of the distillation process, it is expected that the liquid hold-up among selected surfaces and inclination should be a minimum.

Variation of liquid hold-up with time in water for horizontal and vertical surfaces is shown in Fig. 7.5. The liquid mass staying on the surfaces increases with time, first by direct condensation and then by coalescence. At the onset of instability, large

Table 7.4 Dropwise condensation parameters of water used in simulations

Water		
Density of liquid, ρ (kg/m^3)	995	
Saturation temperature and pressure (K, kPa)	303, 4.247	
Latent heat of condensation, h_{fg} (kJ/kg)	2426	
θ_{avg} and hysteresis (°)	110; 17	
Nucleation site density (cm^{-2})	106	
Subcooling, ΔT (K)	1	
Drop radius at nucleation r_{min} (μm)	0.018	
Capillary length r_{cap} (mm)	2.69	
r_{max} (horizontal), r_{crit} (vertical) (mm)	Horizontal	3.59
	Vertical	0.878
Cycle time (s)	Horizontal	417
	Vertical	101
Radius of drop averaged over space and time, R_{avg} (μm)	Horizontal (size of substrate)	5.886 (20 × 20)
	Vertical (size of substrate)	4.673 (20 × 20)
Interfacial resistance, $2/h_{int}$	–	1.14×10^{-4}
Conductive resistance, $(1 - \cos\theta_{avg})R_{avg}/k$	–	0.114×10^{-4}
Heat transfer coefficient averaged over space and time, h (kW/m^2 K)	Horizontal (size of substrate)	10.01 (20 × 20)
	Vertical (size of substrate)	20.35 (20 × 20)

drops are drained away and the liquid hold-up sharply decreases. Fresh nucleation is then initiated, and there is a gradual buildup of mass once again. There is, however, a well-defined average for water, both for the horizontal and vertical surfaces.

Wall Shear Stress and Wall Heat Flux

Including the coalescence timescales discussed in Chap. 3, the entire cycle of the condensation process has been numerically simulated in the present work. Apart from the condensation patterns discussed in previous sections, two other quantities of interest are the wall heat flux and wall shear stress. The surface-averaged heat flux and wall shear stress as well as local values at the coalescence sites as a function of time are presented in Figs. 7.6 and 7.7. The individual effect of a finite time coalescence process on the overall condensation pattern was found to be small. However, the fluxes and stresses during coalescence are substantial as seen from the scales adopted in Figs. 7.6 and 7.7. Since the overall cycle time is of the order of minutes, a millisecond scale coalescence process was not seen to alter surface-averaged values. However, their magnitudes are large enough to be of concern, as discussed below:

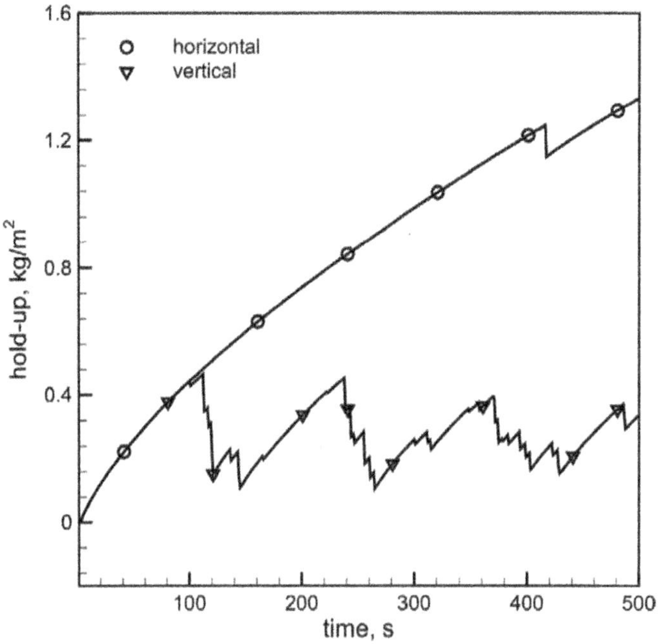

Fig. 7.5 Variation of liquid hold-up with time during condensation of water vapor for horizontal and vertical surfaces of size 20 × 20 mm

Coalescence events occur at several sites, often simultaneously, and maximum values alone are reported in Figs. 7.6 and 7.7. A total of 500 s of the condensation process is presented in the plots. In Figs. 7.6 and 7.7, heat flux and wall shear stress during water condensation over horizontal and vertical surfaces of size 20 × 20 mm are compared.

For a horizontal surface, a condensed drop falls off due to gravitational instability and fluid motion is restricted to the coalescence events. Hence, wall shear stress is mostly zero except for those occasional instants of time when it is large for very short-time durations (~ a few ms). For a vertical surface, wall shear is generated post drop instability when it slides down the surface. Coalescence-induced shear stress is superimposed over this value. Textured surfaces are often coated with a promoter layer and the coating has a finite yield strength, being of the order of 38–1160 mPa (Kim et al. 2006). Wall shear stress generated by coalescence and shown in Fig. 7.7 is greater than this value and indicates the possibility of the coating being worn out. Similarly, large though momentary wall heat fluxes (Fig. 7.6) suggest thermal non-uniformity in terms of the surface temperature that will interrupt the condensation process itself.

Variation of wall heat flux with time in water for both horizontal and vertical surfaces is shown in Fig. 7.6. The first and the second rows provide data for direct condensation where the effect of momentary coalescence is averaged over the entire surface. Peaks in wall heat flux are seen immediately after instability. Owing to

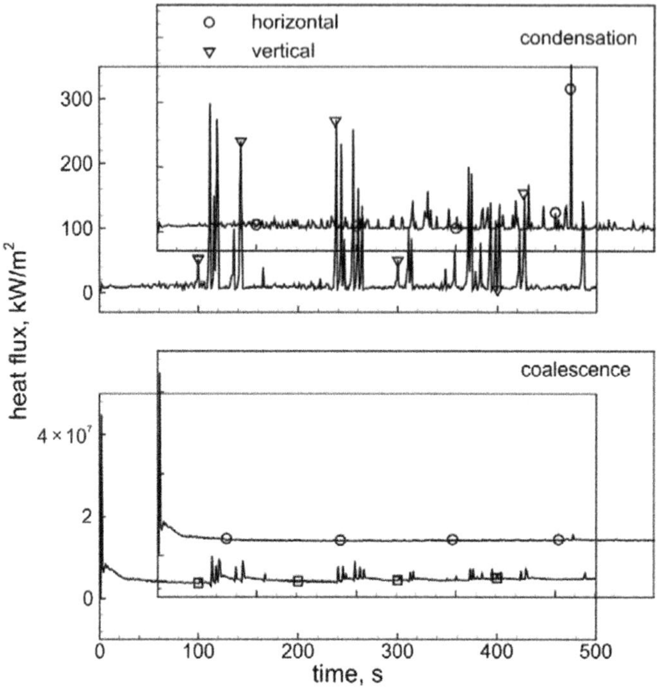

Fig. 7.6 Variation of wall heat flux with time during condensation of water vapor for horizontal and vertical surfaces of size 20 × 20 mm

sliding motion, a vertical surface will reveal a larger number of nucleating drops. A greater portion of drops is carried away from the smaller surfaces because of sliding motion. Larger surfaces may have instability at multiple locations and yet, a smaller portion of the drops will be drained away. Peaks in wall heat flux based on direct condensation as well as coalescence are proportional to this fractional number undergoing instability. Hence, larger fluctuations are seen on smaller surfaces than the larger surfaces for both horizontal and vertical configurations.

Wall heat fluxes generated during coalescence of drops of water are shown in the third and fourth data set rows of Fig. 7.6. In the context of Fig. 7.6, peak wall fluxes for direct condensation are greater for horizontal surfaces compared to the vertical. Orientation does not affect the coalescence fluxes significantly, except that frequent peaks are seen on larger surfaces.

Variation of wall shear stress with time in water for both horizontal and vertical surfaces is shown in Fig. 7.7. The first two rows correspond to wall shear stress generated during coalescence of drops of water on horizontal and vertical surfaces, respectively. The third row of Fig. 7.7 shows wall shear stress generated during the sliding of drops of water on vertical surfaces. The corresponding shear stresses for a horizontal surface are zero since the unstable drops fall off.

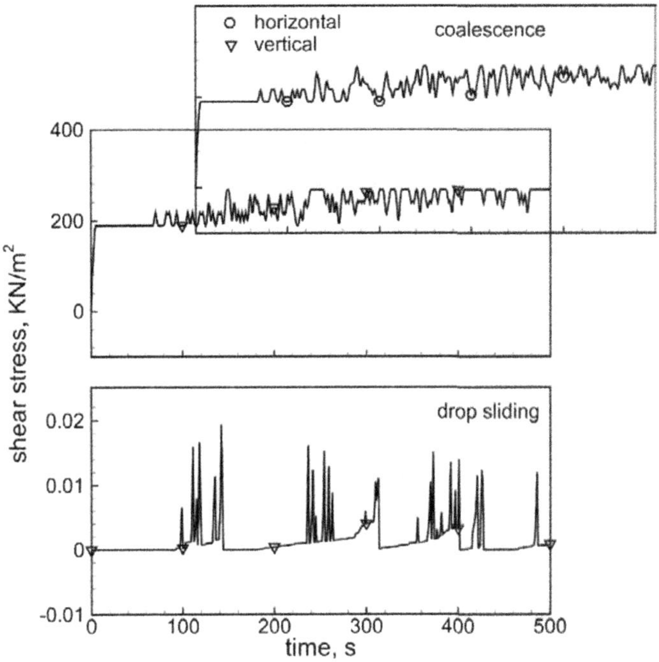

Fig. 7.7 Variation of wall shear stress with time during condensation of water vapor for horizontal and vertical surfaces of size 20 × 20 mm

At instability, drops on larger vertical surfaces must travel a longer distance than a smaller surface. Hence, shear prevails for a longer duration, generating larger wall shear stresses on larger surfaces. Maximum wall shear stress induced during sliding of drops on a vertical surface is 0.02 kN/m^2. Thus, wall shear stresses based on drop sliding show an increasing trend with an increase in surface area though the differences are small.

7.8 Conclusions

A previously developed hierarchical model of dropwise condensation has been extended by including the details of coalescence dynamics. The process starts from direct condensation at nucleation sites over the substrate. It is followed by growth arising from direct condensation as well as coalescence and slide-off or fall-off due to gravity. The process is cyclic and spatially distributed over the substrate. Quantities of interest, including instantaneous condensation patterns, wall shear stress, and wall heat flux have been predicted. Characteristic velocity and timescales of coalescence determined using scale analysis are used to extend the condensation model and determine local instantaneous wall shear stresses and heat transfer rates. The present work reports data related to condensation of water vapor underneath a near-horizontal surface and on a vertical surface. The following conclusions have been drawn from the study:

1. The condensation cycle time of water is smaller on a vertical surface compared to the horizontal. The maximum drop size is smaller for the vertical, the hold-up is smaller, and the average heat transfer coefficient is greater than the horizontal.
2. Heat flux diminishes with time during drop growth, fluctuates with time during the coalescence events as well as drop instability, attaining a peak just afterward. Since coalescence occurs occasionally over a small area, the contribution of the heat flux peaks to the overall substrate-level heat flux is small.
3. Wall shear is generated for a horizontal surface only during coalescence. For a vertical surface, it is seen during coalescence as well as drop instability. The coalescence-induced wall shear stress is substantial. This result is significant because the applicable wall shear stress can tear the surface coating. The contribution of coalescence to the surface-averaged wall shear stress is, however, quite small.

7.9 Closure

A comprehensive mathematical model of dropwise condensation for horizontal and vertical surfaces with the incorporation of length, velocity, and timescales associated with the coalescence of pendant and sessile drops on textured surfaces is developed. The dropwise condensation process starts at the scale of the nucleating drop and progresses towards the scale of the substrate. Various sub-processes are individually modeled to address the entire mathematical model of dropwise condensation (Fig. 7.2). A C++ program is written to carry out simulations on a high-performance computing machine.

Chapter 8
Simulation in a Parallel Environment

Praveen Somwanshi and K. Muralidhar

8.1 Introduction

The model developed in the present study (Chaps. 6 and 7) simulates growth of liquid drops from the vapor phase, one at a time. At the start of the calculation, the number of drops to be tagged is the product of the nucleation site density and the substrate area. This number increases with the size of the substrate area and performing simulations on large real-life surfaces is computationally intensive. In the condensation process, drops of thermodynamically stable minimum radius of a few millimeters are grown following the growth equation until they reach a drop of critical or maximum radius of a few millimeters. Real surface dimensions may span hundreds of millimeters, thus creating a wide range of length scales in the process. In addition, instability, sliding motion, and coalescence timescales are in the range of μs–ms while the overall cycle time can span a few hundred seconds. Thus, the dropwise condensation model is truly multiscale, involving a hierarchy of scales in space and time. Consequently, simulation is computationally intensive.

Even after drops have grown and coalesced, the number of nucleation sites for computations is of the order of 50% of the initial value at the start of condensation. It is thus necessary to examine the possibility of accelerating the computation process. There are multiple methods to address the problems demanding high computational power. Parallel computing is one of them and is discussed further.

8.2 High-Performance Computing Systems

High-performance computing systems at the authors' institute consist of two variants, namely HPC2010 and HPC2013.

S. Khandekar, K. Muralidhar, *Drop Dynamics and Dropwise Condensation on Textured Surfaces*, Mechanical Engineering Series,
https://doi.org/10.1007/978-3-030-48461-3_8

HPC2010 machine consists of 376 nodes. It has initial ratings for R_{peak} (theoretical peak performance) as 34.50 teraflops (1 teraflop floating point operations per second) and R_{max} (achieved peak performance in standardized computation) as 29.01 teraflops. It is made up of Intel Xeon X5570 2.93 GHz 2 CPU-Nehelam (8-cores per node) on HP-Proliant-BL-280c-Gen6 servers with 48 GB of RAM per node. The nodes are connected by Qlogic QDR InfiniBand federated switches which provide throughput of 40 Gbps. It writes at a speed of 5 GBps having a storage capacity of 100 TB. The cluster has additional 96 nodes consisting of Intel Xeon ES-2670 2.6 GHz 2 CPU-Sandy-Bridge (16-cores per node) on HP-Proliant-SL-230s-Gen8 servers with 64 GB of RAM per node which enhances the rating of HPC2010 by 31 teraflops.

HPC2013 machine consists of 901 nodes. It has initial ratings for R_{peak} as 359.6 teraflops and R_{max} as 344.317 teraflops. It is made up of Intel Xeon E5-2670V 2.5 GHz 2 CPU-IvyBridge (20-cores per node) on HP-Proliant-SL-230s-Gen8 servers with 128 GB of RAM per node E5-2670v2x10 core2.5 GHz. The nodes are connected by Mellanox FDR switches based on InfiniBand chassis which provides a throughput of 56 Gbps. It can write and read at a speed of ~23 and ~15 GBps, respectively, having a storage capacity of 500 TB.

These high-performance computing machines have been used for parallel computation in the present study.

8.3 Parallel Computing

Conventional computer programs are written in a manner wherein the instructions are executed one after the other in a sequential manner. The time needed by a simulation is the sum of the times needed by each of the instructions. The simulation time can be reduced by converting the sequential to an equivalent parallel version of the computer code.

Parallel computing is a form of computation wherein several instructions in a computer program are allowed to run simultaneously in parallel. Availability of multiple processors is the only requirement for establishing a parallel computing paradigm. It can be established on a single computer having multiple processors or on a cluster of many computers connected by a network.

Parallelization of computer programs is widely carried out using OpenMP or MPI framework. These approaches are discussed in the following sections:

8.4 Parallelization using OpenMP

OpenMP (Open Multi-Processing) is a standard application programming interface for writing shared memory parallel applications in C, C++, and Fortran. Parallelization of existing sequential codes using OpenMP is easily implemented allowing incremental parallelization. It is suitable for multi-core architectures, is highly portable, and widely used for parallelization.

The OpenMP model of parallel execution is based on a Fork and Join architecture. The master thread works on the sequential portion of the task. It distributes the parallel portion of the task into the available number of worker threads as a "Fork." The worker threads receive their portion of the task and complete the assigned work. The master thread then collects the portion of the work accomplished by the worker threads. Later, the master thread continues the work on the sequential portion of the program. The entire model can be simulated either on the master or on the worker threads depending on the nature of the numerical algorithm. The compiler directives are embedded in the code to achieve parallelization using OpenMP.

8.4.1 Parallelization using OpenMP Paradigm

In the present work, the computational effort required for generating the time-resolved dropwise condensation patterns are distributed over individual processors using the notion of domain decomposition. Here, splitting the space domain into smaller contiguous parts is one of the methods available for parallelization of the computer code using OpenMP. The total condensing substrate is divided into any number of subdomains, matched with the total number of processors available on the computer. The union of all such subdomain level calculations represents growth for the entire substrate, over a given time step.

8.4.2 Validation of OpenMP Enabled Parallel Code

Figure 8.1 shows condensation patterns of bismuth over a vertical surface at instability with 1, 3, and 9 subdomains. The input parameters such as thermophysical properties and nucleation site density (10^5 cm^{-2}) are identical in these calculations. The random site locations for each of the three simulations are also identical. The only difference during computation is in terms of the number of domains. Computations are performed by dividing the total collection of drops into 1, 3, and 9 portions, respectively. An individual drop grows by way of continuous vapor condensation and sporadic coalescence with its neighbor. Instability is observed once the drop radius exceeds the criticality criteria. For the three simulations, the slide-off instability for a vertical surface is seen to occur around 176 s. Figure 8.1 shows that the condensation patterns established using 1, 3, and 9 domains are identical. However, for the computing system available at the authors' institute, the OpenMP mode of parallelization was not scalable (to $>$ 16 processors). Hence, it has not been used for simulation of dropwise condensation on larger surfaces.

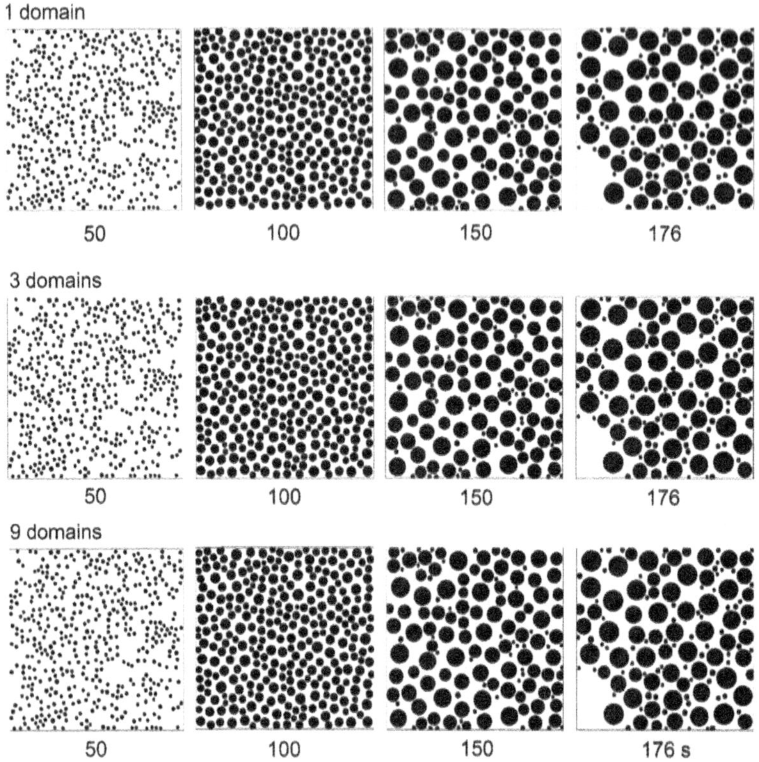

Fig. 8.1 OpenMP implementation: Condensation patterns of bismuth over a 10 × 10 mm vertical surface with 1, 3, and 9 subdomains around the instant of instability when a sufficiently heavy drop slides downward. Starting from a freshly nucleated surface, the first instant of instability is seen to occur at around 176 s for each simulation. On 1, 3, and 9 subdomains, the CPU times required for 200 time steps were 798, 522, and 411 min, respectively

8.4.3 Parallelization using MPI

MPI (Message-Passing Interface) is a language specification to be followed during message passing between the processors. The exchange of data from the address space of one processor to the other is facilitated by MPI. Specific features include collective operations, operations on remote memory, creation of dynamic processes, and input and output operations in parallel. Portability and easy usage are additional advantages of MPI. The goal of the Message-Passing Interface is to establish a practical, portable, efficient, and flexible standard for information exchange during large-scale computing.

The MPI execution model is different from the OpenMP. The unit of parallelism is a process in MPI while it is a thread in OpenMP. Each process has its own address space. The communication between two processors is established through message passing operations such as MPI_send () and MPI_recv (). The data structure of a parallel program consists of a collection of independent portions of data from various processors. The total number of processors is to be specified at the start of MPI among which one is the master and the rest are worker processes. The master processor controls the whole simulation by instructing the worker processors in accordance with the requirements of the mathematical model.

8.4.4 Parallelization of the Computer Code using MPI Paradigm

Domain decomposition, where the space domain is split into smaller parts is one of the methods available for computer code parallelization using MPI. In this approach, the computational effort is distributed to the individual processors, each one representing a subdomain. For this purpose, the total condensing substrate is divided into subdomains by using the available number of processors in a multi-processor architecture. Computations are performed for each time step within individual domains in a sequential manner. The union of all such calculations represents growth for the entire substrate, for a given time step.

All steps in the condensation model are not equally amenable to parallelization. The drops continuously grow until they reach the moment of gravitational instability. Drops can either slide-off or fall-off at criticality, intersecting multiple domains during their motion. This difficulty is resolved by carrying out computations related to instability in a sequential manner on a common (master) processor. On the other hand, the effectiveness of parallelization is nonlinearly dependent on the number of nucleation sites. Hence, for the smaller substrate size (10×10 mm) and nucleation site density (10^5 cm^{-2}) considered in this chapter, the speed-up is substantial; CPU times are provided in the captions of Figs. 8.1 and 8.2. For larger surfaces and a higher nucleation site density considered in later chapters, the computational advantage (speed-up) with respect to the number of processors was found to be less than 100%. Speed-up using MPI was also found to be substantially higher than the OpenMP architecture.

Every domain has one or two common interfaces with its immediate neighbor. The computer code developed in this study needs an access to the data set of the adjacent domains. In addition, 25% of the adjacent domain is added as an overlapping area at the common interfaces. Computations in the overlapping areas are performed on the master processor in a sequential manner. This approach eliminates loss of information that may occur at sharp interfaces, particularly when drops encounter coalescence.

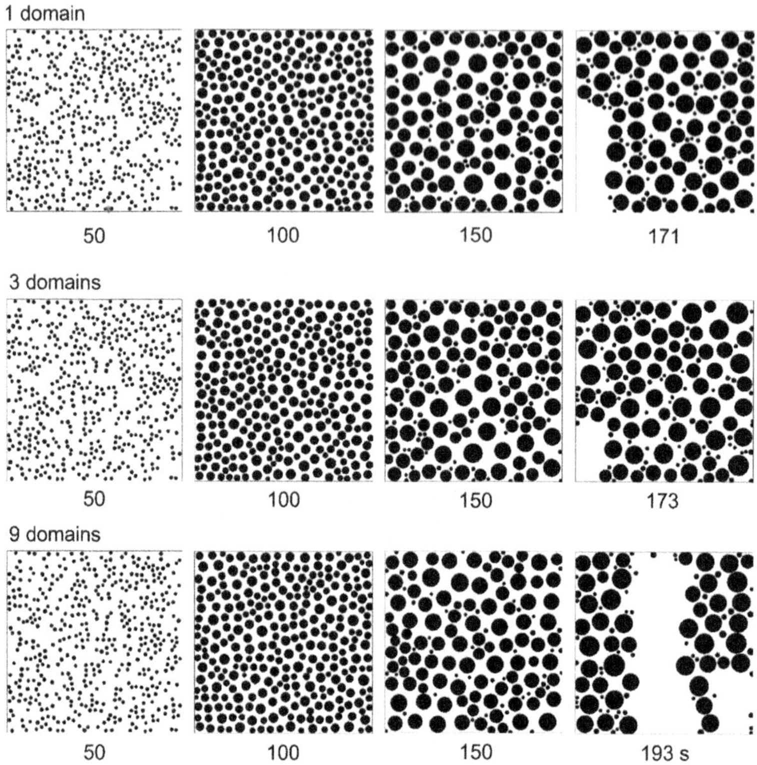

Fig. 8.2 Condensation patterns of bismuth over a 10 × 10 mm vertical surface with 1, 3, and 9 subdomains around the instant of instability when a sufficiently heavy drop slides downwards. Starting from a freshly nucleated surface, the first instant of instability is seen to occur at around 171–193 s for each simulation. CPU times required for 1, 3, and 9 subdomains for 200 time steps were found to be 1171, 135, and 18 min, respectively

8.4.5 Validation of MPI Enabled Parallel Code

Condensation patterns of bismuth over a vertical surface at instability with 1, 3, and 9 subdomains are shown in Fig. 8.2. The input parameters such as thermophysical properties and nucleation site density (10^5 cm^{-2}) are kept constant. The random site locations for each of the three simulations are identical, but different from the simulations using OpenMP. The only difference during computation is in terms of the number of domains. Computations are performed by distributing the total collection of drops into 1, 3, and 9 portions, respectively. Drop grows by way of continuous vapor condensation and sporadic coalescence. Instability is observed once the drop radius exceeds the criticality criteria. The slide-off instability for a vertical surface is seen around 171–193 s. The condensation patterns established using one and three are identical. Condensation pattern arising from 9 domains is similar but departs from the original because of the overlap treatment at the common interfaces. Condensation patterns were determined with up to 400 processors. The patterns consistently retained similarity and the time instant of instability was close to the full domain value. For larger surfaces (30 $\times$ 30 mm and 50 $\times$ 50 mm), a nucleation site density of 10^6 cm^{-2} considered for bismuth and water in other chapters, the match in the time of instability between the sequential and parallel computations was between 1% and 3%.

8.5 Closure

Parallelization of computer programs based on shared memory and distributed memory architectures are discussed. Details of the implementation of OpenMP and MPI frameworks in computer programs are presented. The OpenMP framework is easier to implement in the computer programs compared to MPI. It is, however, not suitable for the simulation of dropwise condensation on large area surfaces, where the number of processors required is quite large. Hence, the MPI framework has been preferred in the present set of simulations.

Chapter 9
Simulation: Dropwise Condensation of Water

Basant Singh Sikarwar, Praveen Somwanshi, K. Muralidhar, and Sameer Khandekar

Nomenclature

C_p	Specific heat at constant pressure, W/kg K
h_{lv}	Latent heat of vaporization, J/kg
h	Heat transfer coefficient, W/m^2 K
k	Thermal conductivity of the condensate, W/m K
N	Number of nucleation sites, cm^{-2}
r	Radius of drop, m; suffix b is for base radius
r_{cap}	Capillary length, $\sqrt{\sigma/g(\rho_l - \rho_v)}$, m
r_{crit}	Size of the drop at instability due to slide-off, m
r_{max}	Size of the drop at instability due to fall-off, m
r_{min}	Radius of thermodynamically smallest drop, m
T	Temperature, K; subscripts w and sat are for wall and saturation
ΔT, $(T_{sat} - T_w)$	Temperature difference between the saturated vapor and condensing wall, K

Dimensionless Quantities

Ja	Jakob Number, $(C_p/h_{lv})_{ref}(T_{sat} - T_w)$
Nu	Nusselt number, hr_{cap}/k
Pr	Prandtl number

S. Khandekar, K. Muralidhar, *Drop Dynamics and Dropwise Condensation on Textured Surfaces*, Mechanical Engineering Series,
https://doi.org/10.1007/978-3-030-48461-3_9

Greek Symbols

α	Inclination angle of the substrate from horizontal, radians
θ	Contact angle, radians or degrees; adv, rcd and avg are advancing, receding and average angles
$\Delta\theta$, $(\theta_{adv} - \theta_{rcd})$	Contact angle hysteresis, °

9.1 Dropwise Condensation of Water Vapor

9.1.1 Effect of Substrate Hydrophobicity

The effect of the hydrophobicity of the substrate towards the condensing liquid is examined in Figs. 9.1, 9.2 and 9.3. Here, drop distribution at selected time instants is pictorially depicted from initial nucleation to the first instance of drop fall-off. Water vapor condenses at 303 K underneath the surface and the degree of subcooling is 5 K. The substrate is horizontal and various contact angles are considered.

Figure 9.1 shows a time sequence of condensation patterns for a contact angle of 90°. Drop diameter at criticality is 4.63 mm and fall-off first occurs at 50.15 min after commencement of condensation. Figure 9.2 shows the corresponding spatio-temporal drop distribution for a contact angle of 120°. Drop diameter at criticality is 3.088 mm and fall-off first occurs at 21.55 min after commencement of condensation. Figure 9.3 shows the spatio-temporal drop distribution for a contact angle of 140°. Drop diameter at criticality is 2.14 mm and fall-off first occurs at 7.25 min after commencement of condensation.

As can be seen, a reduction in wettability increases the contact angle and leads to a smaller base circle of the drop and, therefore, smaller surface forces retaining the drop against gravity. Thus, two effects are clearly visible: (a) The droplet volume at the time of fall-off is smaller (Figs. 9.1, 9.2, and 9.3); (b) with increasing contact angle, the drops achieve fall-off criticality earlier in the cycle.

The area of coverage by drops for various contact angles is shown in Fig. 9.4a. The hydrophobicity of substrates decreases the area of coverage. Therefore, highly hydrophobic substrates (higher contact angle) have higher available exposed nucleation sites density at any given time of condensation (Fig. 9.4b). The size and population of maximum diameter drops have a significant impact on dropwise condensation, due to the limitations imposed by the diffusional resistance of the liquid. The effect of substrate hydrophobicity on the surface-averaged heat transfer rate during dropwise condensation is shown in Fig. 9.4c. The apparent contact angles clearly show an effect on heat transfer. The size of the drop at fall-off as well as the time required for fall-off decrease as the hydrophobicity of the substrate increases. All other conditions remaining unchanged, the fall-off time for a pendant drop is seen to be a linear function of the contact angle (Fig. 9.4d).

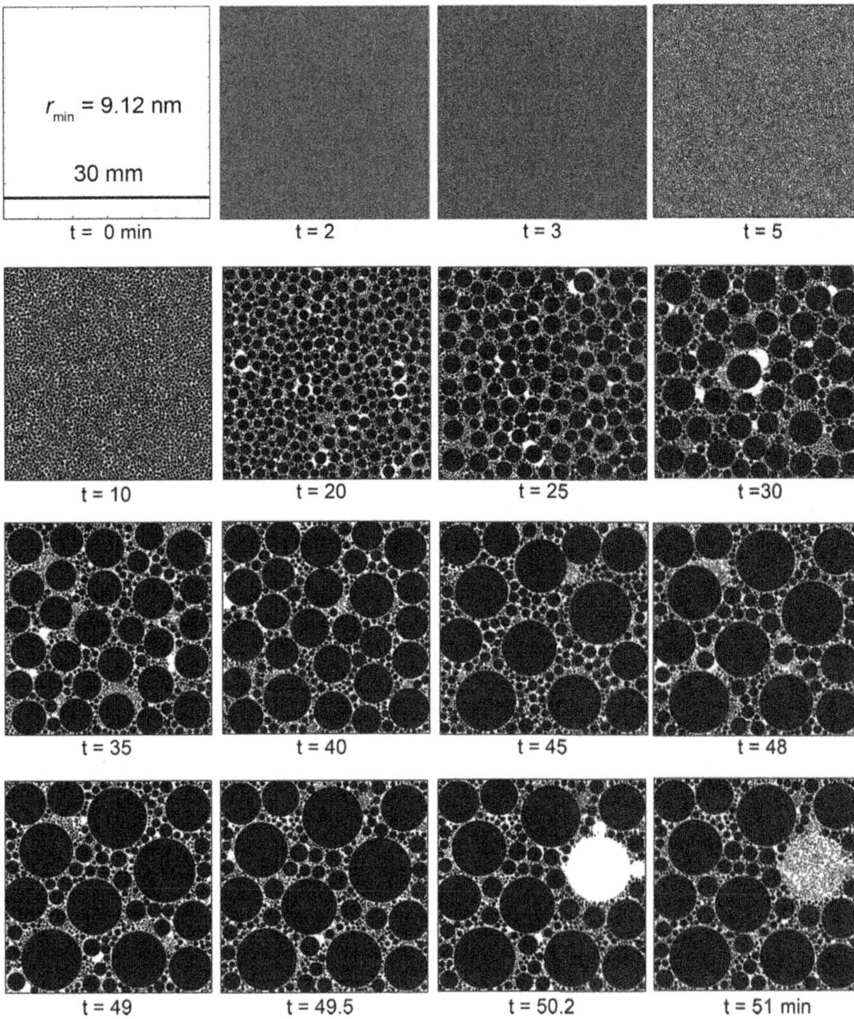

Fig. 9.1 Drop distribution from the start to the first fall-off during dropwise condensation of water vapor at 303 K with 5 K subcooling underneath a horizontal substrate of contact angle 90°

For a given initial nucleation site density, heat transfer can be increased by having a contact angle higher than the 90°. Accordingly, the substrates having higher hydrophobicity result in the condensate having drop size distributions towards the smaller diameter (Figs. 9.1, 9.2, and 9.3). It results in lowering the overall diffusion resistance offered by the condensing drops. In addition, increased hydrophobicity generates many nucleation sites, at any given time. The nucleation sites available for nucleation are shown in Fig. 9.4b. Initially, it decreases according to a power law but after reaching a dynamic steady-state it varies quasi periodically due to coalescence and fall-off. The frequency of drop fall-off, size of the minimum drop and size of the maximum drops, for substrates having various degree of hydrophobicity, during water vapor condensation are summarized in Table 9.1.

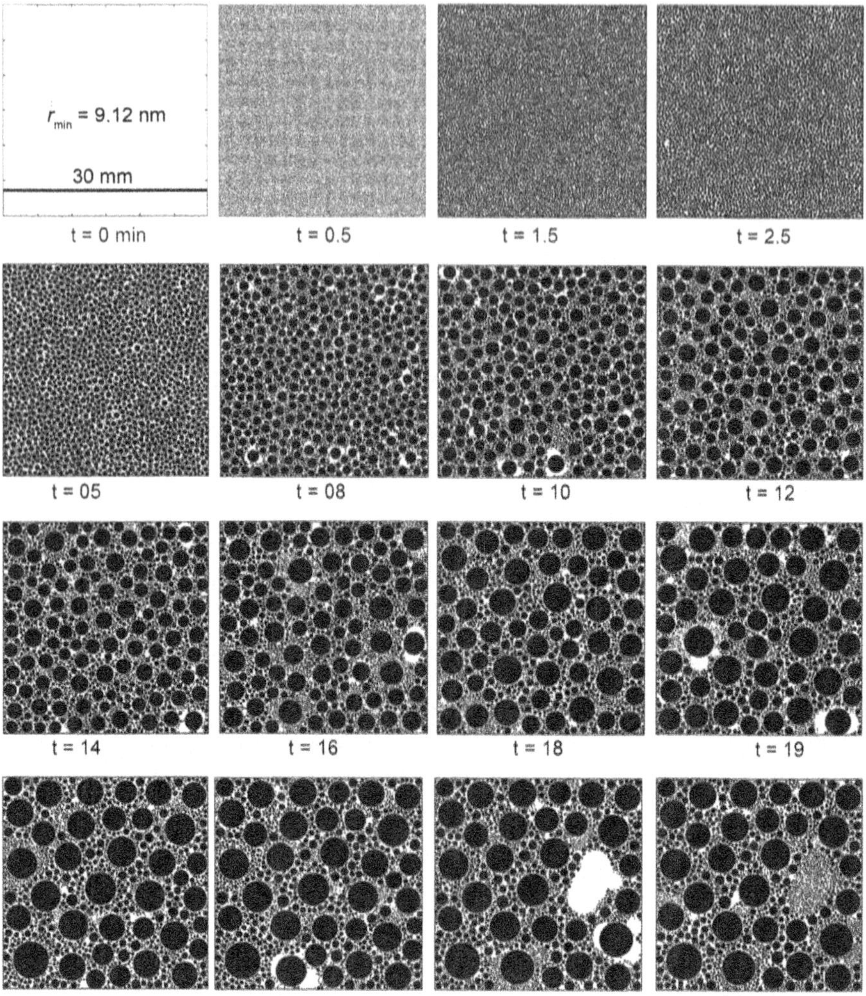

Fig. 9.2 Drop distribution from the start to the first fall-off during dropwise condensation of water vapor at 303 K with 5 K subcooling underneath a horizontal substrate of contact angle 120°

As given in Table 9.1, the substrate hydrophobicity decreases the critical size (of fall-off), hence resulting in a higher population of small drops. Therefore, one concludes that a substrate having a high hydrophobicity with the condensate fluid is desirable in dropwise condensation. Figure 9.5a shows the frequency (namely, the number of drops) as a function of the drop radius, 10 min after commencement of condensation for a contact angle of 90°, saturation temperature of 303 and 5 K subcooling. At later times, drops of higher sizes are to be seen. For the present simulation, the fall-off time of the first drop was approximately 50.2 min. Figure 9.5b

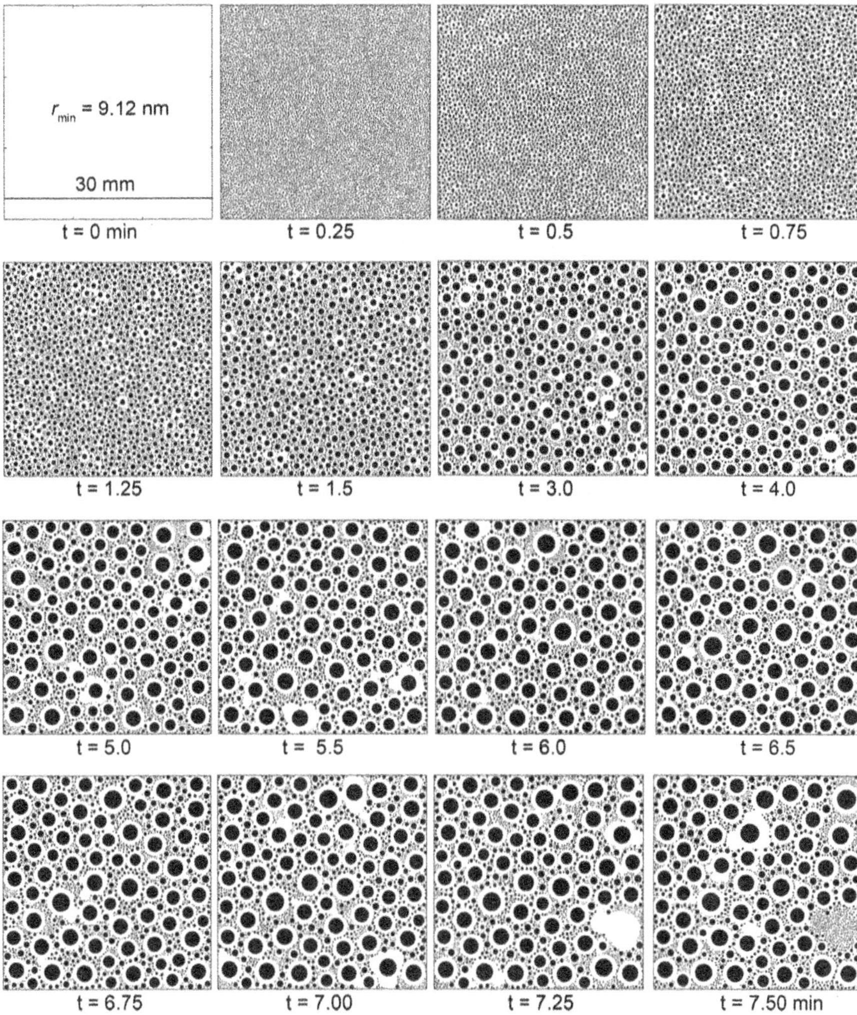

Fig. 9.3 Drop distribution from the start to the first fall-off during dropwise condensation of water vapor at 303 K with 5 K subcooling underneath a horizontal substrate of contact angle 140°

shows that very small droplets nucleate on the substrate at 50 min, immediately before the first drop falls-off at 50.2 min.

9.1.2 Effect of Substrate Inclination

Figures 9.6, 9.7, and 9.8 show the distribution of drops arising from water vapor condensation at 303 and 5 K subcooling from initial nucleation to the first slide-off underneath substrates of various orientation (30°, 60°, and 90°). The pictorial views

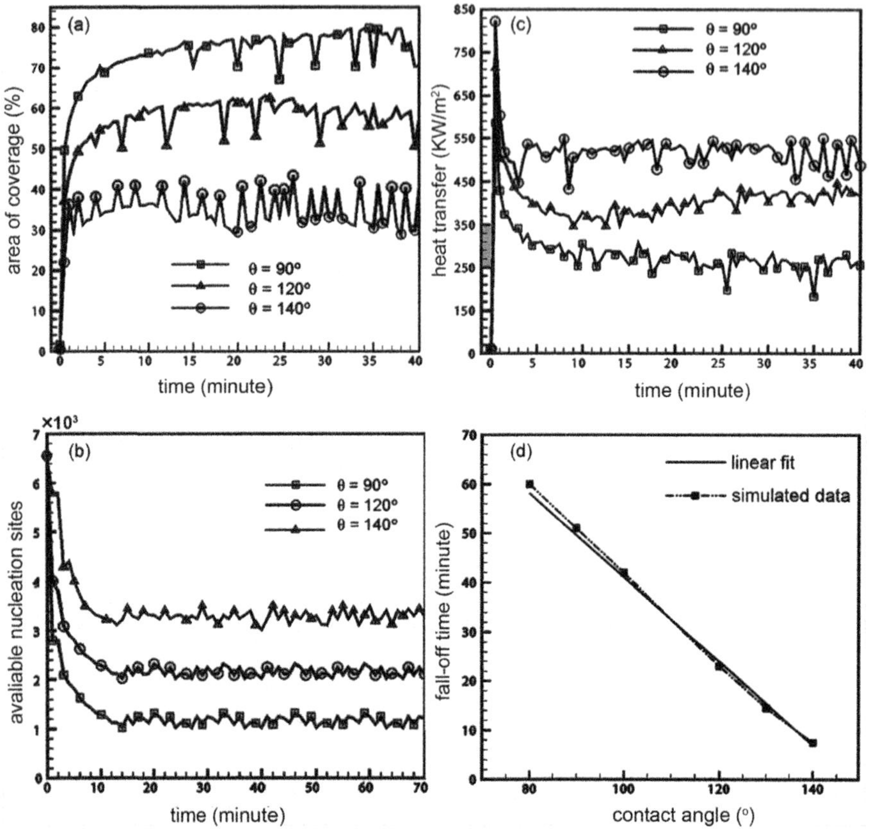

Fig. 9.4 (**a**) Area of coverage as a function of time, (**b**) available nucleation sites over the substrate of size 30 × 30 mm with respect to time for various contact angles at 303 K and degree of 5 K subcooling, (**c**) fluctuations in heat transfer rate on a substrate with respect to time and (**d**) fall-off time of a drop as function of the contact angle

Table 9.1 Results summarizing parameters of dropwise condensation of water vapor at 303 and 5 K subcooling after reaching a quasi-steady state

Contact angle (°)	Radius		Initial nucleation sites (cm^{-2})	Available nucleation sites on a 30 × 30 mm area	First fall-off (min)	Frequency (s^{-1})	Heat transfer (kW/m^2)
	r_{min} (mm)	r_{max} (mm)					
90	9.1×10^{-6}	4.64	10^6	1122	50.2	360	270
			10^7	1137	38.5	187	282
120	9.1×10^{-6}	3.08	10^6	1977	21.5	242	395
			10^7	2015	17.8	155	512
140	9.1×10^{-6}	2.14	10^6	3656	7.2	189	525
			10^7	3684	6.7	85	580

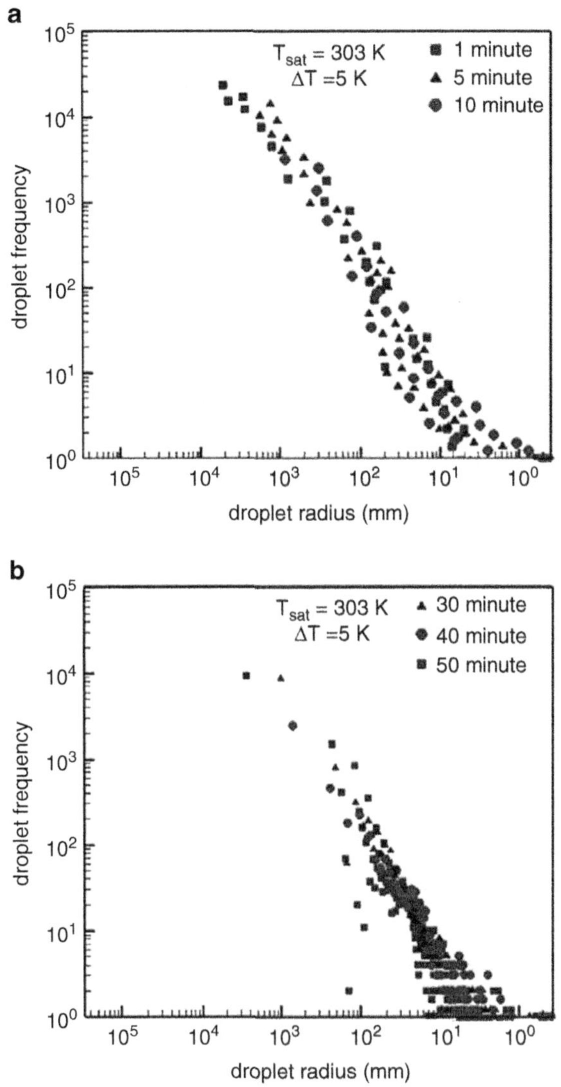

Fig. 9.5 (**a**, **b**) Temporal variation in drop size distribution for condensing water vapor underneath a horizontal substrate for contact angle 90°. For clarity, data for 1–10 min are separately plotted from the data for 30–50 min. The fall-off time for the first drop was equal to 50.2 min in this simulation

of condensations have site density 10^6 cm^{-2}, substrate size 30 × 30 mm for substrates of inclination 30° and 60° and 20 × 20 mm for an inclination of 90°. For these simulations, the advancing and receding angles are taken as 118.5° and 101.5°, yielding an average contact angle of 110° and a contact angle hysteresis of 17°.

Condensation patterns on various inclinations show broad similarity. The point of difference is the size of the drop at slide-off and the average cycle time for slide-off from the substrate. These quantities decrease as the angle of inclination with respect to the horizontal increases.

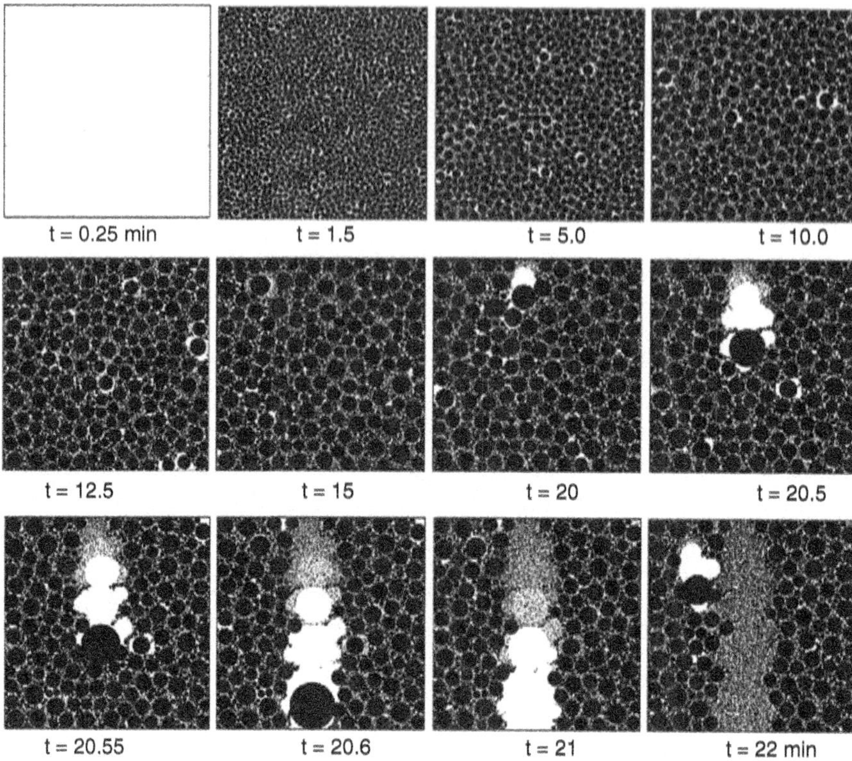

Fig. 9.6 Drop distribution from the start to the first slide-off during dropwise condensation of water vapor at 303 K with 5 K subcooling underneath an inclined substrate. The angle of inclination is 30°, and the size of the substrate is 30 × 30 mm

The variation of the critical size of the droplet with respect to substrate orientation is depicted in Fig. 9.9. The critical drop size decreases with substrate hydrophobicity and inclination, causing a reduction in the cycle time and hence results in more frequent instances of renucleation. The reduction in the cycle-averaged drop size is an important factor in increasing heat transfer from strongly hydrophobic surfaces.

For ease of calculation, the data of Fig. 9.9 is correlated for various surfaces by

$$\frac{r_{\text{crit}}}{r_{\text{cap}}} = \left(2.1612 - 0.7699\theta_{\text{avg}}\right)(\Delta\theta)^{0.5}\alpha^{-0.4266} \tag{9.1}$$

The correlation coefficient of Eq. (9.1) is 99.8%; it simplifies dropwise condensation calculations within the hierarchical model and considerably reduces the computational time.

The heat transfer coefficient of dropwise condensation for a given saturation temperature and degree of subcooling is dependent on the contact angle, contact angle hysteresis, and substrate orientation, which in turn affect the criticality of

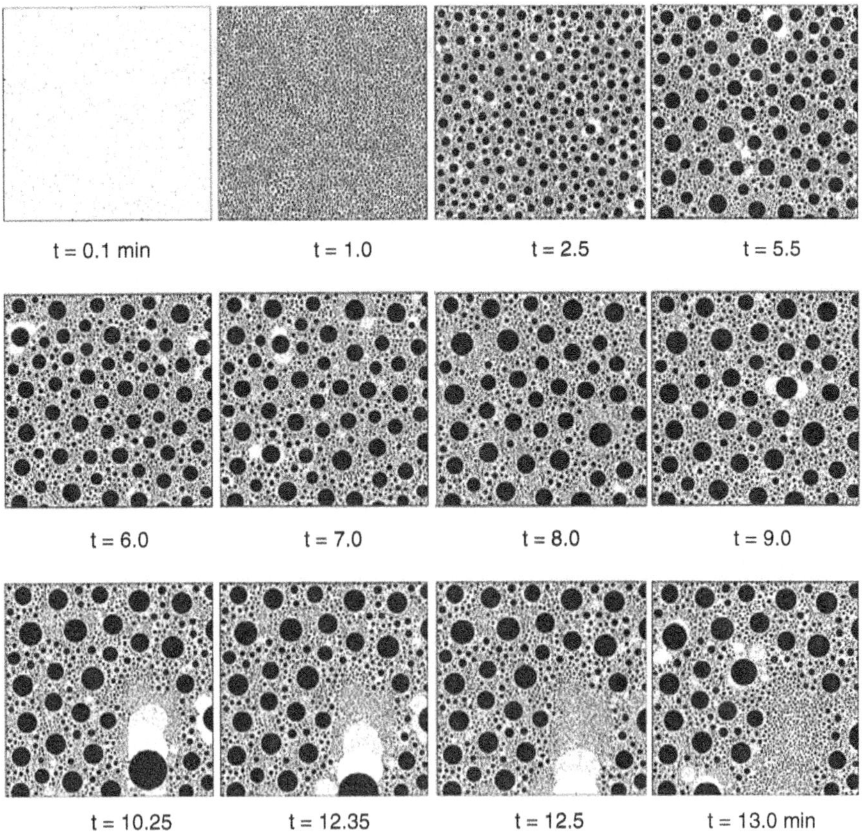

Fig. 9.7 Drop distribution from the start to the first slide-off during dropwise condensation of water vapor at 303 K with 5 K subcooling underneath an inclined substrate. The angle of inclination is 60°, and the size of the substrate is 30 × 30 mm

sliding. The effect of critical radius on heat transfer at 303 and 5 K subcooling is depicted in Fig. 9.10a. It is seen that heat transfer coefficient increases as radius of departure from the substrate decreases. The correlation for heat transfer coefficient is

$$h = 0.19\, r_{\mathrm{crit}}^{-1.2} \tag{9.2}$$

where h is in unit of MW/m^2 K and r_{crit} is in mm.

Figure 9.10b shows the variation of heat transfer coefficient with respect to the angle of inclination of the substrate. Numerically obtained data of water condensing underneath various inclined (0–90°) substrates is shown for an average contact angle of 110°, contact angle hysteresis of 17° and 5 K subcooling at various saturation temperatures. The result exhibits a heat transfer coefficient for vertical substrate that is 40–50% higher than the horizontal substrate.

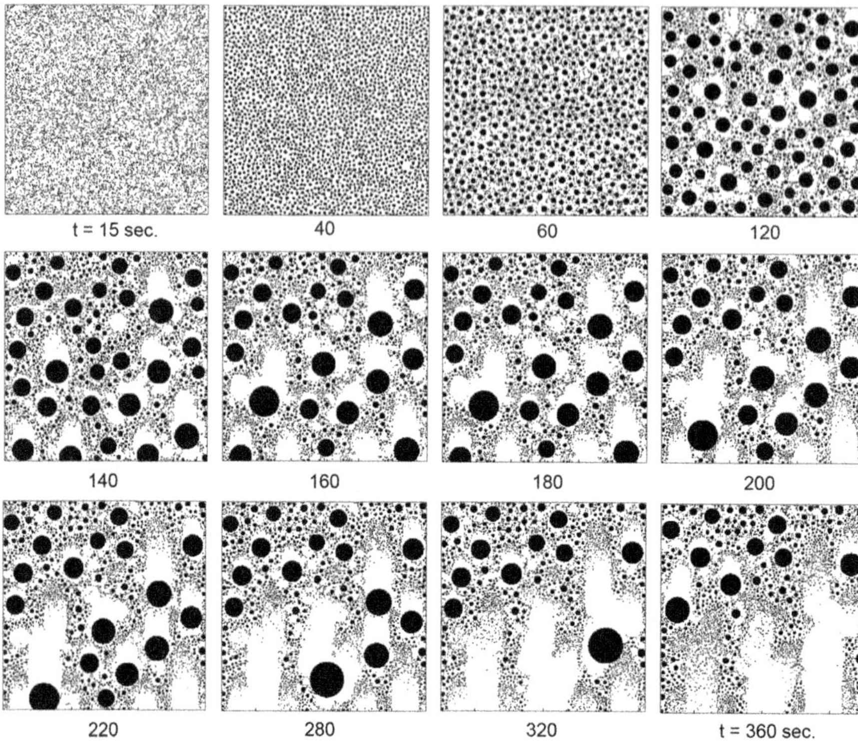

Fig. 9.8 Drop distribution from the start to the first slide-off during dropwise condensation of water vapor at 303 K with 5 K subcooling underneath an inclined substrate. The angle of inclination is 90°, and the size of substrate is 20 × 20 mm

Fig. 9.9 Critical base radius of the drop at instability (r_{crit}/r_{cap}) for a pendant arrangement plotted as a function of substrate orientation (α). Contact angle hysteresis is a parameter while the average contact angle is 105°

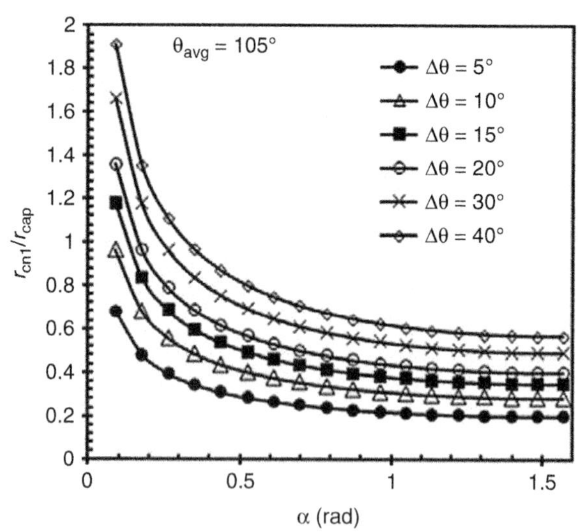

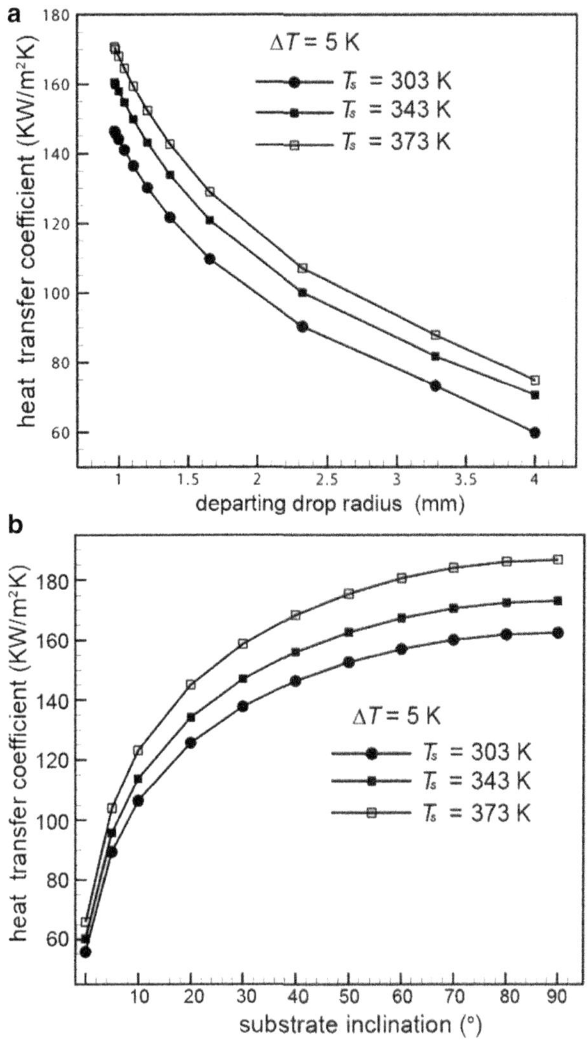

Fig. 9.10 (**a**) Dependence of heat transfer on the departure drop radius. (**b**) Effect of substrate inclination on heat transfer coefficient

The effect of substrate inclination on the temporal distribution of area coverage of drops is presented in Fig. 9.11a. Inclination of the substrate facilitates easier movement of drops by sliding, leading to a sweeping action. Therefore, the effective steady-state coverage is smaller for inclined substrates, changing from 76.1% for a horizontal substrate, 71.2% for 5° inclination, and 67.4% for the substrate with 10° inclination. At the instant of the first fall-off (for the horizontal substrate) and the first slide-off (for the inclined substrate), Fig. 9.11b depicts the drop size distribution as a function of radius, for various inclination angles. The distribution follows a power law with the negative slope increasing with angle, reflecting the repeated appearance of small drops at fresh nucleation.

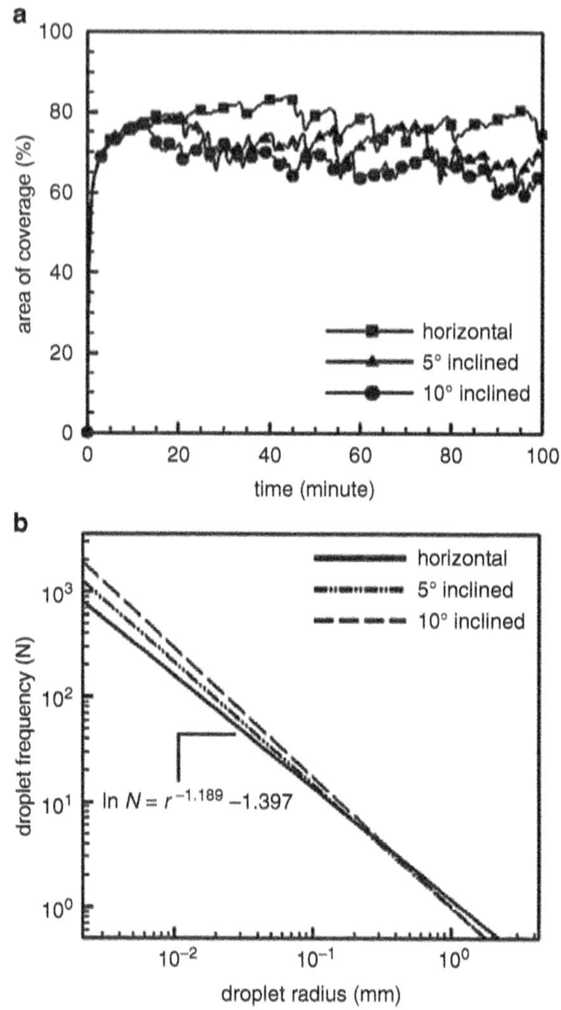

Fig. 9.11 Effect of substrate inclination. (**a**) Temporal variation of area coverage of drops during condensation of water in the pendant mode. (**b**) Drop size distribution just before fall-off (for horizontal substrate) or slide-off (inclined substrate). For this simulation, $\theta_{adv} = 118.5°$ and $\theta_{rcd} = 101.5°$. $T_s = 303\text{K}$, 5 K subcooling

9.1.3 Effect of Saturation Temperature and Subcooling

For a given degree of subcooling ($\Delta T_{sat} = 5$ K), the effect of saturation temperature on drop departure time underneath a horizontal substrate is shown in Fig. 9.12. Increasing the saturation temperature reduces the fall-off time and hence the size of the largest drop, indicating an increase in the overall heat transfer coefficient. The diffusional thermal resistance within the drop is a major limiting factor of condensation heat transfer. Hence, increasing the saturation temperature increases the thermal conductance of the water drop. A marginal increase in the overall resistance is also noticed due to a reduction in the interfacial heat transfer coefficient; it

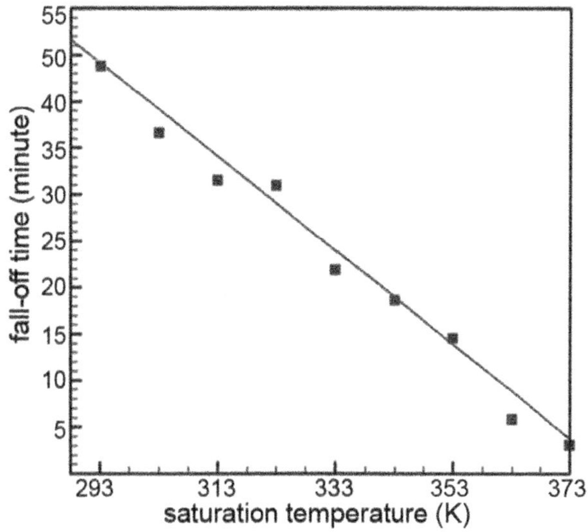

Fig. 9.12 Variation in drop departure time (time required for first fall-off) on a horizontal substrate with respect to the saturation temperature. Fluid employed is water, 5 K subcooling, contact angle = 110°, nucleation site density = 10^6 cm^{-2}. For a given nucleation site density, the fall time has an uncertainty of ± 3 min, depending on the random assignment of initial droplet centers on the substrate

essentially proves to be inconsequential as the overall thermal resistance is dominated by conduction resistance of the droplet.

Figure 9.13 shows the effect of saturation temperature and degree of subcooling on the heat transfer coefficient of dropwise condensation underneath horizontal and vertical substrates. The data show a tendency of increasing heat transfer coefficient with increasing saturation temperature and degree of subcooling. This is caused mainly by the decrease in the interfacial resistance at high saturation temperature of water vapor.

Increasing the degree of subcooling increases the density of active nucleation sites on the condensing substrate. On the basis of numerical data, the heat transfer coefficient (kW/m^2 K) is empirically correlated with the critical radius of drop (mm), degrees of subcooling and saturation temperature (both in units of °C) for water vapor condensation underneath an inclined substrate with a nucleation site density of 10^6 cm^{-2} by

$$h = (0.42\Delta T + 6.4){T_{sat}}^{0.75}\, {r_{crit}}^{-1.18} \tag{9.3}$$

Here, the critical radius depends on contact angle, contact angle hysteresis, and angle of inclination of substrate with respect to the horizontal. It is preferable to cast this correlation in dimensionless form, applicable for all the inclinations, saturation temperature, and degree of subcooling. The heat transfer correlation for condensation of water vapor underneath an inclined substrate is

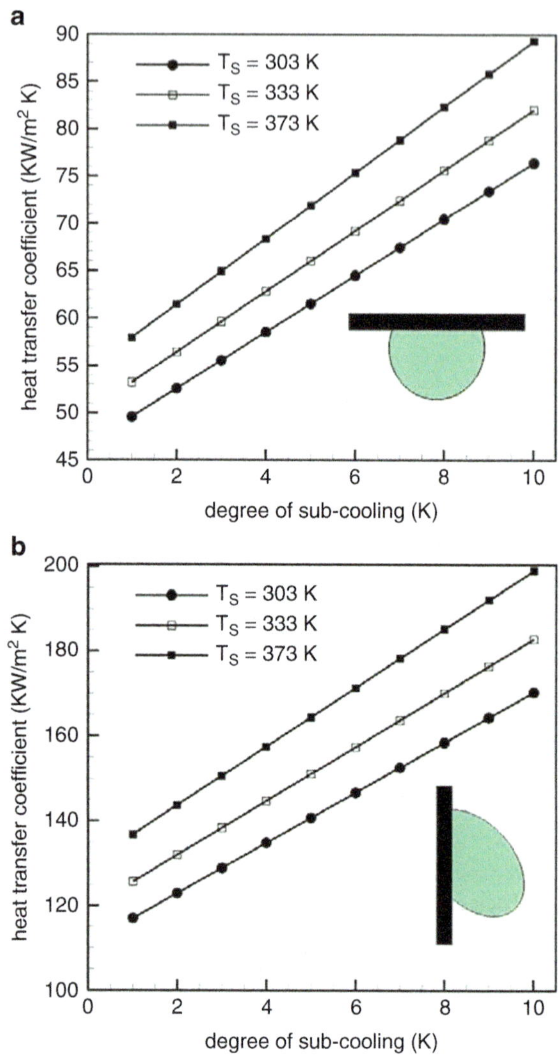

Fig. 9.13 Dependence of heat transfer coefficient on saturation temperature and degree of subcooling. (**a**) Horizontal substrate, (**b**) vertical substrate. Condensing fluid is water average, contact angle is 110°, contact angle hysteresis is 17° and nucleation sites density is 10^6 cm^{-2}

$$\mathrm{Nu} = \left(8.54 \times 10^3\ \mathrm{Ja} + 240\right)\left(\frac{T_{\mathrm{sat}}}{T_{\mathrm{ref}}}\right)^{0.75}\left(\frac{r_{\mathrm{crit}}}{r_{\mathrm{cap}}}\right)^{-1.18} \tag{9.4}$$

Equation (9.4) has a correlation coefficient of 98.5%. Standard reference values used are properties of water at the normal boiling point.

9.1.4 Effect of Nucleation Site Density

Increasing the density of nucleation sites leads to a large overall heat transfer in dropwise condensation. This effect arises mainly from a reduction in the size of the drop before coalescence. Early coalescence allows virgin spaces for new initial drops, causing a high population of small drops. One can conclude that a surface providing a higher nucleation sites is desirable for dropwise condensation. The number of nucleation sites is chosen as a parameter for condensation underneath a horizontal substrate, the working fluid being water. The effect of initial nucleation site density on heat transfer is shown in Fig. 9.14 for a contact angle of 110° and a saturation temperature of 303 K.

As the number of nucleation sites increases per unit area, many small drops nucleate on exposure of the surface to vapor, i.e., the average drop size within a cycle decreases. The conduction resistance is thus lowered, leading to an increase in average heat transfer coefficient.

9.1.5 Effect of Promoter Layer Thickness

Dropwise condensation of water underneath metal surfaces is rarely observed in natural conditions. It is generally promoted with a suitable coating. An understanding of the role of coatings is critical not only because it determines the surface wettability, but it adds an extra thermal resistance. The nucleation density is also dependent on the promoter layer. The mathematical model of the present work is utilized for designing and quantifying the effect of the coating.

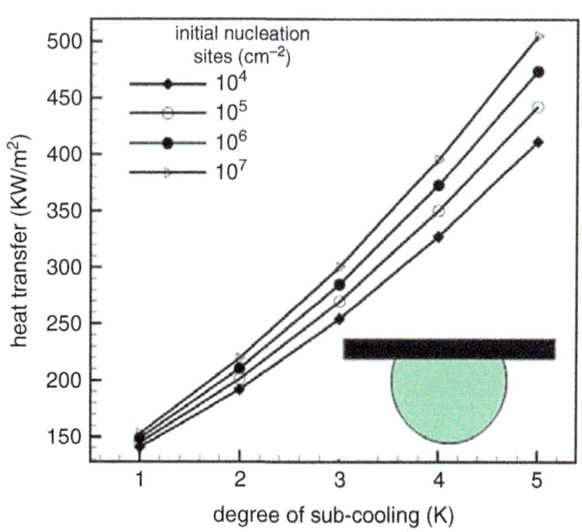

Fig. 9.14 Overall heat transfer rate as a function of subcooling with nucleation site density as a parameter

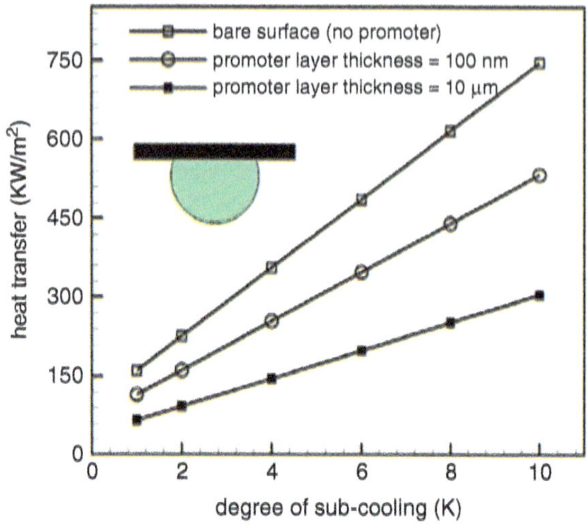

Fig. 9.15 Overall heat transfer as a function of the degree of subcooling with varying thickness of the promoter layer underneath a horizontal substrate. Fluid considered is water at 303 K, contact angle of 100° and nucleation site density of 10^6 cm^{-2}

Figure 9.15 presents the overall heat flux for coatings of varying thicknesses. The computation was carried out when the layer with the thermal conductivity of 0.28 W/m K creates a 100° contact angle. The overall heat flux is significantly influenced by the thickness. When the 10 μm thick promoter thickness is reduced to 100 nm without any change in the contact angle and nucleation density, heat transfer improves by a factor of 1.75. If the same hydrophobicity can be achieved without any promoter, the condensing surface can produce 1.4 times the heat transfer of the 100 nm thick promoter and 2.45 times that of the 1 μm promoter, respectively. This result shows that a redundantly thick coating results in a significant degradation of heat transfer.

9.1.6 Effect of Wettability Gradient

The effectiveness of dropwise condensation is improved by moving the liquid drop that grows on or underneath a solid substrate. The droplet moving over the surface wipes other droplets off. Consequently, more unexposed area is available where smaller droplets can form again. This process of wiping and formation of new small droplets exhibits low heat transfer resistance and is the reason for a large heat transfer coefficient.

Literature (Lee et al. 1998; Daniel et al. 2001; Liao et al. 2006; Zhu et al. 2009; Pratap et al. 2008) suggests various ways of controlling the motion of droplets. A simple approach for mobilization of drops is to incline the surface with respect to horizontal, wherein the gravitational body force is responsible for the droplet motion. Alternatively, one can introduce a variation of surface tension gradient on the substrate. Surface tension can be varied as follows: (1) applying a large temperature gradient on the substrate, in which Marangoni effect leads to drop motion and

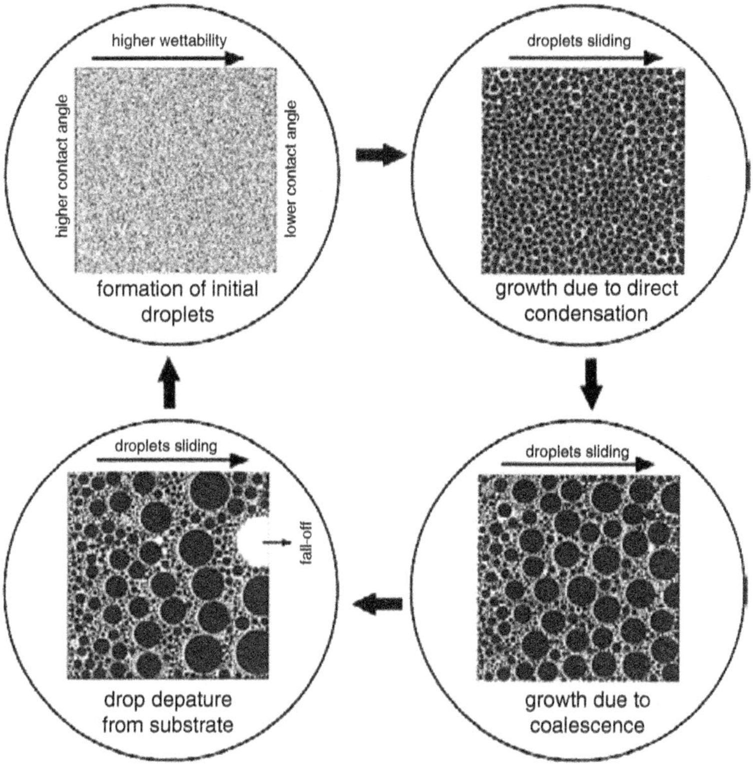

Fig. 9.16 Cycle of dropwise condensation observed in water vapor condensation underneath a horizontal substrate with unidirectional wettability gradient

(2) movement of micro-droplets underneath a horizontal surface by a variable-surface-energy coating, which creates a wettability gradient. To facilitate drop motion by artificially forming a wettability or surface energy gradient on the surface by suitable chemical treatment is a promising technique for drops motion as compared to applying a temperature gradient on the substrate. It is quite possible on copper and glass surfaces by depositing organic long chain monolayers (Subramanian et al. 2005; Pratap et al. 2008).

Daniel et al. (2001) and Bonner (2010) reported from experiments that the condensation on a wettability gradient surface is quite large as compared to a horizontal substrate without wettability gradient. Against this background, dropwise condensation of water underneath a horizontal surface with unidirectional constant wettability gradient is numerically simulated by the mathematical model. Figure 9.16 shows features of the condensation cycle underneath a horizontal substrate with wettability gradient. These are like those of an inclined surface. The points of difference for a graded surface are: (1) drops shift towards the higher wettability side and hence drops of all the sizes are in motion, (2) the velocity of drops depends

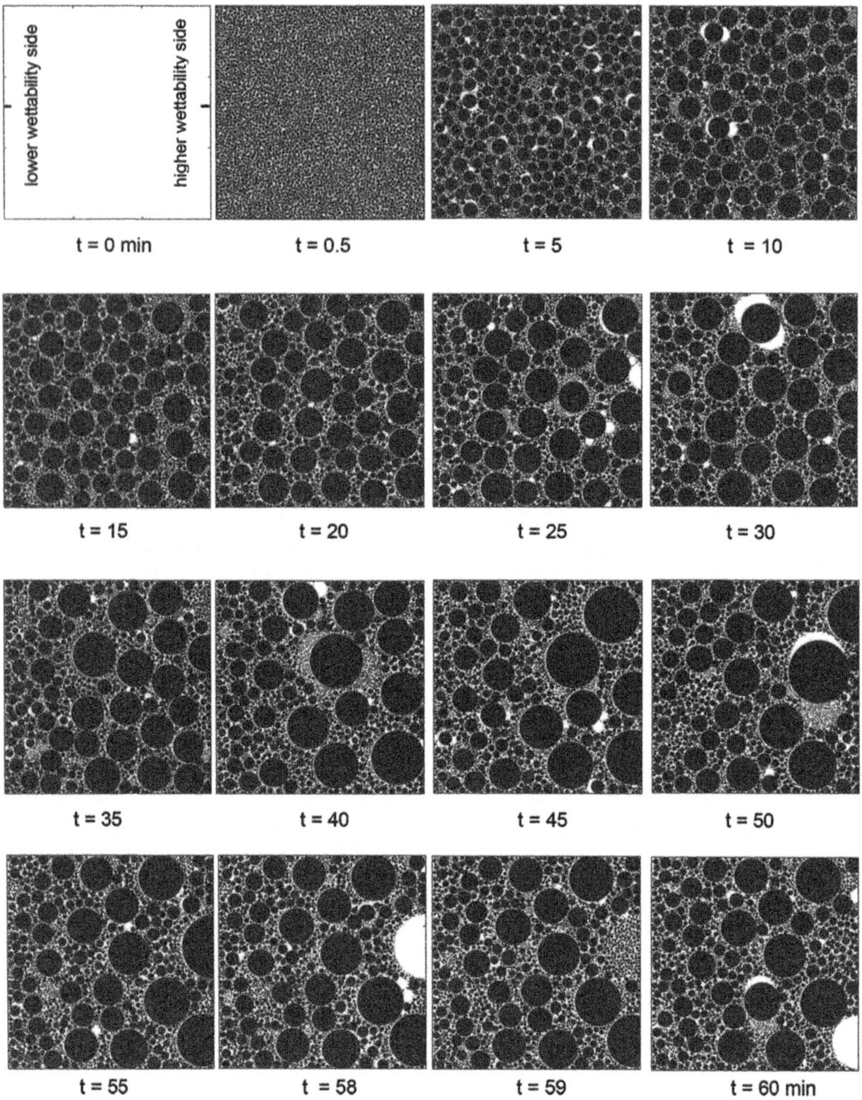

Fig. 9.17 Distribution in drops in dropwise condensation of water vapor underneath a horizontal substrate with wettability gradient 0.33° mm^{-1}. Lower wettability side has a contact angle of 100°, size of substrate is 30 × 30 mm, nucleation site density 10^6 cm^{-2} at saturation temperature 303 and 5 K subcooling

on their size and position on the substrate, and (3) growth and sliding occur simultaneously.

Figures 9.17 and 9.18 show the drop distribution from initial nucleation to dynamic steady-state at selected instants of time underneath horizontal substrates

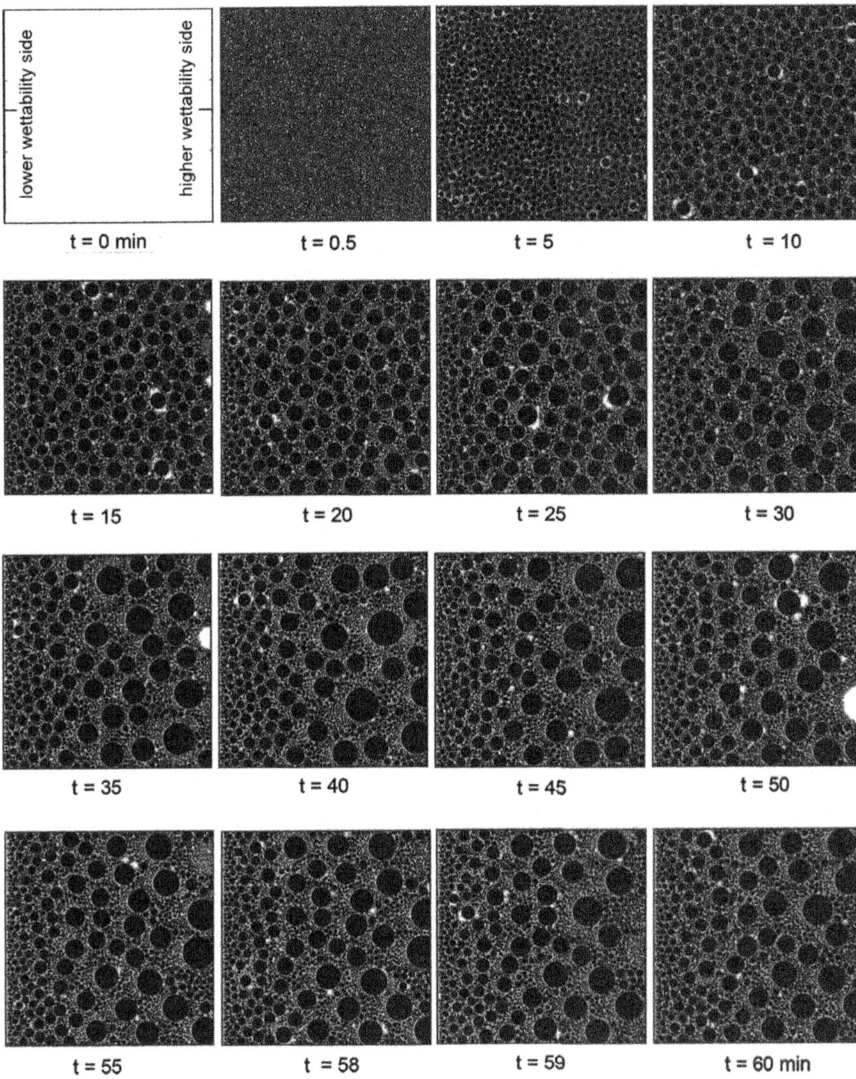

Fig. 9.18 Distribution in drops in dropwise condensation of water vapor underneath a horizontal substrate of wettability gradient 0.33° mm^{-1}. Lower wettability side has a contact angle of 110°, size of substrate is 30 × 30 mm, nucleation site density 10^6 cm^{-2} at saturation temperature 303 and 5 K subcooling

with constant wettability gradient. Figure 9.17 shows a condensation pattern of water vapor at a saturation temperature of 303 and 5 K subcooling underside of substrate having contact angles of 100° and 90°. Figure 9.18 shows the condensation of water vapor at under similar conditions as in Fig. 9.17 but with contact angles of 110° and 100°. In both the surfaces, the wettability gradient is 0.33° mm^{-1}. As drops grow, they become unstable and move towards the higher wettability side of the substrate. Therefore, on an average, larger drops are present at the higher wettability side. The patterns of drops underneath a wettability gradient follow approximately the same trend as those underneath an inclined substrate. The point of difference is that there is no critical size for drop instability. Every drop becomes unstable due to surface tension difference at the three-phase contact line. Gravity, viscous, and surface tension forces are important to determine the terminal velocity of the drop. Hence, the sliding velocity depends on the size and position of the drop underneath the substrate. Fall-off is observed as a rule, on the higher wettability side of the substrate. Figures 9.17, 9.18, and 9.19 also depict small drops present at the lower wettability side of the substrate. On the higher wettability side, the driving force of drop becomes small, the drop cannot move, and it reaches criticality of fall-off (Fig. 9.17, 58 min). These results reveal that the micro-drop size can be moved as the hydrophobicity of wettability gradient substrate increases.

Figure 9.20 compares dropwise condensation patterns of water vapor underneath horizontal and inclined non-graded substrates with a horizontal graded substrate. Spatial distribution of drops at an instant just before the first drop leaves the surface on a graded substrate, first slide-off from an inclined substrate and the first fall-off from a horizontal substrate, respectively, are compared in Fig. 9.20a.

As wettability gradient induces motion to the drops of every size, there exists an exposed virgin area behind every droplet on the graded substrate, as shown in Fig. 9.20a. Hence, the fraction of total area exposed for fresh condensation tends to be greater for a graded surface when compared to the other two configurations. Figure 9.20b shows the frequency of occurrence of a drop of a given radius, namely the histogram, on the three substrates, at an instant just before slide-off or fall-off criticality is attained. Drops slide-off for a graded surface as well as for the inclined. Drops fall-off from a horizontal surface. From Fig. 9.20b, it is clear that a graded substrate has a larger number of smaller sized drops as compared to the other two. Largest drops are formed on a horizontal substrate before they fall-off. This eventually leads to a slower condensation rate on the horizontal substrate; in this regard, the graded substrate shows a clear promise from a perspective of heat transfer enhancement.

Figure 9.20c presents the area of coverage created by the footprints of the drops on the substrate, as a function of time. As soon as the virgin substrate is exposed to vapor flux, direct condensation is initiated, and the area coverage of drops increases rapidly. Later, coalescence dominates direct condensation, eventually leading to droplet criticality. The cycle is then established, and the area coverage tends to stabilize. On a horizontal substrate, only a fall-off criticality is possible while on a graded substrate, a slide-off criticality is usually achieved first. During sliding motion, a droplet may fall-off in transit due to increase in its weight. A quasi-steady-state is eventually reached, after which the area coverage oscillates around an average value.

Results shown in Figs. 9.17, 9.18, 9.19, and 9.20 reveal that area coverage is smaller for the graded surface, making the exposed area greater than the other two surfaces considered. Further, drop instability in the form of a slide-off event is

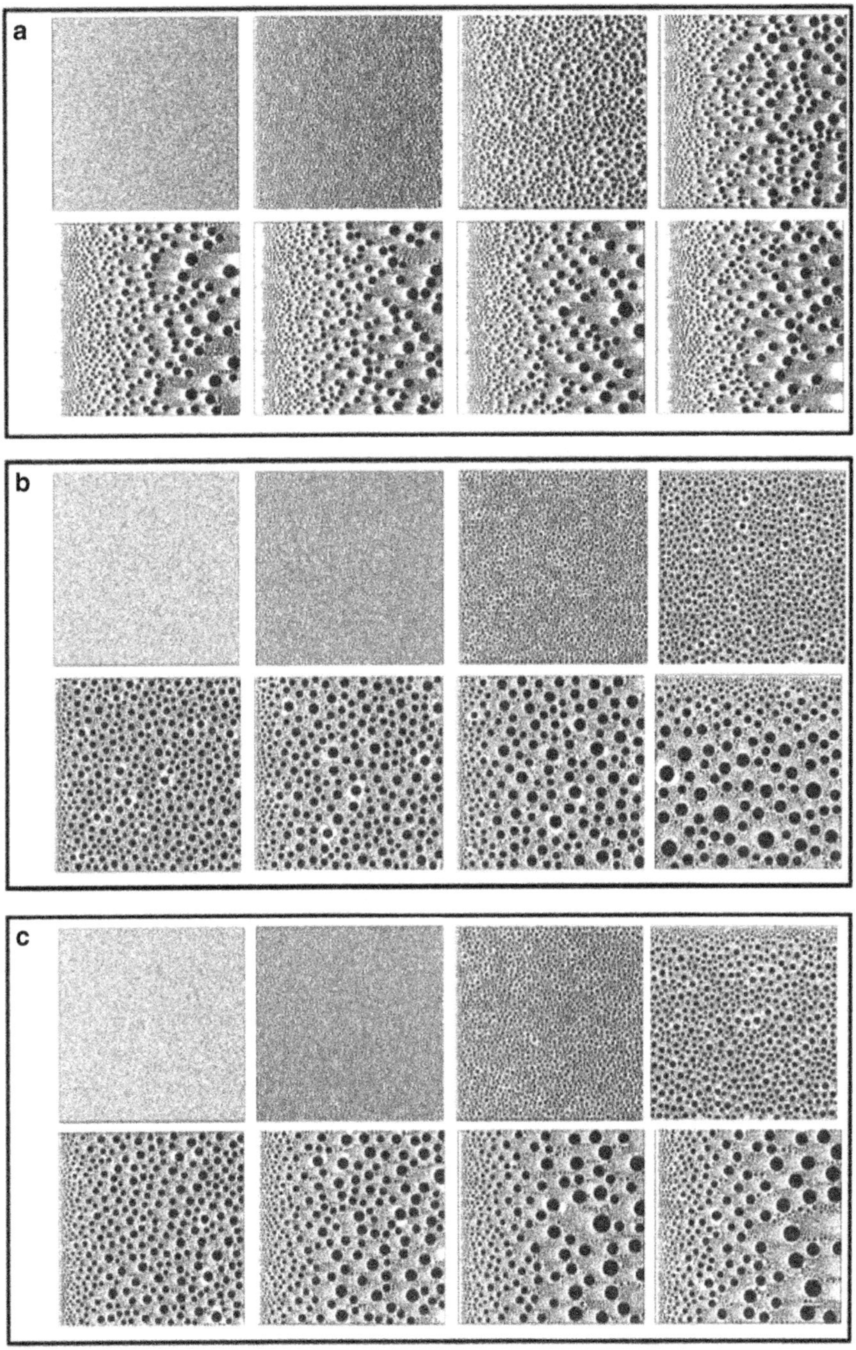

Fig. 9.19 Drop distribution condensed water vapor underneath a horizontal substrate with a wettability gradient. Contact angles are (**a**) 130° and 90°, (**b**) 130° and 100°, and (**c**) 130° and 120°

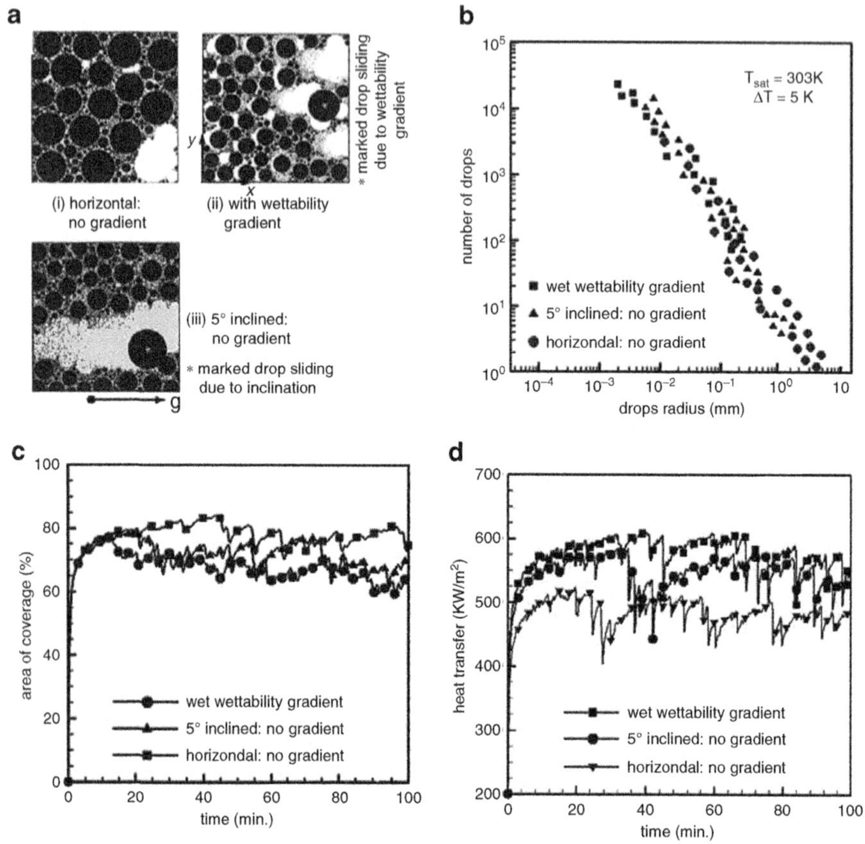

Fig. 9.20 Dropwise condensation of water vapor at 303 K with 5 K subcooling underneath surfaces of various textures. (**a**) Drop distribution underneath various substrates soon after instability. (**b**) Number of drops as a function of drop radius, just before the largest drop leaves the surface. (**c**) Effect of the choice of the substrate on percentage area of coverage. (**d**) Variation of heat transfer coefficient with time over various substrates

relatively early on the graded surface. As a direct consequence, heat transfer coefficient is expected to be higher for a surface with variable wettability. Heat transfer rates computed on these surfaces were found to be 450 (horizontal), 520 (inclined), and 540 (graded horizontal) in units of kW/m^2, with a subcooling of 5 K and a condensation temperature of 303 K.

9.2 Closure

Water vapor condensation underneath horizontal, inclined, and wettability gradient surfaces have been studied by numerical simulation. The effects of contact angle, its hysteresis, inclination of the substrate, thermophysical properties of the working fluid, and saturation temperature of condensation are investigated. Based on numerical data, heat transfer coefficients of water vapor condensation are correlated. In order, the horizontal, inclined, and the graded surface experience (a) larger to smaller drop sizes, (b) longer to shorter cycle times, and (c) lower to higher heat transfer coefficients. Validation of the mathematical model against experiments is discussed in the following chapter.

Chapter 10
Dropwise Condensation of Bismuth on Horizontal and Vertical Surfaces

Praveen Somwanshi, K. Muralidhar, and Sameer Khandekar

Nomenclature

A	Surface area, m^2
C_p	Specific heat of condensate at constant pressure, J/kgK
h	Heat transfer coefficient, $q''/(T_{sat} - T_w)$, W/m^2K
h_{int}	Interfacial heat transfer coefficient, W/m^2K
h_{lv}	Latent heat of vaporization, J/kg
k	Thermal conductivity of condensate, W/mK
k_{coat}	Thermal conductivity of coating, W/mK
m_{avg}	Mass of the condensate averaged over space and time, kg
M	Molecular weight, kg/kmol
N	Nucleation site density, cm^{-2}
p	Pressure, N/m^2
q	Surface heat transfer, W
q''	Surface heat flux, W/m^2
r_{avg}	Radius of drop averaged over space and time, m
r_b	Base radius of drop, m
r_{crit}	Radius of drop at instability due to slide-off, m
r_{max}	Radius of drop at instability due to fall-off, m
r_{min}	Minimum radius of the thermodynamically stable drop, m
R	Base radius of coalesced drop under equilibrium conditions, m
R_{avg}	Base radius of drop averaged over space and time, m
t	Time, s; suffix IS is inertia-surface tension; IV is inertia-viscous; VS is viscous-surface tension
Δt	Time step, s
T_{avg}	Average temperature of condensation $= (T_{sat} + T_w)/2$

S. Khandekar, K. Muralidhar, *Drop Dynamics and Dropwise Condensation on Textured Surfaces*, Mechanical Engineering Series,
https://doi.org/10.1007/978-3-030-48461-3_10

T_{sat}	Saturation temperature in vapor phase, K
T_w	Substrate temperature, K
ΔT	Degree of subcooling, $(T_{sat} - T_w)$, K
U	Terminal velocity of sliding drop or relative velocity between the wall and sliding drop; also characteristic velocity of drop coalescence, m/s
v_l, v_v	Specific volume, m^3; l for liquid and v for vapor
V	Volume of liquid drop, m^3

Dimensionless Quantities

Bo	Bond number, $\frac{\rho g R^2}{\sigma}$
$C_f, \overline{C_f}$	Local and surface-averaged skin friction coefficient, $\tau_w/(1/2)\rho U^2$
Fr	Froude number, $\frac{U^2}{gR}$
$(\mathrm{Nu})_{sd}$	Local Nusselt number of single sliding drop, $h\, r_b/k$
Oh	Ohnesorge number, $\frac{\mu}{\sqrt{\rho R \sigma}}$
Pr	Prandtl number, $\mu C_p/k$
Re	Reynolds number, $\rho UR/\mu$
We	Weber number, $\frac{\rho U^2 R}{\sigma}$

Greek Symbols

α	Inclination angle, radians
μ	Dynamic viscosity of liquid phase, Pa-s
ρ	Liquid density, kg/m^3; l for liquid and v for vapor
σ	Surface tension coefficient at liquid-vapor interface, N/m
$\tau_w, \overline{\tau}_w$	Local and average wall shear stresses, N/m^2
θ	Contact angle, degrees
θ_{adv}	Advancing contact angle, degrees
θ_{rcd}	Receding contact angle, degrees
$\Delta\theta$	Contact angle hysteresis, $(\theta_{adv} - \theta_{rcd})$, degrees
θ_{avg}	Average contact angle, $^\circ$

10.1 Introduction

As discussed in Chaps. 5–7, condensation occurs on a solid wall when the surface temperature falls below the local saturation temperature of the adjoining mass of vapor. On specially treated surfaces, liquid droplets will appear at specific nucleation sites. As the phase change process proceeds, these droplets grow by direct condensation from the vapor phase, coalesce with neighboring drops, and may fall off or start to slide down the substrate. The process is cyclic and drop instability will prevent the formation of a liquid film. Such a phase change process, termed dropwise condensation (Carey 2008), is heterogeneous. Vapor condenses in the form of discrete liquid drops on or underneath a cold solid substrate.

Dropwise condensation can be sustained only on specially textured surfaces (Rose 2002; Vemuri and Kim 2006; Rausch et al. 2008). The heat transfer coefficient during dropwise condensation can be quite high, for example, up to 30 times greater than the filmwise mode, when tested with Langmuir-Blodgett surfaces (Rose 2002) and 5–20 times better when a promoter layer is used (Koch et al. 1998a, b). Leaching of the coated or textured substrate can alter its wettability characteristics and result in its aging. Hence, apart from heat fluxes, it is important to estimate wall shear stresses created during drop movement.

Drops are expected to form at individual nucleation sites, while the area between the drops remains inactive with respect to condensation (Leach et al. 2006). The diameter of the smallest drop at nucleation can be estimated from thermodynamic considerations (Carey 2008). Subsequently, the drop grows by direct condensation at a rate determined by the conduction resistance through the drop, interfacial heat transfer coefficient, and the available temperature difference. Larger drops also grow by coalescence with their neighbors. When a certain size is reached, drops become gravitationally unstable, fall off, or slide along the surface, wiping other drops along their path. Fresh nucleation sites are thus revealed, and the condensation process repeats in a cyclic manner.

A multiscale dropwise condensation model, starting from the atomic scale, progressing towards the growth of droplets, coalescence, and drop instability has been reported by various authors (Rose 2002; Vemuri and Kim 2006; Rausch et al. 2008) for condensation of water vapor. The model developed by the authors, as described in Chap. 6, is comprehensive and the deformed drop shape is determined by the two-circle approximation. In addition, the process of coalescence is taken to be instantaneous.

In this chapter, the original model is extended by including the characteristics of coalescence of the droplets. While condensation of bismuth is the primary focus, results for water are also generated for comparative purpose. The effect of drop coalescence on wall shear stress and heat transfer rates in dropwise condensation is determined. In addition, the mathematical model is parallelized using MPI to run for large surface areas on a high-performance computing system.

The choice of bismuth as the working fluid is motivated by the range of applications where liquid metals are encountered. These include the use of liquid sodium in fast breeder reactors and in distillation processes required to produce high purity precious metals such as gold. Thermophysical properties of bismuth and water at the selected saturation pressure and temperature are given in Table 10.1. Density of water is substantially smaller than bismuth; it has higher latent heat of vaporization, lower surface tension and viscosity. Thermal conductivity of bismuth is substantially higher than that of water, thus giving rise to a small Prandtl number. Differences in thermophysical properties lead to significant changes in the drop sizes, cycle time, and heat fluxes for bismuth relative to water. Even under near-vacuum conditions, bismuth evaporates at a relatively high temperature when compared to water. The

Table 10.1 Thermophysical properties of bismuth and water

Property	Bismuth	Water
Density, ρ (kg/m^{-3})	9904.4	995
Latent heat of vaporization, h_{lv} (kJ/kg)	178.9	2426
Specific volume, υ (m^3/kg)	0.0001 (l), 327.86 (v)	0.001 (l), 29.74 (v)
Surface tension, σ (N/m)	0.371	0.0709
Thermal conductivity, k (W/mK)	13.44	0.62
Molecular weight, M (kg/kmol)	208.98	18
Dynamic viscosity, μ (Pa-s)	1.34×10^{-3}	0.769×10^{-3}
Advancing contact angle, θ_{adv} (deg)	118.5	118.5
Receding contact angle, θ_{rcd} (deg)	101.5	101.5
Saturation temperature, T_{sat} (K)	635	303
Wall temperature, T_w (K)	630	298
Prandtl number (−)	0.014	5.182

degree of subcooling between the saturated vapor and the substrate is taken as 5 °C. Since the temperature interval is small, a constant thermophysical property approximation at the average temperature is utilized. At high levels of subcooling, drops are expected to form a liquid film on the substrate.

Liquid bismuth has been reported to display non-Newtonian behavior under certain circumstances. In the present work, bismuth is taken to be Newtonian with a definite viscosity for the following reason. Fluid motion in dropwise condensation is seen during (a) the instability of the liquid drop, either during its sliding motion or fall-off and (b) coalescence. Both events occur for less than 0.1% of the overall condensation cycle time. The coalescence characteristics are determined by a variety of properties such as surface tension, contact angle, density, and volume. In the context of small drops, surface tension and contact angle are central to the initial pressure difference and is unaffected by non-Newtonian behavior. Reynolds numbers during drop coalescence in the range of 400–500 may momentarily appear and non-Newtonian effects should be separately studied for this short duration. Viscosity and shear rates of bismuth in the liquid phase appear only during the late decay of the coalescence-generated velocities and during the sliding of drops over a vertical surface. They do not affect the peak velocities arising initially from inertia-surface tension coupling. Reynolds number during sliding motion can be in excess of 1000 (Sikarwar et al. 2013a, b) while Prandtl number is small (Table 10.1). The wall heat flux in the liquid metal is strongly affected by the large thermal diffusivity and is less sensitive to fluid motion. For these reasons, Newtonian fluid model for bismuth is expected to be satisfactory. Since condensation pattern and wall heat fluxes are determined by events that occur for a majority of the time, non-Newtonian behavior of bismuth has not been considered in the mathematical model.

10.1.1 Mathematical Modeling of Dropwise Condensation

Models of the dropwise condensation process have been reported in the literature with varying levels of approximations (Vemuri and Kim 2006; Kim and Kim 2011).

The initial mathematical model of dropwise condensation developed by the authors is described in detail elsewhere (Khandekar and Muralidhar 2014). The model proceeds from nucleation of drops to their growth and instability but treats coalescence to be an instantaneous process. The model, comprising these four steps, is schematically shown in Fig. 10.1. The model input includes nucleation site density, contact angle and hysteresis, interfacial properties, thermodynamics of phase change, and the thermophysical properties of the condensing phase. Thermophysical properties of the vapor and liquid phases are evaluated at the average of the substrate and saturation temperatures (Table 10.1). The model predicts the instantaneous condensation pattern, the mass of condensate leaving the surface, and wall heat flux. Additional quantities of interest are the cycle time of instability, liquid holdup, instantaneous and time-averaged heat fluxes, and wall shear stress. Validation studies against experiments of Rose (2002) have been discussed by the authors elsewhere (Khandekar and Muralidhar 2014; also see Chap. 3).

The instantaneous space-averaged heat transfer coefficient over an area A of the substrate during dropwise condensation is estimated in terms of the mass of the vapor condensing over the surface per unit time m_{avg} and the subcooling ΔT. Shear

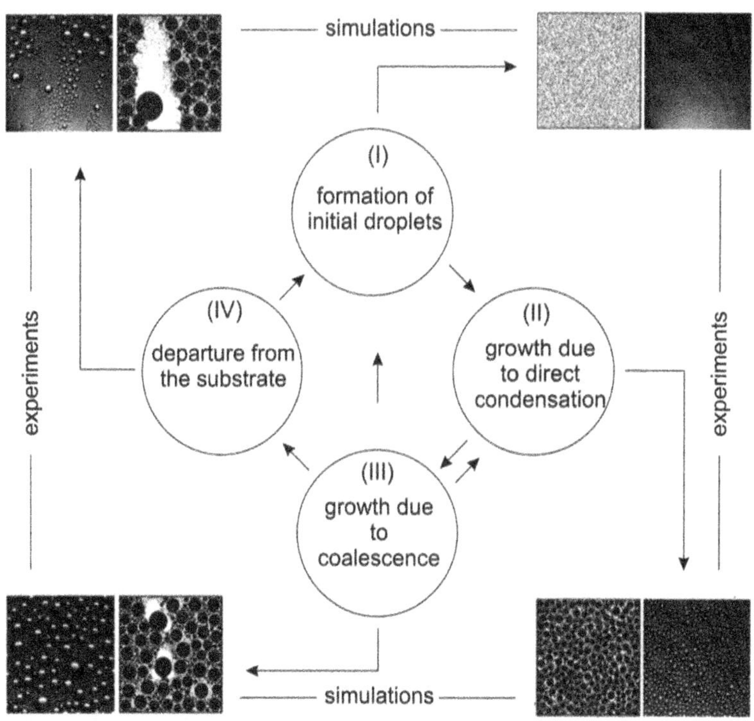

Fig. 10.1 Schematic drawing of a mathematical model representing the dropwise condensation cycle. The model represents the condensation process in four steps, namely nucleation, growth by direct condensation of vapor, growth by droplet coalescence, and drop instability. Experimentally recorded images in water vapor (left) are shown jointly with those obtained from simulation

forces are generated by each drop when it begins to slide over the substrate. The mass of the droplet can increase during its travel on the substrate. These are discrete events and are cumulatively determined over the cycle time of condensation. The calculation of space and time-averaged shear stresses on the substrate arising from drop movement is described in Chaps. 6 and 7. The determination of wall shear stress during coalescence is described in Chap. 3.

The condensation model of the present study extends the past work of Sikarwar et al. (2013a, b) in two significant respects. First, coalescence is not treated as instantaneous. Velocity scale and timescale are estimated from non-dimensional analysis and in turn, used to estimate additional heat flux as well as the wall shear stress. Secondly, simulations are adapted to a high-performance computing system using domain decomposition that permits the use of MPI to run the computer code in parallel. The second step permits simulations over surfaces as large as 50 × 50 mm in a reasonable amount of time. In addition, the choice of bismuth is novel since it is a model liquid metal with clearly prescribed thermophysical properties but has not been studied in the context of dropwise condensation.

10.1.2 Scale Analysis of Drop Coalescence

The discussion here is for the merger of two small liquid drops over textured surfaces that are hydrophobic with respect to the wetting phase. The properties of the textured surface appear in the model through the specification of the contact angle. Contact angles are a source of uncertainty in modeling dropwise condensation. Superhydrophobic surfaces with contact angles greater than 140° may be prepared only over small areas. For this reason, lower contact angles that, nevertheless, make the surface hydrophobic and promote dropwise condensation have been utilized in the present study.

For a horizontal surface, the equilibrium static angle is taken to be 110° while for a vertical surface, advancing and receding angles are 118.5° and 101.5°, respectively. These values are expected for bismuth condensation on a chemically textured copper substrate. In the condensation context, coalescence takes place over a horizontal surface in the pendant configuration. The drops are composed of the same liquid while coalescence takes place in a vapor environment. The drops are not pinned at the three-phase contact line and the footprint evolves with time. In the following discussion, scale analysis developed for a horizontal configuration is taken to be applicable for a vertical surface as well.

A large number of forces appear in the flow field with relative magnitudes that change with time. The forces can be due to surface tension, gravity, viscosity, and inertia. These primary forces are used to estimate the internal pressure of the drops. Drop size plays an important role in coalescence dynamics apart from fluid and interfacial properties. For drops of smaller volume, surface tension would be more important than inertia and gravity. Gravity forces lower internal pressure in pendant drops and hence the resulting velocities, in comparison to the sessile. In the present

Table 10.2 Magnitudes of dimensionless parameters estimated for bismuth and water under ambient conditions. Velocity scale U is taken as $(\sigma/\rho R)^{0.5}$, where R is the base radius of the coalesced drop at equilibrium

Volume (μL) (U, (m/s))	Base radius, (mm)	Bo	We	Fr	Re	Oh
Bismuth						
0.05(0.39)	0.237	1.437×10^{-3}	1	68.130	696.034	0.015
0.5(0.27)	0.510	0.979×10^{-3}	1	14.678	1021.640	0.068
1.5(0.22)	0.736	0.815×10^{-3}	1	7.057	1226.920	0.142
2.5(0.20)	0.872	0.749×10^{-3}	1	5.020	1335.960	0.199
7.5(0.17)	1.258	0.623×10^{-3}	1	2.413	1604.400	0.414
Water						
0.05(0.54)	0.237	0.008	1	127.882	161.022	0.62×10^{-3}
0.5(0.37)	0.510	0.037	1	27.551	236.348	4.23×10^{-3}
1.5(0.31)	0.736	0.075	1	13.245	283.839	3.52×10^{-3}
2.5(0.28)	0.872	0.106	1	9.422	309.063	3.24×10^{-3}
7.5(0.24)	1.258	0.221	1	4.530	371.165	2.69×10^{-3}

study, drops coalesce in vapor environment. The density difference between water and the ambient air is taken practically to be the density of the liquid. Surface tension is considered at liquid-vapor interface while a contact line force arises at the solid-liquid-vapor three-phase contact line.

The approach adopted here for scale analysis is as follows. Dimensionless parameters are defined using a pair of forces at a time, leading further to velocity and timescales. The realization of these scales in experiments is then examined. Using the experimental data as the yardstick, applicable forces during coalescence are then identified.

Instead of resolving the spatio-temporal distribution of the flow field, each regime of coalescence may be represented by its characteristic length, velocity, and timescale. This approach is quite suitable from a modeling perspective. Coalescence is taken to introduce additional velocity, length, and timescales that are suitably incorporated in the condensation cycle. Non-dimensional quantities for equal drop volumes of bismuth and water are summarized in Table 10.2. The corresponding timescales are given in Table 10.3. The viscous-surface tension timescale is relevant closer to the instant of coalescence while viscous-inertial timescale is suggestive of the drop evolution closer to final equilibrium. The intermediate timescale (in ms) represents the appearance of large velocities jointly with significant deformation of the drop interface and is defined by a balance of inertia forces and surface tension. This timescale is of greatest interest to dropwise condensation.

Table 10.3 Timescales estimated for coalescence of bismuth and water drops under ambient conditions; IS inertial-surface tension; VS viscous-surface tension; IV inertia-viscous

Volume(μL)	Base radius(mm)	t_{IS}(ms)	t_{VS}(ms)	t_{IV}(ms)
Bismuth				
0.05	0.237	0.595	0.855×10^{-3}	0.414×10^{3}
0.5	0.510	1.882	1.842×10^{-3}	1.923×10^{3}
1.5	0.736	3.260	2.657×10^{-3}	4.000×10^{3}
2.5	0.872	4.208	3.150×10^{-3}	5.622×10^{3}
7.5	1.258	7.289	4.543×10^{-3}	11.695×10^{3}
Water				
0.05	0.237	0.434	2.70×10^{-3}	0.070×10^{3}
0.5	0.510	1.374	5.81×10^{-3}	0.325×10^{3}
1.5	0.736	2.379	8.38×10^{-3}	0.675×10^{3}
2.5	0.872	3.072	9.94×10^{-3}	0.949×10^{3}
7.5	1.258	5.32	1.43×10^{-3}	1.975×10^{3}

,

10.2 Results and Discussion

Instantaneous condensation patterns of bismuth on vertical and horizontal surfaces are discussed in this chapter. Simulations on realistic meaningful surfaces are prohibitively expensive. As an alternative, simulations are carried out over substrates of increasing sizes. The sensitivity of the condensation pattern to substrate size is the examined. Condensation statistics of bismuth and water are compared.

Three parameters that strongly influence simulation are the nucleation site density, surface hydrophobicity, and the accommodation coefficient. For the present discussion, density of the nucleation sites at which liquid drops nucleate (with a radius of r_{min}) is chosen as 10^6 cm^{-2}; this value is quite commonly used in the condensation of water. For bismuth, nucleation site density data is surface-specific and has not been extensively tabulated. To facilitate comparison with water, this value of 10^6 cm^{-2} has been retained. The surface hydrophobicity is characterized by the equilibrium contact angle. In this context, it may be noted that fabrication of stable large-area superhydrophobic surfaces continues to be a major challenge. Such surfaces are invariably prepared by coating a substrate. Specialized coatings that make the surface superhydrophobic ($\theta_{equilibrium} > 150°$) tend to wear out at fairly low shear stresses and have limited shelf-life. For this reason, a hydrophobic surface with a modest equilibrium contact angle of 110° (hysteresis 17°) has been selected for analysis. The third factor, namely the accommodation coefficient (σ) determines the interfacial heat transfer coefficient. Following the discussion in the literature, values of $\sigma = 0.05$ for water (Carey 2008) and 0.67 for bismuth (Pukha et al. 2005) have been used in the present study. Though uncertainty levels in these values is quite high, apart from their dependence on pressure, temperature, and surface conditions, the dropwise condensation statistics are shown in later discussions of this chapter to be less sensitive to their choice.

A quantity of importance to be extracted from simulation is liquid holdup over the substrate. A larger average holdup indicates a greater heat transfer resistance and is undesirable in many applications. Similarly, a larger holdup indicates smaller productivity of the distillate and is equally undesirable. It is to be expected that the holdup would be greater for a horizontal surface compared to a vertical, being related to the cycle time and the drop size at instability. The degree of subcooling employed in simulations for bismuth as well as water is 5 K. The saturation temperature of condensing bismuth vapor is taken to be 635 K while that of water is 303 K.

Near-horizontal surfaces, with up to 10° inclination from the horizontal, and truly horizontal surfaces show instability mainly due to fall-off of large drops, and their respective condensation patterns were found to be indistinguishable. The results discussed in the following paragraphs use a 10° orientation for the near-horizontal surface but the surface, for simplicity, is addressed as horizontal.

10.2.1 Condensation Patterns of Bismuth on Vertical and Horizontal Surfaces

Spatio-temporal drop distributions, from initial nucleation to drop instability, are shown in Fig. 10.2a, b for condensation of bismuth vapor. The underside of a horizontal surface and a vertical surface are individually studied. Surface dimensions of 10 × 10 mm, 30 × 30 mm, and 50 × 50 mm are compared. The respective images are scaled in Fig. 10.2a, b for uniformity. As a result, the size of drops on a surface of size 50 × 50 mm is portrayed to be smaller than on a surface of size 10 × 10 mm. Sizes of equipment of interest are larger than the dimensions studied here, but a truly multiscale simulation starting from nucleation is computationally intractable on a device scale. The approach adopted in the present work is to examine surfaces of increasing dimensions and the influence of confinement on the condensation dynamics.

The simulation begins with drops of minimum radius placed at the nucleation sites. Subsequently, they are grown over a time step by direct condensation of the surrounding vapor. Partially grown drops are coalesced with their neighbors if they touch each other. Otherwise, they continue to grow by direct condensation. Fall-off on a near-horizontal and slide-off on a vertical surface are observed if the gravitational instability criteria are fulfilled. During sliding motion, the size of the sliding drop increases by coalescence with those in its path. Instability exposes the surface to fresh nucleation and the entire cycle of nucleation to instability is resumed. When large drops merge or slide off/fall off, fresh sites are exposed, and small drops are placed at these locations. As a result, the relative population of small drops increases. Just ahead of criticality, the average drop diameter over the surface is the greatest. The combined effect of direct condensation, coalescence, and instability creates drop

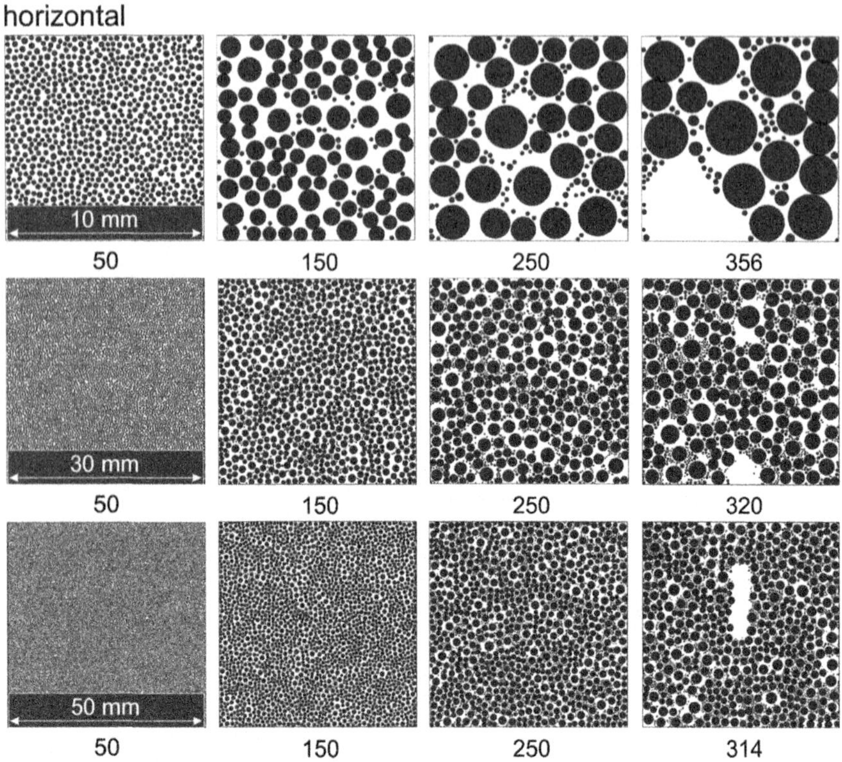

Fig. 10.2 (**a**) Comparison of full surface condensation patterns of bismuth vapor over horizontal surfaces of size 10 × 10 mm in the first row, 30 × 30 mm in the second row, 50 × 50 mm in the third. Time stamps show various stages of growth of drops and instability. The time instant of instability is seen to be insensitive to the domain size. (**b**) Comparison of full surface condensation patterns of bismuth vapor over vertical surfaces of size 10 × 10 mm in the first row, 30 × 30 mm in the second row, 50 × 50 mm in the third. Time stamps show various stages of growth of drops and instability. The time instant of instability is seen to be insensitive to the domain size

population of varying sizes over the surface (Fig. 10.2a, b). These features, as discussed here, are seen in horizontal as well as vertical surfaces of the three increasing substrate sizes.

The size of a typical drop increases with time till criticality, by direct condensation and/or the coalescence process. Since the conduction resistance scales with the drop size, the relative magnitudes of conduction and convection resistance change with the drop radius, and hence with time. At any time instant, a collection of small and large drops of varying sizes prevail over the surface. Soon after instability of a large drop, the number of small drops increases due to fresh nucleation over the virgin substrate created by the movement of the unstable drop. As a part of the condensation cycle, the drop radius, and its interfacial area will increase with time. Since latent heat release and interfacial heat transfer coefficient both refer to the same

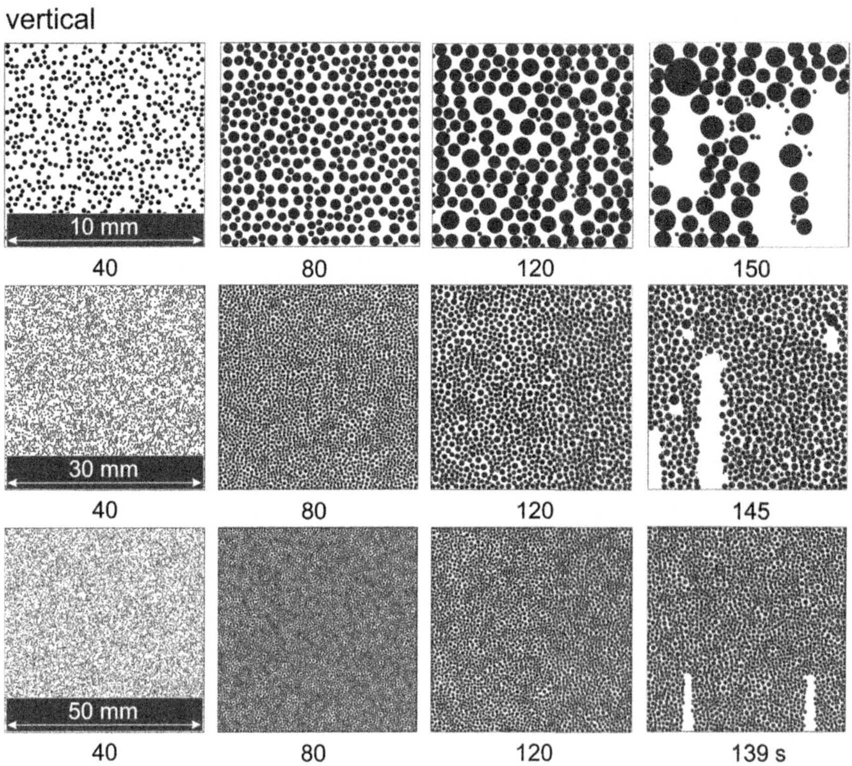

Fig. 10.2 (continued)

vapor-liquid interface area, interfacial resistance becomes a constant, independent of the heat transfer area. In contrast, conduction resistance is small for small drops and scales with drop size. Expressions for these resistances are discussed in earlier chapters, and their numerical estimates are summarized in Table 10.4. Conduction resistance plays an important role relative to interfacial resistance in water and the total resistance changes with drops size. Thus, thermal resistance in water is minimum when the population of small drops is large and attains a maximum just before instability. For a given level of subcooling, heat flux passing through the substrate per unit time is large when the surface is populated mainly by small drops and is small when some drops have grown to greater size. Specifically, heat fluxes are large at fresh nucleation and will attain a minimum ahead of instability. In bismuth, the roles of conduction and interfacial resistances are reversed and hence, the total resistance is less sensitive to drop size.

Including finite time coalescence of drops on horizontal and vertical surfaces will further increase the wall heat flux over the value that is predicted with the assumption of instantaneous coalescence, for which the associated heat flux is zero. The coalescence event, however, lasts for a few milliseconds wherein the coalesced resultant drop grows to a finite size.

Table 10.4 Comparison of dropwise condensation parameters of bismuth and water

		Bismuth	Water
θ_{avg} and hysteresis (degrees)		110; 17	
Nucleation site density (cm^{-2})		10^6	
Subcooling, ΔT (K)		5	
Drop radius at nucleation r_{min} (μm)		0.052	0.0035
Capillary length r_{cap} (mm)		1.94	2.69
r_{max} (horizontal), r_{crit} (vertical) (mm)	Horizontal	2.44	3.59
	Vertical	0.6	0.878
Cycle time (s)	Horizontal	313–356	73
	Vertical	138–149	21
Radius of drop averaged over space and time, R_{avg} (μm)	Horizontal (size of substrate) (mm × mm)	2.449 (10 × 10) 2.199 (30 × 30) 2.198 (50 × 50)	8.009 (30 × 30)
	Vertical (size of substrate) (mm × mm)	2.370 (10 × 10) 2.368 (30 × 30) 2.147 (50 × 50)	7.775 (30 × 30)
Interfacial resistance, $2/h_{int}$ ($K\text{-}m^2/W$)	–	51.44×10^{-4}	1.1×10^{-4}
Conduction resistance, $(1 - \cos\theta_{avg}) R_{avg}/k$ ($K\text{-}m^2/W$)	–	0.00219×10^{-4}	0.173×10^{-4}
Heat transfer coefficient averaged over space and time, h (kW/m^2K) with standard deviation	Horizontal (size of substrate) (mm × mm)	2.72; 16.8 (10 × 10) 3.44; 10.2 (30 × 30) 2.97; 6.36 (50 × 50)	15.88; 2.78 (30 × 30)
	Vertical (size of substrate) (mm × mm)	1.48; 5.27 (10 × 10) 2.005; 3.64 (30 × 30) 2.01; 2.79 (50 × 50)	22.8; 1.84 (30 × 30)

For a vertical surface, shear stresses are generated during the sliding motion of the drop. For a strictly horizontal surface, fall-off is taken to be instantaneous and shear stresses are absent. Coalescence-induced velocity generates shear stresses for both surface orientations which could be comparable or even greater than that arising from drop sliding motion.

Referring to Fig. 10.2a, b as well as Table 10.4, the size of the drop at criticality is larger for horizontal surfaces than the vertical. This is because gravity has to overcome the average contact angle for a horizontal surface while it has to overcome the advancing angle-to-receding angle difference for a vertical surface. Hence, the retention of mass is larger for a horizontal surface when compared to the vertical. It may be realized that drop growth rates are limited by the sum of conduction and vapor-liquid interfacial resistance and diminish for larger drops. In addition, since the drop grows to a larger size over a horizontal surface, the condensation cycle time is clearly greater here when compared to the vertical.

Figure 10.2a, b show that the vertical surface experiences sliding instability, as expected, for the three surfaces studied for the condensation of bismuth vapor. For a near-horizontal surface, it is mainly fall-off though a short signature of slide-off is visible for a horizontal surface of 50 × 50 mm dimensions. Comparing the time instants of commencement of instability, it is seen that the surface dimensions play a role. The moment of instability is earlier on larger surfaces when compared to the smaller; it is a trend common to horizontal as well as vertical surfaces. The first instant of instability is in the range of 314–356 s for horizontal surfaces and 139–150 s for vertical surfaces over the range of dimensions considered. The proximity of the substrate boundaries is seen to stabilize large drops by limiting their growth rate, thus contributing to higher cycle times for smaller surfaces.

Condensation parameters of bismuth and water are compared in Table 10.4. Thermodynamically stable minimum radius of bismuth is 15 times greater than water. The capillary radii are of similar magnitude, showing that density and surface tension change in similar proportions. Accordingly, the maximum radius of liquid bismuth at the instant of instability for horizontal surfaces is smaller than water only by a factor of 1.5. The critical radius of bismuth at the instant of instability for vertical surfaces is smaller than water by a factor of 1.3. The average radius of a drop for bismuth during the condensation process is determined from a histogram analysis of the instantaneous condensation patterns; it is smaller for bismuth than water for horizontal and vertical surfaces by factors of 4 and 3.5, respectively.

Table 10.4 also shows that the cycle time for a vertical surface is smaller than the horizontal. Cycle time of bismuth is larger than water for horizontal and vertical surfaces by factors of five and seven, respectively. The products of density and latent heat of vaporization are similar for the two fluids. Yet, Table 10.4 shows that the heat transfer coefficient of bismuth is smaller than water for horizontal and vertical surfaces by factors of 4.6 and 11.4, respectively, for reasons discussed below.

Thermal resistance, defined as a ratio of the temperature difference and the wall heat flux at the scale of a single drop arises from conduction through the drops, and the interfacial heat transfer coefficient, the two terms appearing in series. Expressions for these resistance components for a single drop, as it evolves in time, are given in Table 10.4. For numerical evaluation and comparison, the representative radius of the drop selected is the average value that is obtained among all drops jointly evolving over the substrate within a condensation cycle. Conduction resistance, being proportional to radius, is small for a small drop and large for a large drop. It reaches a maximum value just before drop instability. A smaller cycle time will experience frequent nucleation and show a preference for smaller drops and hence a smaller overall conduction resistance. However, the interfacial resistance is a constant irrespective of the size of the drop, as noted earlier. The total thermal resistance is the sum of conduction as well as interfacial resistance which is convective in origin. In addition, the interfacial heat transfer coefficient in water is much larger than for bismuth, Table 10.4. In terms of the sum of the conduction and interfacial thermal resistance, the minimum to maximum variation for water is much greater when compared to bismuth. Accordingly, in water, one can see a strong correlation among small drop sizes, small cycle time, and large heat transfer coefficient. Specifically, heat transfer coefficient for water is higher for a vertical surface when compared to the horizontal, while the average drop size is smaller.

In bismuth, interfacial resistance is certainly dominant and is a constant irrespective of the drop size. The heat transfer coefficient during bismuth condensation is smaller than in water. While the peaks attained in heat fluxes are smaller and frequent in water, Figs. 10.6 and 10.7 show strong individual peaks in heat fluxes for bismuth. Accordingly, Table 10.4 also includes the time-averaged heat transfer coefficient along with its standard deviation. In water, standard deviation in heat transfer coefficient data is small, relative to the time averaged value. However, including standard deviation in the heat transfer coefficient, such a clear difference between horizontal and vertical surfaces for bismuth condensation cannot be discerned.

Condensation patterns at the instant of instability for horizontal and vertical surfaces are shown in Fig. 10.3 for the three sizes considered. Here, the central 10×10 mm region is extracted for discussion. These patterns are located away from the boundaries, and the question is, the extent to which the substrate size influences the condensation process at the center. Figure 10.3 also includes the time

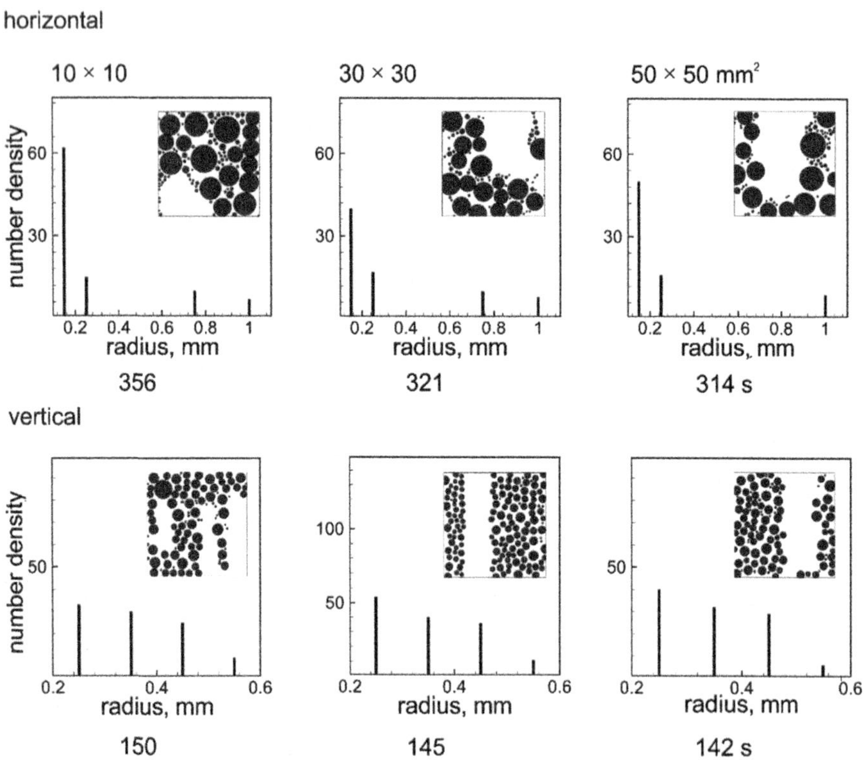

Fig. 10.3 Comparison of condensation patterns and corresponding time for bismuth vapor at the occurrence of first instability (INSET). The number density data is shown for the central 10×10 mm region, for the three substrate sizes (10×10 mm, 30×30 mm, 50×50 mm), both for horizontal and vertical substrates. At the onset of instability, $r_{max} = 2.44$ mm for horizontal substrate and $r_{crit} = 0.6$ mm for the vertical substrate

instant of instability. For a horizontal surface, instability occurs due to droplet fall-off while it involves sliding motion for the vertical substrate. As discussed with reference to Fig. 10.2a, b, instability occurs earlier for a vertical surface as compared to the near-horizontal. The time instant of instability changes with the size of the substrate, but the influence is marginal. However, a consistent trend of boundary-induced stabilization on smaller surfaces and early instability over a larger surface are seen.

The number density distribution in terms of drop sizes at selected instants of time is shown in Fig. 10.3, where the population of small drops is often larger than that of large drops. The skewness in this ratio is smaller for a vertical surface where very large drops fall off first due to gravitational instability; this is not so for horizontal substrates. The data of Fig. 10.3, combined with the liquid holdup curves (Chap. 9) in Figs. 10.4 and 10.5, indirectly provide the number density variation with time.

Figure 10.3 also shows the drop size distribution over the surface at the moment of instability, in a number density versus radius plot. A radius interval of 0.1 mm is used for this purpose. Soon after the time instant of instability, a number of small fresh drops are nucleated. The corresponding number is very large, and the drop radii are small. The data points corresponding to very small drops are common to all the graphs and are not shown for clarity. The critical radius of instability is greater for the horizontal surface when compared to the vertical (Table 10.4), explaining the scales selected for the x-axis. For a horizontal surface, a few large drops are to be seen with the population of the largest drops (~ 1 mm) being smaller than the smallest drops (~0.2 mm). This trend of decreasing number density with radius is seen on all the three substrates. For a vertical surface, the pattern at instability shows

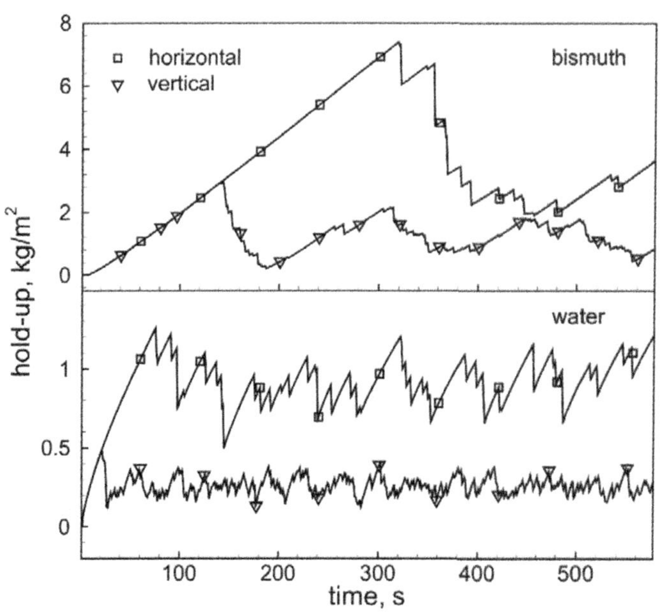

Fig. 10.4 Variation of liquid bismuth holdup in mass units with time compared with water for a surface of size 30 × 30 mm; horizontal and vertical orientations are both considered

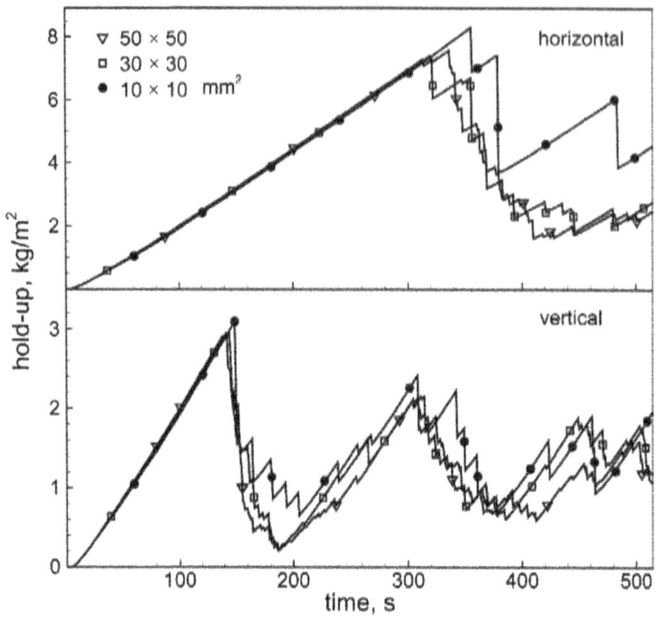

Fig. 10.5 Variation of liquid holdup with time in bismuth for horizontal and vertical surfaces of size 10 × 10 mm, 30 × 30 mm, 50 × 50 mm

a vertical sweep, along which, a gravitationally stable drop has been swept away. Subsequently, a large number of small nucleating drops appear. This creates a large number of drops having small radii, in contrast to large drops that are prone to instability and get eventually wiped out from the surface. Thus, the number density trend as a function of the characteristic radius has, once again, a negative slope for the vertical surface.

10.2.2 Liquid Holdup

From an application viewpoint, liquid holdup over the surface is a quantity of great importance. It is the difference between the amount of vapor condensed and the amount drained away. Since the condensate is the productive aspect of the distillation process, liquid holdup for the selected surface and inclination should be a minimum.

Instantaneous liquid holdup as a function of time is shown in Fig. 10.4. Here, bismuth is compared with water for a surface of size 30 × 30 mm, for both horizontal and vertical configurations. The liquid mass staying on the surfaces increases with time, first by direct condensation and then by coalescence. At the onset of instability, large drops are drained away and the liquid holdup sharply decreases. Fresh nucleation is then initiated and there is a gradual build-up of mass once again. There is however a well-defined average for the fluids considered.

Two major observations to emerge from Fig. 10.4 are the following. Holdup (units of kg/m^2) on a horizontal surface is greater than the vertical surface, for both, water and bismuth. This trend makes vertical surfaces as preferred over the horizontal for condensate recovery applications. Average holdup for bismuth is larger than water by a factor of 3.8 (for horizontal) and 4.8 (for vertical). Looking at vertical and horizontal substrates individually, it is noted that the average radii of droplet are comparable for the two fluids, respectively. The condensation cycle time of water (from droplet nucleation to occurrence of instability and its repetition during dynamic steady state) for both horizontal and vertical substrates is smaller than bismuth, in conformity with the data of Table 10.4. The density ratio between bismuth and water is of the order of 10. These effects collectively explain the enhanced holdup factor of bismuth to water (= 4.8) for vertical substrates. The average radius of a bismuth drop is smaller than water by a factor of four (Table 10.4) and correlates with the slower onset of instability and an increase in cycle time.

Variation of liquid holdup with time in bismuth for horizontal and vertical surfaces of size 10 × 10 mm, 30 × 30 mm, and 50 × 50 mm is shown in Fig. 10.5. The plots are quite similar, indicating that the condensation cycle broadly scales over the dimensions considered. Smaller differences can, however, be detected for reasons discussed here. The average holdup is higher for a smaller surface (such as 10 × 10 mm) since an unstable drop drags only a smaller fraction of liquid with it. For larger surfaces, instabilities are initiated at multiple locations and a substantial portion of the liquid condensed on the surface travels with the sliding drop. There is also an effect of the average size of the liquid drop. The drop radius influences conduction resistance and is larger for larger drops, as noted earlier. The interfacial heat transfer coefficient is not influenced by the size of the drop but is small for bismuth vapor condensing over a liquid drop. Jointly, interfacial resistance dominates conduction resistance even for large drops. The combined resistance is larger, leading to a slower growth rate on a larger surface. It leads to higher density of larger drops on the larger surface. Hence, following instability, a greater amount of liquid mass held on the surface is removed, leading to a sharper reduction in the average holdup. These details are minor, and Fig. 10.5 shows the overall effect of domain size to be marginal.

10.2.3 Wall Shear Stress and Wall Heat Flux

Apart from the condensation patterns discussed in previous sections, two other quantities of interest are the wall heat flux and wall shear stress. The surface-averaged heat flux and wall shear stress as well as local values at the coalescence sites as a function of time are presented in Figs. 10.6, 10.7, 10.8, and 10.9. The effect of the individual finite time coalescence events on the overall condensation pattern

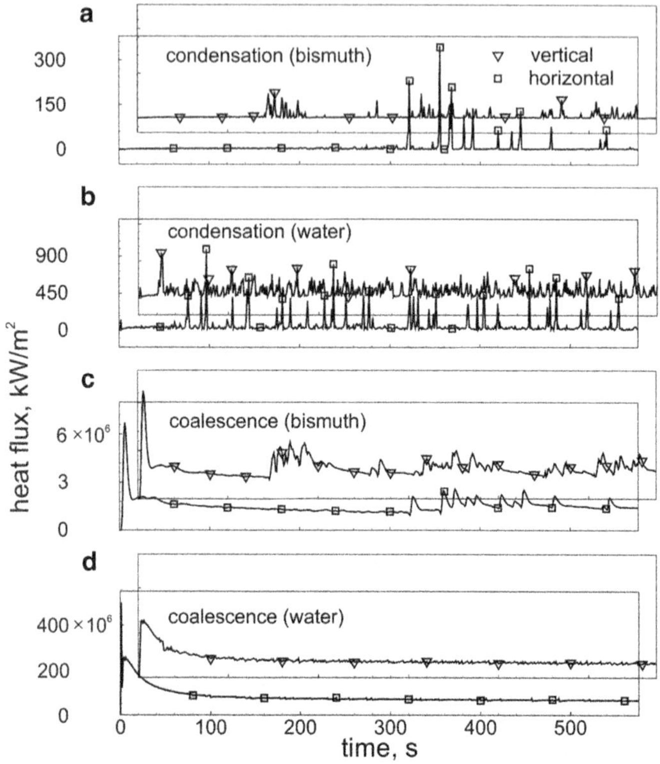

Fig. 10.6 Variation of wall heat flux with time for horizontal and vertical surfaces of size 30 × 30 mm (**a**) due to direct condensation in bismuth, (**b**) due to direct condensation in water, (**c**) due to coalescence in bismuth, and (**d**) due to coalescence in water. Note the difference in the wall heat flux scales

was found to be small. However, the fluxes and stresses during coalescence are substantial, as seen from the scales adopted in Figs. 10.6, 10.7, 10.8, and 10.9. Since the overall cycle time is of the order of minutes, a millisecond-scale coalescence process was not seen to alter the surface-averaged values. However, their magnitudes are large enough to be of concern, as discussed below.

Coalescence events occur at several sites, often simultaneously, and maximum values alone are reported in the Figs. 10.6, 10.7, 10.8, and 10.9. A total of 500 s of the condensation process is presented in the plots. In Figs. 10.6 and 10.8, bismuth is compared with water. In Figs. 10.7 and 10.9, heat flux and wall shear stress during bismuth condensation over surfaces of three different areas are compared.

Wall heat fluxes arising from (Case (i)) direct condensation (Figs. 10.6a, b and 10.7a, b) and (Case (ii)) coalescence (Figs. 10.6c, d and 10.7c) have been distinguished in the presented data. Note the scale difference in these figures in the heat flux values. Prominent and visible variability in time in the former (Case (i)) arises from drop instability where the drop size is large as it attains the critical size for

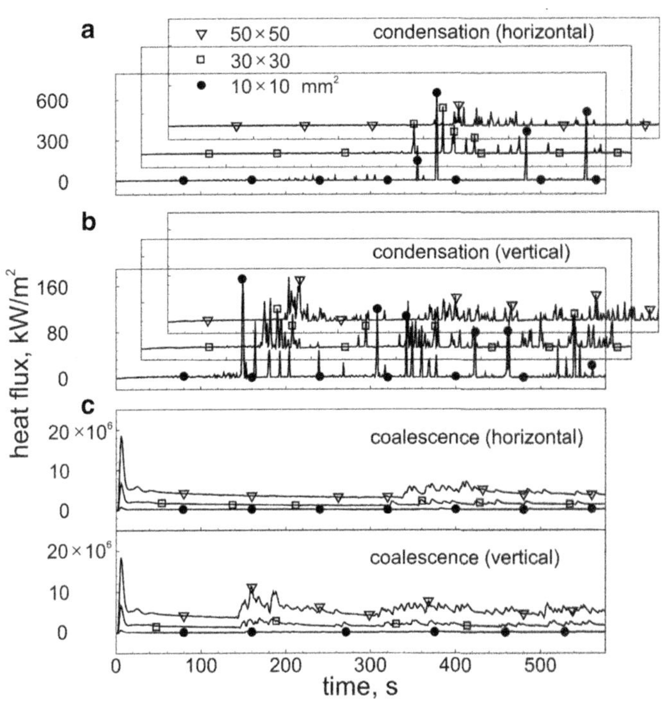

Fig. 10.7 Variation of wall heat flux with time in bismuth for surfaces of size 10 × 10 mm, 30 × 30 mm, 50 × 50 mm (**a**) due to direct condensation for horizontal orientation, (**b**) due to direct condensation for vertical orientation, and (**d**) due to coalescence for both horizontal and vertical orientation

sliding or fall off. For a horizontal surface, the drop size at fall-off instability is larger than slide-off instability of the vertical. Hence, the associated peaks in heat flux data are larger for Case (i). It is also to be emphasized that these peaks in heat flux originate primarily from a collection of nascent drops that nucleate immediately after instability of a larger drop. In Case (ii), coalescence events are spatially distributed and the extent of variability in time of the coalescence events and the resulting peak in heat flux is smaller. In general, the dropwise condensation cycle (from nucleation to instability and its periodicity) has a shorter overall time period for the vertical surface. A higher degree of variability, including coalescence, is visible as compared to the horizontal substrate.

Similar explanation can be offered for the wall shear stresses though with a difference. Wall shear is zero during the growth phase of a drop. For a horizontal surface, it is non-zero only during coalescence (Figs. 10.8a, b and 10.9a, b). For a vertical surface, it is non-zero during instability as well as coalescence. Multiple coalescence events spread over the surface give rise to a near-uniform distribution of wall shear stress. Selected peaks occurring due to drop instability arise only for the vertical surface as seen in Figs. 10.8c and 10.9c.

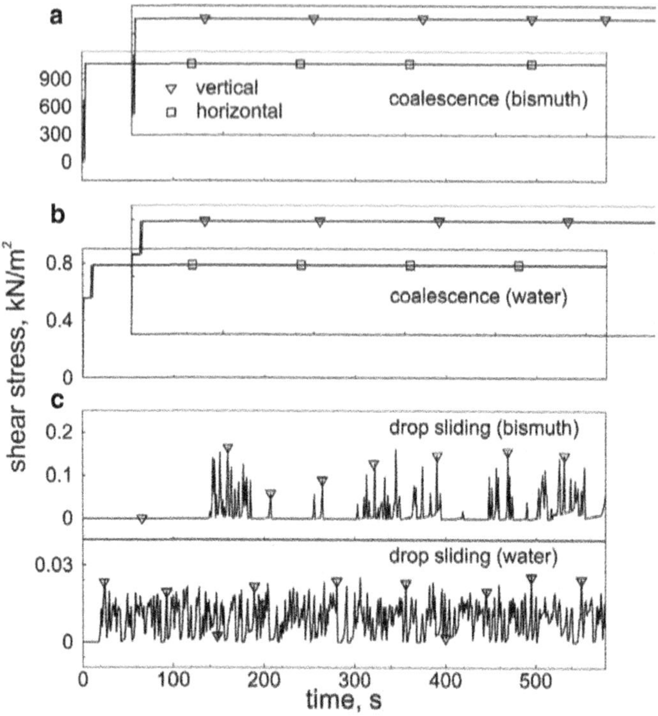

Fig. 10.8 Variation of wall shear stress with time for surface of size 30 × 30 mm (**a**) due to coalescence in bismuth for both horizontal and vertical orientation, (**b**) due to coalescence in water for both horizontal and vertical orientation, and (**c**) due to drop sliding for vertical orientation for bismuth and water

For a horizontal surface, a condensed drop falls off due to gravitational instability and fluid motion is restricted to the coalescence events. Hence, wall shear stress is mostly zero except for those occasional instants of time when it is large for very short time durations (~ a few ms). For a vertical surface, wall shear is generated post drop instability when it slides down the surface. Coalescence-induced shear stress is superimposed over this value.

Textured surfaces are often coated with a promoter layer and the coating has a finite yield strength, being of the order of 38–1160 mPa (Kim et al. 2006). These yield stress values are typically applicable for organic monolayer coatings that are mainly reported in the context of condensation of water vapor. It is evident that such a coating technology on substrates, applicable for the case of water, may not withstand instantaneous shear stress values during condensation of other fluids, including metal vapor. Such coatings are prone to leaching out due to continuous cycles of condensation occurring over it, as is evident from Figs. 10.8 and 10.9. Similarly, large, though momentary, wall heat fluxes (Figs. 10.6 and 10.7) suggest thermal non-uniformity in terms of surface temperature that will affect the condensation process itself.

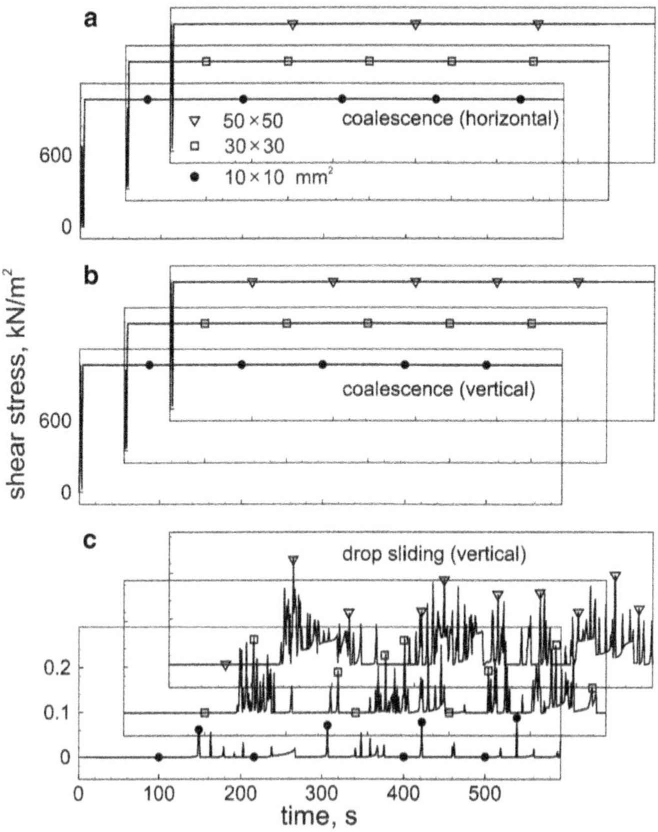

Fig. 10.9 Variation of wall shear stress with time in bismuth for surfaces of size 10 × 10 mm, 30 × 30 mm, 50 × 50 mm (**a**) due to coalescence for horizontal orientation, (**b**) due to coalescence for vertical orientation, and (**c**) due to drop sliding for vertical orientation

Direct condensation occurs at all drops, small and large from the vapor phase over the selected time step. The mass of liquid condensed over this time step multiplied by the latent heat release and divided by the sum of drop footprint areas is numerically equal to the average instantaneous wall heat flux. This is the part of heat flux associated with direct condensation. Heat flux related to coalescence arises only at those sites where two drops merge and are significant when the merging drops are physically large. Hence, the second component of heat flux is obtained by tracking coalescence events occurring at selected sites over the substrate.

Variations of wall heat fluxes with time in bismuth and water, for both horizontal and vertical surfaces of size 30 × 30 mm, are shown in Fig. 10.6. Figure 10.6a, b correspond to direct condensation with the effect of local coalescence averaged out over the entire surface. Individual peaks of heat flux correspond to droplet instability. In comparison, heat flux through stationary drops is quite small and is not visible in Fig. 10.6. Peaks in wall heat flux generated due to direct condensation are seen to be larger for a horizontal, compared to a vertical surface for both water and bismuth (Fig. 10.6a, b). Cycle time is smaller for a vertical surface compared to the horizontal, as seen in the appearance of frequent multiple peaks for water and bismuth. In addition, a larger portion of the drops is drained away from the vertical surface by sliding motion than horizontal surface during instability. Peaks in heat fluxes arise from the freshly nucleated drops, post instability. These factors combine to achieve peak heat fluxes for a horizontal surface greater than the vertical. Secondly, Fig. 10.6 shows that fluxes in water are uniformly greater than for bismuth. The product of density and latent heat of condensation for water is larger than bismuth by a factor of 1.36. Thus, latent heat is not a major reason that differentiates the wall heat flux between water and bismuth. The underlying reasons are a lower cycle time and a lower minimum drop size for water relative to bismuth, as summarized in Table 10.4. Heat fluxes arising from coalescence are reported in Fig. 10.6c, d. The peaks in water are frequent but the magnitude is smaller because of a lower Reynolds number during coalescence (Table 10.1).

Variation of wall heat flux with time in bismuth, for both horizontal and vertical surfaces of size 10 × 10 mm, 30 × 30 mm, and 50 × 50 mm are shown in Fig. 10.7. Figure 10.7a, b provide data for direct condensation where the effect of momentary coalescence is averaged over the entire surface. The effect of surface area can be understood in the following manner. The total number of drops experiencing peak wall heat flux depends on the size of the surface. For a 50 × 50 mm surface, the number of such drops undergoing instability is the greatest, while the 10 × 10 mm surface has lowest number of drops simultaneously undergoing instability. Peaks in wall heat flux are seen immediately after instability. Owing to sliding motion, a vertical surface will reveal a larger number of nucleating drops. A greater portion of drops is carried away from the smaller surfaces because of sliding motion. Larger surfaces may have instability at multiple locations and yet, a smaller portion of the drops will be drained away. Peaks in wall heat flux based on direct condensation, as well as coalescence, are proportional to this fractional number undergoing instability. Hence, larger fluctuations are seen on smaller surfaces than the larger surfaces for both horizontal and vertical configurations.

Wall heat fluxes generated during coalescence of drops of bismuth are shown in the Fig. 10.7c. For horizontal surfaces of sizes 10 × 10 mm, 30 × 30 mm, and 50 × 50 mm, these values are 0.16×10^6 kW/m^2, 1.64×10^6, and 4.86×10^6 kW/m^2, respectively. Average values of wall heat flux generated during coalescence of drops of bismuth on the three vertical surfaces are 0.24×10^6 kW/m^2, 2.3×10^6 kW/m^2, and 6.77×10^6 kW/m^2, respectively. Wall heat flux, both based on direct condensation and coalescence, shows increasing trends with an increase in surface area though the differences are small. As discussed in the context of Fig. 10.6, peak wall fluxes for direct condensation are greater for horizontal surfaces compared to the vertical. Orientation does not affect the coalescence fluxes significantly (Fig. 10.7). It is also seen that frequent coalescence peaks are seen on larger surfaces as compared to the smaller ones. Coalescence-related heat fluxes show momentary large peaks. However, the drop footprint is small and their contribution to the cumulative surface-averaged value of the net heat flux is small. Thus, the average value of the heat flux is principally governed by the direct vapor condensation process over the substrate.

Variation of wall shear stress with time in bismuth and water for both horizontal and vertical surfaces of size 30 × 30 mm is shown in Fig. 10.8. The instantaneous wall shear stress induced during coalescence of any two pair of droplets is estimated using centroidal analysis of Chap. 3. It is multiplied by the footprint area of the coalesced drop. At any instant of time, several such pairs may be coalescing on the substrate. The net shear force generated due to all such occurrences of coalescence events divided by the footprint area of all coalesced drops yield the average shear stress contribution due to droplet coalescence. This data is reported in Fig. 10.8a, b for bismuth and water, respectively.

The effect of substrate size on wall shear stress is shown in Fig. 10.9 for bismuth. Substrate areas, 10 × 10 mm, 30 × 30 mm, and 50 × 50 mm, respectively, are considered. Figure 10.9a, b correspond to wall shear stresses generated during coalescence. These values are, in general, quite large, because they are associated with a small area under the coalescing drops. Since the coalescence events are spatially distributed, wall shear stress associated with coalescence for the substrate as a whole is nearly constant in time (both in Figs. 10.8a, b and 10.9a, b).

Figures 10.8c and 10.9c present data for shear stresses generated during the sliding motion of individual droplets after they become unstable over a vertical substrate. This computation is carried out from a numerically determined Reynolds number correlation described in the previous work of the authors (Khandekar and Muralidhar 2014). For a horizontal surface, this contribution to shear stress is zero as unstable drops do not slide but actually fall off from the substrate. For vertical surfaces, sliding velocities which are generated are usually small, the drop size continues to become larger as the sliding drop gains mass. The process is infrequent relative to coalescence. Hence, surface-averaged wall shear stress due to sliding motion of unstable drops is substantially smaller than the coalescence-based wall shear stress. Instability events are more frequent in water when compared to bismuth, as seen in cycle times listed in Table 10.4. Accordingly, a larger number of momentary peaks are visible for water (Fig. 10.8c). For coalescence as well as in sliding motion, the shear stress generated in water is much smaller than bismuth, a result that follows from a larger dynamic viscosity of bismuth. From Fig. 10.9c, it is also noted that at instability, drops on larger vertical surfaces have to travel a longer distance than a smaller surface. Hence, shear prevails for a longer duration, generating larger wall shear stresses on larger surfaces. The maximum wall shear stress induced during sliding of drops on vertical surfaces of sizes 10 × 10 mm, 30 × 30 mm, 50 × 50 mm are 0.1 kW/m^2, 0.16 kW/m^2, 0.23 kN/m^2, respectively. Thus, wall shear stresses based on drop sliding show an increasing trend with an increase in surface area though the differences are small.

Coalescence-based wall shear stress is directly proportional to characteristic velocity and dynamic viscosity and inversely proportional to the base radius. Estimated velocity for bismuth is larger than in water by a factor of 1.44 and the average radius of a bismuth drop is smaller than water by a factor of 4 (Table 10.4). Viscosity of bismuth is larger than water by a factor of 175 (Table 10.1). These parameters combine to generate larger wall shear stress in bismuth relative to water. During sliding motion, bismuth drops acquire a much higher terminal velocity (Table 10.2) and accordingly, shear stresses are again higher, relative to water.

10.3 Closure

A hierarchical model of dropwise condensation described in Chaps. 6 and 7 has been extended by including the details of droplet coalescence dynamics. The process starts from direct condensation at nucleation sites over the cold substrate. It is followed by growth arising from direct condensation as well as coalescence and slide off or fall off due to gravity, depending on the substrate orientation. The process is cyclic and spatially distributed over the substrate. Quantities of interest, including instantaneous condensation patterns, wall shear stress, and wall heat flux, have been predicted. Characteristic velocity and timescales of coalescence determined using scale analysis are used to extend the condensation model and determine local wall shear stresses and heat transfer rates. The present work reports data related to condensation of bismuth vapor underneath a near-horizontal surface and on a vertical surface. These are further compared with the condensation of water vapor on a hydrophobic surface. Changes in condensation parameters with the overall surface area of the substrate are investigated. Instantaneous wall heat fluxes and shear stresses arising from coalescence events are seen to be substantial. However, the surface and time-averaged heat flux data is barely affected. Similarly, wall shear stress data is only marginally influenced. However, large instantaneous shear stresses have a significant impact on the life of the coated surface and should be considered in analysis.

Relative to water, liquid bismuth has a higher density, higher thermal conductivity, higher surface tension with vapor, higher viscosity but a much smaller latent heat of vaporization, thus generating several contrasts with it. Cycle time, minimum radius of drop, holdup, and vapor-liquid interfacial resistance of liquid bismuth are larger than water for both horizontal and vertical surfaces. Maximum radius of drop, average radius of drop, conductance resistance, and average heat transfer coefficient of liquid bismuth are smaller than water for both horizontal and vertical surfaces. As a result, wall heat flux based on direct condensation as well as coalescence is substantially larger for water than liquid bismuth for both horizontal and vertical surfaces. The average heat transfer coefficient in bismuth is smaller than for water. Fluctuations in heat fluxes in water are small but frequent while in bismuth, the local peaks arising from drop instability are larger and occasional. Wall shear stresses generated during coalescence as well as sliding of drops are larger for liquid bismuth than water for both horizontal and vertical surfaces.

Part III
Dropwise Condensation Experiments

Chapter 11
Dropwise Condensation: Experiments

Basant Singh Sikarwar, K. Muralidhar, and Sameer Khandekar

Nomenclature

θ_{adv}; θ_{rcd}	Advancing and receding angles of the liquid drop on the chosen substrate, degree
$h_c(\alpha)$	Heat transfer coefficient for a substrate at an angle α, W/m^{2}K
$h_c(90°)$	Heat transfer coefficient for a vertical substrate, W/m^{2}K
r_{min}	Minimum radius of the liquid drop at nucleation, m
t	Time, s
T_{sat}, T_w, ΔT	Saturation temperature of pure vapor, wall temperature, and degree of subcooling $T_{sat} - T_w$, K

11.1 Introduction

Experimental determination of the heat transfer coefficient during dropwise condensation is a difficult task because of the many intricacies involved in the process. The driving temperature difference is small, essentially resulting in a high heat transfer coefficient. Further, uncertainties associated with the microscale sub-structure of contact line shapes and motions, dynamic temperature variations below the condensing drops, effect of roughness and inhomogeneity of the substrate structure, control of true boundary conditions, microscale instrumentation, and transport dynamics of coalescence, merger, wipe-off, renucleation cycles, and the leaching rates of the promoter layer add to the difficulty in conducting repeatable experiments. Very high heat transfer rates (and therefore a very low temperature differential) coupled with the above factors also hinder generation of repeatable experimental data. Consequently, many conflicting experimental results have been published over the years, some results showing considerable scatter (Fig. 11.1).

S. Khandekar, K. Muralidhar, *Drop Dynamics and Dropwise Condensation on Textured Surfaces*, Mechanical Engineering Series,
https://doi.org/10.1007/978-3-030-48461-3_11

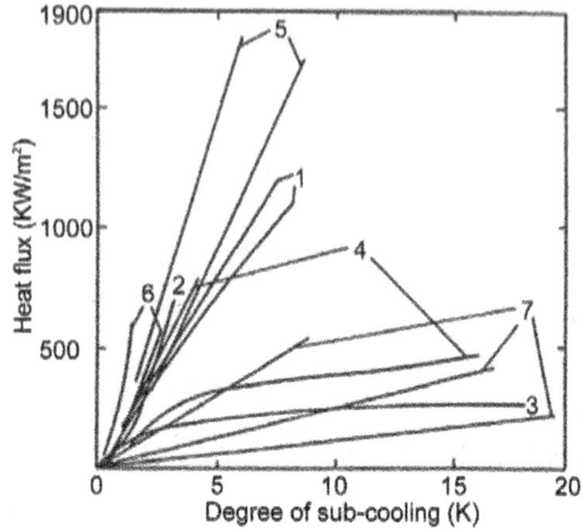

Fig. 11.1 Experimental results on dropwise condensation (water, ~1 bar) Stephan (1992). *1* Hampson and Özisik (1952), curve for two different promoters; *2* Wenzel (1957); *3* Welch and Westwater (1961); *4* Kast (1965); *5* Le Fevre and Rose (1965); *6* Tanner et al. (1968); *7* Griffith and Lee (1967)

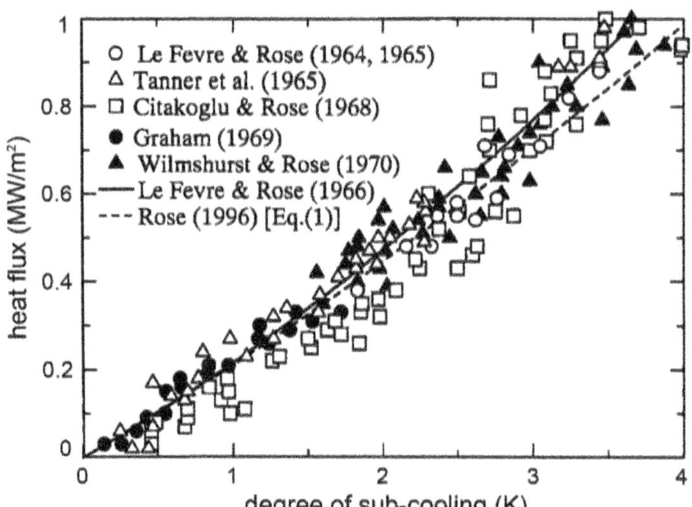

Fig. 11.2 Dropwise condensation of steam on a copper surface (vertical) at atmospheric pressure compared with the theory of Le Fevre and Rose (1966). (Adapted from Rose et al. 1999)

Improved experimental techniques have led to reproducible and reliable experiment data to an extent; see, for instance, Le Fevre and Rose (1964, 1965) and Citakoglu and Rose (1968a, b). Several authors (Tanasawa 1991; Stephan 1992) reported that the heat transfer coefficient of dropwise condensation for steam at an atmospheric pressure, under the normal gravitational acceleration and on a vertical copper surface is about 230 ± 30 kW/m^{2}K in the heat flux range of 0.1–1 MW/m^2, provided there is no effect of non-condensing gases and steam is approximately quiescent. Heat transfer coefficient of dropwise condensation of steam at atmospheric pressure has been summarized by Rose et al. (1999) (Fig. 11.2). Several representative measurements shown in the figure are close to the theory Le Fevre and

Rose (1966). Data for other vapors and measurement under other thermal conditions are still scarce. Despite sustained research over the past three decades, the prediction of the correct heat transfer rate during dropwise condensation over a surface remains a challenge (Tanasawa 1991; Stephan 1992; Rose 2002), mainly due to lack of knowledge of the local transport mechanisms of drop formation.

With the advent of newer coating/manufacturing and nanoscale fabrication techniques, promoting long-term sustainability of dropwise condensation by physical texturing and/or chemical coating holds considerable prospect for enhancing heat transfer. An example of enhanced performance of compact steam condensers having chemically coated flow passages of only a few millimeters width is demonstrated by Majumdar and Mezic (1999). It is necessary to understand the effect of various parameters on heat transfer during dropwise condensation, as reviewed next.

Marto et al. (1986) tested several polymer gold and silver coatings for sustaining dropwise condensation of steam and reported that the heat transfer coefficient in dropwise condensation is as high as six times when compared to the filmwise. Zhao et al. (1996) reported heat transfer coefficient on Langmuir-Blodgett treated surface to be than 30 times more than that of filmwise condensation on a bare surface. Koch et al. (1998a, b) showed the effect of hydrophobicity on heat transfer coefficient on a chemically textured vertical substrate. Heat transfer coefficient was found to decrease with an increase in wettability.

Ma and Wang (1999) proposed that the heat transfer coefficient increases with the increase in the surface free energy difference between the condensate and the condensing substrate. Hence, surface modifications for promoting dropwise condensation by silanation and ion implantation are of particular interest. These would yield continuous dropwise condensation along with a high heat transfer coefficient. Leipertz and Choi (2000) reported heat transfer rates on several metallic substrates (copper, titanium, aluminum, high-grade steel, and hastelloy) treated by ion implantation. Ions considered were nitrogen, oxygen, and carbon ions, with varying ion density. Das et al. (2000a, b) applied an organic self-assembled monolayer coating to enhance the dropwise condensation, the corresponding increase in the heat transfer coefficient being a factor of four.

Vemuri et al. (2006) performed a condensing experiment over various coated substrates and reported long-term sustainability and enhancement of heat transfer coefficient. The authors coated a copper substrate with self-assembled monolayers (SAMs) of *n*-octadecyl mercaptan and stearic acid. An increase in heat transfer coefficient by a factor of three was reported as compared to a bare copper substrate. Ma et al. (2008) experimentally studied dropwise condensation on a vertical plate for a variety of non-condensable gas (NCG) concentration, saturation pressure, and surface subcooling. A fluorocarbon coating was applied to promote dropwise condensation. Departure of drops was inferred as the dominant factor for the steam–air condensation heat transfer enhancement.

Rausch et al. (2007, 2010a, b) observed that the heat transfer coefficient on an ion-implantation surface is more than five times than that of filmwise condensation. Ion implanted metallic substrates have stable condensation as well as high heat transfer coefficient over a long time duration. Chen et al. (2009) experimentally investigated the effects of various chemical coatings and their long-term durability on the dropwise mode of condensation. A reduction in the heat transfer coefficient was seen with the elapsed condensation time, suggesting possible leaching of the chemical coating. Dietz et al. (2010) investigated droplet departure frequency using electron microscopy to understand enhancement of dropwise condensation on superhydrophobic surfaces. A reduction in drop departure size shifts the drop size distribution to smaller radii, which may enhance the heat transfer rate.

Ma et al. (2012) investigated experimentally the heat transfer characteristics in the presence of a non-condensable gas (NCG) on superhydrophobic and hydrophobic surfaces including the wetting mode evolution on the roughness-induced superhydrophobic surface. Superhydrophobic surfaces with high contact angle ($>150°$) and low contact angle hysteresis ($<5°$) were seen to be an ideal condensing surface to promote dropwise condensation of water and enhance heat transfer. With increasing NCG concentration, the droplet undergoes transition from the Wenzel to Cassie-Baxter mode.

Miljkovic et al. (2012) studied the effect of droplet morphology on heat transfer during dropwise condensation on superhydrophobic nanostructured surfaces. These surfaces were designed to be Cassie stable and favored the formation of suspended droplets on the top of the nanostructures as compared to the partially wetting droplets which locally wetted the base. Cassie stable droplets were seen to have minimal contact line pinning and promoted passive droplet shedding at sizes smaller than the characteristic capillary length. However, the gas films underneath such droplets significantly hindered the overall heat and mass transfer performance.

11.1.1 Thermophysical Properties of Condensate

Several sets of results are available for dropwise condensation of steam on copper at atmospheric pressure. Fewer results are available on the heat transfer coefficient at pressures lower than one atmosphere (Tanner et al. 1968; Graham 1969; Wilmshurst and Rose 1970; Tsuruta 1993; Hatamiya and Tanaka 1986). They show a tendency of decreasing heat transfer coefficient with decreasing pressure (Fig. 11.3).

Condensation of other vapors in the form of drops has been reported by several authors. Wilmshurst and Rose (1974) performed condensation experiments of aniline and nitrobenzene on PTFE-coated substrate. Stylianou and Rose (1983) reported condensation of ethylene-glycol on a copper substrate. Utaka et al. (1987, 1994) performed a condensation experiment with propylene-glycol, ethylene-glycol, and glycerol vapors on copper substrate using a monolayer type promoter below atmospheric pressure. Quantitatively, the heat transfer characteristics for organic vapors differ from those for steam due to wide variation of physical properties. Owing to the lower liquid thermal conductivity of the organic fluids, relatively low heat transfer coefficients are to be seen in comparison with steam. For moderate subcooling, the heat transfer coefficient for dropwise condensation is significantly larger than for film condensation. It can also be seen that the surface subcooling ranges of ideal

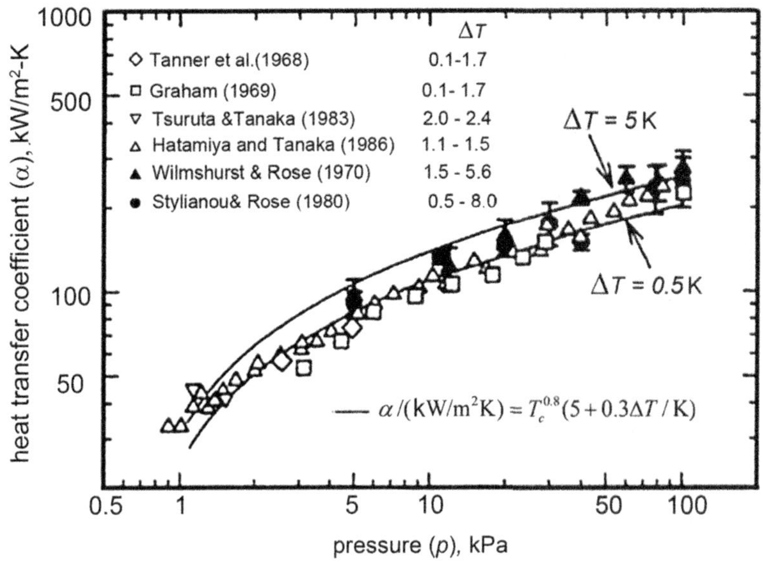

Fig. 11.3 Dependence of heat transfer coefficient on pressure for steam condensation on copper substrate in the form of drops (Rose et al. 1999)

dropwise condensation differ widely, depending on the choice of the fluid. For a fluid of higher surface tension such as glycerol, dropwise condensation is maintained for larger surface subcooling, compared to a lower surface tension liquid such as propylene-glycol.

Dropwise condensations of low Prandtl number vapors are scarcely presented in the literature. In many situations, a singular behavior is observed for low Prandtl number systems, for example, liquid metals where $\mathrm{Pr} \approx 0.01$. Moreover, condensation of liquid metals also plays an important role in many engineering processes. Only a few researchers have considered dropwise condensation of metal vapors. Bakulin et al. (1967) reported the effect of a non-condensable phase on temperature drop at the liquid-vapor interface during the dropwise condensation of sodium, potassium, and lithium. Interfacial resistance to mass transfer at the liquid-vapor interface was seen to play an important role in the condensation of metal vapors. Rose (1972) modified the previously reported theory of dropwise condensation and showed that the degree of subcooling affects heat transfer in mercury though the effect is less than in water. Necmi and Rose (1977) measured vapor-to-condensing surface temperature difference and the corresponding heat flux for various vapor pressures during dropwise condensation of mercury on a vertical substrate. Niknejad and Rose (1984) compared the experimental data of mercury with their own theory of dropwise condensation developed for water and found significant differences. Literature on dropwise condensation of other metal vapors such as sodium, potassium, and bismuth is not available though the liquid phases of these substances have rather large surface tension.

Many researchers (Takeyama and Shimizu 1974; Tanasawa and Utaka 1983; Tanasawa 1991; Rose 2002; Rose 2004; Lan et al. 2009) reported increase in heat flux along with an increase in the surface subcooling. At a higher subcooling, the rate of drop nucleation may increase because the minimum radius of drop must be smaller.

11.1.2 Physico-Chemical Properties of Substrate

The substrate hydrophobicity contact angle hysteresis and state of drop (Cassie versus Wenzel) on or underneath a substrate depend on its physico-chemical properties with respect to the condensing fluid. These parameters play an important role in dropwise condensation. Many researchers (Lee et al. 1998; Lara and Holtzapple 2011; Baojin et al. 2011) reported that high contact angle (>150°) and low contact angle hysteresis (<5°) is an ideal combination for a condensing surface. Neeumaqnn et al. (1978) reported that heat-transfer during dropwise condensation of water vapor strongly depends on contact angle hysteresis. This is because the surface conductance increases with decreasing contact angle hysteresis. Kim and Kim (2011) reported a strong effect of contact angle on the heat transfer rate. A large contact angle leads to the enhancement of heat transfer. Miljkovic et al. (2012) reported that the heat transfer coefficient of dropwise condensation depends on the morphology of droplets on the substrate. The initial growth rates of partially wetting droplets (Wenzel) were six times larger than the suspended droplets (Cassie). Experimental results showed that partially wetting droplets (Wenzel) had four to six times higher heat transfer rates than the suspended droplets (Cassie).

Although the heat transfer coefficient of dropwise condensation of vapors are strong functions of physico-chemical properties such as contact angle and contact angle hysteresis, there is no correlation linking the heat transfer coefficient with contact angle, contact angle hysteresis, and droplet morphology on the substrate. Such a relationship has been examined in this monograph.

11.1.3 Substrate Having a Wettability Gradient

In dropwise condensation, liquid droplets forming on a subcooled non-wetting surface are removed from the surface by gravitational forces when the droplets reach a critical mass. The dependence on gravity for liquid removal limits the utilization of dropwise condensation in low gravity aerospace applications and horizontal surfaces. Various authors (Zhao and Beysens 1995; Daniel et al. 2001; Liao et al. 2006) have applied a novel passive technique based on surface energy gradient in the condensing surface to remove droplets. Daniel et al. (2001) observed the random movements of droplets to be biased towards the more wettable side of the surface. Powered by the energies of coalescence and directed by the forces of the chemical gradient, small drops (0.1–0.3 mm) display speeds that are faster than those of typical Marangoni flows. Wettability gradient on a horizontal substrate has implications for passively enhancing heat transfer in heat exchangers and heat pipes. Bonner (2009) verified experimentally that a wettability gradient substrate has high heat transfer compared to a horizontal substrate. Gu et al. (2005) and Bonner III (2010) enhanced heat transfer of a condensing system by creating energy gradient on the condensing substrate.

11.1.4 Substrate Orientation

The study of orientation of the cold substrate is important in dropwise condensation and enhancement of heat transfer. Many researchers (Citakoglu and Rose 1968b; Izumi et al. 2004; Leipertz and Fröba 2006) have reported high rate of water vapor condensation on vertical substrates for a given degree of subcooling. Leipertz and Fröba (2008) reported the following correlation for the heat transfer coefficient in dropwise condensation as a function of the inclination of the substrate

$$h_c(\alpha) = h_c(90^\circ) \cdot [\sin \alpha]^\kappa, \tag{11.1}$$

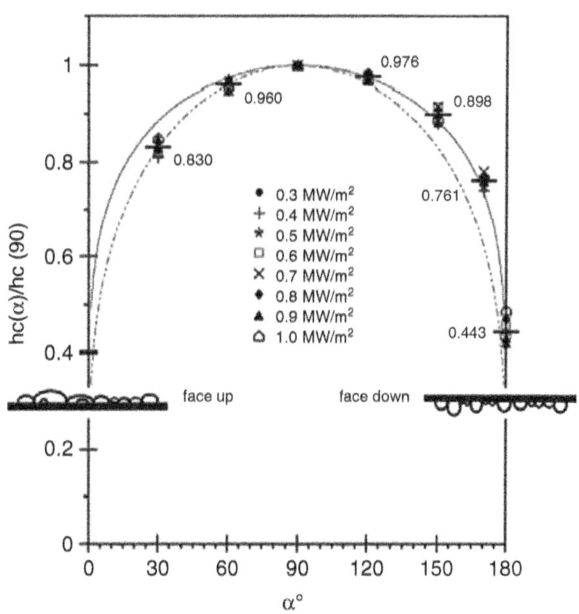

Fig. 11.4 Variation of heat transfer coefficient with respect to angle of inclination of the substrate (Leipertz and Fröba 2008)

where $h_c(\alpha)$ is the heat transfer coefficient of dropwise condensation at angle α and $h_c(90°)$ corresponds to that of a vertical substrate. In Fig. 11.4, the value of k is ~0.270 for the dashed line and ~0.176 for the solid line. The angle of inclination is defined to be 0° for the horizontal surface with the drops on the upper side of the substrate (sessile mode) and 90° for a vertical. From 90° onward, drops form on the lower side of the substrate and the 180° horizontal substrate refers to the pendant mode of dropwise condensation.

The heat transfer data for water vapor in dropwise condensation with respect to orientation (sessile and pendant) are shown in Fig. 11.4. It is clear that the pendant mode over a horizontal substrate yields a higher heat transfer coefficient as compared to the sessile. Heat transfer coefficient is the highest for a vertical substrate and decreases with increasing inclination. For an inclined substrate, the surface is swept clean of drops and this renewal of the growth process is responsible for a higher heat transfer coefficient. In contrast, drops over a horizontal surface become large and fall off by gravity in the pendant mode or spread over the substrate and cover it by a layer of the condensate liquid, in case of sessile drops. In both the cases, the surface is not regularly refreshed by fresh condensation, resulting in a lower heat transfer coefficient. Tanasawa et al. (1976) measured the dependence of heat transfer coefficient on the departing drop diameter. Authors reported that the heat transfer coefficient is proportional to the departing drop diameter to the power of about ~0.3. Lawal and Brown (1982) and Briscoe and Galvin (1991b) reported that a pendant drop is less stable as compared to a sessile drop on an inclined substrate, suggesting that heat transfer during dropwise condensation underneath an inclined substrate is marginally better than its counterpart above the surface. Therefore, surface orientation is an important parameter in the enhancement of heat transfer coefficient in dropwise condensation. Surface modification techniques generally used to develop surfaces favorable for dropwise condensation are discussed in Chap. 11, and the experimental study performed on silanized glass substrate is presented here.

11.2 Experiments on Chemically Textured Surfaces

The experimental apparatus was designed to study dropwise condensation under controlled conditions on the underside of a cold substrate and is schematically shown in Fig. 11.5. The setup primarily consisted of the main cylindrical stainless steel vacuum chamber (better than 10^{-5} mbar abs.) of inner diameter 180 mm and length 120 mm (Fig. 11.5a, b). It was closed from the two ends by specially designed flanges. The lower flange was fitted with a $\lambda/4$ optical viewing window. Typical photographs of condensing droplets are shown in Fig. 11.5c. In addition, the optical window also had an annular space around, wherein the working fluid inventory (distilled and deionized water) was stored. A circular, 1.5 mm thick mica strip heater (OD = 70 mm, ID = 40 mm) was attached outside the annular space to give the necessary heat input, as shown in the cut section of the experimental setup in Fig. 11.5d. The upper end of the main vacuum chamber was closed with a

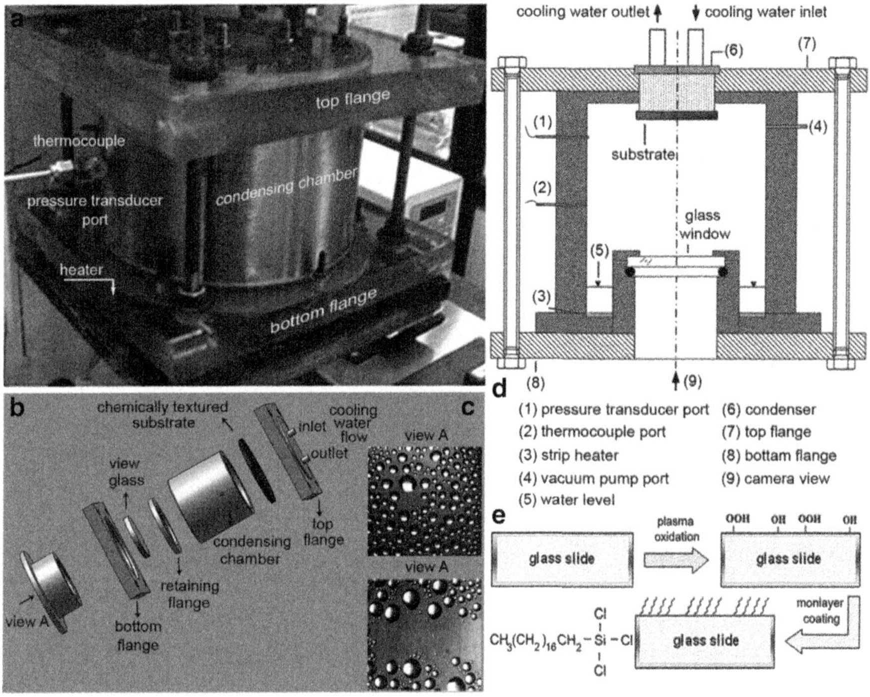

Fig. 11.5 Details of the experimental setup to study dropwise condensation under controlled conditions underneath a substrate. (**a**) Photograph shows the details of the main condensing chamber. (**b**) Exploded view of the condensing chamber showing all the components. (**c**) Typical images of the condensing droplets at two different times, as captured from View-A. (**d**) Cross-sectional view of condensing chamber. (**e**) Schematic diagram explaining the CVD process

polycarbonate square flange with an inbuilt cavity wherein cold water was circulated to maintain constant temperature boundary conditions. The condenser capacity was at least 20 times that of the maximum expected heat transfer rate. The chemically coated glass substrate of 100 × 100 mm was integrated on the upper flange (Fig. 11.5d). Connections for evacuation, pressure transducer, and temperature sensors were provided on the main condensing chamber wall. The temperature of the condensing vapor was measured with one K-type thermocouple (Omega®, 0.5 mm diameter) of accuracy ±0.2 °C after calibration. It was placed centrally in the chamber at a distance of 25 mm from its side wall. The condensing chamber pressure was measured by an absolute pressure transducer (Honeywell, accuracy 0.1% FS, NIST traceable calibration, range 0–1.2 bar). Online data acquisition was carried out with 16-bit PCI-4351 card (National Instruments®). The entire assembly could be tilted to any desired inclination between 0°–50°. A color CCD video Camera (Basler® A202KC with 1024 × 1024 pixels at 100 fps) was used to capture the images of the drops forming on the underside of the chemically textured substrate (View A, Fig. 11.5b). Length scales were calibrated by imaging a grid with known periodicity. A diffuse white light source symmetrically placed around the camera was directed on the substrate from the optical window on the bottom flange so as to maintain a near-parallel and symmetric beam on the droplets ensuring proper contrast level for subsequent edge detection.

11.2.1 Experimental Methodology

Dropwise condensation of distilled and deionized water, underneath a horizontal substrate and an inclined substrate having various inclination (10°, 15°, and 30°), was carried out underneath a glass substrate which was coated with octyl-decyl-tri-chloro-silane ($C_{18}H_{37}C_{13}Si$). Complete details of glass silanation process is presented in Chap. 12. The chamber temperature was maintained at 27 °C in all the experiments with cold substrate maintained at 22 °C. The static contact angle of water drop placed on the chemically textured substrate was measured to be 96° ± 0.5° for droplet volume range of 50–100 μL. Dropwise condensation was achieved at the desired saturation pressure by controlling the coolant temperature and the heat throughput. Once quasi-steady state was reached, the correspondence between the saturation pressure and the condensing vapor chamber temperature was continuously monitored. The high-quality video images recorded were digitally processed (using Image-J® software) to get the relevant parameters of interest, i.e., area of coverage, droplet size distribution, fall-off/slide-off, and coalescence events. The primary steps in finding the area of coverage were: (a) Digital image acquisition (b) Contrast thresholding and pinning to reduce pixel noise (c) Droplet detection with geometry attributes (d) Measurement of total digitized pixel area covered by the droplets and (e) Finding the area of coverage by dividing the total pixel area of all the droplets by the total area of the acquired image. Droplets below a diameter of around 0.1 mm could not be resolved with the imaging hardware used. The image processing software was first tested against benchmark images.

The experimental process was simulated by the mathematical model for both horizontal and inclined arrangement of the substrate. After validation, simulations were performed for the range of parameters not covered in the experiments. Here, the effect of the static contact angle, nucleation site density, thermophysical properties of the working fluid, physico-chemical properties of the liquid-substrate, and the angle of inclination of the substrate are considered.

11.2.2 Experimental Validation of Computational Model

Experimental results of condensation patterns, and the corresponding predictions of numerical simulation for water vapor at a saturation temperature of 27 °C and subcooling of 5 °C, are compared both in qualitative and quantitative terms underneath a horizontal and an inclined substrate. Nucleation sites density is taken to be 10^6 cm^{-2} in the simulation.

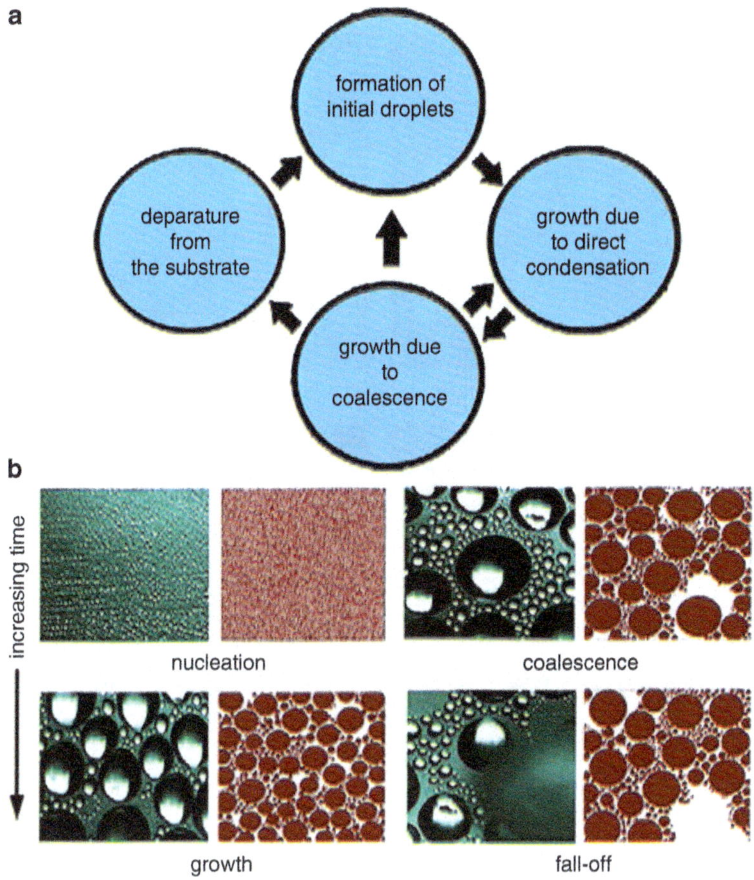

Fig. 11.6 (**a**) The cycle of major physical processes observed in the pendant mode of dropwise condensation on a horizontal substrate. (**b**) Qualitative comparison of experimental images of dropwise condensation on silanated glass substrate of area 25 × 25 mm (coated with octyl-decyl-tri-chloro-silane, $C_{18}H_{37}C_{13}Si$) with corresponding images generated by simulation. The hazy patch seen in the top-left section of the last experimental image is due to the fact that the droplet has fallen on the viewing glass through which images are recorded

Horizontal Substrate

Figure 11.6a schematically depicts the major observable processes of dropwise condensation underneath a horizontal substrate. These are nucleation, growth, coalescence, and fall-off of droplets. Figure 11.6 visually and qualitatively highlights these processes, as observed experimentally (View-A in Fig. 11.5b) and captured in the computer simulation. The statistical nature of the overall process, with multiple generations of droplets in different stages of their respective growth phase and present simultaneously on the substrate, is clearly visible. Contrary to the perfect

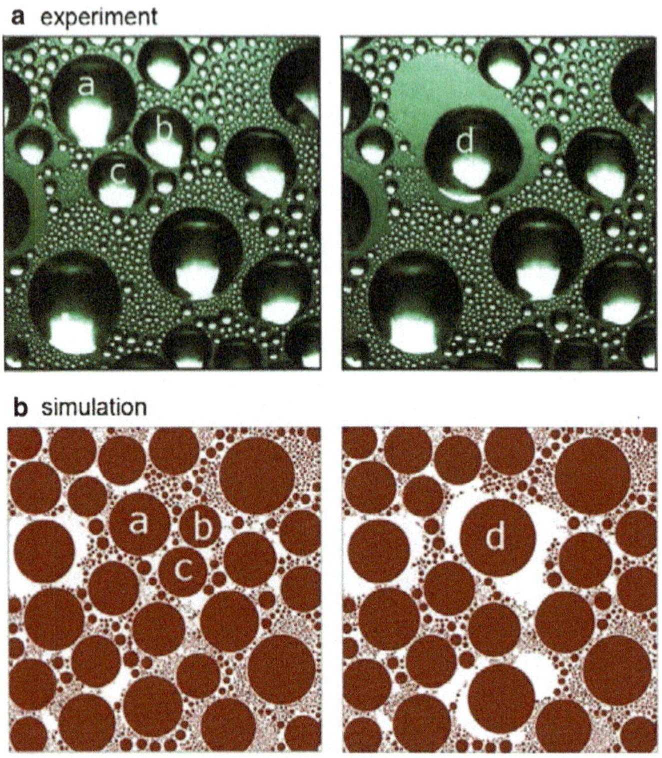

Fig. 11.7 Sequence of two images observed during experiment and corresponding simulation, showing coalescence of three droplets a, b, and c over a horizontal substrate, resulting in the formation of composite drop d

circular footprints of the droplet bases assumed in the simulation, local phenomena such as pinning of the contact line, capillary waves, contact line inertia during droplet merging, and the dynamics of the liquid-vapor interface cause deviations that are observable in the experiments. Specifically, droplet pinning, and the non-circular base of the footprint, can be clearly seen in the experimental images. Thus, the mathematical model can be further refined to cover local disturbances. However, major phenomena related to dropwise condensation underneath horizontal substrates are well simulated by the model.

Figure 11.7 depicts coalescence of three drops (marked a, b, and c) as observed during the experiment and revealed in the simulation. In the simulation, the center of the new resulting drop (after coalescence, i.e., drop-d) is determined by a mass weighted average of centroid of constituent droplets before coalescence (i.e., droplets a, b, and c).

The assumption that the coalesced volume takes up the weighted center of mass of the original droplets is vindicated by this representative comparison. The merger

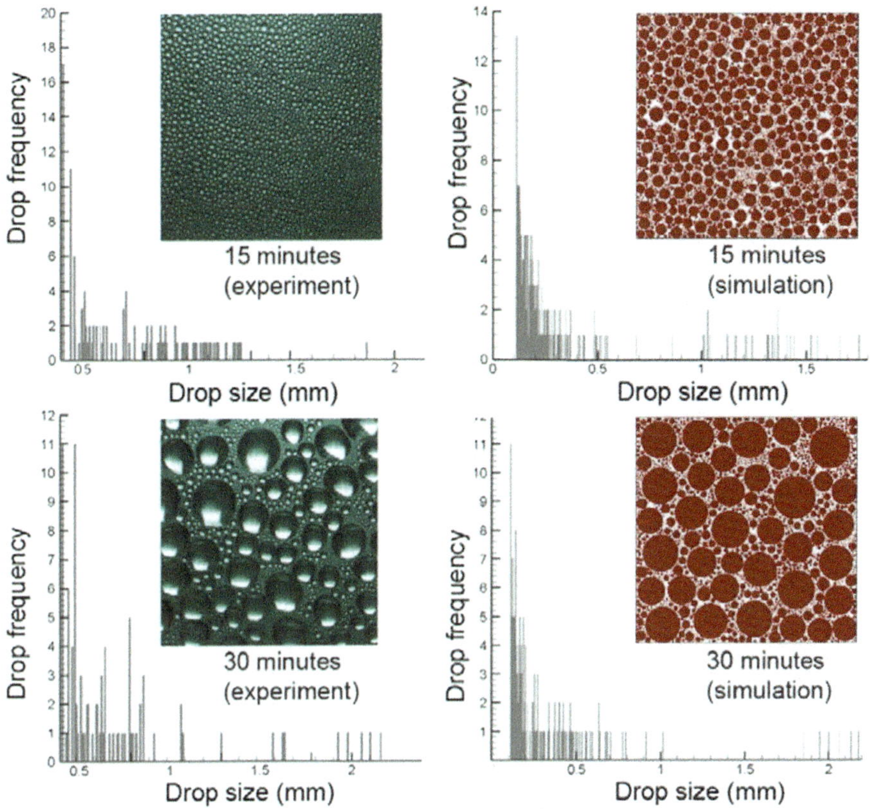

Fig. 11.8 Visual and statistical comparison of experimental and simulated spatial drop distribution patterns and the corresponding histograms of droplet frequency at the dynamic steady-state

results in the exposure of virgin areas around the drop where renucleation of the new generation droplets will commence. Droplet mergers bring about near-instantaneous changes in the total area coverage as well as the drop size distribution. A closer look at the edges of the droplets during experiments, especially the larger droplets, also reveals that the shapes of their bases are not exactly circular, with local pinning phenomenon of the contact line at certain locations (e.g., see drop d in the experimental image). As the droplets merge, experimental images show that it takes a certain finite time (of the order of 0.1–300 ms, depending on the respective sizes of the coalescing droplets) for the surface and body forces to redistribute the fluid in the coalesced drop and come to the state of minimum possible energy level; the new contact line shrinks and tends to be as circular as possible in a finite relaxation time; local pinning can distort its circularity.

Figure 11.8 shows the spatial drop size distributions underneath the horizontal substrate, at 15 min and 30 min, respectively, after the commencement of the condensation process. No fall-off has yet taken place. The strong temporal variation

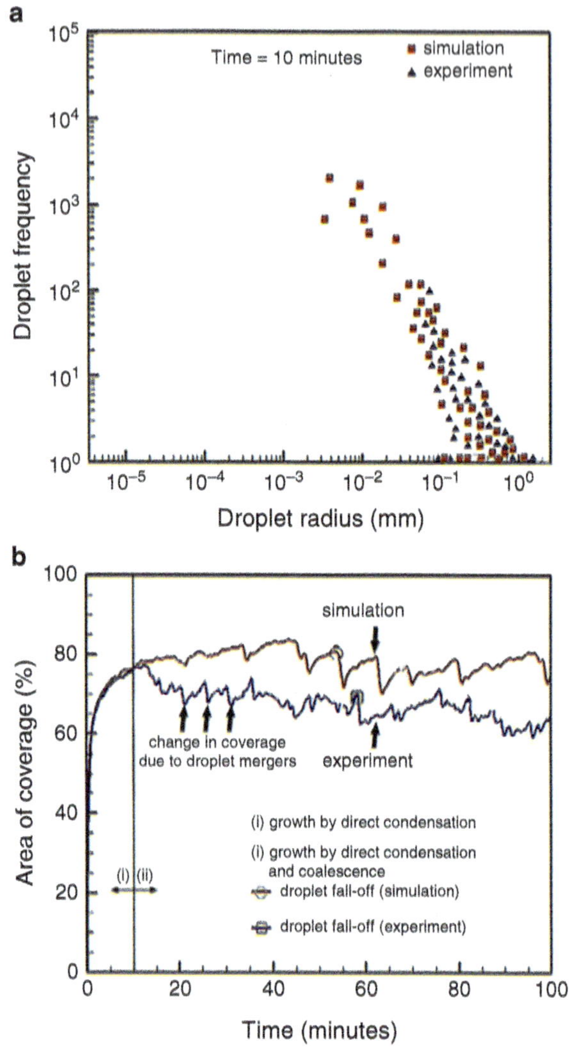

Fig. 11.9 (**a**) Drop size distribution from experiments and simulation at a time of 10 min after the commencement of dropwise condensation. (**b**) Time-wise variation of the area coverage of droplets over the substrate

of size distribution of droplets is clearly visible. As can be seen, after a 15-min interval, the distribution shows moderately sized droplets with the maximum diameter of ~2.0 mm. As the time progresses, droplets merge exposing virgin areas; an increase in number density of very small droplets (below ~0.5 mm) is clearly visible at 30 min. In addition, the number density of larger droplets (greater than ~2.0 mm) has increased substantially. The simulated histograms are denser than the experimental counterpart due to the loss of information in experimental data during image processing of droplets below ~0.1 mm. For the same reason, the experimental and simulation histograms of the 15 min data are more dissimilar than at 30 min. Initially, as condensation commences, the number of smaller sized droplets is quite large. At later times, droplets of higher diameter are greater in number and are captured well by the digital camera. In the latter part of the process, the growth is chiefly dominated by coalescence and the number density distribution shifts towards larger sized drops.

Figure 11.9a compares the experimental and simulated droplet frequency plotted as a function of the drop radius, 10 min after commencement of the condensation process. The experimental fall-off time for the first drop was approximately 58–62 min while the simulation predicted a number in the range of 48–54 min. It is clear that drops whose radius is less than ~0.1 mm have not been recorded by the camera. The corresponding range of drop sizes that could be included in the simulation is 10^{-3} to 1.0 mm. Although the order of magnitude of $r_{\min}$ $(t = 0)$ is ~10^{-4} mm, nearly all the original drops have since grown to the order of 10^{-3} mm at 10 min, mostly by direct condensation. Droplet coalescence has not yet started, as can be seen in Fig. 11.9b, where the temporal change in area coverage of drops is presented. Initially, there is a rapid increase in the coverage and later approaches a dynamic quasi-steady state. Two distinct zones clearly seen in the experimental and simulation data are: (a) growth due to direct condensation in the initial period and (b) growth due to coalescence. Large local fluctuations in area coverage represent time instants when drops either coalesce to form larger drops or a large drop falls off/slides off. The fact that smaller drops could not be accounted due to imaging limitations explains the higher values of coverage area in simulation (73.1%) as compared to experimental data (64.5%).

Figure 11.10 shows the complete sequence of experimental and simulated drop distribution, from the appearance of drops of minimum radius to the formation of drops of critical radius, underneath a horizontal substrate of 30 × 30 mm area. The first image is at a time instant of 1 min and thereafter the images are at approximately 10 min intervals. The last image is presented at 65 min for the experiment and 52 min for the simulation, respectively. For this experiment, the first fall-off occurred at 59.5 min while in the corresponding simulation, the first instance of fall-off was observed at 51 min and 10 s. This discrepancy may arise due to the following factors: (a) Non-condensable gases in the experimental chamber can deteriorate the heat transfer coefficient and delay the drop growth rate. (b) Inexact nucleation site density. (c) The local effects of pinning and contact line dynamics can lead to higher frictional stresses that enhance surface forces and delay fall-off. The comparisons in Figs. 11.7 and 11.10 for a horizontal substrate, however, show that the simulator satisfactorily captures the major processes of dropwise condensation, both from a qualitative and a quantitative standpoint.

Inclined Substrate

Various attributes of dropwise condensation of water at saturation temperature of 27 °C ($\Delta T_{\text{sat}} = 5$ °C) underneath an inclined substrate (15° from horizontal; $\theta_{\text{adv}} = 111°$, $\theta_{\text{rcd}} = 81°$), recorded in experiments and observed in numerical simulation are shown in Figs. 11.11, 11.12, 11.13, and 11.14.

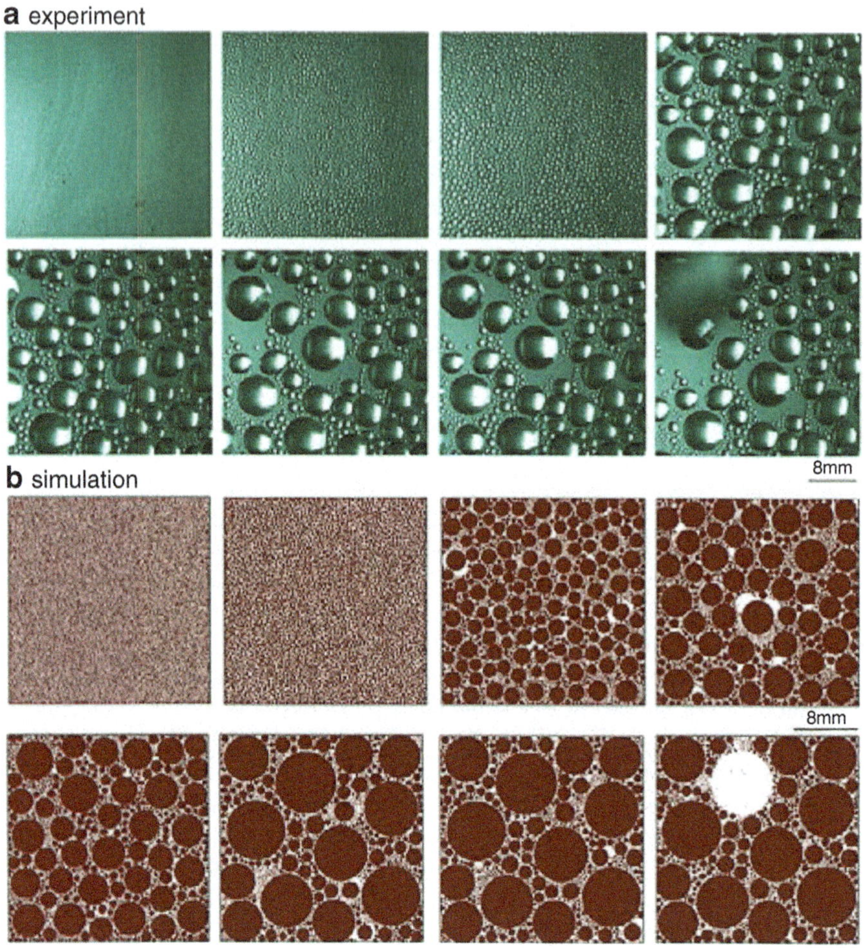

Fig. 11.10 Comparison of experiments and simulation for the complete sequence of dropwise condensation process, from the appearance of drops of minimum radius to the drops of critical radius underneath a horizontal silanated glass substrate of 30 × 30 mm area

Major physical processes observed on an inclined substrate are similar to those of the horizontal substrate, except that the simple fall-off mechanism is replaced by a more complex combination of slide-off and fall-off, as shown in Fig. 11.11. On an inclined substrate, a critically sized sliding droplet, while sweeping other droplets on its path, may either (a) reach the end of the substrate without falling off or, (b) may acquire enough mass to be pulled in the downward direction, thus falling off from the substrate, before actually reaching the edge of the substrate. The scenario realized will depend on the rate of growth of the drop, coalescence, and the length of the substrate itself. The other physical processes of nucleation, direct condensation growth, coalescence, and merger dynamics are quite similar to that of horizontal

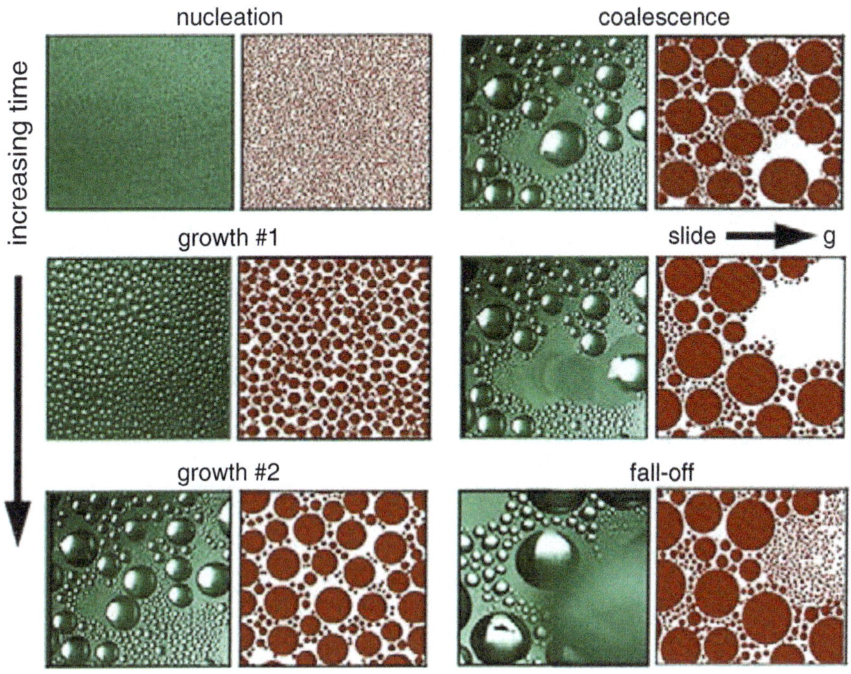

Fig. 11.11 Cycle of individual processes in dropwise condensation of vapor over an inclined cold substrate. (**a**) Cycle of individual sub-processes which constitute dropwise condensation. (**b**) Qualitative depiction of the footprints of the droplets during the cyclic growth process (for water, subcooling $\Delta T_{sat} = 5$ °C, average contact angle = 96°, nucleation site density = 10^6 cm^{-2})

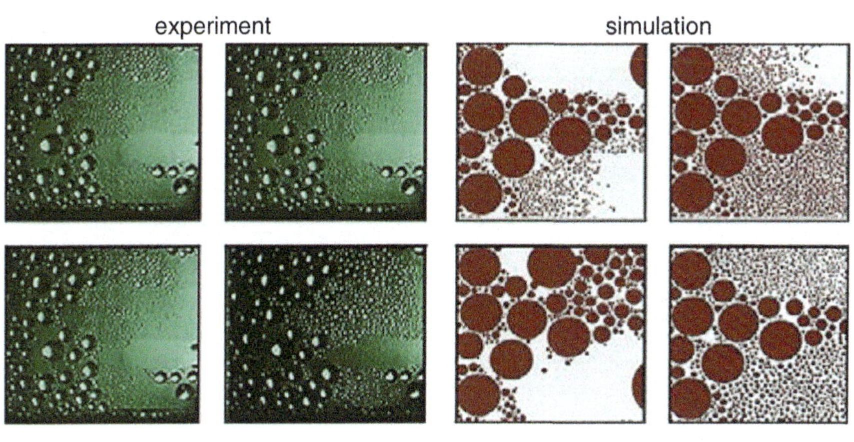

Fig. 11.12 Various stages of droplet condensation on the inclined substrate (15° from the horizontal) recorded during experiments and simulation. The commencement of sliding and sweeping actions of the drop as it gathers mass during transit and renucleation of the virgin exposed surface, when the sweeping action is complete are clearly seen

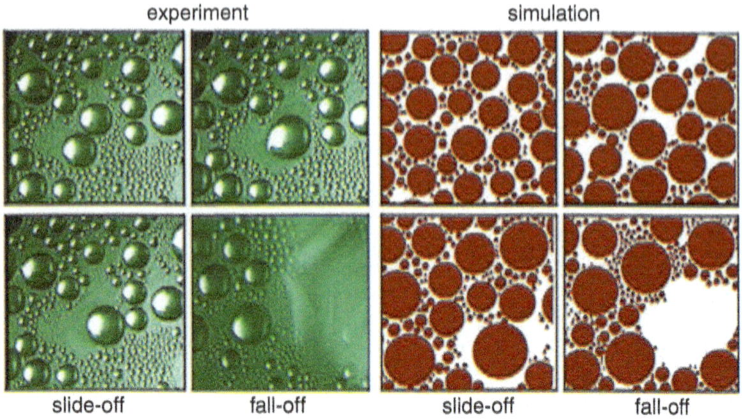

Fig. 11.13 Temporal stages of droplet condensation on the inclined substrate (5°) recorded during experiments and simulation. The sliding drop gathers mass during transit and reaches criticality of fall-off within the physical domain

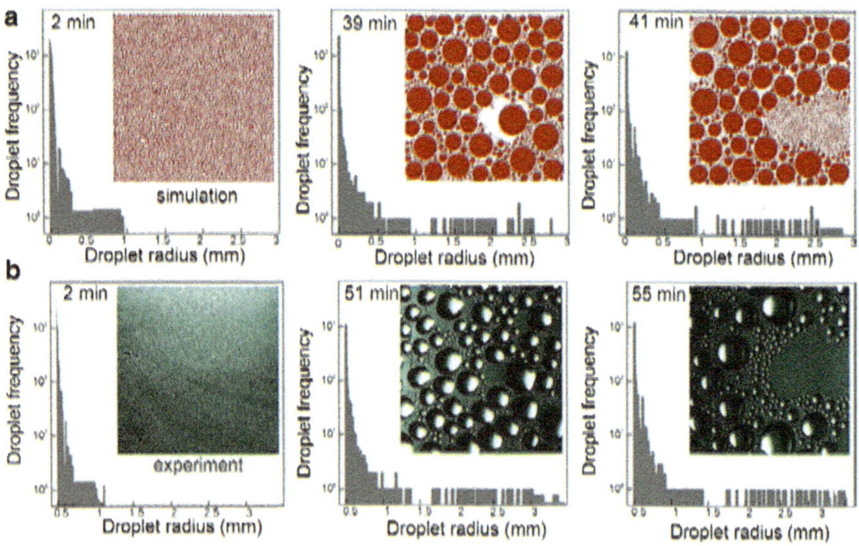

Fig. 11.14 Size distribution of drops condensing underneath an inclined (15° from the horizontal) silanated glass substrate of size 25 × 25 mm as recorded in the simulation and in experiments (**a**) at time = 2 min from the commencement of dropwise condensation, (**b**) at critical state of slide-off, and (**c**) just after a complete sweeping action is completed by a sliding drop

substrate. The fact that the gravity vector now acts at an angle to the growing droplets leads to unsymmetrical drop deformation. The contact angle hysteresis plays a role in the static force balance.

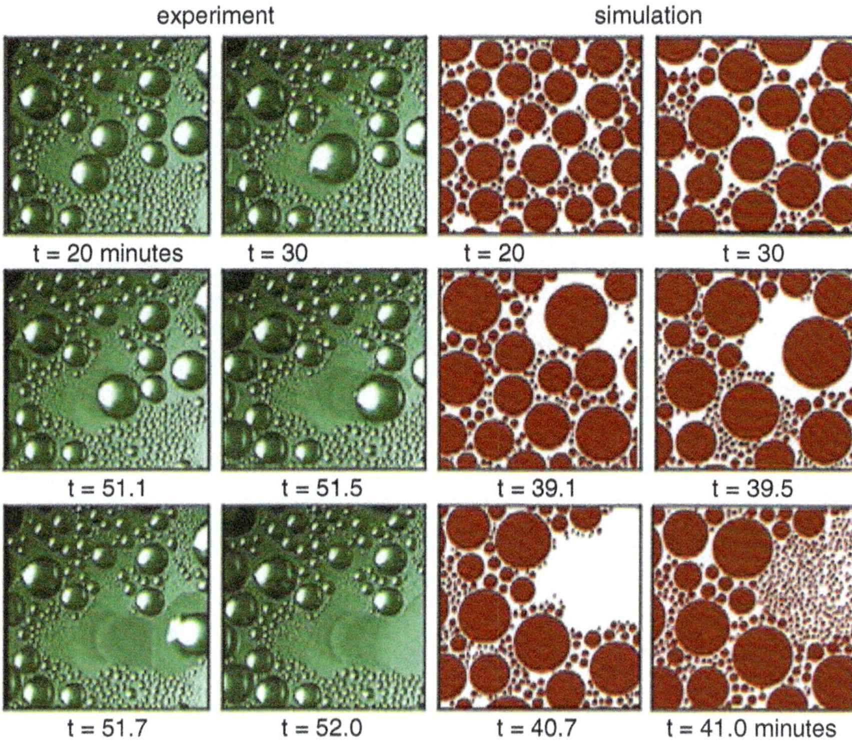

Fig. 11.15 Temporal stages of droplet condensation on an inclined substrate (15°) recorded during experiments and simulation. The commencement of sliding and sweeping actions of the drop as it gathers mass during transit and renucleation of the virgin exposed surface, when the sweeping action is complete, are clearly seen

On the inclined surface, critical sized droplets first begin to slide, rather than fall-off, as observed underneath a horizontal substrate. Criticality is achieved by direct condensation or alternatively, by coalescence with the adjoining drops. Thus, depending on the length of the substrate and timescales of direct growth and growth due to coalescence, there are various possibilities observed during the experiment underneath an inclined substrate. These include the following:

Slide-off criticality is achieved, and during the entire slide-off on the substrate, fall-off criticality is not achieved.

Slide-off criticality is achieved, and during the slide-off underneath the substrate, fall-off critically is also achieved before the droplet traverses the complete substrate length scale. Both these possibilities have been incorporated in the simulation (Figs. 11.12 and 11.13).

Figures 11.12 and 11.13 depict the complex sequence of slide-off, rapidly followed by sweeping, fall-off, and renucleation. After the first instance of slide-off, it is interesting to note that the subsequent slide-offs and sweeping actions occur

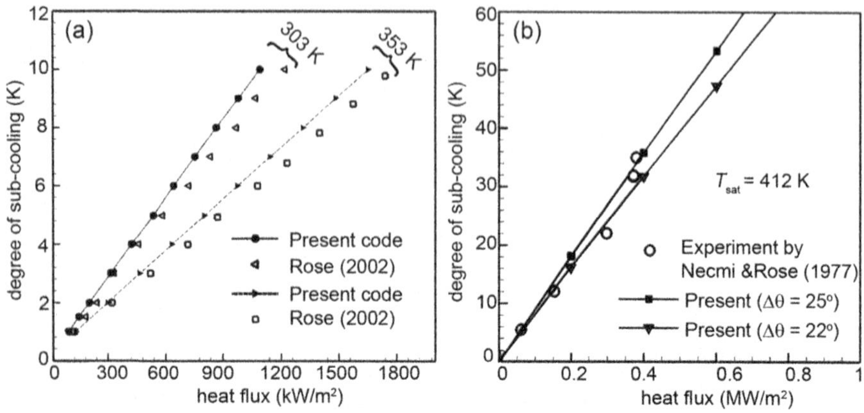

Fig. 11.16 Comparison of average substrate heat flux obtained by simulation with experimental data reported in the literature. (**a**) For water vapor condensation: Variation of heat flux, underneath a horizontal substrate, at $T_{sat} = 30$ °C and $T_{sat} = 50$ °C, respectively, with subcooling. (**b**) For mercury vapor condensation: Average substrate heat flux plotted as a function of degree of subcooling for dropwise condensation of mercury over a vertical plate at saturation temperature of 139 °C and contact angle hysteresis ($\Delta\theta$) = 22° and 25°

at a greater frequency. The mathematical model satisfactorily captures these processes.

Figure 11.14 depicts the experimental images and histograms of droplet frequency along with the corresponding simulation data for a hydrophobic surface of 15° inclination. The critical stage of slide-off is also pictorially compared; a discrepancy in the actual time of slide-off in experiments as opposed to simulation is again observed. Soon after the slide-off, virgin areas are created, fresh nucleation sites are exposed and renucleation commences (Fig. 11.14a, b).

The complete temporal sequence of events on the inclined substrate is shown in Fig. 11.15. Unlike a horizontal substrate, the drop dynamics on an inclined substrate is unique because the criticality of droplet motion and the series of events soon thereafter (sweeping and/or fall-off) happen extremely quickly leading to a sudden reduction in area coverage. Moreover, repeated removal of drops leads to the time-averaged area of coverage being smaller for the inclined substrate when compared to the horizontal. At the instant of the first slide-off, the area coverage is 58.8% in simulation and 49.5% from experiments. The discrepancy is again attributed to the loss of data pertaining to small sized droplets during experimental observations. It is clear from Fig. 11.15 that drop slide-off underneath the inclined substrate occurs earlier than the corresponding time instant of fall-off underneath a horizontal substrate.

The variation of the average substrate heat flux for dropwise condensation of water with respect to the degree of subcooling ($T_{sat} - T_w$) at condensation temperatures 30 °C and 50 °C, respectively, on a horizontal chemically textured substrate is shown in Fig. 11.16a. The comparison of the present simulation with the theory put forward by Le Fevre and Rose (1966) for dropwise condensation of water on a monolayer promoter layer, as reported by Rose (2002) is also shown. The model results are further validated against experiments for condensation of mercury vapor on a vertical surface (Fig. 11.16b). The experimental data for the surface-averaged wall heat flux as a function of the vapor to surface temperature difference is adopted from the work of Necmi and Rose (1977). Figure 11.16 shows good overall agreement for all levels of subcooling. Hence, one may conclude that the model is robust and applicable to liquids, having a wide range of Prandtl numbers, condensing on substrates with various orientations and conditions.

11.3 Closure

Experiments for dropwise condensation on the measurement of heat transfer coefficient are reviewed. Dropwise condensation experiments on chemically coated glass substrate are performed, and results of condensation patterns and the corresponding predictions of numerical simulation for water vapor are compared. The prediction of the model is in fair agreement with the experimental data of condensation of water vapor. Average heat flux as a function of degree of subcooling for water and mercury are compared. Although there is some discrepancy in the data obtained, major phenomena related to dropwise condensation underneath horizontal substrates are well simulated by the mathematical model.

Chapter 12
Surface Preparation: Some Techniques

Mahesh Kumar Yadav, Praveen Somwanshi, Sameer Khandekar, Sanghamitro Chatterjee, Mohit Gonga, K. Muralidhar, and Sudeep Bhattacharjee

Nomenclature

r Roughness factor, –
r_c Critical radius for transition between Wenzel and Cassie-Baxter states, m

Greek Symbols

ϕ_s Solid surface area with respect to total surface area, –
γ Surface tension, N/m
θ Contact angle, °
θ^* Wenzel contact angle, °
$\theta_\#$ Cassie-Baxter contact angle, °

Subscripts and Superscripts

g Gas phase
l Liquid phase
s Solid phase

Abbreviations

AAO Anodic aluminum oxide
AFM Atomic force microscopy
FESEM Field emission scanning electron microscopy
HA Hard anodization

S. Khandekar, K. Muralidhar, *Drop Dynamics and Dropwise Condensation on Textured Surfaces*, Mechanical Engineering Series,
https://doi.org/10.1007/978-3-030-48461-3_12

HMS	Di-methyl-chloro-silane
MA	Mild anodization
OTS	Octadecyl-tri-chloro-silane
RMS	Root mean square
SAM	Self-assembled monolayer

12.1 Introduction

Condensation of vapor on a solid surface can occur in either dropwise mode or filmwise mode. A mixed-mode of condensation is also possible where dropwise and filmwise mode of condensation transpires concurrently. This has been already discussed in earlier chapters. How to achieve the dropwise condensation in practical situations is discussed in this chapter. An important property generally used to recognize the mode of condensation over a solid surface is called "wettability." It defines the spreading or contracting characteristics of the fluid to the solid, depending on forces between a fluid and a solid surface in an immiscible medium (Extrand 2016). Here, spreading means flow of fluid in the form of thin films, whereas contacting means coiling of fluid in the form of drops. From the molecular point of view, wettability is determined by a balance between the cohesive intramolecular interaction between the liquid molecules and the adhesive interaction between the liquid and the solid surfaces in contact (Israelachvili 2011; de Gennes et al. 2004; de Gennes 1985). Accordingly, wettability is mainly governed by micro/nanostructuring of the solid surface, chemical compositions of the fluid and the solid surface. The range of wetting varies from complete-wetting surfaces (contact angle ~0°; superhydrophilic) to non-wetting surfaces (contact angle → 180°; superhydrophobic) (Xin and Hao 2010). A particular value of the wettability defines the contact angle at the intersection point of the solid-fluid interface (Extrand 2016; de Gennes et al. 2004). Depending on the value of contact angle (typically more than 120°), dropwise shape of the fluid can be realized over the solid surface. In fact, this phenomenon is highly involved in many natural processes, such as self-cleaning of plant leaves, feathers of birds, and legs/back/wings of insects (Abdulhussein et al. 2016, Shin et al. 2016). In practical applications, the solid surfaces are engineered to modify the forces acting at the solid-fluid interface to get the desired wetting characteristics or surface properties. Some of the examples of such specially engineered surfaces include application in biomedical field (in blood vessel replacement and controllable drug delivery), microfluidics, dropwise condensation, self-cleaning, lubrication, fuel cell transport, and textiles industries (Zhao et al. 2010; Bhushan et al. 2009; Prabhu et al. 2009; Furstner et al. 2005; Onda et al. 1996).

Specially engineered surfaces, which promotes fluid-flow in the form of drops over the solid surface, found applications in many practical applications. It is vital to comprehend underlying concepts of wettability to understand the requirements/constraints to engineer surfaces for a preferred mode of condensation in practical applications. Therefore, the basics of wettability are discussed next before we move to the available surface engineering techniques to comprehend the dropwise mode of condensation.

The wetting characteristics or equilibrium shape of the liquid on the solid surface is governed by forces acting at the three-phase contact line in the solid plane, i.e., at an intersection point of the solid, liquid, and the surrounding environment. Force balance at the interaction point is given by the famous Young's equation (Bormashenko 2015):

$$\gamma_{sg} - \gamma_{sl} = \gamma_{lg} \cos \theta \tag{12.1}$$

Here, γ denotes the surface tensions between the two phases indicated by the subscripts s, g, and l (representing solid, gas, and liquid, respectively). θ is the Young contact angle formed between the liquid-gas interface and the solid surface.

While deriving the Eq. (12.1), it is assumed that the surface where the liquid interacts is perfectly smooth and chemically homogenous. Besides, the mass of the liquid is small enough to neglect the effect of gravitational force.

Equation (12.1) clearly shows that contact angle will vanish when the surface tension of the solid-gas interface becomes equal to the sum of the surface tensions of the liquid-solid and liquid-gas interfaces. This is the case where surface wetted easily and called a high energy surface. The complete wetting of the surface is achievable by tuning surface forces, as confirmed by the thermodynamic phase transition of the partial wetting surface into the fully wetting surface at a characteristic temperature. Contrary to this, the contact angle may increase as high as 180° in theory for lower (negligible) energy surface (Wang et al. 2015a, b). However, it is practically not feasible due to the finite surface energy of a material. Therefore, along with lower energy surface, roughness, and/or texture on the surface at micro/nanolevel becomes important to reach this extreme contact angle of nearly 180° (Ramiasa-MacGregor et al. 2016).

The wetting characteristics of the rough and/or textured surface are usually described by two important equations: Wenzel and Cassie-Baxter. In the Wenzel state, the liquid follows the rough surface, and the contact angle is linked to the Young's equation as follows (Quere 2008; Wenzel 1936):

$$\theta^* = r \cdot \cos \theta, \tag{12.2}$$

where θ^* is the Wenzel contact angle, and r is a roughness factor defined as the ratio of rough to plane surface area.

Equation (12.2) clearly shows that roughness greater than 1 ($r > 1$) improves the hydrophilicity and hydrophobicity for $\theta < 90°$ and $\theta > 90°$, respectively (Ramiasa-MacGregor et al. 2016). However, for $\theta > 90°$, the liquid in the form of a drop will sit over the top of the asperities of the solid surface with air trapped in the gap below the liquid. For such a "fakir" drop sitting on the heterogeneous surface, called the Cassie-Baxter state, the contact angle is assumed to be average between angles on the solid surface and on the air (180°). The contact angle with a fraction of solid surface area ϕ_s and a fraction of the air $(1 - \phi_s)$ is given as follows (Abdulhussein et al. 2016; Milne and Amirfazli 2012; Cassie and Baxter 1944):

$$\cos\,\theta_{\#} = -1 + \phi_s(\cos\,\theta + 1), \tag{12.3}$$

where $\theta_{\#}$ is the Cassie-Baxter contact angle.

Equation (12.3) shows that $\theta_{\#}$ increases as ϕ_s decreases, theoretically achieving $\theta_{\#} = 180°$ for $\phi_s = 0$, i.e., the liquid and the solid surface is physically in contact with the trapped air only. However, the roughness factor "r" in Eq. (12.2) also decreases with a decrease in ϕ_s, promoting the Wenzel state (Wang et al. 2015a, b; Quere 2008).

It is clear from the above discussion that wettability can be easily tuned by adjusting the three surface tension forces for ideal smooth and homogenous surfaces (Eq. 12.1). However, for real rough surfaces, two different states of the liquid on the solid surface are achievable. In the first state where liquid follows the solid surface, i.e., the Wenzel state, higher wettability is achieved due to increased interaction between the liquid and the solid. In the second state, where the liquid sits over the top of the asperities of the solid surface with trapped air below, i.e., the Cassie-Baxter state, lower wettability is seen due to minimal surface contact between the interacting liquid and the solid. There is a critical roughness "r_c" which separates these two states is deduced from the intersection of Eqs. (12.2) and (12.3) (Bormashenko 2015),

$$r_c = (\phi_s - 1)/\cos\,\theta + \phi_s \tag{12.4}$$

The critical radius "r_c" above this value favors the Cassie-Baxter state, and below this, the Wenzel state. Therefore, different wettability can be achieved on a particular surface by tuning the surface energy. There are several ways to modify the surface energy, and thereby controlling the wettability of solid surfaces. These include a coating of low surface energy materials, chemical modification, and engineering of surface roughness/surface texturing (Phan et al. 2009; Gorman et al. 1995; Sondag-Huethorst and Fokkink 1994). Basics of surface preparation techniques are deliberated in next section, followed by detailed discussion on five distinct techniques to promote the dropwise condensation, i.e., glass silanization, chemical texturing on copper, anodization of aluminum, laser machining, and ion implantation.

12.2 Surface Preparation Techniques

In process equipment, dropwise condensation can be realized by suitably treating the condensing surface. The treatment will ensure partial wetting of the surface by the condensate liquid in the sense that the contact angle greater than 90° is achievable.

The wetting characteristics of condensate over the cold substrate can be broadly controlled by two different means:

(a) Modify the surface of the substrate
(b) Alter the condensing vapor chemically, say by injecting a chemical which promotes non-wetting behavior

Other methods that rely on changing the pH value of the condensate can be used so that dropwise condensation can be controlled by using an electrical potential (electro-wetting), changing the condensing temperature, and other techniques. Among these methods, substrate modification has emerged as the most popular and effective strategy.

A good drop promoter technique should be long lasting, involving low surface energy, low contact angle hysteresis, and low thermal resistance. The method should be easy to apply, nontoxic, and must be compatible with the system in which it is used, i.e., it should not impair the proper functioning of the other parts of the system. Superhydrophobicity appears ideal to promote continued dropwise condensation which results in rapid removal of condensate drops; however, such promotion has not been reported on engineered surfaces. For any technique used for promoting dropwise condensation, the longevity of the textured surface is critical. With the advent of newer manufacturing, coating, and nanoscale fabrication techniques, surface treatment of the substrate holds considerable prospect in terms of providing the required long-term sustainability of dropwise condensation.

There have been two generic methods that can be used to modify the wettability of the substrate. The first one is chemical grafting or adsorbing molecules with wetting characteristics of their own (chemical texturing). The second is to texture the surface by altering the surface topography/roughness by patterning, called physical texturing. Roughening a surface will increase wettability, in general, unless special patterns of the right scale are employed. In contrast, chemical coatings have gained prominence because of the larger choices available and are reviewed here.

12.2.1 Chemical Texturing

Chemical texturing can be created by coatings, such as organic compounds, with hydrophobic groups (Blackman et al. 1957; Watson et al. 1962; Ma and Wang 1999; Vemuri et al. 2006), inorganic compounds (Erb 1965; Erb and Thelen 1965; Erb and Thelen 1966; Zhao et al. 1996), polymers (Marto et al. 1986; Zhang et al. 1986; Mori et al. 1991, Ma et al. 2002), or special surface alloys (Erb 1973; Zhao et al. 1991; Koch et al. 1997; Ma et al. 2000a, b). Other coating materials include, for example, Teflon (Stylianou and Rose 1980). These surfaces are created by preparing a weak solution of Teflon (AF1600) in FC-75. The samples are dip-coated in this solution with different pulling speeds to achieve the desired film thickness. After dip-coating, the samples are annealed in a furnace for ~10–30 min at temperatures ranging from 100 °C to 300 °C (Ma and Wang 1999; Ma et al. 2000b).

Though simple in concept, such surfaces suffer from long-time sustainability issues that do not allow application to real-life, large-scale processes. Leaching by the motion of drops over the surface can also result in degradation of the coating. Amorphous hydrogenated carbon films (a-C:H) with diamond-like mechanical properties have been modified by adding new elements to the film, e.g., silicon or fluorine (Grischke et al. 1994), reducing its surface energy. These coatings have been studied for their heat transfer characteristics. Such coatings are mechanically and chemically stable but introduce an additional thermal resistance. This drawback can be overcome by other surface modifications which do not form an additional layer. Ion-implantation is an example that has been tested successfully by Zhao et al. (1990), Zhao et al. (1991), Leipertz and Choi (2000), and Zhao and Burnside (1994).

12.2.2 Physical Texturing

Fabrication of hydrophobic surfaces by physical texturing is, in principle, quite simple. It can be generated by creating a suitable roughness. A review of the subject (Nakajima et al. 2001) reveals a wide range of methods for producing roughness on substrates. Figure 12.1 shows some methods available to create roughness distribution on the substrate. Pros and cons of these techniques are summarized in Table 12.1.

Most researchers (Sommers and Jacobi 2006; Sommers and Jacobi 2008) reported hydrophobic and superhydrophobic surfaces produced by etching and lithography. Some (Lau et al. 2003; Chen et al. 2007; Hsieh et al. 2008; Boreyko and Chen 2009) reported continuous dropwise condensation on a superhydrophobic surface with short carbon nanotubes deposited on micro-machined posts, a two-tier

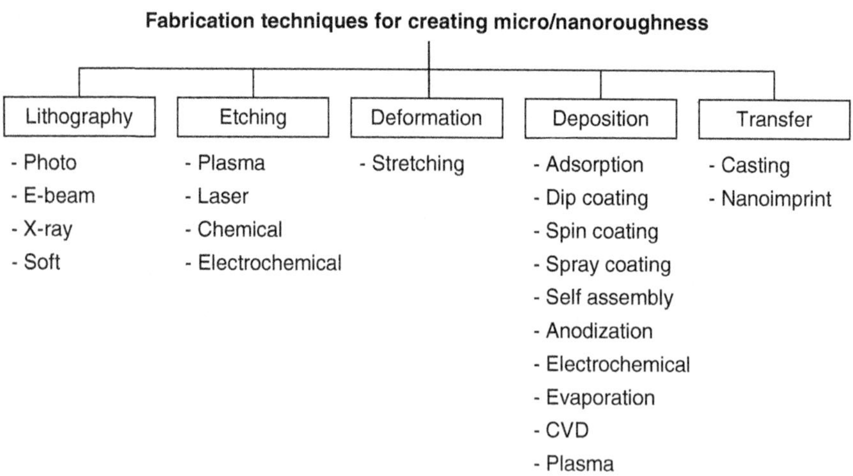

Fig. 12.1 Typical methods to fabricate micro/nano-roughened surfaces (Bhushan and Jung 2011)

Table 12.1 Pros and cons of various surface fabrication techniques

Techniques	Pros	Cons
Lithography	Accuracy	Large area, slow process, high cost
Etching	Fast	Contamination, less control
Deposition	Flexibility, cheap	High temperature, less control
Self-assembly	Flexibility, cheap	Require suitable precursor

texture mimicking lotus leaves. Surface preparation by physical texturing as reported in the literature is reviewed below.

Jessensky et al. (2003) introduced a new technique for the fabrication of a superhydrophobic surface by anodic oxidation of metals such as aluminum, titanium, tungsten, and hafnium. These metals may all be anodically oxidized when put into an electrolyte. Such surface treatments are common in the industry, as anodization of aluminum and titanium creates hard scratch-resistance and protects the underlying surface from further oxidation. When refined to a specific processing regime, anodization results in the formation of a highly ordered nanostructure.

Miwa et al. (2000) prepared various superhydrophobic films of different surface roughness. The relationships between the sliding angle, the contact angle, and the surface structures were investigated. Sliding angle of water droplet was seen to decrease with increasing contact angle. Microstructures revealed that the surface texture traps air and assists in the preparation of low-sliding-angle surfaces.

Sommers and Jacobi (2006) describe photolithographic techniques to obtain micro-patterns on aluminum surfaces with parallel grooves, 30 μm wide and tens of microns in depth. Experimental data show that a droplet placed on the micro-grooved aluminum surface using a micro-syringe exhibits an increased apparent contact angle. For droplets condensed on these etched surfaces, more than a 50% reduction in the volume needed for the onset of sliding is obtained.

Liu et al. (2006) prepared micro/nanoscale binary-structured composite particles of silica/fluoropolymer by using an emulsion-mediated sol-gel process. It is composited on various substrates by using simple spray-coating or spin-coating methods to create superhydrophobic surfaces. Results show that the static contact angle of water on the substrate is larger than 150°.

Boreyko and Chen (2009) generated a superhydrophobic substrate, composed of two-tier roughness with carbon nanotubes deposited on silicon micro-pillars and coated with hexadecanethiol. Continuous dropwise condensation was spontaneously formed on a superhydrophobic surface without any external forces. Spontaneous drop removal resulting from the surface energy released upon drop coalescence led to an out-of-plane jumping motion of the coalesced drops at a speed as high as 1 m/s.

Dietz et al. (2010) reported a novel technology to achieve superhydrophobic coating with microscopic roughness on a copper surface. A layer of verdigris was grown on fresh pure copper surface by exposing copper to air and a mist of acetic acid solution. The coating was oxidized to CuO. A self-assembled monolayer coating of n-octadecanethiol was obtained on the outermost surface. Results showed that the static contact angle of a water drop was $153.1 \pm 1.7°$.

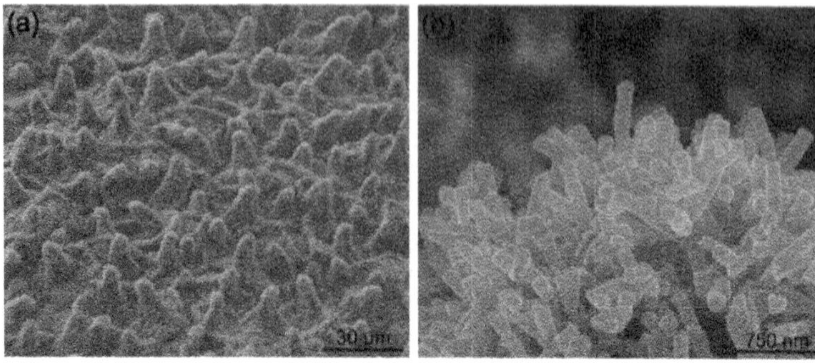

Fig. 12.2 SEM images of a lotus leaf, Nelumbo nucifera. (**a**) The surface is covered with hierarchical roughness so that the micro-bumps and the basal area are decorated with nano-protrusions. (**b**) Randomly oriented nano-cylinders that cover a micro-bump (Cha et al. 2010)

Cha et al. (2010) fabricated six different surfaces, one natural and five artificial. As a natural hydrophobic surface, a lotus leaf, Nelumbo nucifera was used. A lotus leaf collected from a local pond was cleaned with an air gun to remove dust particles. The leaf surface was covered with hydrophobic epicuticular wax crystals, and its water repellency was enhanced by the intrinsic surface structure. A scanning electron microscope of a lotus leaf surface is shown in Fig. 12.2.

Artificial surfaces were prepared with bare silicon wafers, single-roughness surfaces with micro-pillar arrays, single-roughness surfaces with nanoscale pillars, hierarchical surfaces with micro-pillars decorated with nano-protrusions only on their tops (surfaces with partial dual roughness), and hierarchical surfaces with nanoscale roughness on both micro-pillar tops and bases (surfaces entirely with dual roughness). The process is schematically shown in Fig. 12.3. Contact angle of water drops on these surfaces varied from 140° to 170°, depending on the micro-pillar density. On micro/nanostructured surfaces, the condensate drops prefer the Cassie state which is thermodynamically more stable than the Wenzel state.

Despite breakthrough in nano-machining, there is hardly any literature that has reported generation of a hydrophobic surface by conventional machining process. Bhutani et al. (2013) used hand lapping process on aluminum and copper substrate to make the surface hydrophobic. Lapping pastes of three different grades were used to produce surface roughness of the order of 0.5, 1.5, and 3.5 μm (RMS). The highest contact angle obtained was 95° on these substrates. McCarthy et al. (2010) reported fabrication and characterization of biomimetic superhydrophobic surfaces synthesized using self-assembly and metallization of the Tobacco Mosaic Virus (TMV) onto micro-pillar arrays. Superhydrophobic surfaces with static contact angles greater than 150°, and droplet hysteresis less than 10°, were seen to resist wetting and exhibit self-cleaning properties.

In the next section, five distinct techniques used for preparing low surface energy surfaces in our laboratory are described in detail.

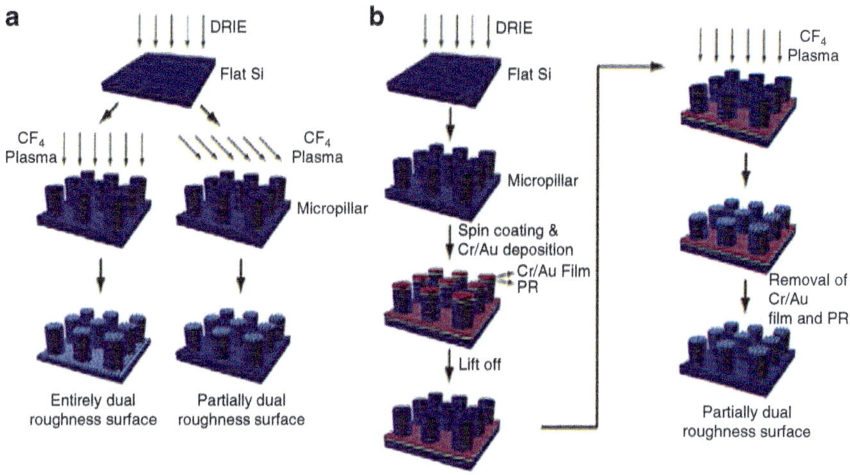

Fig. 12.3 Fabrication process of dual-roughness surfaces. (**a**) Fabrication of the surface with dual roughness with the direct incidence of CF_4 plasma. (**b**) Fabrication of the surface with partially dual roughness via masking the basal area with a Cr/Au layer (Cha et al. 2010)

12.3 Surface Modification Techniques

Five different techniques used in our laboratory to engineer the surface properties/ wettability to promote the dropwise condensation is discussed hereunder.

12.3.1 Silanization of Glass

Silanization is one of the methods of modification of surface energy in which grafting of long aliphatic chains on silica is done using tri-chloro-silane. The hydrophobic surface is created by coating the Self-Assembled Monolayers (SAMs) on the substrate (Fig. 12.4a). Octadecyl-tri-chloro-silane (OTS), di-methyl-chloro-silane (HMS), trichloro-silane (MTS), and propyl-tri-chloro-silane (HTS) are generally used as SAMs (Fig. 12.4b). Among the SAMs, Octa-decyl-tri-chloro-silane (OTS) was found to yield the best quality surface for dropwise condensation because of the smallest contact angle hysteresis. To deposit OTS on a surface, the cleaned substrates were kept in a solution of 60 mL bi-cyclo-hexane, 35 drops carbon-tetrachloride and 20 drops OTS. During this time, the OTS molecules bond covalently on silicon dioxide substrates. After a few minutes, the substrates were taken out of the silane solution and rinsed with chloroform.

To coat the surface with di-methyl-chloro-silane (HMS), trichloro-silane (MTS), and propyl-tri-chloro-silane (HTS), the cleaned samples were kept in a desiccator together with a small quantity of the desired silane. Silane vaporizes in the closed environment of the desiccator and gets deposited on the substrate. After 15–20 min

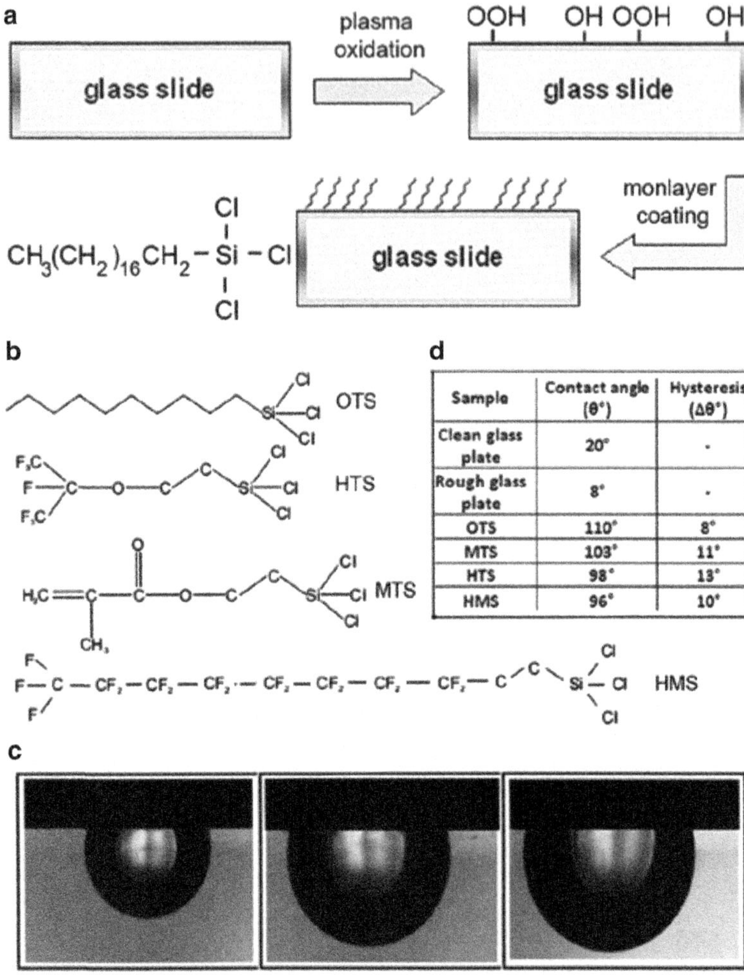

Fig. 12.4 (**a**) Schematic diagram explaining the chemical vapor deposition process. (**b**) Representations of various self-assembled monolayer (silane) molecules. (**c**) Image of pendant drop of volume 5 μL, 10 μL, and 15 μL, respectively, on HMS textured substrate. (**d**) Measurement data

of evaporation, a silane monolayer gets bonded covalently with the oxide surface (Genzer and Efimenko 2000). After taking out the samples from the desiccators, they are rinsed by chloroform. Co-evaporation of various silanes can also be carried out to achieve intermediate surface energy but at the cost of a higher contact angle hysteresis.

The preparation of the hydrophobic substrate by self-assembled monolayers uses the chemical vapor deposition process. The samples were first cleaned by sonicating them in ethanol, acetone, and toluene bath for 3–10 min, respectively. The substrate was dried with compressed nitrogen gas carefully while changing from one solvent to another. Subsequently, the samples were cleaned by an oxygen plasma torch followed by a dry CO_2 snow-jet (Cras et al. 1999). The samples were kept in Piranha solution (50% H_2O_2 and 50% H_2SO_4 by volume) for 2–4 h. Piranha solution is highly oxidizing and requires special care in handling. Surface energy modification was accomplished by coating samples with OTS. Thereafter, the substrate was washed in distilled water and dried in nitrogen. Nascent oxygen released when sulfuric acid reacted with hydrogen peroxide cleaned the surface.

Experimental setup used for chemical vapor deposition consisted of a vacuum pump (rotary vane rougher pump coupled with diffusion pump, ultimate vacuum level ~10^{-5} mbar), plasma oxidizer (with RF generator of 6–18 W power, frequency 8–12 MHz) and a desiccator. Inside the reactor, which was maintained at low vacuum pressure, the high frequency oscillating electromagnetic field ionized the silane molecules forming plasma. This interacted with the glass substrate by removing organic contamination from its surface. The high energy plasma particles combined with the contaminants to form carbon dioxide (CO_2) or methane (CH_4). The silanation process is explained schematically in Fig. 12.4a. Figure 12.4c, d shows the image of a pendant drop of varying volumes (5 μL, 10 μL, and 15 μL, respectively) on chemically textured substrates along with the measurement data.

12.3.2 Copper Texturing

It can be inferred from Eq. (12.2) that the contact angle is strongly influenced by roughness of the surface. Copper showed more reliability than the other metals/non-metals for chemical texturing and hence particularly used for chemical texturing. The chemical texturing modifies the surface morphology at the nanoscale. Hence, ultimate care has to be taken during the pre-processing of the surface. A copper test surface (99% pure) is taken for grinding using a fine grade grinding wheel. The surface is cleaned using neutral liquid detergent (Labolene®-Fischer Scientific) and flushed with deionized water immediately after the grinding to remove the debris and burr associated. Subsequently, the copper surface is polished using emery papers, starting from a grit size of 600–2500. In the next step, the polished copper surface is cleaned using liquid detergent and flushed with deionized water. This cleaned surface is kept in an ultra-sonication bath for 20–30 min and subsequently dried in nitrogen gas. The cleaned copper surface is immersed in aqueous $AgNO_3$ (0.01 mol dm^{-3}) for 20 s at room temperature, then washed with deionized water and dried again in nitrogen gas. Chemical products of the reaction are

$$Cu\ (s) + 2AgNO_3\ (aq.) \rightarrow Cu(NO_3)_2\ (aq.) + 2Ag\ (s) \qquad (12.5)$$

Chemically treated copper surface is subsequently immersed in solution (1 mmol-dm^{-3}) of 3, 3, 4, 4, 5, 5, 6, 6, 7, 7, 8, 8, 9, 9, 10, 10, 10—heptadecafluoro-1-decanethiol ($CF_3(CF_2)_7CH_2CH_2SH$) in CH_2Cl_2 for 5 min and dried with nitrogen

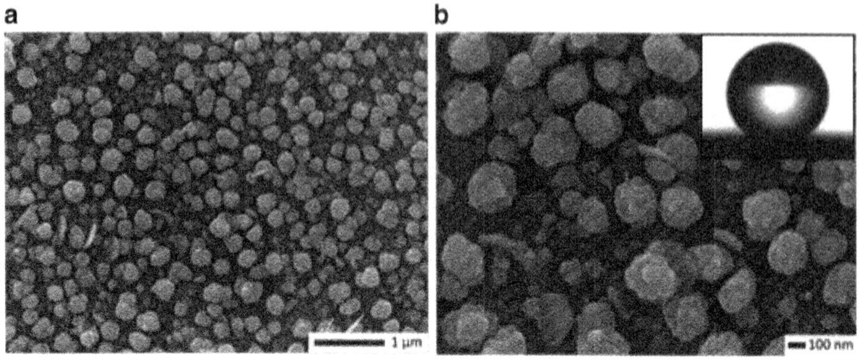

Fig. 12.5 FESEM images of a hydrophobic copper substrate as-prepared at (**a**) low (×20,000). (**b**) High (×50,000) magnification. Inset shows drop of 4.0 μL volume on the textured copper surface having contact angle 150°

gas. The chemical texturing procedure adopted here is similar to Larmour et al. (2007).

Figure 12.5 shows the field emission scanning electron microscopy images (FESEM; JSM-7100F, JEOL) at low (Fig. 12.5a) and high magnifications (Fig. 12.5b). It is clearly seen that extrusions similar to micro-flowers of diameter, with size ranging from 5 to 580 nm, are uniformly grown because of oxidation of the copper surface. These extrusions are acting as a rough surface on which liquid drop sits like the Cassie-Baxter state with gaps between these extrusions filled with air.

The image of a static water drop of 4 μL on the textured surface, forming a contact angle of 150°, is shown in the inset of Fig. 12.5b. Therefore, we can say that non-wetting characteristics of the copper surface, which helps in realizing dropwise condensation, is significantly improved by creating micro/nanotexture on the surface.

12.3.3 Anodization of Aluminum

Anodization of aluminum to generate nanoporous structures had been extensively investigated and used in many applications (Arya et al. 2016; Singh et al. 2013; Poinern et al. 2011; Keller et al. 1953). This is mainly because of the formation of highly ordered Anodic Aluminum Oxide (AAO) at specified parameters (voltage, current density, type of electrolyte, temperature, etc.) and outstanding properties of AAO such as chemical and thermal stability and high surface area (Md Jani et al. 2013). AAO is an electrochemical process carried out in a mildly acidic solution such as oxalic, phosphoric, or sulfuric acid by applying a positive voltage to produce nanoporous structures. The mechanism of pore formation is still not very clear (Wang et al. 2012). One of the generally accepted explanations is electrical field-assisted oxide dissolution mechanism, where it is assumed that dissolution of the oxide at oxide-electrolyte interface forms an equilibrium with anodic formation of the oxide at the oxide-metal interface, allowing a steady-state growth of the pores (Diggle et al. 1968; Hoar and Mott 1959). It is believed that pores are formed in three stages (Wang et al. 2012; Parkhutik and Shershulsky 1992): (a) initial formation of barrier oxide layer, (b) start of nucleation in the outer regions of the barrier oxide, and (c) propagation of porous structure through the barrier oxide layer via electrical field-assisted oxide dissolution. Ideally, porous cylindrical structures are formed after prolonged anodization in packed hexagonal array to minimize the energy of the system (Pratap 2015). It is also shown that stress-driven interface diffusion and interaction of neighboring cells of the nanopores undergoing volume expansion also help in the self-organization of the formed nanopores (Liao et al. 2016; Houser and Hebert 2009).

Further, the Masuda group in their pioneering work had shown that highly ordered nanopores structures can be formed via two-step anodization process (Masuda and Fukuda 1995) and different pore arrangements (square and triangle) using shallow depressions of particular shape on the aluminum surface (Masuda et al. 2001). Aluminum anodized in the first step is etched off to get a nanosized pre-textured aluminum surface. The nanosized pre-textures act as pore initiation sites in the second anodization steps carried out at the same parameters, resulting in the formation of highly ordered nanopores (Masuda and Fukuda 1995). These studies made it feasible to produce nanopores of desired properties by efficiently controlling their size, shape, and pattern within typical ranges of pore diameter (10–400 nm), interpore distance (50–600 nm), thickness of porous layer (10 nm–150 μm), pore aspect ratio (10–5000), pore density (10^9 to 10^{11} cm^{-2}), and porosity (5–50%) (Kasi et al. 2010; Lee and Wong 2009; Lee et al. 2008).

The anodization of aluminum is categorized as: mild anodization (MA) and hard anodization (HA), depending on the applied voltage (Oh 2010; Rajendra et al. 2005). However, self-ordered AAOs were obtained only for limited specified applied voltage depending on the type of electrolyte used (Masuda et al. 1998; Li et al. 1998a, b; Masuda and Fukuda 1995). For demonstration purposes, 40 V for MA and 165 V for HA, with oxalic and phosphoric acids as electrolytes, respectively, are considered here. These voltages have also been chosen to have significantly different interpore distance, i.e., different size of nanostructures (~100 and 500 nm) (Nielsch et al. 2002; Li et al. 1998a, b; Masuda et al. 1997), and used two-step anodization approach to develop ordered porous structures. Additionally, anodization is performed on commercial grade aluminum using a two-step anodization approach. The particular steps followed in the anodization process is detailed underneath.

In the case of MA, ultra-sonically cleaned and deionized (DI) water washed aluminum sheet (having a linear type of intrusion, see inset Fig. 12.6a) is anodized in 0.3 M oxalic acid solution at 40 V for 12 h in the first anodization step. The solution is continuously stirred using a magnetic stirrer at 300 rpm and maintained at a temperature of 5 °C by circulating cooling water using high heat removal capacity water bath (Julabo®-FL-2506, accuracy ±0.1 °C). The anodized aluminum is etched off at 60 °C for 2 h in a solution of 6% phosphoric acid, 2% chromic acid (by weight) made in DI water. The second anodization step is then performed under the same

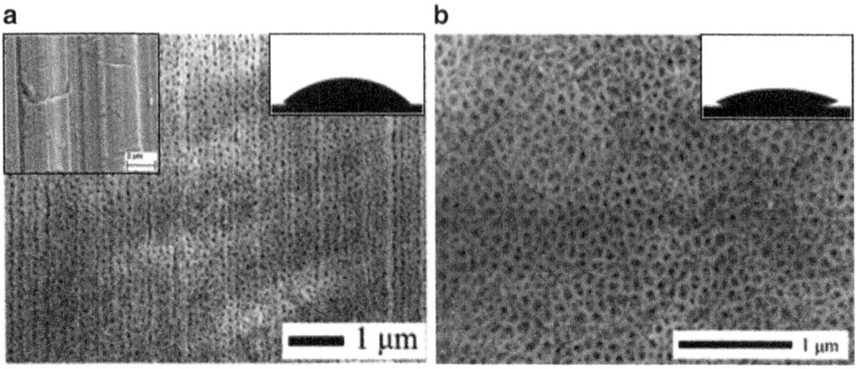

Fig. 12.6 Prepared hydrophilic surface from commercial grade aluminum by its MA at 40 V for 12 h in 0.3 M oxalic acid solution, physically changing surface texture in a specified linear pattern at micro/nanoscale. Figures show FESEM images of the MA surface: (**a**) after first anodization and (**b**) after second anodization. Inset shows FESEM images of commercial-grade aluminum surface and photos of 8 μL water droplet sitting on the anodized aluminum surface having a static contact angle of 45° and 28.5° in (**a**) and (**b**), respectively

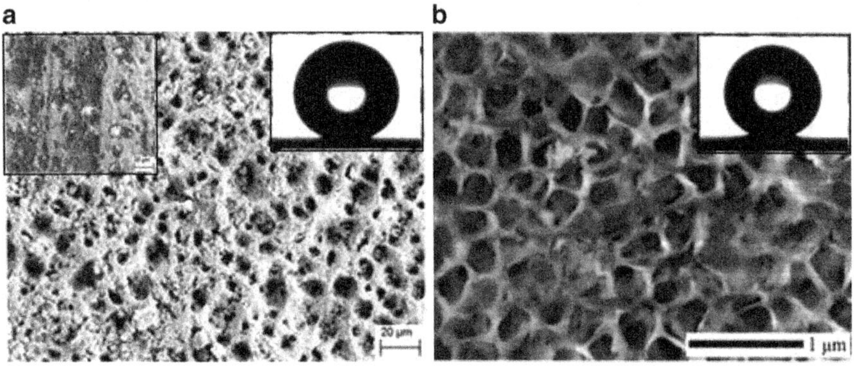

Fig. 12.7 Prepared superhydrophobic surface from commercial grade aluminum by its HA at 165 V for 2 h in 0.5 M phosphoric acid solution, physically changing surface texture in a hexagonal pattern at micro/nanoscale and coating with a low free surface energy chemical OTS. Figures show FESEM images of the HA surface: (**a**) after first anodization and (**b**) after second anodization. Inset shows FESEM images of commercial-grade aluminum surface and photos of 8 μL water droplet on the anodized aluminum having a static contact angle of 152° and 159° in (**a**) and (**b**), respectively

parameters for 12 h. The formed nanopores are then enlarged in 5% phosphoric acid (by weight) at 30 °C for 20 min.

In the case of HA, ultra-sonically cleaned grind-finished aluminum surface (see inset Fig. 12.7a) is first etched in a 0.16 M sodium hydroxide aqueous solution at 70 °C for 3 min. Further, the aluminum sheet covered with a layer of polymer on one side to prevent its oxidation to enable effective heat removal is anodized in 0.5 M phosphoric acid solution at 165 V and 5 °C for 2 h in the first anodization step. The dried anodized plate after the first step is kept in a 5% phosphoric acid solution (by weight) at room temperature for 30 min. The treated plate is again dried at 60 °C for 20 min and then coated in a 0.06 M OTS solution made in hexane in a cooled room (10 °C) for 2 h. The anodized surface in the first step is then ultra-sonically washed in acetone for 5 min and further etched off at 60 °C for 1 h in a solution of 6% phosphoric acid, 2% chromic acid (by weight), and DI water. The second step anodization is performed on the treated surface under the same parameters for 2 h, and all subsequent steps were repeated.

Two commercial grade aluminum sheets, one having a linear type of pattern and other grind-finishing, are anodized at MA (40 V) and HA (165 V) conditions in a two-step anodization, respectively. The representative FESEM images after the first-step and second-step anodization are shown in Figs. 12.6 and 12.7, respectively. Figure 12.6a shows the approximate linear type of pattern due to initial linear intrusion on the plate. However, after etching the anodized aluminum for 2 h, a developing self-ordering pore configuration is seen in second-step anodization (Fig. 12.6b). A static contact angle of 45° is obtained for 8 μL water droplet after first anodization at 40 V in oxalic acid (Fig. 12.6a) with an interpore distance of ~100 nm, corresponding to $\approx$ 2.5 μm-V^{-1} removal rate reported in the literature (Lee et al. 2006; Li et al. 1998a, b; Masuda and Fukuda 1995). After the second-step anodization, much-improved wettability with a static contact angle of 28° is obtained for the same volume of a water droplet (Fig. 12.6b). This can be attributed to higher ordered and deep nanostructures formation after second anodization resulting in improved wettability.

For HA at 165 V in phosphoric acid followed by OTS chemical coating, hexagonal type of pattern is observed with 412.5 nm interpore distance (Lee et al. 2006; Li et al. 1998a, b; Masuda and Fukuda 1995). A static contact angle of 152° is obtained on the first anodized surface for an 8 μL water droplet (Fig. 12.7a). After the second anodization, the static contact angle has increased to 159° for the same volume of water droplet (Fig. 12.7b). It clearly shows that superhydrophobic surfaces with commercial-grade aluminum sheets can be fabricated via two-step HA and subsequently coating the nanoporous surface with a low free surface energy chemical such as OTS.

It is shown that different wettability characteristics on the aluminum surface can be obtained by altering physical texturing at nanoscale through anodic oxidation and subsequent low surface energy coating on the formed nanopores. It is also noted that lower interpore distance (~100 nm) improves the wettability and large interpore distance (~400 nm) with a subsequent coating of low free surface energy material results in the formation of the non-wetting surface. Accordingly, the surface can be engineered to achieve a particular contact angle, and as a result, a specific mode of condensation over the surface.

12.3.4 Laser Machining

Laser patterning or micro-machining is another promising method to modify surface textures. Ultrashort laser (fs or ns) is commonly used to simultaneously develop microstructure patterns of micro/nanoscale on various materials like metals (Long et al. 2015a, b; Wu et al. 2009), semiconductors (Baldacchini et al. 2006), glasses (Ahsan et al. 2013), silicon (Yong et al. 2013; Zhang et al. 2012), and polymer (De Marco et al. 2012). The important features of surface texturing using laser machining is that it is a non-contact method, causes minimal distortion of bulk material, precise, and have the flexibility to prepare complex patterns by integrating it with a digital control three-dimensional translation stages.

The nanostructures formed on the surface by local ablation of the material using a laser beam is termed as laser-induced periodic surface structures (LIPSS). The material heated by low energy density laser (which is above the ablation threshold, known as gentle ablation) evaporate or sublimate the material, and forms smooth nano-rippled structures (Moradi et al. 2013; Jiang and Tsai 2003). However, high energy density laser (below melting threshold) converts the material into plasma forms, dominated by thermal vaporization (called as strong ablation) (Mannion et al. 2004). The depth of material removed is typically hundreds of nanometer in this region, forming rough nano-rippled structures (Mannion et al. 2004; Jiang and Tsai 2003). The ablation rate of the material and hence the diameter and height of the developed microstructures depend on the incident laser fluence, type of incident beam, number of repetitions, and of course on the material used (Moradi et al. 2013). An increase in laser fluence, scanning speed, and laser pulses per unit area increase the diameter and height of the micro/nanopillars formed on the material surface (Jagdheesh et al. 2011). Moreover, high fluence at the center of the Gaussian beam (usually used in laser micro/nanotexturing) results in a higher material removal rate at the center of the ripples compared to its edges. However, amplitude and periodicity of ripples remain equal or less than the wavelength of the laser beam, maybe because of interference and scattering of incident laser radiation (Bauerle 1996).

It is observed that the wettability of the laser textured surface changes with time when exposed to air (Long et al. 2015a, b; Kietzig et al. 2009). Several mechanisms such as creation of hydrophobic groups due to adsorbed atmospheric organic matter, decomposition of carbon-di-oxide into carbon with active magnetite (Long et al. 2015a, b; Bizi-bandoki et al. 2013), and partial de-oxidation of material oxide (e.g., CuO and ZnO; formed due to oxidation of copper and zinc in atmospheric oxygen during laser ablation) into Cu_2O have been suggested in the literature (Ta et al. 2015). Therefore, the range of wettability is observed over laser-ablated surfaces over time, and this effect is extremely reproducible.

In this technique, micro/nanostructures on the copper surface are developed using a solid-state pulsed ytterbium fiber laser. Initially, designing and drawing of a required array to be machined on the substrate (pillar, 50×50 μm; channel width and depth, 50 and 53 μm) were drawn. A cleaned copper substrate using ethanol followed by a sonication bath for 20 min was textured using high speed and

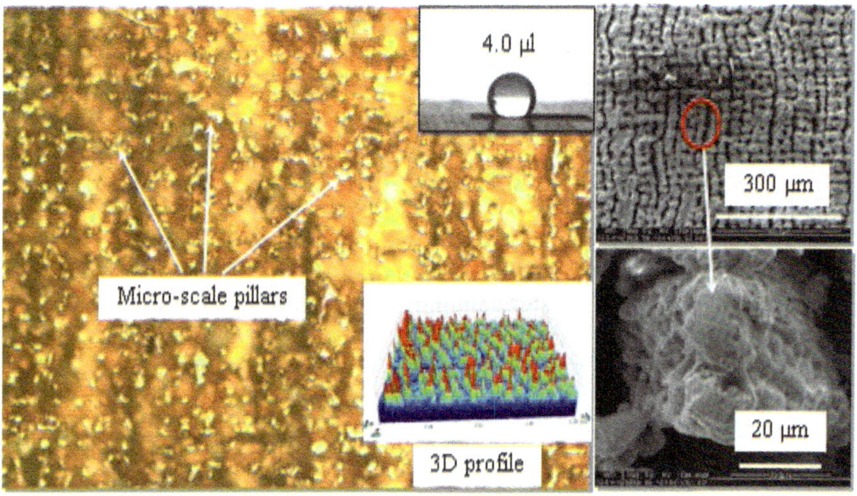

Fig. 12.8 (**a**) Optical image and 3D profile (2–3 µm narrow pillars) of the laser textured copper substrate. Inset shows 4 µL water drop on the textured copper substrate having contact angle 135°. (**b**) FESEM images of the substrate showing surface grain-like structures

precision laser (21 W, 6.8 mm/s, double run) with the drawn array as input (Gogna 2016). Subsequently, unwanted debris is removed using high-pressure air followed by a sonication bath in ethanol for 20 min.

Digital and FESEM images of the physical textured copper surface are shown in Fig. 12.8. A digital image and a three-dimensional profile of textured surface having 2–3 µm pillars are shown in Fig. 12.8a, and FESEM images in Fig. 12.8b show developed grain-like structures on the textured surface. It is observed that the contact angle increased from 30° to 40° on the untreated copper surface to 135° (inset of Fig. 12.8a) after the texturing. Such a high value of contact angle is sufficient to appreciate condensation in dropwise mode over the engineered surface.

12.3.5 Ion Implantation

Another important method of controlling wettability is by ion beam irradiation of surfaces, which has become an emerging field of research. Uniform ion irradiation having energies ranging from few eV to MeV on solid surfaces can cause pattern formation, smoothing (Anzenberg et al. 2012; Madi et al. 2008), and can modify the electrical (Khakshouri and Duffy 2009; Alkemade 2006; Hansen et al. 2006; Ziberi et al. 2005) and optical properties (Martin et al. 2005; Schmuki et al. 1998) of materials. Previous studies in our laboratory also showed changes in electrical sheet resistance and Debye temperature by low energy ion beam irradiation (Chowdhury and Bhattacharjee 2013; Chowdhury and Bhattacharjee 2011). However, the research on the application of ion beams for controlling the wettability of materials is relatively new, and not much work has been done.

Keller et al. (2011) were among the pioneer researchers in this direction. They presented a method of tuning hydrophobicity of atomically flat mica surfaces by 100 eV Ar^+ irradiation. Due to extremely low energies, only negligible roughening of the surface was observed, and the chemical composition remained intact. However, ion irradiation caused preferential removal of the outermost K^+ ions from the surface, leading to exposure of the underlying aluminosilicate sheets. The irradiated surface thus exhibited an enhanced chemical reactivity towards hydrocarbons, resulting in the absorption of thin hydrocarbon film from the environment.

Kumar et al. (2014) performed 200 keV Ar^+ oblique (60° with respect to the surface normal) ion beam irradiation on Si (100) surfaces. The authors observed that ripple patterns oriented perpendicular to the direction of the ion beam is generated as the beam fluence varies from 3×10^{17} to 3×10^{18} ions/cm^2. The calculated value of RMS roughness increases exponentially with beam fluence, but the wavelength of the ripples stays almost constant. The contact angle of water on Si was found to decrease from 76° to 50° due to an increase in surface roughness. It is as per the Wenzel law (Wenzel 1936), which states that an initially hydrophilic surface becomes more hydrophilic with an increase in surface roughness.

The studies of the wettability of the ripple patterned Si surface was investigated by Garg et al. (2016), where they performed 60 keV Ar^+ irradiation on Si at two irradiation angles, namely 0° and 60°. They found that high energy ion irradiation causes amorphization of the topmost Si surface. This amorphous Si layer has a lower surface energy (1050 mJ/m^2) than crystalline Si (2512 mJ/m^2). The reduction of surface energy caused a transition from hydrophilic to hydrophobic nature of the surface. For the case of a 60° irradiation angle, ripple patterns are developed. The observed behavior was attributed to surface anisotropy and inhomogeneous distribution of implanted Ar atoms near the surface.

Singh et al. (2016a, b) performed 100 MeV Ag ion irradiation on thin films of Ag deposited on a quartz substrate. The contact angle of water on fused quartz and Ag is 62° (in the hydrophilic regime) and 95° (in the hydrophilic-hydrophilic transition regime), respectively. Irradiation with a low dose (1×10^{13} ions/cm^2) leads to the development of rough Ag surfaces that enhance its hydrophobicity according to Wenzel law. But at higher fluences of these swift heavy ion irradiation (1×10^{14} ions/cm^2), sputtering dominates due to high energy ion bombardment that leads to decrease the coverage area of Ag and causes exposure of underlying quartz substrate. This process can create a chemically inhomogeneous surface (a two-component-Ag/Quartz system) and be consistent with the macroscopic Cassie-Baxter law (Cassie and Baxter 1944).

The dynamic wettability of nanostructured surfaces was studied by Ramos et al. (2003). Single crystals of $LiNbO_3$ were bombarded by 500 MeV ^{208}Pb ions, and randomly distributed defects were created. The defect concentration was found to be proportional to the irradiation fluence. The average contact angle was found to be constant for these nano-tailored surfaces. It was observed that for defect density lower than a critical value, the contact angle hysteresis increases linearly with defect concentration, indicating that the defects pin the contact line individually, which supports the theoretical prediction by Joanny and de Gennes (1984).

The above studies were mostly done with high energy ion beams, and thus, the tuneable wettability was done primarily by surface nanostructuring and chemisorption. But low energy ion beams of energy <1 keV does not induce appreciable roughness on hard metallic surfaces. Their interaction with matter remains confined within some subsurface atomic layers (~50 nm). Inert gas molecules can be implanted in the subsurface region of the irradiated samples (Chowdhury and Bhattacharjee 2013). Due to their inert nature, these atoms do not make any chemical bond with the host atoms, and thus, the metallic properties of the irradiated samples remain intact. However, they can alter the wettability. It is well known that metals are characterized by high surface energy, and the dispersive component of the surface energy dominates wettability (Cognard 1984; Schrader 1984). As inert gas molecules have very low dispersive surface energy, their presence near the surface can reduce the surface free energy by an appreciable amount. These implanted impurities stay inside the metallic network in the atomic form without forming any macroscopic patch work of different chemical composition at the surface. Due to thier inert nature, chemical compunds are not formed by interaction with the host atoms. In this manner, low energy ion bombardment causes an atomically heterogeneous system that improves non-wetting properties of the surface and helps in realizing the dropwise mode of condensation over the solid surface.

In this technique, thin metal films of copper, aluminum (coated on a glass substrate), and gold (coated over a silicon substrate) are uniformly irradiated using beams of 0.5 keV Ar^+ with a fluence of ion beam ranging 1.2×10^{15} to 5.2×10^{16} ions/cm^2. The thickness of thin metal films is reduced to ~140–175 nm (measured by a stylus profilometer) from initially ~200 nm after the uniform ion irradiation at the highest fluence (Chatterjee et al. 2017). Representative images of the untreated and ion-implanted gold film (captured using Atomic Force Microscopy (AFM)) are shown in Fig. 12.9a, b, respectively. In the images, variations in RMS roughness values of untreated and treated surfaces are shown. Besides, the static and the dynamic contact angle of deionized water on the irradiated films is measured (Fig. 12.9c, d, respectively). It is clearly seen that the increased surface roughness or fluence of irradiated Ar^+ ions improves both, the static and the dynamic contact angle.

12.4 Closure

Dropwise mode of condensation is realized on surfaces with low free surface energy or non-wetting surfaces. Theoretically, the wettability of surfaces can be altered in two ways: (a) coating lower surface energy materials, and (b) micro/nanostructuring on the surface. In practical applications, a combination of (a) and (b) approaches are adopted to develop such surfaces. Five techniques, which employ the above approaches and used in our laboratory, i.e., glass silanization, chemical texturing,

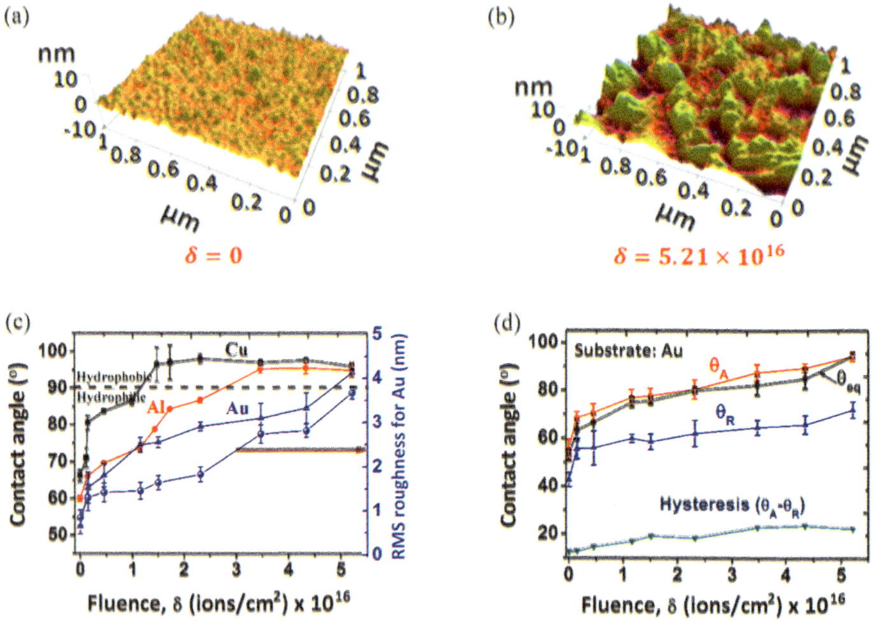

Fig. 12.9 (**a**, **b**) Present surface roughness, measured using atomic force microscopy, on pristine and ion-irradiated gold (Au) films, respectively. Variation of static and dynamic contact angle of deionized water with the fluence of Ar^+ irradiations is shown in (**c**) and (**d**), respectively (Chatterjee et al. 2017)

anodization of aluminum, laser machining, and ion implantation, are delineated in this chapter. In all the techniques, the developed surfaces have shown non-wetting properties or higher contact angles, and therefore, are suitable for practicing dropwise condensation.

Chapter 13
Measurement of Condensation Heat Transfer

Mahesh K. Yadav, Maneesh Punetha, Abhinav Bhanawat, Sameer Khandekar, and K. Muralidhar

Nomenclature

$f(t)$	Unknown transient wall heat flux, W/m^2
J	Sensitivity matrix, –
k	Thermal conductivity, W/mK
L	Thickness of the medium, m
N	Number of sine or cosine terms, –
P	Parameters to be estimated, –
q	Heat flux, W/m^2
S	Sum of squared errors, –
t	Time, s
T	Temperature, K
x	Local length, m
Y	Experimental temperatures, K

Greek Symbols

α	Thermal diffusivity, m^2/s
θ	Contact angle, °
ξ	Perturbation, –
μ	Damping parameter, –
ω	Angular frequency, Hz
Ω	Diagonal matrix, –

S. Khandekar, K. Muralidhar, *Drop Dynamics and Dropwise Condensation on Textured Surfaces*, Mechanical Engineering Series,
https://doi.org/10.1007/978-3-030-48461-3_13

Subscripts and Superscripts

0	Condition at time $t = 0$
a	Advancing angle
i, j, k	Summation/iteration indices
m	Heat conduction material
r	Receding angle
s	Sensor material
sat	Saturation condition
x	Length variables

Abbreviations

HFS	Heat flux sensor
HTC	Heat transfer coefficient
IHC	Inverse heat conduction

13.1 Introduction

Condensation in the form of drops are highly sought in engineering applications as high heat transfer are realizable at exceptionally low temperature difference. In engineering systems, condensation on the solid surface is of specific interest. A significant advancement in understanding of the surface condensation at micro- and nanoscale in recent time (Enright et al. 2014), and the development of various mathematical/numerical models also supports implementing dropwise condensation in real systems. However, actual implementation of the dropwise condensation in any thermal system will require long-term sustainable operation as well as verifiable knowledge of heat transfer bounds in the range of desired operating parameters. Various techniques adopted to promote dropwise condensation are already discussed at length in the previous chapter. In this chapter, we will focus on identifying/developing suitable instrumentation for estimating the steady-state/transient condensation flux. A real challenge lies in estimating the transient condensation flux in a non-intrusive manner and evaluating the low temperature gradients, while minimizing the uncertainty in the measurement in actual experiments. Additionally, the dependency of condensation processes on surface morphology makes the overall process highly complex. Surface morphology, including physical and chemical texturing strongly affects the number of nucleation sites, contact line motion, local pinning dynamics, precursor layer interactions, and wettability transitions. The presence of non-condensable gases, which is mostly the case in practical applications, further complicate the analysis (Punetha and Khandekar 2017; Yadav et al. 2016; Agrawal et al. 2015). Experimental studies become essential to appreciate complexity of the process, support developing suitable models for predicting the condensation flux, and much needed for the validation of numerical models. Details on numerical modeling, associated difficulties, and necessity of experimental data for validation is already provided in Part A and Part B, and available in (Xie et al. 2018; Qi et al. 2015; Sikarwar et al. 2012, 2013a, b). In this chapter, heat transfer characteristics under different orientation of the surface resulting in different modes of dropwise condensation, i.e., sessile mode, pendant mode, and condensation at different orientation of the test surfaces will be studied through dedicated experiments. As thermo-hydrodynamics of condensation heat transfer are closely woven, hydrodynamics of the drop condensation is also presented via digital imaging for a comprehensive understanding. Additionally, surface treated to promote dropwise condensation generally loses their properties over time and end up with endorsing filmwise condensation heat transfer. This puts an additional requirement on the measurement instrument used for dropwise condensation heat transfer to have capabilities/applicability range to estimate filmwise condensation heat transfer. Accordingly, the instrumentation suggested in this chapter is also verified to estimate the heat transfer in filmwise condensation. Overall, readers will be introduced to the complexity involved in the measurement of condensation heat transfer, and experimental protocol to follow for truthful estimation through controlled experiments.

In this section, the basics of filmwise and dropwise condensation heat transfer, and difficulties in the measurement of the condensation fluxes are discussed. Thereafter, use of conventional heat flux sensors (HFSs) in estimating the condensation heat transfer is scrutinized. Later on, a fully noninvasive inverse heat conduction (IHC) measurement technique, based on measured temperature data at internal locations inside the test surface is proposed. Finally, applicability of IHC model to filmwise and dropwise condensation is shown through two experimental case studies. Dropwise condensation experiments at different inclination angles of the test surface are performed and inferences emanating from this study is provided. The difficulties associated with temperature and condensation heat transfer measurements, and experimental philosophy to minimize intrusion in the measurement data is highlighted throughout the chapter.

13.2 Condensation Modes: Dropwise and Filmwise

Dropwise and filmwise mode of condensation are significantly different in the way they are realized over the surface. A thin film of condensate gets formed over the entire test surface in case of filmwise mode, whereas dropwise condensation occurs in a quasi-cyclic fashion. This way, the dropwise condensation process is dynamic in nature and generally renders order of magnitude higher Heat Transfer Coefficient (HTC) than the filmwise condensation (Khandekar and Muralidhar 2013). Out of the two, which mode of condensation is going to prevail mainly depends on the wettability of the surface. Wettability can be seen as spreading or contracting characteristics of the fluid with respect to the solid substrate. It is governed by

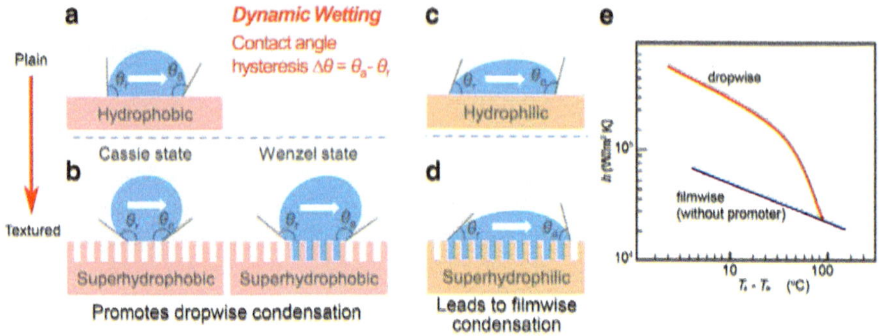

Fig. 13.1 (**a–d**) Show different wetting characteristics of plain and textured surfaces. Typical values of heat flux during the dropwise and the filmwise mode of condensation is compared in (**e**). ((**a–d**) Are reprinted from Wen et al. (2018))

competing forces at the three-phase contact line, namely surface tension at the liquid surface, the free surface energy of the solid surface, and interaction with the surrounding. The force balance is given by Young's equation (Israelachvili 2011; Carey 2008; de Gennes et al. 2004). By tuning these forces, one can control and achieve the desired condensation pattern. In general, the working fluid (vapor in the present study) and the atmospheric conditions are application specific, and largely fixed. It is the free surface energy of the solid surface, which is mainly altered to achieve the desired mode of the condensation. Generally, the adopted methodology for this is to either provide a thin-film coating on the surface of some suitable/desired low surface energy material or micro/nanoscale texturing on the surface of interest (Kim et al. 2018; Wen et al. 2018; Bisetto et al. 2014). How these altered surfaces can lead to different condensation rates can be understood through Fig. 13.1. Dynamic wetting characteristics on plain and textured surfaces are shown in the figure. In situations where the surface tension force of the condensate liquid is dominant over the free surface energy of the surface, condensed liquid tends to shrink and draw together in a nearly spherical droplet shape (Cho et al. 2016; Yan et al. 2011). This is a typical situation where the contact angle is greater than 90° (Fig. 13.1a, b), and a dropwise mode of condensation is usually realized. In contrast, cases where the free surface energy of the solid subjugates the surface tension force of the liquid, condensation heat transfer via filmwise mode becomes dominant (see Fig. 13.1c, d, Wen et al. 2018). As the condensate liquid moves over the surface, the pinning effect on the surface may lead to different contact angles at advancing and receding fronts, known as advancing and receding contact angles (Kim et al. 2018; Somwanshi et al. 2018; Israelachvili 2011). These are depicted as "θ_a and θ_r," in Fig. 13.1. Exploration of condensation phenomena becomes more complex for condensation over a textured surface where two states, Wenzel and Cassie (Yan et al. 2011), are encountered depending on whether the liquid penetrates through the micro-textures on the condensing surface or not. This is shown in Fig. 13.1b for condensation on a superhydrophobic surface. Contrary to this, only a Wenzel state is feasible over a textured hydrophilic surface, as noted in Fig. 13.1d. It is emphasized that the different hydrodynamics characteristics are feasible for the condensate solely by altering the surface properties. This also results in order of magnitude difference in the heat transfer rates for dropwise and filmwise condensation (particularly for small values of wall-subcooling) (Khandekar and Muralidhar 2013), as evident from Fig. 13.1e.

13.3 Thermo-Hydrodynamics of Condensation Heat Transfer

Heat transfer rates during dropwise and filmwise condensation are significantly different, as noted earlier. To convincingly capture the heat transfer characteristics, it is essential to understand the subtle nuances of thermo-hydrodynamics during the condensation process in the dropwise and the filmwise modes. This is shown through Fig. 13.2, where the heat transfer mechanism in dropwise and filmwise condensation is shown. Dropwise condensation occurs in a cyclic fashion, as shown in Fig. 13.2a, b for condensation on a plane vertical and underneath the horizontal surfaces, respectively. A dropwise condensation cycle, in general, consists of four sub-processes: nucleation, growth, coalescence, sliding, or fall-off (Sikarwar et al. 2011, 2012). Recently, droplets jumping away from the test surface after the coalescence, as shown in Fig. 13.2a, have also been explored (Oh et al. 2017). As drops sit on the test surface and add resistance to subsequent condensation, hydrodynamics of the droplets can provide substantial insight into the heat transfer rates. However, different scales of droplet sizes over the course of a condensation cycle, typically from some nanometers at the nucleation, to the range of millimeters during the slide-off process for condensation on a vertical surface (Fig. 13.2a), makes the analysis complex. A similar quasi-cyclic process is followed for condensation underneath a test surface, with the main difference being gravity force pulling the drops away from the test surface in this configuration, affecting their stability criteria. This leads to comparatively larger sized drops at fall-off or slide-off situations, as compared to condensation on the vertical surface, as is visible from Fig. 13.2a, b (typically 1.5–2.5 mm to 5–10 mm for condensation on vertical and horizontal surfaces, respectively (Sikarwar 2013)). In this way, the cycle time as well as the condensation rate also depends on the orientation of the surface. A typical range of the cycle time for condensation on the vertical and the horizontal surfaces are in the range of few seconds and some tenths of seconds, respectively. Only quasi-steady-state is achievable in case of dropwise condensation; hence, transient measurements are highly desirable. As a result, a measurement system based on the assumption of steady-state heat transfer is incapable of appreciating the inherent transience of the problem and is unsuitable for measurement. A reliable measurement system must have response time faster than the cycle time to truly capture the quasi-cyclic variation in heat fluxes. In fact, quick response time is preferred to distinctly quantify heat transfer rates at each sub-process. A measurement sensor/

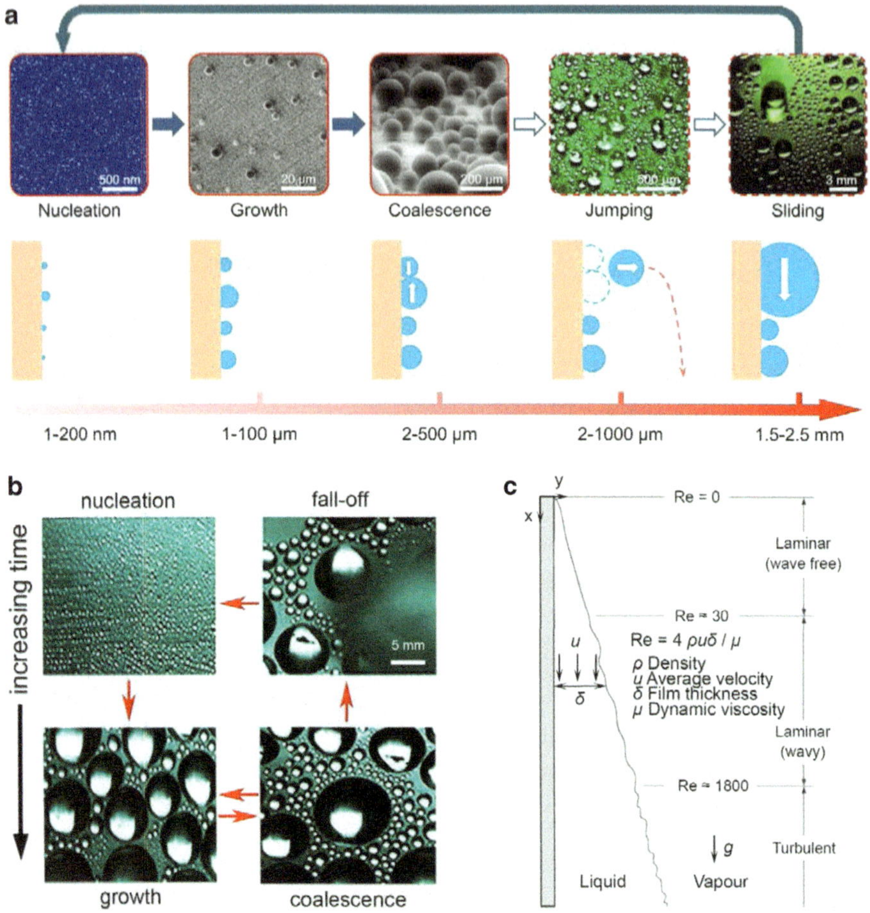

Fig. 13.2 Dropwise condensation cycle and multiscale characteristics for vapor condensation on (**a**) a vertical plane surface, and (**b**) underneath a horizontal surface (pendant mode of condensation). (**c**) Shows the filmwise condensation over a vertical surface and three regimes of flow inside the liquid film. (Reprinted from (**a**) Wen et al. (2018), and (**b**) Sikarwar (2013))

system which has minimum response delay and phase-lag is therefore favored for such heat transfer problems.

The thermo-hydrodynamics of heat transfer during filmwise condensation is quite distinct from the dropwise case. A thin liquid film gets formed over the surface during the filmwise mode of condensation. This film is continuously removed from the surface by gravity force. Depending on the film Reynolds number, generally three different flow regimes are realized (Cengel and Ghajar 2017). These include wave-free laminar, wavy laminar, and turbulent, as shown in Fig. 13.2e. It can be seen from the figure that the film thickness keeps on increasing as one moves downward on the surface, and as a result, resistance to the heat transfer also increases. Contrary to this, waves and ripples developed over the liquid-vapor interface results in a chaotic movement which eventually end up in promoting the heat transfer (Swartz and Yao 2017). In this way, condensate film which forms over the surface, particularly over a long surface, is highly unordered, and as a consequence, spatio-temporal variation in heat flux is experienced over the entire surface (Collier and Thome 1994; Stephan 1992). Similar to the dropwise condensation, a non-intrusive measurement is needed in this case as well, to avoid alteration in the heat transfer due to alteration of the surface morphology. The difference being that dynamics of the heat transfer in filmwise condensation is mainly governed by waviness on the condensate film rather than the condensation cycle time in the dropwise mode.

From the above discussion, it becomes clear that condensation phenomena are considerably dependent on the surface texturing and the free surface energy of the solid material. Therefore, any physical sensor which requires mounting at the surface of interest, i.e., the surface where condensation is going to occur, must have coherent surface properties (surface morphology, and thermal properties) for realistic measurements of the ensuing heat flux. Failure to do so will result in unaccounted measurement errors. In fact, the very nature of the condensation process may get fundamentally altered at the local level. Therefore, a measurement system, which do not alter the surface characteristics, is particularly sought in the condensation process. A dynamic and cyclic/quasi-cyclic nature of the dropwise condensation process also requires a faster response system for transient measurements. The complications involved in the condensation heat flux measurement are highlighted and how to accurately estimate the heat flux is presented. In this process, use of commercially available heat flux sensors, and inverse heat conduction technique based on temperature inversion are explored (described in next sections). Finally, suitability of the adopted instrumentation in actual thermal systems is enforced through a real-time experimental study. Experiments are performed for filmwise and dropwise mode of condensation, and thermo-hydrodynamics of the process is elucidated through recorded heat transfer data and digital images.

13.4 Limitations of Conventional Heat Flux Sensors

The most common way of measuring heat flux is by attaching an HFS at the point of interest. Many methods are available to determine heat flux under different conditions, which are already detailed (Childs et al. 1999; Diller 1993). One of the commonly used conventional HFS, a differential type sensor, is picked to scrutinize its suitability in estimating the condensation flux. In a differential type HFS, the measured temperature gradient across a pair of thermocouples embedded inside a solid is correlated with the heat flux (Diller 1993). A single unit of thermocouple junction in a thermopile HFS is shown in Fig. 13.3a. In general, the output of a single thermocouple junction is in the range of a few millivolts. Hence, a number of such thermocouple junctions are arranged in series in an actual thermopile HFS to improve its sensitivity (Fig. 13.3b). A typical thermopile HFS package is composed

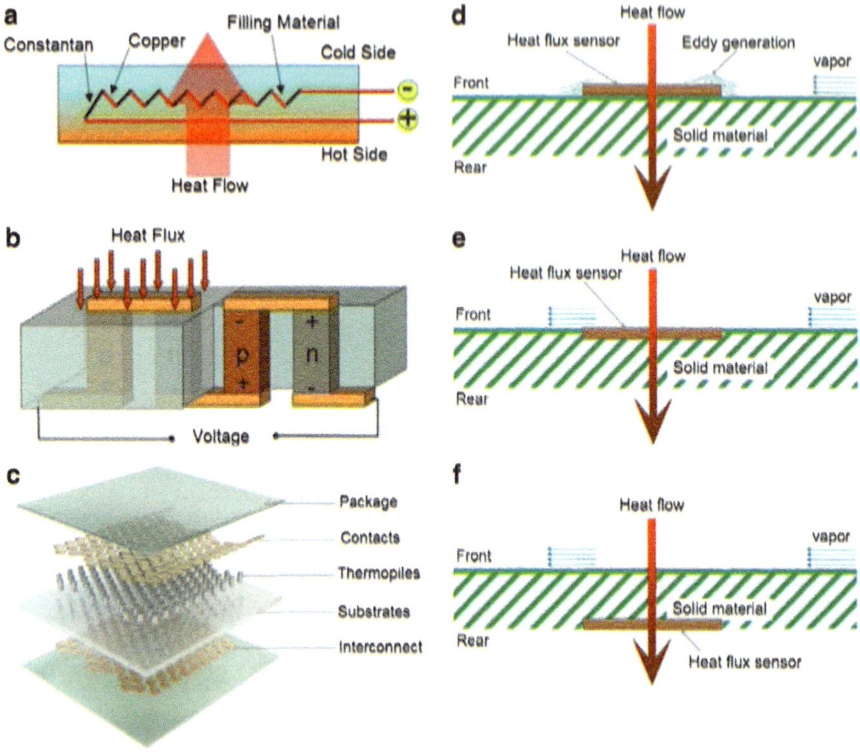

Fig. 13.3 (**a**) A single unit of thermocouple junction in a thermopile HFS, (**b**) multiple thermocouple junction units arranged in series to improve sensitivity (output voltage per unit heat flux) of a thermopile HFS, (**c**) packaging of a typical thermopile HFS. (Adapted from http://www.greenteg.com and http://www.azosensors.com/). (**d**, **e**) Show non-flush and flush-mounted HFS, respectively, at the front side of the solid material (the surface where condensation is envisaged), whereas (**f**) shows flush-mounted HFS at the rear side of the solid material. ((**a–c**) Are reprinted from Singh et al. (2016a, b))

of a layer of thermopiles sandwiched between the substrates. This is depicted in Fig. 13.3c, where different layers inside a thermopile HFS are shown. With the advancement in the capabilities of microscale fabrication, the sensors are continuously miniaturized and intrusion due to the mounting of the sensors is reduced. As a result, the use of these sensors is continuously growing. The representative mounting arrangements of the sensor are shown in Fig. 13.3d–f. Two types of arrangements are possible: non-flush and flush-mounted HFS. A non-flush mounting arrangement, shown in Fig. 13.3d, considerably disturbs the velocity and thermal boundary layer and thereby largely alters the heat transfer (Childs et al. 1999). In convective conditions, generation of eddies and highly turbulent behavior near the edges of the sensor makes it extremely difficult to deduce the correct heat transfer rate. As a result, this mounting arrangement is not appropriate for heat flux measurements in a convective environment for any heat transfer application, let aside the surface texture-dependent condensation flux. Non-flush mounting arrangements, at the front side (the surface where condensation is envisaged) and at the rear side of the surface are shown in Fig. 13.3e and Fig. 13.3f, respectively. Whether these mounting arrangements of the sensors can serve as a potential measurement system for condensation heat transfer or not is examined through a numerical case study. A one-dimensional heat transfer is assumed through a solid substrate and appropriateness of the sensor in estimating the heat flux is discussed in this section. The emphasis here is to understand the nuances of mounting a thermopile HFS with respect to errors in the subsequent heat flux measurement. The representative values of thermal and geometrical properties (for the heat conducting material and the sensor) and boundary conditions in the numerical simulation are selected to genuinely demonstrate the measurement capabilities of a thermopile-based HFS for condensation application. For example, one-dimensional heat conduction is considered in a heat conducting medium having thermal diffusivity seven times larger than the HFS material (corresponding to diffusivities ratios of stainless steel to a typical HFS material). A constant heat transfer coefficient condition ($h = 4$ kW/m^{2}K, $T = 373$ K) at one boundary and a constant temperature condition ($T = 298$ K) on the other boundary is considered in present study, which is typical of a case undergoing condensation heat transfer.

Transient heat flux recorded for the case of a heat flux monitor (a numerical representative of the actual HFS) at the front and the rear surfaces are shown in Fig. 13.4a, b, respectively. The true value of heat fluxes at the front surface (case when no heat flux sensor is mounted on the surface) is compared with the heat flux recorded from a heat flux monitor mounted at the front and the rear surfaces, and one estimated from the Fourier law of heat conduction. The true value of the temperatures recorded from the two temperature monitors (mounted at known locations inside the solid medium) is used to obtain the Fourier heat flux. It is seen from Fig. 13.4a, b that mounting a sensor in itself reduces the heat transfer through the solid medium. Isotherms of the two cases also show localized heating and cooling at the front and rear side of the sensor, respectively. Although transient heat flux estimated from the sensor mounted at the front surface has a bias error, the overall trend is appropriately followed. Therefore, proper calibration of the sensor mounted at the front surface can correctly represent the ensuing transient heat flux (Singh et al. 2016a, b). Furthermore, a stationary underestimation or bias error in the heat flux is found at steady-state due to added thermal resistance by the mounting of the sensor. Contrary to this, an inherent systematic delay in the measurement of rear mounted HFS makes it unsuitable for transient measurements (Fig. 13.4b). Similarly, Fourier law of heat conduction is also unable to delineate the time dependence of the problem. In summary, the front mounted sensor is suitable for transient measurement of heat flux only after a proper calibration whereas the rear mounted sensor and the Fourier law of heat conduction can only be used for steady-state measurements with proper error analysis.

How steady-state heat flux at the front surface gets affected due to mounting a sensor on commonly used heat conducting materials is shown in Fig. 13.4c. A large deviation (error) found in surface heat flux indicates that the thermal impedance by

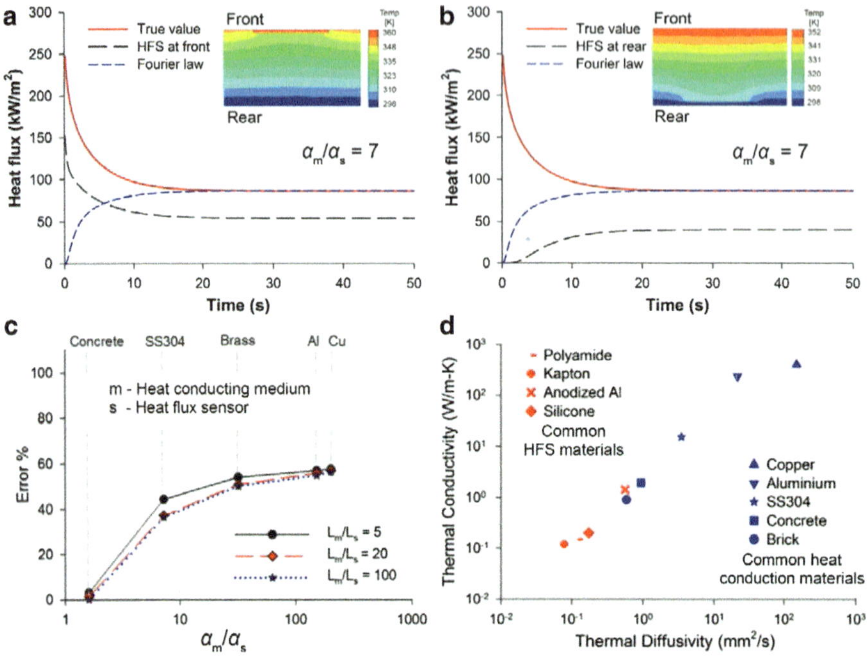

Fig. 13.4 Transient heat flux for (**a**) HFS mounted at the front surface and (**b**) HFS mounted at the rear surface. Respective isotherms are also shown in the figures (**a**) and (**b**) for clear depiction. (**c**) Shows the error percentage in measured heat fluxes using an uncalibrated HFS mounted at the front surface for heat conduction through commonly used materials in the engineering applications. In (**d**), relative comparisons of thermal properties of commonly used HFS and heat conducting materials at 25 °C are shown. ((**c**, **d**) Are reprinted from Singh et al. (2016a, b))

the sensor is considerably larger than the heat conducting material of similar thickness (Singh et al. 2016a, b). This can be explained in Fig. 13.4d where thermal conductivities and diffusivities of commonly used heat conduction material and HFS material are compared. The thermal properties of heat conducting material generally used in engineering applications are shown by blue symbols, whereas the properties of HFS materials are shown using red symbols. As can be seen, an order of magnitude difference exists between the thermal diffusivities of the two materials (higher diffusivity ratios of heat conducting material to the sensor material, i.e., α_m/α_s). Therefore, one must be careful while measuring heat flux using a sensor which has significantly different thermal properties than the heat conduction medium. The heat conduction material and the sensor material should be chosen such that their thermal properties match as closely as possible to reduce the intrusion due to the mounting of the sensor. The error percentage in estimated heat flux for different ratio of thickness of the heat conducting material to the sensor material (L_m/L_s) is also shown in Fig. 13.4c. The error percentage does not change much with a change in relative thickness of the heat conducting and the sensor material, and predominantly depends on the ratio of thermal diffusivities.

In summary, the sensor mounted at the front surface, although suitable for transient measurement, largely affects the heat flux estimation. Furthermore, it also alters the surface morphology at the location of interest and hence unsuited for measuring heat flux of a surface texture-dependent phenomenon such as surface condensation. Fourier law of heat conduction, although it gives correct results at steady-state condition, is erroneous in predicting the transient heat flux. This erroneous measurement is a manifestation of diffusional lag in the temperature measurement. If, by some mechanism, this diffusional lag can be accounted for, we can very well predict the transient heat flux through temperature data from two thermocouples embedded at known locations. This is discussed next, where inversion of temperature using IHC technique is adopted to evaluate the condensation flux.

13.5 Inverse Heat Conduction (IHC) Technique

The inverse heat conduction (IHC) technique is based on an estimation of a cause (unknown quantities) with the knowledge of its effect. In the context of heat transfer problems, the cause can be an unknown boundary condition or thermophysical properties of the medium or a combination of these, whereas effect can be a temperature distribution inside the medium or heat conduction rate (Beck et al. 1985). Our interest here is to estimate the condensation heat flux at the surface, a cause, based on its effect on temperature at an internal location (recorded using micro-thermocouples). The idea is to obtain the heat flux at the solid surface where condensation is occurring such that temporal temperature distribution matches with the recorded values at a particular location(s) in the solid medium. This way, it is an optimization problem to find the correct surface heat flux by inverting the recorded temperature data (Özisik and Orlande 2000). Obviously, the estimation accuracy will depend on trueness of the recorded temperature data and the optimization algorithm. Most of the IHC problems are highly sensitive to measurement error and generally ill-posed in nature. Therefore, it is essential to have a precise measurement system for recording the temperature data, and also a robust solution algorithm for careful estimation of the desired unknown quantities. These aspects are discussed herewith.

13.5.1 Measurement of Temperature

Measurement of the temperature record using thermocouples is one of the simplest and most commonly used techniques. A thermocouple is a small bead/junction of dissimilar metals and requires mounting at the same location where the temperature data need to be quantified. It works on the principal of the Seebeck effect, in which a temperature difference at the junction of two dissimilar metals produces a voltage difference. Mostly, the voltage difference is in the range of some microvolts to few millivolts. A calibration equation is then used to estimate the actual value of temperature. Thermocouples, which need mounting at the measurement points/locations and generate only a small voltage across the junction, must be used in accordance with best practice guidelines (ASTM Committee E20 1993). Particularly in IHC applications, where a small deviation in the measurement data can lead to erroneous results, accuracy becomes much more critical. Therefore, good practices which are followed during the measurement of the temperature data from the embedded thermocouples are highlighted here in the following points.

(a) Typically, voltage signal output from thermocouples could be a fraction of an mV and use of shielded cable is highly recommended to avoid noise pick-up from the external electric field. A low-pass filter should be used to remove any out-of-phase signal. It is always a good practice to have proper grounding for the measurement junction.
(b) Choosing the right type of thermocouple as per the requirement is a must for any measurement. In the present condensation studies where the temperature is supposed to vary between 20 °C and 150 °C, a K-type thermocouple is chosen. A fast response type measurement system is also necessitated for condensation applications to faithfully capture transience of the problem, particularly for dropwise condensation where heat flux may vary in a cyclic fashion. To minimize inertia of the thermocouple, a micro-thermocouple (bead diameter of 76.2 μm; sheath diameter of 1 mm; Omega®) is preferred. Assuming no contact resistance between the solid medium and the bead, it will result in $\approx$ 5 ms of response time at 20 °C for this thermocouple (Singh et al. 2017). This time constant is good enough to faithfully record the rapid temperature variation inside the solid medium. Ideally, as per the Nyquist criterion, a measurement frequency must be at least twice the frequency of the signal which one would like to analyze.
(c) Proper calibration of the thermocouple is at the heart of the measurement accuracy. This is carried out against a standard Pt100 resistance thermometer (NIST traceable calibration curve) by performing steady-state measurements using a constant temperature water bath (Julabo® ME-26, temperature stability: $\pm$0.1 °C). Based on the recorded temperature data from the two thermocouples, i.e., K-type and Pt-100 type, a linear calibration fit with $\pm$0.3 °C accuracy is achieved.
(d) In an IHC estimation, precisely knowing the measurement locations of the sensor is highly desirable. Besides, temperature data available from embedded thermocouple closer to the boundary of interest are needed for enhanced sensitivity to the measurement (Beck and Woodbury 2016; Beck et al. 1985). High-Precision electro-discharge machining is used to drill the millimeter size holes of 1 mm with a tolerance of +0.05 to +0.15 mm at a known depth away from the surface of interest for mounting the micro-thermocouples. System level calibration was proposed by Battaglia et al. (2001) and the idea was followed by Löhle et al. (2013). Additionally, in situ calibration for the measurement location through the inverse model itself has been explored by Chen et al. (2018). Although this step is not incorporated in the present study, it is a significant

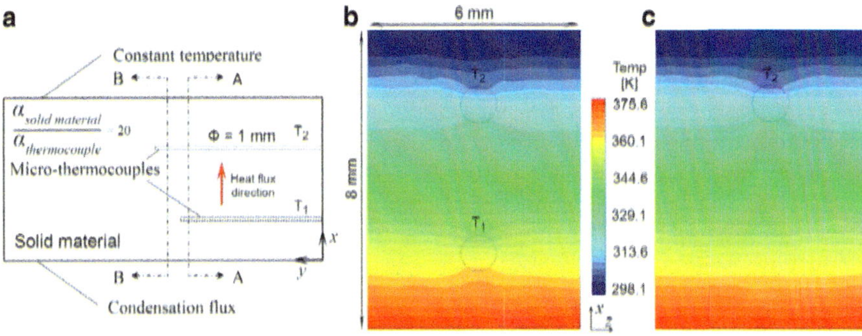

Fig. 13.5 (**a**) Shows a typical mounting arrangement inside the solid medium. Two boundary conditions, condensation flux and constant temperature, are also marked. Isotherms at locations A-A and B-B in figures (**b**) and (**c**), respectively, show intrusion due to the mounting of the thermocouples

advancement to minimize error propagation due to uncertainty in the thermocouple position.

(e) Embedding a thermocouple inside a solid medium intrudes with the heat transfer process and may result in erroneous measurements. This possibility is explained in Fig. 13.5. A typical geometrical configuration along with mounting of two micro-thermocouples (T_1 and T_2) inside the solid medium is shown in Fig. 13.5a. Three-dimensional simulation at steady-state is performed in ANSYS® CFX to analyze this situation. A representative condensation heat flux condition on one side ($q = 120$ kW/m^2) and a constant temperature condition ($T = 298$ K) on the other side is prescribed. The constant sink temperature considered here is noticeably lower than the temperature on the surface where condensation takes place. All other boundaries are taken to be insulated. A representative thermal diffusivity of the base material and effective thermal diffusivity of the thermocouple material, including the sheath are taken to be in the ratio α (base material)/α (thermocouple) = 20.

An enlarged view of isotherms obtained in the x-z plane at two locations A-A and B-B in the y-direction is shown in Fig. 13.5b, c. The locations are chosen such that intrusion caused by mounting of the micro-thermocouples in all the three directions can be seen. Lower thermal diffusivity of the thermocouple material in comparison of the substrate results in a lower heat transfer through the embedded length of the thermocouples. Consequently, marginal accumulation of heat before the thermocouple T_1 and three-dimensional heat transfer nearer to the thermocouple location is observed in Fig. 13.5b. However, these are localized effects and diminish just beyond the thermocouple location, specifically the size of the bead diameter in x- and z-directions. Similarly, no intrusion effect of mounting the thermocouple T_1 is seen in the y-direction at section B-B in Fig. 13.5c. These results are presented for mounting a micro-size thermocouple with one order of magnitude difference in the thermal diffusivities of the test surface and the thermocouple. If thermal diffusivity ratio is larger than the one considered here, or size of the thermocouple is bigger than few millimeters, such a numerical exercise will have to be performed to get an overall idea and range of the thermal field affected by mounting the thermocouple, and accordingly optimize the measurement locations.

(f) Selecting a proper direction of the drill-holes and installation of the thermocouples is important for reliable measurements. It is found that the thermocouples installation parallel to the local isotherms provide accurate estimation of temperature (Fahrni et al. 2018; Attia et al. 2002). It is also known that contact pressure between the heat conducting material and the thermocouple tip affects the measurement accuracy and therefore sufficient contact pressure should be applied for improved measurement accuracy (ASTM Committee E20 1993).

(g) With all the precautions mentioned above, it is necessary to verify recorded temperature from the thermocouples for their trueness in a real application. This is carried out to ensure that the thermocouples work properly after mounting and there is a healthy and mechanically tight contact between the thermocouple bead and the solid material. A thin layer of high thermal conductivity paste is used between the thermocouple bead and the solid material to guarantee a tight contact (Singh et al. 2017). Use of the thermal paste and mounting of the thermocouples can also introduce error in temperature data. Therefore, in situ calibration of the thermocouple under controlled conditions is mandatory before their final use. In situ calibration can be performed by comparing the recorded temperature data from a thermocouple for a standard one-dimensional heat transfer problem (say, by employing a constant heat flux condition on one boundary using a precise heating unit and a constant temperature on the opposite boundary) with theoretical values of temperature at the same location. Such an exercise should be performed under similar conditions as expected during the actual application. If the variation in the recorded and the respective theoretical temperature data is significant, a new calibration equation should be developed to get accurate measurement data during the actual experiment (Löhle et al. 2013).

The above-mentioned procedures have been adopted in the present work to obtain temperature data in the two case studies discussed in the next section. Estimation of the condensation heat flux at the surface using the temperature data from the embedded thermocouple(s) is further discussed.

13.5.2 Mathematical Model

A mathematical model is developed to get the condensation heat flux from the temperature data measured in the interior of the solid substrate. In this model, one-dimensional transient heat conduction in the Cartesian coordinate system is considered. Governing energy equation, boundary conditions, and initial condition considered are as stated below.

Energy equation:

$$\frac{\partial T}{\partial t} = \alpha \frac{\partial^2 T}{\partial x^2}, 0 < x < L_x, t > 0 \tag{13.1}$$

The boundary conditions are specified as follows. At the front surface, time-dependent heat flux occurring due to condensation of vapor is symbolically prescribed. Its exact form is as yet undetermined. Hence,

$$q = -k\frac{\partial T}{\partial x} = f(t) \text{ at } x = 0, t > 0, \tag{13.2}$$

where $f(t)$ is the unknown heat flux released during vapor condensation at $x = 0$.

At an internal location L_x, temperature is measured by the sensor and is prescribed as a dynamically varying condition,

$$T = T_x \text{ at } x = L_x, t > 0, \tag{13.3}$$

where T_x is temperature variation recorded by the thermocouple at location $x = L_x$.

The initial condition within the substrate is

$$T = T_0 \text{ at } t = 0 \text{ over } 0 < x < L_x \tag{13.4}$$

The primary objective here is to estimate the unknown condensation heat flux at $x = 0$, using additional information such as recorded temperature data from the thermocouple at an internal location in the solid medium. Optimization for the unknown condensation flux is performed using the Levenberg-Marquardt algorithm, as described next.

13.5.3 Solution Algorithms

As stated at the start of this section, IHC problems are ill-posed and require a special solution algorithm. A variety of algorithms have been developed for this purpose (Chang et al. 2017). These can be broadly classified as parameter estimation and function estimation. As specificities of the solution algorithms are not obligatory to this chapter, details of various solution algorithms are not discussed. Interested readers may go through the available literature for comprehensive minutiae on the subject (Chang et al. 2017; Özisik and Orlande 2000; Beck et al. 1985). Levenberg-Marquardt algorithm, a parameter estimation approach, is used for inverse estimation in this study, and elaborated here. The algorithm is particularly chosen for its applicability in both linear and nonlinear parameter estimation problems, and better stability for assumed values of the parameters at the start (Cui et al. 2016). In this algorithm, function $f(t)$, as considered in Eq. (13.2), is assumed to be a function of unspecified coefficients and known trial functions. A truncated Fourier series is used for this purpose

$$f(t) = P_0 + \sum_{j=1}^{N} P_{cj} \cos(j\omega t) + \sum_{i=1}^{N} P_{si} \sin(j\omega t), \tag{13.5}$$

where $P_{0,}$ $P_{cj,}$ and P_{si} are constant parameters to be obtained by an optimization approach. Sines and cosines are the trial functions, the total number of sine and cosine terms is 2 N, and ω represents angular frequency.

The optimization algorithm, which is based on minimizing the sum of squares of errors over the measured data, is solved for a particular parameter "P" in Eq. (13.5) using the following expression (Özisik and Orlande 2000):

$$P^{k+1} = P^k + \left[\left(J^k\right)^T J^k + \mu^k \Omega^k\right]^{-1} \left(J^k\right)^T \left[Y - T\left(P^k\right)\right], \tag{13.6}$$

where P^k is any one of the parameters in Eq. (13.5), Y is the temperature obtained from the thermocouple at the interior location, $T(P^k)$ is the estimated temperature at this location obtained by solving the direct problem (Eqs. 13.1–13.5), $J(P^k)$ is a sensitivity matrix, μ^k is a damping parameter and Ω^k is a diagonal matrix (Singh et al. 2017). Quantities $J(P^k)$ and Ω^k are calculated from their definitions

$$J\left(P^k\right) = \frac{\partial T^T\left(P^k\right)}{\partial P^k} \tag{13.7}$$

$$\Omega^k = \mathrm{diag}\left[\left(J^k\right)^T J^k\right] \tag{13.8}$$

In the sensitivity matrix $J(P^k)$, its entries, namely the sensitivity coefficients are evaluated as

$$J_{ij} = \frac{T_i\left(P_1, P_2, \ldots, P_j + \varepsilon P_j, \ldots, P_N\right) - T_i\left(P_1, P_2, \ldots, P_j, \ldots, P_N\right)}{\varepsilon P_j}, \tag{13.9}$$

where ε is a small perturbation in parameter(s) P_j and taken between 10^{-3} and 10^{-5}.

Using Levenberg-Marquardt algorithm, parameters in Eq. (13.5) are evaluated iteratively in two steps: initially, the first parameter P_0 is calculated, and after that, all the remaining parameters are estimated (Yadav et al. 2018). The following stopping criteria are used to derive the optimized values of the parameters

$$S\left(P^{k+1}\right) < \xi_1 \tag{13.10}$$

$$\left|\left(J^k\right)^T\left[Y - T\left(P^k\right)\right]\right| < \xi_2 \tag{13.11}$$

$$\left[P^{k+1} - P^k\right] < \xi_3, \tag{13.12}$$

where ξ_1, ξ_2, and ξ_3 are user defined tolerances.

For details on the Levenberg-Marquardt algorithm, refer to Özisik and Orlande (2000). The inversion methodology adopted here has been extensively validated and benchmarked against numerical as well as experimental data in Yadav et al. (2018) and Singh et al. (2017). Specificities of the inversion model are: (a) use of dynamically varying recorded temperature from a thermocouple as boundary condition (refer to Eq. 13.3) eliminates the requirement of any knowledge of the boundary condition on the front surface exposed to the condensation heat flux, and (b) use of truncated Fourier series as a generalized trial function makes the model suitable for estimating any functional variation in heat flux at the front boundary. A measurement frequency of 1 Hz is used in the reported experiments. A suitable interpolation of temperature data is performed wherever found necessary to satisfy numerical stability criterion in the IHC model. A total of 30 terms of sines and cosines are considered in the truncated Fourier series (Eq. 13.5) and used as the trial functions. The temperature data from thermocouple closer to the unknown boundary (T_1, T_3, and T_5 as shown in Fig. 13.6c is used for inversion (owing to higher measurement sensitivity for data closer to the boundary where one wants to estimate the prevailing thermal condition) whereas data from thermocouples T_2, T_4, and T_6 are used for dynamically varying boundary conditions.

13.6 Case Studies

Two case studies are performed to show the applicability of the IHC model in estimating heat flux in the filmwise and the dropwise modes of condensation. An experimental setup developed for this purpose is shown in Fig. 13.6. A three-dimensional CAD model of the setup, including a two-stage diffuser and a nozzle unit, is shown in Fig. 13.6a. A photo of the adequately insulated and well-instrumented test section installed at Indian Institute of Technology (IIT) Kanpur, India, is shown in Fig. 13.6b (Yadav 2019).

The primary components of the setup include: the test section, a two-stage diffuser and a nozzle sections/unit to maintain proper inlet and outlet conditions, respectively, a high-resolution DSLR camera and a data acquisition system. The test section of the facility is a rectangular box of 300 × 250 × 200 mm. It has three viewing windows (two rectangular and one circular), one on each wall, with a cooling chamber and test plate on the fourth wall. Two test plates, each having an active cooled area of 100 × 100 mm (cooled using an array of impinging water jets with a heat transfer coefficient ≈ 25 kW/m^{2}K and made of stainless steel), are used in the experiments. Different measurement locations on the first test plate are shown

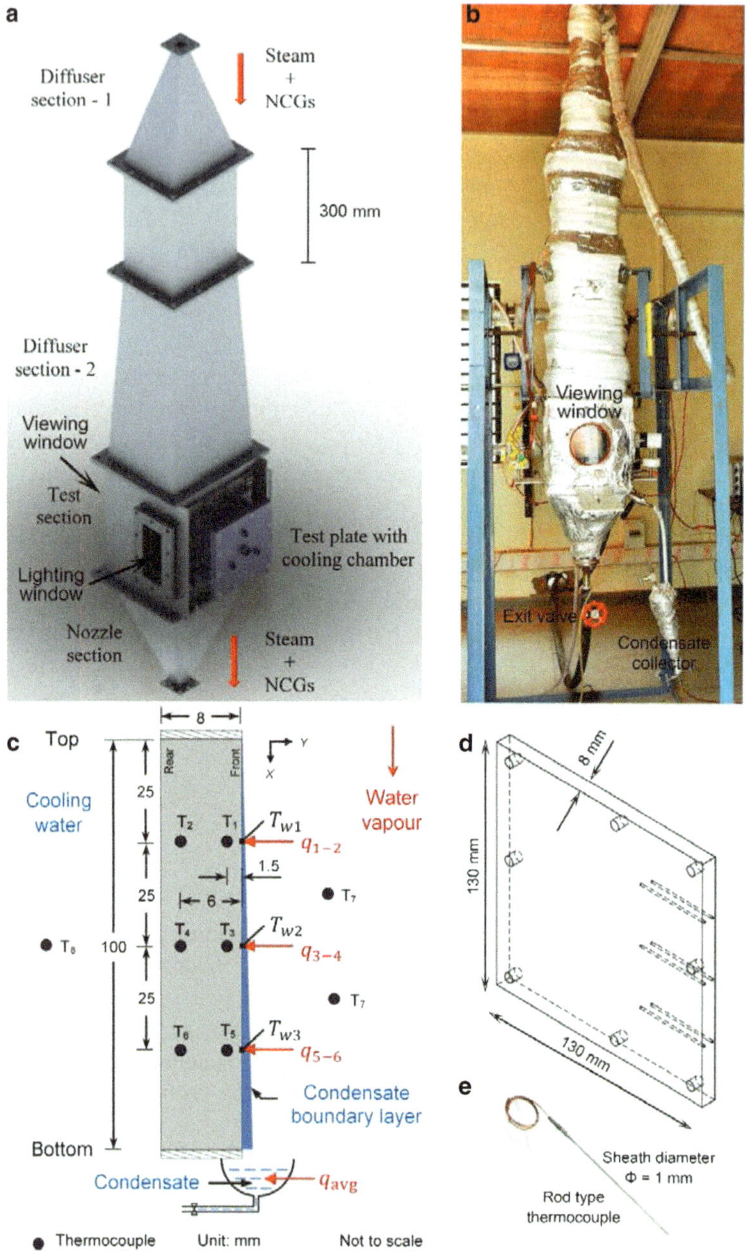

Fig. 13.6 Experimental facility developed for the condensation experiments. (**a**) A three-dimensional CAD model of the test section with a two-stage diffuser and a nozzle section unit, respectively. (**b**) Shows an adequately insulated and instrumented experimental setup. Different measurement locations for the temperature inside the test surface are shown in (**c**). An isometric view of the test surface with the unequal depth of the temperature sensor is highlighted in (**d**), whereas a rod-type micro-thermocouple is shown in (**e**)

in Fig. 13.6c. Three thermocouples pairs, namely, T_1–T_2, T_3–T_4, and T_5–T_6, at vertically equidistance locations (25 mm between any two pairs) are mounted along the thickness (8 mm) of the test plate. The thermocouples T_1, T_3, T_5 and T_2, T_4, T_6 are mounted at 1.5 ± 0.1 mm and 6.5 ± 0.1 mm depth from the front surface, respectively. All the geometrical parameters in the second test plate are similar to the first one, except the plate thickness. A thickness of 20 mm is considered for the second plate with thermocouples mounted at 5.0 ± 0.2 mm and 10.0 ± 0.2 mm depths from the front surface. Different depth of the thermocouples is considered to allow discrete diffusional lag inside the test plate (diffusion time for 1 mm and 5 mm depths are 60 ms and 1.5 s, respectively).

Millimeter size holes of 1.1 ± 0.05 mm size with unequal depths are made on the test plate for insertion of the thermocouples (Fig. 13.6d). Rod type thermocouple having 1 mm diameter (Fig. 13.6e) are installed in the holes to obtain temperature variation while minimizing obstruction in heat flow. Two more thermocouples, T_7 and T_8, are installed to measure the bulk and the cooling water temperatures, respectively. A pressure gauge is also installed on the test section to record the bulk pressure. Auxiliary systems such as a diesel fired boiler unit, pressure reducing station, and a steam separator unit is used to generate the saturated steam at desired pressures. A properly insulated flexible piping arrangement is used to supply the steam towards the test section through the diffuser unit. A flexible piping arrangement allows the whole experimental unit to rotate at fixed pivot and facilitates condensation studies at different inclination angles from sessile to pendant mode. Steam condenses on the front side of the test surface (Fig. 13.6c) and collected in a conical shape condensate collector (Fig. 13.6b). In order to continuously remove the heat released during the condensation process and maintain the test surface at the desired temperature, an impinging array of cooling water jets mechanism integrated with a constant temperature water bath is employed. The complete details of the experimental setup are available in (Yadav 2019). The steam condensation experiments are performed at different operating parameters, and temperature variation at specified locations inside the test surface are recorded using the micro-thermocouples. Selected results for the estimation of the filmwise and the dropwise condensation heat transfer are presented in the following subsections.

13.6.1 Case Study 1: Filmwise Condensation

Two types of experiments are performed for filmwise condensation: (a) condensation on the 8 mm thick test surface with the test section filled with nearly quiescent steam, and (b) condensation of uniformly flowing steam (with velocity ≈ 0.1 m/s and flow Reynolds number of ≈ 500) on the 20 mm thick test surface. Heat flux results obtained from the IHC method using temperature data recorded at location 50 mm away from the top of the test surface (data from T_3 and T_4 thermocouples) are shown in Fig. 13.7. The chosen location is representative only, and other measurement locations can also be considered.

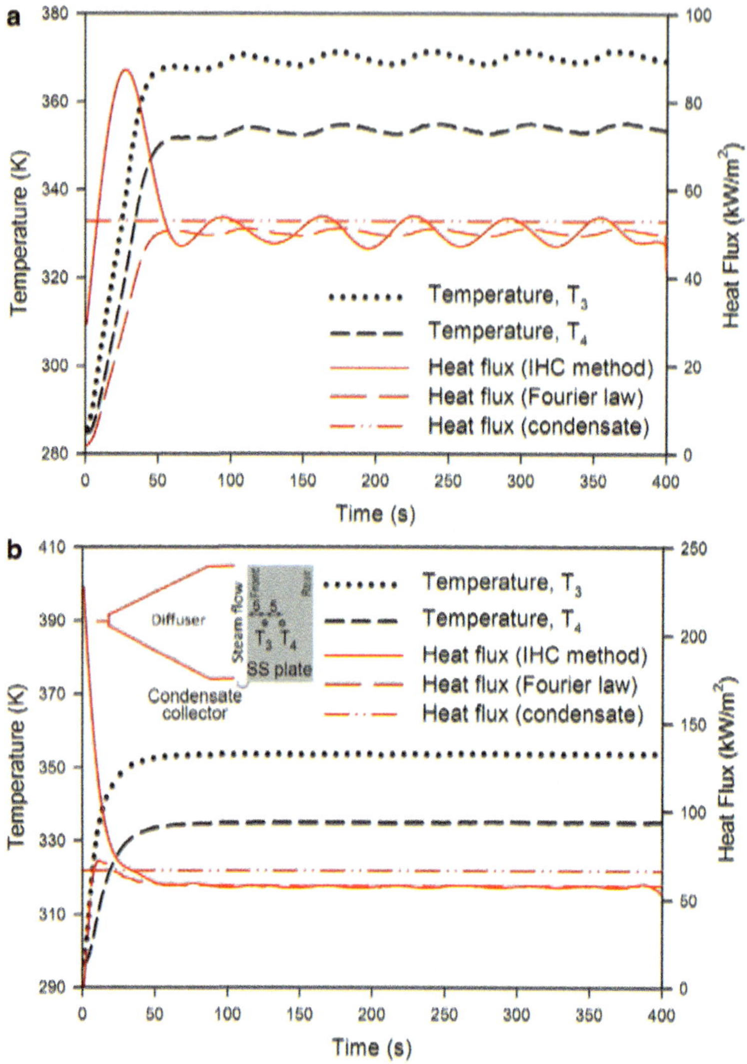

Fig. 13.7 Shows results for filmwise condensation of the saturated steam on the vertical test surface. (**a**) Temperatures and the heat fluxes for condensation at the bulk pressure of 1.2 bar abs. under stagnant condition (the test section filled with steam) and cooling water temperature of 289 K. Results for condensation of uniformly flowing steam on the vertical test surface in an open atmosphere and cooling water temperature of 293 K is shown in (**b**). A schematic of the test plate with a typical arrangement for uniformly flowing steam over the test surface is shown in the inset. ((**a**) is reprinted from Yadav et al. (2018))

The experiment in (a) is performed at 2 bar absolute bulk pressure of steam ($T_{sat} = 393$ K) and cooling water temperature of 303 K. Corresponding temperature variation recorded from thermocouples T_3 and T_4 are shown in Fig. 13.7a. The condensation flux on the test surface estimated using the IHC model (with temperature recorded from T_3 thermocouple as input in the inversion model), Fourier law of heat conduction, and average heat flux obtained via condensate gravimetric measurement are also shown in the figure. In an ideal situation where no non-condensable gas is present at the start of the experiment, condensation flux on the test surface commences with a maxima, owing to the highest temperature gradient. However, presence of the air (a non-condensable gas) during the initial phase of the experiment significantly hampers the overall heat transfer rate. As the steam coming inside the test section expels the trapped air as well as heats the bulk space, highest heat transfer is achieved after this initial time of $\approx$ 30–40 s (Fig. 13.7a). Thereafter, development of a condensate thin film over the test surface provides additional thermal resistance to the condensation heat flux. This results in a substantial reduction in heat transfer. Finally, once the condensate boundary layer gets fully developed, a nearly constant heat flux is obtained.

The fluctuations in heat fluxes which are owing to the fluctuating temperature variation at locations T_3 and T_4 (due to on/off toggle relay control of the boiler unit between pressure limits of 2.0 ± 0.15 bar) are captured by the IHC model, but with a minor phase-shift compared to temperature data. A surge tank, used to suppress fluctuation in the steam supply after the boiler unit, is intentionally not used here to see whether the IHC model is capable in capturing the cyclic variation in the heat fluxes. It is found that the IHC model is able to estimate condensation heat fluxes during transient as well as at steady-state whereas heat flux obtained from the Fourier law is suitable for steady-state measurements only. A nearly matching heat transfer at steady-state from the IHC model, Fourier law of heat conduction and condensate gravimetry further confirms the trueness of the measurement.

As noted before, condensation of steam flowing over the vertical plate has been studied in (b). The experiment has been carried out with a cooling water temperature of 293 K. The resulting temperature at two locations, T_3 and T_4, and estimated heat fluxes from the IHC model, Fourier law of heat conduction, and condensate gravimetry are shown in Fig. 13.7b. A test plate of 20 mm thickness is used here, and the measurement locations are shown in INSET of Fig. 13.7b. A large temperature gradient at the start (initial $\simeq$ 50 s) results in significantly large heat flux. This variation is correctly predicted by the IHC model, whereas the Fourier law of heat conduction is found to be suitable for steady-state measurements only. Similar to the experiment in (a), the heat flux obtained from the IHC model, Fourier law and estimated via the collected condensate mass at steady-state matches reasonably well.

Through these two experimental studies, it is shown that the IHC model can predict the condensation flux using the internal temperature data. The estimation of different variations in heat fluxes, viz. suddenly increasing and decreasing, and cyclic variations, are accurately captured. A continuous condensate film is formed over the surface in these experiments, and as a result, heat flux variations here are not as dynamic in nature as in the case of dropwise condensation. How the IHC model will fare when used for estimating the dynamically varying heat flux in case of the dropwise condensation is explored in the next application.

13.6.2 Case Study 2: Dropwise Condensation

Dropwise condensation experiments are performed on 8 mm thick test plate at 15 different surface orientations: sessile mode (condensation over the horizontal test surface), pendant mode (condensation underneath the horizontal test surface), and various angles in between these two asymptotic limits. All the experiments are performed at 1.2 bar abs. bulk pressure of steam and 289 K cooling water temperature. Temperature variations during the experiments are recorded using embedded thermocouples at the quasi-steady-state condition, whereas the condensation cycle is obtained from the recorded videos during the experiments. A viewing window (Fig. 13.6b) is used for this purpose and videos are recorded using a DSLR camera (Nikon® D7100) at 60 fps. A complete condensation cycle showing timescale at different stages (A-G) for the sessile mode of condensation is shown in Fig. 13.8. The different stages are: A—initiation of nucleation, B—growth of the condensate drops, C—two unequal size drops just before merging, D—bigger size drop just after coalescing, E—restructuring of the bigger drop through balancing of liquid interface force (surface tension force), surface force (free surface energy of the test surface), and gravitational force, F—restructured large size drops, and finally G—slide-off from the surface to provide fresh surface for starting the next cycle. A close look of the timescale suggests that coalescence process (D) took place in few milliseconds ($\approx$ 4 ms in Fig. 13.8), whereas all other processes carry over up to few hundreds of

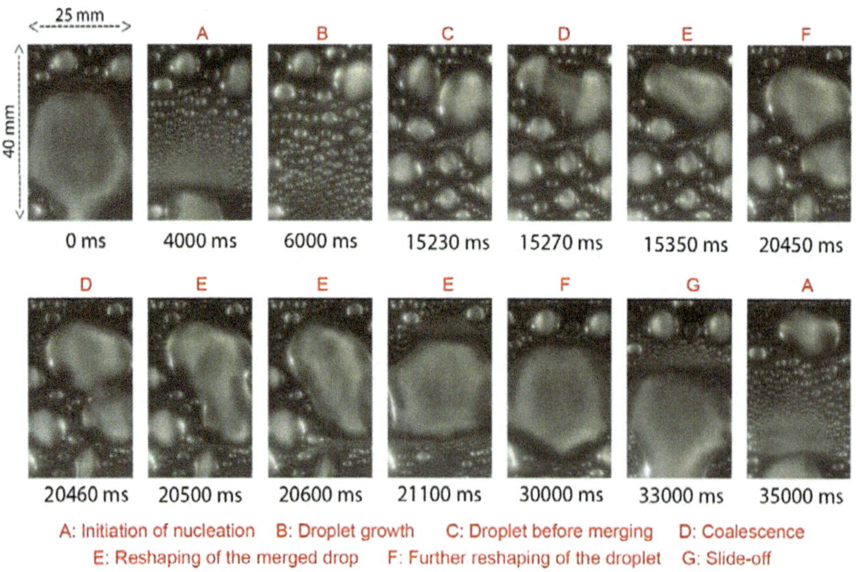

Fig. 13.8 Drop cycle for condensation of saturated steam on a vertical surface (i.e., 0°) at a bulk pressure of 1.2 bar abs. and cooling water temperature of 289 K under stagnant condition. The time when the coalesced drop is just about to get swept is taken as the reference time (0 ms). The time estimated for each snapshot with this reference point is shown below the snap

milliseconds. The snapshots also confirm the different stages of condensation cycle presented in earlier chapters. In other cases where gravity assists sliding of the drops over the surface (condensation on a vertical surface) or pull the drops against the surface (pendant mode of condensation), the timescale will be different. This timescale will also depend on the surface properties (particularly surface roughness and morphology), and temperature difference between the surface and the bulk steam, and the dynamic contact angle of the drop. Depending on requirement of the resolution of heat transfer, one needs to choose the appropriate sampling frequency (as per the Nyquist criteria) and a suitable measurement system to support the chosen sampling frequency accordingly.

A sampling frequency of 1 Hz is used to record the temperature data in the experiments reported here. These temperature data at known locations (T_1 and T_2, Fig. 13.6) are inverted using the inversion methodology and estimated heat flux for dropwise condensation on the vertical surface, sessile, and pendant modes are shown in Fig. 13.9. A lower heat flux with smoother cyclic variation is seen for the sessile mode of condensation. In this mode, drops keep growing over the surface until they spread over the surface due to gravity force and then slide away from the surface owing to hydrostatic force. Unlike this mode of condensation, gravity pulls the drops against the surface in pendant mode of the condensation. As drop(s) falls from the surface at local location(s), condensate boundary layer gets distributed and results in

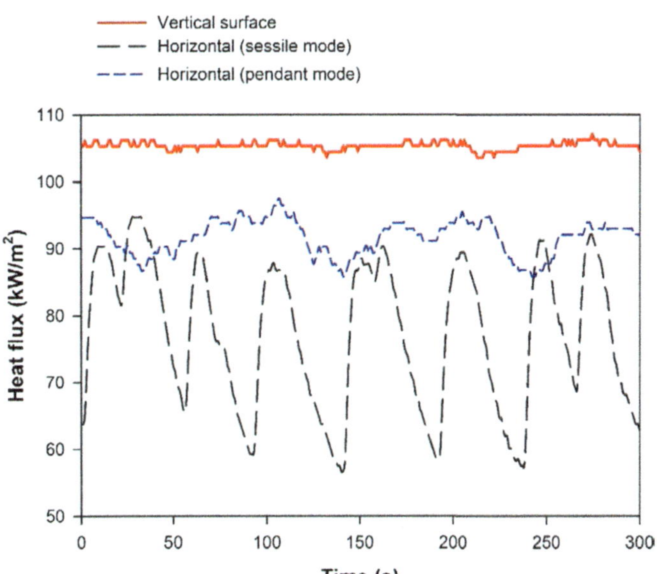

Fig. 13.9 Shows temporal heat flux variation for dropwise condensation of saturated steam at a bulk pressure of 1.2 bar abs. and cooling water temperature of 289 K under stagnant condition. The heat flux variation is shown for three orientations of the test surface, i.e., vertical surface, and over and underneath the horizontal test surface (representative of sessile and pendant modes of condensations, respectively)

significant fluctuations in the ensuing fluxes (Fig. 13.9). This also leads to higher heat flux in this case as compared to sessile mode of condensation. In case of condensation on the vertical surface, smaller drops favored by the gravitational force rapidly falls down from the surface. Due to the lower measurement frequency, a cyclic variation in the estimated heat flux from the IHC model could not be identified in this case. Therefore, it becomes important to identify the cycle time for a dropwise condensation process and adopt an appropriate measurement methodology to record the temperature data. Finally, the condensation flux can be estimated using temperature data in the IHC model.

From the above discussion, it is clear that condensation cycle as well as the related heat flux variations strongly depends on orientation of the surface where condensation takes place. This will also affect the condensation heat flux for various surface inclinations. Accordingly, dropwise condensation experiments are performed at different inclination angles to quantify the heat flux dependence on orientation. The experiments are performed for a number of inclination angles ($\theta = 0°$, $\pm 15°$, $\pm 30°$, $\pm 45°$, $\pm 60°$, $\pm 75°$, $\pm 87°$, and $\pm 90°$) at a cooling water temperatures of 289 K and a bulk pressure of 1.2 bar abs. inclination angle "θ" is measured here from the vertical surface (considered as 0°) with "+" direction representing sessile mode and "−" direction as pendant mode of condensation. The time-averaged heat flux at three locations from the top of the test surface (corresponding to a pair of thermocouples, i.e., T_1–T_2, T_3–T_4, and T_5–T_6) is estimated using the Fourier law of heat conduction. Although dropwise condensation is essentially a transient process with cyclic variation, the time-averaged heat flux values are equally important for designing a thermal system based on this phase change process. Therefore, time-averaged heat flux over a minimum of ten condensation cycles at each inclination angles is reported. The average heat fluxes are evaluated at locations 25, 50, and 75 mm away from the top of the test surface, and termed as HF1, HF2, and HF3, respectively. The variation of heat fluxes at the three locations with inclination angles is shown in Fig. 13.10a. It can be seen from Fig. 13.10a that at each of the three locations, the highest heat flux is corresponding to the case of vertical plate, i.e., at $\theta = 0°$. Thereafter, the heat flux decreases as the test surface is inclined to face upward or downward at some inclination angle from the vertical. The heat flux values for the inclined plate are found to be similar for the sessile ($+\theta$) and pendant ($-\theta$) mode of condensation at lower inclination angles (for $\theta < +60°$ and $\theta > -60°$) of the test surface. However, at higher inclination angles (for $\theta > +60°$ and $\theta < -60°$), heat flux decreases considerably for both the modes, albeit for different reasons (explained in next paragraphs). Moreover, the magnitudes of heat flux at +90° (horizontal surface, sessile mode) are found to be lower as compared to −90° (horizontal surface, pendant mode). These results are discussed below.

For sessile mode of condensation, as the inclination angle is increased from 0° to +60°, the heat flux decreases slightly as compared to the vertical case. This is because of a small decrease in the effective gravitational force ($g\cos\theta$, as $\cos\theta \approx 1$ for lower values of θ) responsible for the sweeping of condensate drops from the surface. As a result, condensate drops are easily swept from the surface by

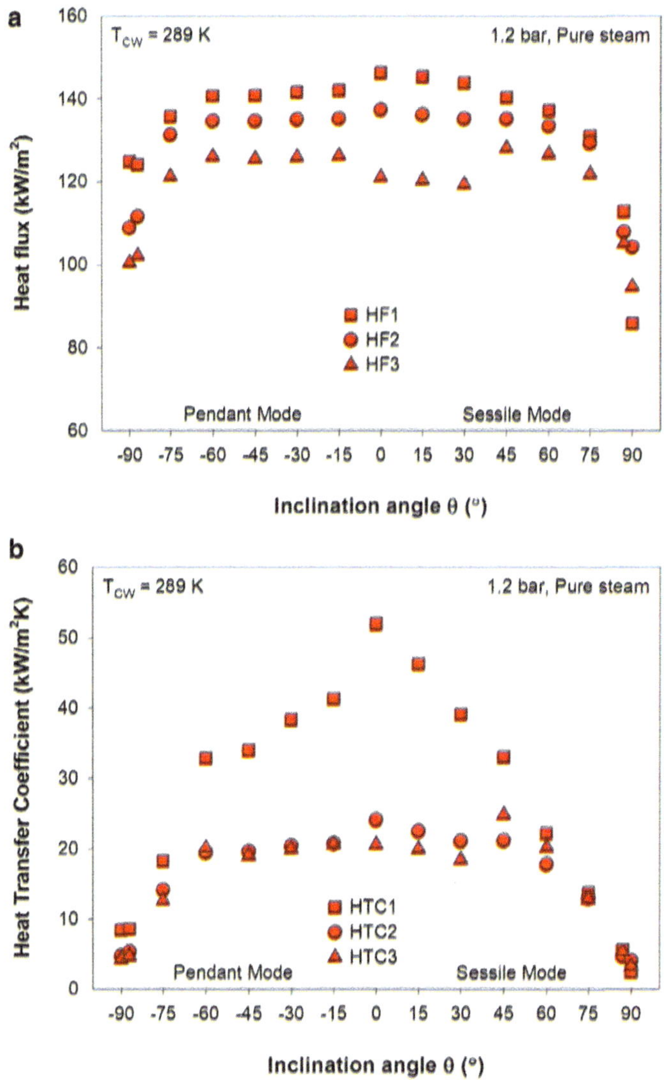

Fig. 13.10 (**a**) Heat fluxes and (**b**) heat transfer coefficients for condensation of pure steam at 1.2 bar abs. bulk pressure and cooling water temperature of 289 K. The variations at different inclination angles for three locations on the test surface is shown

the gravitational force and decrease in overall heat transfer is not very significant. However, at inclination angles higher than +60°, the effective gravitational force reduction becomes significant. This leads to an increase in the time for which a drop remains over the test surface. It also results in an increase in their size due to further condensation and more importantly merging of the adjacent drops. These large drops cover substantial area on the surface, which slows the further condensation. This reduction in heat transfer can be attributed to increased thermal resistance and unavailability of fresh nucleation sites. In fact, at $\theta = +90°$, where the test surface is completely horizontal and there is no effective gravitational force to remove the condensate, the coalesced drops become so large that they lead to liquid holdup over the complete surface. The large coalesced drops are only removed when hydrostatic pressure force inside the thick film overcomes surface tension of the condensate liquid. Therefore, a significant decrease in the heat flux is obtained in Fig. 13.10a for the case of dropwise condensation over horizontal test surface.

For the pendant mode of condensation, as the inclination angle is decreased from 0° to −60°, the heat flux decreases slightly, with magnitudes similar to those obtained in the sessile mode at the respective inclination angles. The drops on the inclined surface are found to slide along the surface and fall off at the edge, rather than fall down at the place where they nucleated. The size of the departing droplets observed in this case is similar to that observed in the sessile mode at low inclination angles. This shows that gravitational effect is negligible compared to the surface tension force up to this range of angles (approximately 60°). Thus, the time-averaged heat transfer in pendant and sessile mode are comparable at a particular inclination angle up to ≈ 60°. However, for inclination angles lower than −60° ($\theta < -60°$), the gravitational force starts to elongate the droplets, which increases the thermal resistance. This results in a significant decrease in the heat flux. At $\theta = -90°$, the plate is completely horizontal and facing downward. In this case, the drops become elongated and keep falling down from random locations. This observation is confirmed by digital imaging during the experiments. The overall heat flux decreases due to the large thermal resistance provided by the increased drop sizes. Still, the magnitude of the heat flux in the pendant mode at $\theta = -90°$ is found to be higher as compared to the sessile mode at $\theta = +90°$ at higher inclination angles. It is because of the significant disturbance caused by the falling droplets in the liquid boundary layer for the pendant mode of condensation. The heat flux variation at the three measurement locations is shown in Fig. 13.10a. A higher heat transfer is obtained for locations nearer the top of test surface as drops sliding over the surface increases the resistance at lower locations on the surface.

The variation of HTC with the inclination angle follows a trend similar to that of the heat flux. The results for HTC at different surface inclination angles are shown in Fig. 13.10b. It is found that the HTC at $\theta = -90°$ (pendant mode on horizonal surface) is higher than that at $\theta = +90°$ (sessile mode on horizontal surface). These results are in-line with the observations reported by Leipertz and Fröba (2008). The authors had reported a reduction in HTC as the inclination angle is increased from the vertical, and a higher HTC in the pendant mode compared to the sessile mode on a horizontal surface. In addition, higher HTCs are obtained at locations closer to the top of the test surface. In fact, for the two lower measurement locations, i.e., 50 and 75 mm away from the top of test surface, nearly identical HTCs are obtained. These results indicate that to get the maximum benefit of dropwise condensation heat transfer in practical applications, smaller sized surfaces in the vertical orientation should be used.

13.7 Closure

Estimation of the heat flux during the filmwise and the dropwise condensation require transient measurements due to temporal variation in the condensate film during filmwise condensation and cyclic nature of dropwise condensation. In addition, phase change heat fluxes are large and result in small temperature differences. Commercially available heat flux sensors are found to be quite unsuitable for such measurements. One needs to be careful in assessing the quality of heat flux data obtained from these sensors. An inverse heat conduction (IHC) technique-based measurement system is a robust and accurate system for evaluating surface heat flux during such phase change phenomena. The specific requirement for this measurement system is an accurate knowledge of temperature at precisely one or more known locations inside the solid medium and assisting hardware to support recording of the data at the desired frequency. Laboratory-scale experiments show the validity of this approach. It is also found that small modular surface arrangement should be used to get the maximum heat transfer in dropwise condensation.

Chapter 14
Measurement of Heat Transfer Rates under a Liquid Drop During Dropwise Condensation

Gagan Bansal, S. Khandekar, and K. Muralidhar

Nomenclature

A	Area, m^2
D	Diameter, m
g	Acceleration due to gravity, m/s^2
H, S, I	Hue, Saturation, Intensity scale, –
h	Heat transfer coefficient, W/m^2K
h_{fg}	Latent heat, J/kg
K_1	Constant, –
k	Thermal conductivity, W/mK
M	Molecular weight, kg/kmol
n	Number density, m^{-2}
P	Pressure, N/m^2
$\dot{Q}$	Heat transfer rate, W
q''	Heat flux, W/m^2
R	Gas constant, J/K-mole
R, G, B	Red, Green, Blue scale, –
T	Temperature, K
$\overline{T}$	Average temperature, K
t	Thickness, m
υ	Specific volume, $m^{3/}kg$
λ	Wavelength of light, m
$\hat{\sigma}$	Accommodation coefficient, –
σ	Surface tension, N/m
ρ	Density, kg/m^3

S. Khandekar, K. Muralidhar, *Drop Dynamics and Dropwise Condensation on Textured Surfaces*, Mechanical Engineering Series,
https://doi.org/10.1007/978-3-030-48461-3_14

Subscripts

avg	Average value
exp	Experimental
liq	Liquid
max	Maximum
min	Minimum
pix	Digitized pixel
sat	Saturation
sub	Substrate
total	Total value
vap	Vapor

14.1 Introduction

When compared to heat transfer through liquid films, significantly higher heat transfer coefficients are achievable in dropwise condensation (Bansal et al. 2009). In an engineering device, the condition of dropwise condensation is realized by appropriately texturing, the surfaces exposed to vapor (Rose 1981, 1998, 2002). There is a renewed interest in fully understanding the dropwise mode of condensation due to the possibility of robust manufacturing of engineering surfaces—namely physical and chemical texturing, as well as surface modification (Leipertz and Fröba 2006; Vemuri et al. 2006; Kananeh et al. 2006). Complementary interest for miniaturization and compactness of heat transfer equipment is also to be seen (Goldstein et al. 2006).

In spite of sustained research in the past decades, many issues related to heat transfer during dropwise condensation remain unresolved (Rose 2002; Ma et al. 2002). This is mainly because small changes in the surface morphology, and hence, surface energy, leads to changes in the droplet distribution, affecting the ensuing thermo-hydrodynamics. In addition, the overall mechanism of formation of a droplet on an engineered surface involves varied length scales, from atomistic orders at early phases of nucleation on the one hand, to scales affected by the body force distribution vis-à-vis the surface tension forces, on the other (Venables 2000). In addition, changes in contact angles and the hysteresis phenomenon of the contact line remain as heuristic parameters in the predictive models. Frequently, sessile or pendant droplets on substrates are modeled as hemispheres, completely neglecting the dynamic contact angles that appear in real-time operation (Collier and Thome 1996). Very high heat transfer rates, and therefore, very low temperature differential, coupled with the above factors also hinder generation of repeatable experimental data. Thus, experiments that concern heat transfer during dropwise condensation are quite complex (Rose 1981, 1998, 2002; Goldstein et al. 2006; Ma et al. 2002).

A common assumption in heat transfer analysis and the interpretation of experimental data is the isothermal nature of the substrate on which condensation is taking place. Quite often, the temperature gradient in the normal direction, recorded in the bulk substrate by suitably located thermocouples, is extrapolated to determine the average temperature of the substrate (Collier and Thome 1996; Carey 2008). Subsequently, average condensation heat transfer coefficients are determined. However, the inherently time-dependent droplet size distribution results in surface temperature fluctuations during the cyclic sequence of vapor condensation, drop formation, coalescence, and their removal. A variation of temperature/heat flux at the base of each condensing drop is obvious when we consider a mechanistic model (Graham and Griffith 1973; Abu-Orabi 1998). Neglecting such crucial details leads to an error in estimating the actual local and average values of heat transfer coefficient.

Although the inherent time dependence of heat transfer in dropwise condensation has been acknowledged in the literature, spatio-temporal determination of temperature fluctuations under experimetal settings is not trivial. Need for such a data generation is essential to relate the droplet ensemble hydrodynamics to the eventual process performance. Conventional thermometry (e.g., with standard micro thermocouples) cannot provide spatial information of temperature distribution. Moreover, for measurements on mini and microscales (as in the case of individual condensing droplets), spatial constraints and sensor intrusion thwarts the acquisition of accurate primary information. Hence, to overcome such limitations, liquid crystal thermography (LCT) has been employed. The spatio-temporal variation of temperature (and therefore the heat throughput via the droplet base area in the normal direction) can be obtained from LCT. This technique allows the determination of thermal behavior of the condensing drops on the scale of an individual droplet.

The fluid considered is pure deionized water condensing under controlled conditions on the underside of a polyethylene substrate. Study of stand-alone polymer substrates as well as polymer-coated metal substrates subjected to dropwise condensation have practical engineering applications for engineering systems ranging from compact polymer heat exchangers, contact lenses, thermal enclosures for horticulture applications, and dew formation on polymer food packages (Beysens 2006; Briscoe et al. 2005; Ma et al. 2002; van der Geld et al. 2001). Videography and high spatial resolution LCT are simultaneously employed to derive information on the condensing droplets. A specially designed experimental setup enables in-situ measurements of condensation over a suitably textured surface (Bansal et al. 2009). Results obtained are compared with the classical one-dimensional heat transfer formulation of dropwise condensation (Carey 2008; Graham and Griffith 1973; Abu-Orabi 1998).

14.2 Experimental Setup

The experimental apparatus is designed to study dropwise condensation under controlled conditions on the underside of a substrate in the pendant mode. It permits simultaneous visualization of the condensing droplets as well as real-time temperature measurement using LCT sheet, as shown in Fig. 14.1. It consists of a cylindrical polycarbonate (Makrolon®) vacuum chamber ($<10^{-6}$ mbar) of 180 mm ID and length 110 mm, closed from the two ends. The lower brass flange is fitted with a $\lambda/4$ optical viewing window for digital videography and photography(view A). There is an annular space around this viewing window for storing the working fluid inventory. A circular 1.5 mm thick, mica strip heater (OD = 70 mm, ID = 40 mm) is attached below this annular space to give the necessary heat power input. The upper end of the condensing chamber is sealed with a polycarbonate flange fitted with a transparent $\lambda/4$ optical glass disk of 100 mm diameter and 8.0 mm thickness. Outside the glass disk, a rectangular section coolant flow passage (200 × 6 mm) is provided. One face of the glass disk can be subjected to constant temperature boundary conditions by varying the temperature of the flowing coolant water.

External connections for evacuating the condensing chamber, and measuring pressure and temperature are provided on the main chamber wall. Condensing vapor temperature is measured with K-type thermocouples (National Instruments®) of accuracy ±0.05 °C, after calibration. The condensing chamber pressure is measured by an absolute pressure transducer (M/s Honeywell®, accuracy 0.1% FS, NIST traceable calibration, Range 0–1.2 bar). Online data acquisition is performed with a 16 bit PCI-4351 card (National Instruments®) card.

A color CCD camera (Basler® A202KC, resolution: 1024 × 1024 pixels) is used to capture the LCT images (View-B) while a Sony Digital CCD camera is used for capturing the optical video images of the condensing surface (View-A). A diffuse white light source with a controllable light intensity output is used for illumination.

14.3 Experimental Methodology

14.3.1 Calibration of TLCs

Liquid crystals supplied by M/s Hallcrest® in micro-encapsulated form, displaying the full RGB color spectrum, are used in this study. These crystals respond to changes in temperature by changing their color. They have chiral (twisted) molecular structures and are optically active mixtures of organic chemicals turning from colorless (black against a black background) to red at a given temperature (called as the event temperature) and, as the temperature is increased, pass through the other colors of the visible spectrum in sequence (orange, yellow, green, blue, violet) before turning colorless (or black) again at a higher temperature, called the clearing

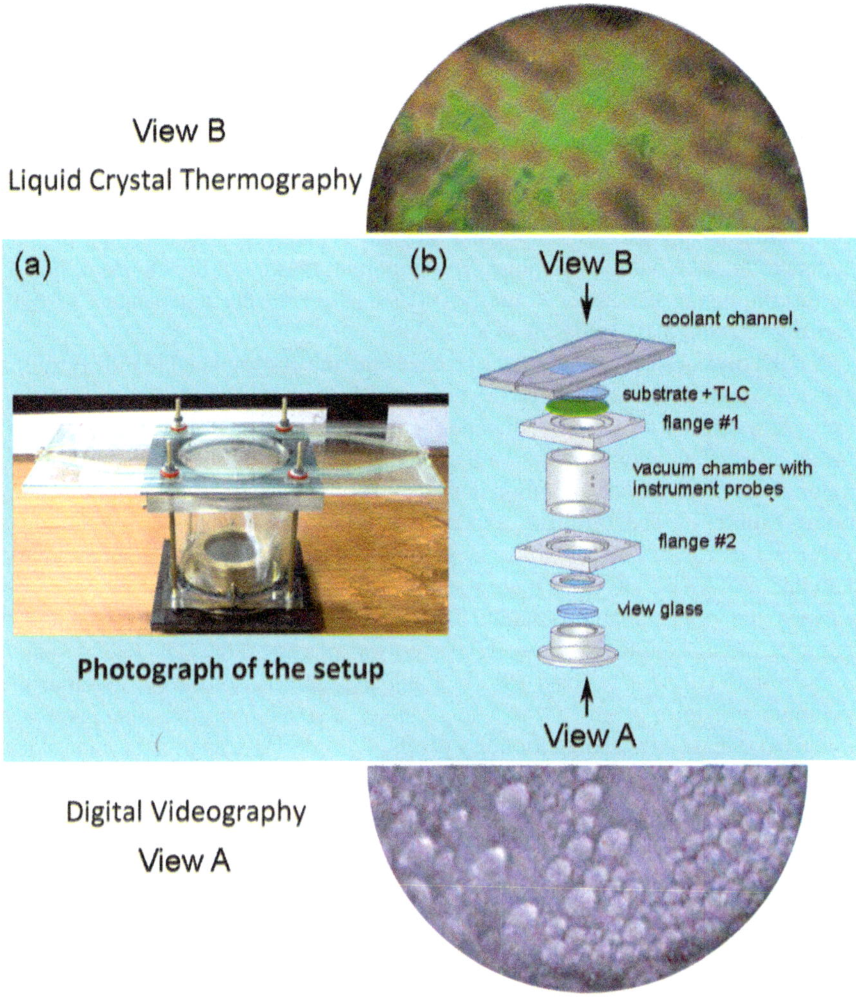

Fig. 14.1 Details of the setup to observe dropwise condensation on the underside of a substrate. (**a**) Photograph shows the details of the main condensing chamber (**b**) Exploded view of the chamber showing all the components. Camera View A from the bottom of the setup gives the photograph of the condensing pendant droplets on the substrate (as shown below) while Camera View B from the top provides the RGB image of the liquid crystal sheet

temperature. The color changes are reversible; on cooling, this color change sequence is reversed. Thus, the local spatial temperature distribution can be captured by a digitizer (e.g., a CCD camera with a frame grabber card) and suitably quantified in 8/16 bit pixel-specific information. This pixelized response, corresponding to the local temperature, is usually available in the form of the three primary colors—red, green, and blue (RGB). To relate these to temperature, the color response needs to be reduced to a single value. This step is accomplished by converting the pixel level RGB values to the hue, saturation, and intensity (HSI) scale and choosing the hue parameter to scale with temperature. The important reason for the choice of hue as the discriminating parameter is the fact that it is practically independent of light intensity or illumination (Stasiek et al. 2006).

A polyethylene sheet (70 μm thick) coated with the encapsulated liquid crystals is attached on the lower surface of the top optical window. Dropwise condensation takes place directly on the polyethylene sheet. Before commencement of the experiments, in-situ calibration was carefully conducted under controlled conditions. It is important to perform the calibration in situ so that identical lighting conditions exist during calibration and the main experiments. Calibration of the LCT sheet was performed as follows. Without the evaporator heater power, water was circulated (kelvimat HAAKE® DC10K20, accuracy ±0.1 °C) at controlled temperature steps between the operational bandwidth of the LCT sheet. A pre-calibrated Pt-100 reference thermocouple (accuracy ±0.05 °C) was placed on the substrate. A region of interest (ROI) was selected for obtaining the calibration curve. Although the minimum size for a given ROI is a single pixel, a size greater than one pixel was utilized to statistically account for noisy pixels. Considering scatter in the hue values, the nominal hue value was estimated based on the median of the hue distribution. A fifth-order polynomial was used to fit the calibration data relating hue and temperature as measured by the thermocouple (Stasiek et al. 2006; Muwanga and Hassan 2006). Figure 14.2 shows a sample calibration curve drawn between hue and the operating substrate temperature; the corresponding recorded RGB images of the ROI by the color CCD camera are also shown. Subsequent experiments have been carried out under identical conditions of illumination and laboratory environmental conditions. Following the procedure given by Hay and Hollingsworth (Hay and Hollingsworth 1998), the overall accuracy of the measurement process is estimated to be 0.42 °C. This is about 6% of the useful range of the LCT and compares well with other similar reported experimental data (Muwanga and Hassan 2006; Höhmann and Stephan 2002). It includes uncertainty in the primary thermocouple sensors, sensitivity of hue calculation from RGB values, and standard error estimate of curve fitting.

14.3.2 Experimental Procedure

The condensing chamber is assembled with a small quantity of liquid inventory kept in the annular space at the lower portion of the apparatus. The chamber is evacuated

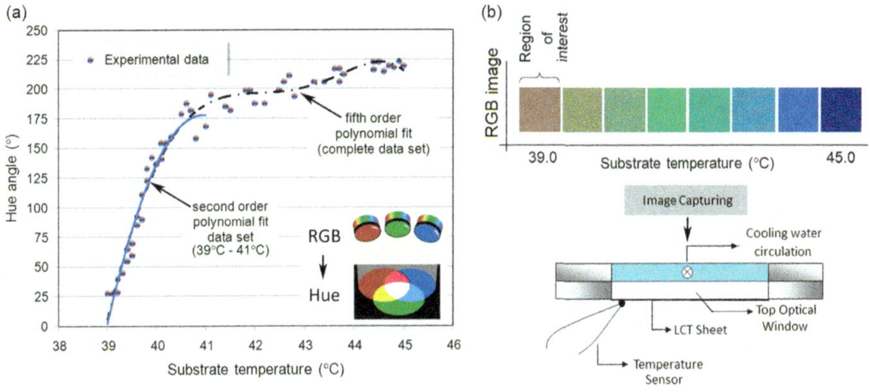

Fig. 14.2 (**a**) Typical calibration curve of the liquid crystal sheet relating the substrate temperature to the hue scale. (**b**) RGB images obtained during the calibration step, as detailed below in the schematic of the setup. These are further processed to get the HSI images

by a turbomolecular vacuum pump. A fraction of water flashes to vapor that fills up the chamber. The evacuation process is continued for several minutes so as to ensure complete removal of non-condensable gases in the chamber. At this stage, it is ensured that the absolute pressure in the chamber is equal to the saturation pressure of the working fluid at the chamber temperature. Any operating temperature-pressure combination of the condensing fluid can be obtained by a suitable choice of the coolant water temperature (flowing in the upper rectangular passage) and the heat input to the chamber by the circular heaters placed below.

Condensation in the form of pendant drops commences almost immediately after evacuation is initiated. Hence, it is not possible to capture the liquid crystal thermographs in the initial phase of nucleation and droplet growth. Nucleated drops grow in size by direct condensation of the vapor and then by coalescence with the neighboring drops. Once the weight of the drop exceeds a certain threshold value, it falls back into the reservoir. An open space is created at such locations, where fresh condensation is initiated. Thus, after the passage of a few minutes, a dynamic steady-state of evaporation and condensation is established in the apparatus. The relevant data, namely LCT images and condensations patterns of drops have been acquired after such a quasi-steady state is reached wherein growth of drops is primarily dominated by coalescence. Condensation on virgin areas occurs only when such areas are created by drops falling off from the substrate.

14.3.3 Data Reduction

The overall data reduction scheme applied to LCT images and condensation patterns at dynamic steady-state obtained during the dropwise condensation process is shown in Fig. 14.3. Part (a) shows the physical model considering a single condensing drop

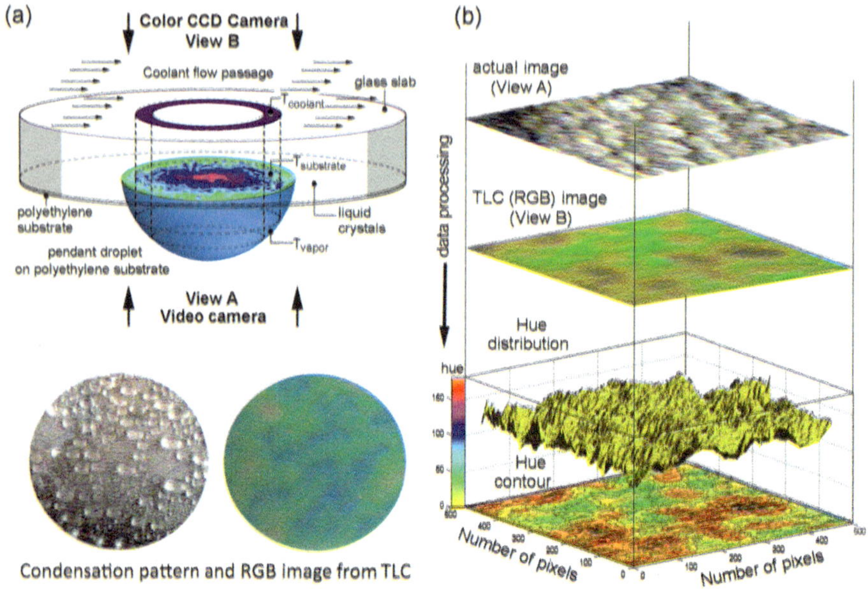

Fig. 14.3 (**a**) Model for the estimation of the local heat transfer coefficient for dropwise condensation in pendant mode. View A provides the direct picture of the drop distribution while View B provides the liquid crystal thermograph. (**b**) Series of operations employed for data reduction to otain the local tempearture field (**c**) Typical photograph of condensing drops obtained from Camera View A and LCT image obtained from Camera View B

on the substrate in the pendant mode; shown below are typical images obtained from the experiments. View A provides the instantaneous photograph of droplets distributed over the cold surface. The number density of drops as a function of drop diameter can be obtained from this digital image. View B is the corresponding liquid crystal thermograph of the entire substrate, as recorded by a CCD camera. The instantaneous spatial temperature distribution on the substrate is known via the LCT calibration curve. The color-to-temperature conversion scheme is delineated in Fig. 14.3b. It shows the digital image (View A), RGB image (View B), the corresponding distribution of hue contours (and therefore, the corresponding surface temperature distribution, from Fig. 14.2) over a selected area of the substrate. Individual condensing droplets and their corresponding base temperature distribution during the condensation process can thus be clearly estimated.

For a given liquid crystal thermograph and its complementary photograph of the condensation pattern, heat transfer calculations can be carried out at the scale of a drop as well as the scale of the area imaged by the camera. The data reduction scheme adopted in this work for the two scales is as follows (Bansal et al. 2009):

At the Substrate Level

- Calculate the net heat flux from the entire substrate to the cooling water supply by assuming a one-dimensional heat transfer approximation.
- Calculate the heat transfer coefficient during dropwise condensation by the net heat transfer rate obtained in step (i) and dividing it by the applicable degree of subcooling ($\overline{T}_{vap} - \overline{T}_{sub}$), where $\overline{T}_{sub}$ is the space-averaged temperature on the polymer substrate, as estimated from the liquid crystal thermograph.

Thus, considering one-dimensional heat transfer through the entire system, the instantaneous local heat flux through a pixelated area on the substrate is

$$q''_{pix} = (\dot{Q}_{pix}/A_{pix}/A_{pix}) = \frac{k_{glass}}{t_{glass}} \cdot (T_{pix} - T_{w}) \tag{14.1}$$

The net heat transfer to the cooling water from the individual droplet and the entire substrate, respectively, can be estimated with

$$\dot{Q}_{drop}\big|_{exp} = \int_{\text{drop base area}} q''_{pix} dA_{pix} \tag{14.2}$$

$$\dot{Q}_{total}\big|_{exp} = \int_{\text{total substrate area}} q''_{pix} dA_{pix} \tag{14.3}$$

The average temperatures of the base area of individual pendant drops and the complete condensing substrate (in the experiments, $A_{sub} = A_{glass}$) are estimated, respectively, by area averaging the local pixel temperatures on the substrate

$$\bar{T}_{sub} = \left(\frac{1}{A}\right) \int_{\text{total substrate area}} T_{pix} dA_{pix} \tag{14.4}$$

$$\bar{T}_{drop} = \left(\frac{1}{A}\right) \int_{\text{drop base area}} T_{pix} dA_{pix} \tag{14.5}$$

The average condensation heat transfer during dropwise condensation is a direct manifestation of the combination of interfacial resistance to condensation of vapor on the drop surface and the diffusional resistance the droplet offers. Knowing the mean condensing vapor temperature inside the chamber, the steady-state average dropwise condensation heat transfer coefficient through the polymer substrate can be estimated by

$$\bar{h}_{exp} = \frac{\dot{Q}_{total}\big|_{exp}}{A_{sub}(\bar{T}_{vap} - \bar{T}_{sub})} \tag{14.6}$$

At the Droplet Level:

- Isolate individual droplets that are formed on the substrate, i.e., through the respective complementary images obtained via Views A and B.
- Calculate heat transfer rate from the base area of the individual droplet to the coolant and compare it with the one-dimensional heat throughput for an individual droplet, with steady-state assumption. The latter accounts for the interfacial heat transfer resistance, diffusion/conduction resistance through the droplet, and the additional constriction resistance due to the droplet curvature. The droplet geometry is assumed to be uniformly hemispherical for all cases, as confirmed with static contact angle measurements of pendant droplets on the surface. Dynamic apparent contact angles were not measured.

A one-dimensional model, which considers heat transfer through individual droplets, has been described by many authors (Carey 2008; Graham and Griffith 1973). Considering interfacial resistance, drop curvature resistance, and diffusional resistance, the average heat transfer through the base of the droplet is (Bansal et al. 2009)

$$\left(\overline{T}_{\text{vap}} - \overline{T}_{\text{drop}}\right) = \frac{2\dot{Q}_{\text{drop}}\big|_{\text{model}}}{h_i \cdot \pi \cdot D_{\text{drop}}^2} + \frac{\left(\overline{T}_{\text{vap}} - \overline{T}_{\text{drop}}\right) D_{\min}}{D_{\text{drop}}} + \frac{\dot{Q}_{\text{drop}}\big|_{\text{model}}}{2k_{\text{liq}} \cdot \pi \cdot D_{\text{drop}}} \tag{14.7}$$

The interface heat transfer coefficient h_i is

$$h_i = \left(\frac{2\hat{\sigma}}{2 - \hat{\sigma}}\right) \frac{h_{\text{fg}}^2}{\overline{T}_{\text{vap}} v_{\text{liq}}} \left(\frac{M}{2\pi R \overline{T}_{\text{vap}}}\right)^{0.5} \tag{14.8}$$

Thus, we can compare Eqs. (14.5) and (14.7) to validate the modeled and experimentally observed heat throughput across an individual droplet. In fact, if the complete range of droplet distribution is also estimated by the digital image analysis of condensation patterns, the total heat throughput passing through the substrate can be obtained as

$$\dot{Q}_{\text{total}}\big|_{\text{model}} = \left(\frac{\pi \cdot \left(\overline{T}_{\text{vap}} - \overline{T}_{\text{sub}}\right)}{2}\right) \int_{D_{\min}}^{D_{\max}} n_D \cdot D^2 \cdot \frac{(1 - D_{\min}/D)}{(1/h_i + D/4k_l)} \cdot dD \tag{14.9}$$

Equation (14.9) can be compared with experimental estimation of the total heat throughput, i.e., Eq. (14.3) above. In the present experiments, due to limitations in optics, the smallest droplet which we could digitally measure with 99% confidence level was of the order of 0.4 mm. It is much larger than the minimum diameter of the drops that can possibly grow for a given wall subcooling (Carey 2008) as given by the following expression

$$D_{\min} = \frac{4 \cdot v_{\text{liq}} \cdot \sigma \cdot T_{\text{s}}}{h_{\text{fg}}(T_{\text{sat}} - T_{\text{s}})} \tag{14.10}$$

Equation (14.10) yields D_{min} of the order of 0.01 mm for an accommodation coefficient of unity (though it is expected to be two orders smaller). Hence, it is clear that a large fraction of small sized drops could not be imaged in the experiment. Since a relatively large portion of the total heat flux is carried by droplets of small diameters, the verification of Eq. (14.9) with respect to Eq. (14.3) could not be conducted. This non-availability of the droplet distribution function, especially in the lower droplet size range, is indeed a generic problem in experimental investigations of dropwise condensation (Rose 1981, 1998, 2002). A common practice that circumvents this shortcoming is to assume a suitable droplet distribution function in such a manner that it fits the experimental heat throughput. In this chapter, we restrict our attention to single droplets in the range of 0.35 mm to D_{max}. The maximum diameter is obtained from the ratio of surface tension to the gravitational body force, i.e., the Bond number, and is defined by

$$D_{max} = K_1 \left(\frac{\sigma}{g(\rho_{liq} - \rho_{vap})} \right)^{0.5} \tag{14.11}$$

For the present experiments, D_{max} was obtained to be 4.73 mm ± 0.02 mm which gives $K_1 = 1.732$ in Eq. (14.11).

14.4 Results and Discussion

Drop shapes, condensation patterns obtained on the substarte, and simultaneous LCT images of the condesing drops are reported. Of interest is the comparison of heat transfer rates calculated from dropwise condensation patterns over the polyethylene LCT surface to those obtained from the LCT images.

14.4.1 Static Contact Angle Measurements

Contact angles of static pendant droplets of water on the polyethylene substrate were measured, as shown in Fig. 14.4. There is considerable static contact angle hysteresis with substrate inclination. The pendant drop shape can be estimated by a two-dimensional representation of the Young-Laplace equation, as outlined in Chap. 1. While the droplet contact line was assumed to be pinned in the calculations, this is not true for real surfaces, as can be seen in the figure when the substrate gets inclined. For a horizontal substrate, the contact angles justify the use of a hemispherical geometry in the one-dimensional heat transfer model. The assumption that dynamic contact angles realized during the actual condensation process would be identical to those values recorded under static conditions may not be true.

14.4.2 Heat Transfer Through Individual Drops

Figure 14.5a shows the liquid crystal thermograph of a single water droplet of diameter 2.96 mm. The hue distribution over the drople base is shown in Fig. 14.5b. Figure 14.5c shows the variation of heat throughput passing through the mid-plane of the single drop, as identified in Fig. 14.5a. Figures 14.6a, b show another example of an adjacent droplet pair and the hue distribution on their base area. Figures 14.6c, d depict the instantaneous heat transfer rate through these two individual drops, marked Drop #1 and Drop #2, along the vertical plane passing through their center, as indicated, under quasi-equilibrium conditions.

Many interesting features are observable from these data.

(a) Isotherms indicating temperature differentials can be clearly seen on the base area of the condensing pendant droplets of water. Thus, for a given temperature difference between the glass plate and the condensing vapor, the varying degree of droplet thermal resistance due to its changing droplet thickness gets manifested as temperature distribution over the droplet base.
(b) The contact line region provides the least resistance to heat flow and thus maximum heat transfer rate appears near this region. This is the zone where the thickness of the droplet is lowest, thus indicating that the ensuing diffusive thermal resistance is the lowest.
(c) An adsorbed liquid thin film exists on the substrate around the vicinity of the three-phase contact line of the condensing droplets which tends to pose a resistance to heat transfer.
(d) The central portion of the drop poses the maximum resistance to heat transfer.
(e) Smaller sized drops have a lower thermal resistance per unit area as compared to the drops of larger diameter.

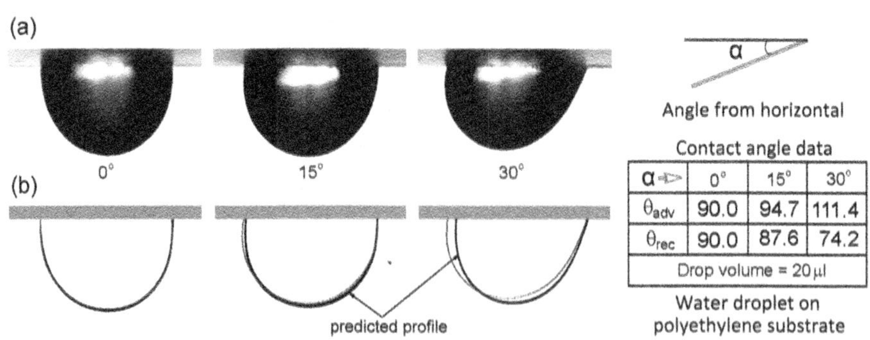

α	0°	15°	30°
θ_{adv}	90.0	94.7	111.4
θ_{rec}	90.0	87.6	74.2
Drop volume = 20 μl			

Fig. 14.4 (**a**) Static apparent contact angle of a pendant water droplet on a polyethylene substrate, which is made to incline from the horizontal (α) to study the effect of contact angle hysteresis, as shown. The advancing and receding contact angles are noted in the table. (**b**) Droplet shapes, generated with the assumption that the advancing liquid front remains pinned on the substrate, are compared with the actual experimentally obtained shapes

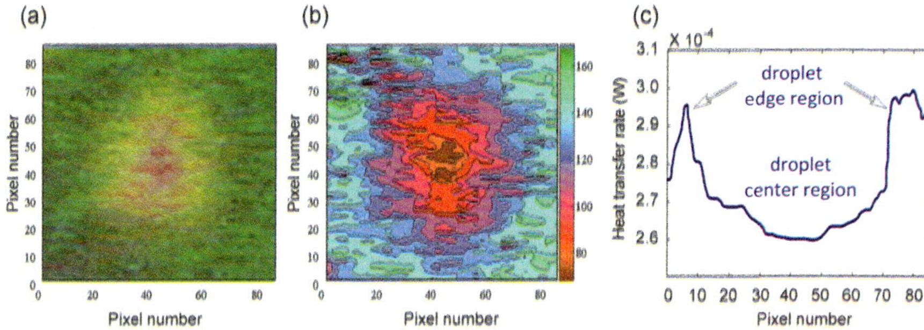

Fig. 14.5 (**a**, **b**) The figures show the LCT-RGB image of an isolated pendant droplet placed below a liquid crystal sheet during dropwise condensation and its corresponding hue contour plot. Images have been recorded after quasi dynamic steady state has been attained. (**c**) The heat transfer rate at a plane passing through the middle of the droplet is shown as a function of position

The above observations agree with the one-dimensional heat transfer model described in Sect. 14.3.3.

Figure 14.7a shows the computed heat transfer rates through individual droplets as a function of drop diameter and degree of subcooling, computed using equations (14.7)–(14.9). The accommodation coefficient for condensation of water has been taken to be unity in the present discussion (see Sect. 5.5.5 for a discussion on the subject). Two bands of data are shown; one set corresponds to results obtained by neglecting the thermal resistance of the polyethylene substrate, while the other set corresponds to the case wherein this thermal resistance is included in the one-dimensional heat transfer model. The degree of subcooling is varied from 0.4 °C to 2 °C. When the thermal resistance due to the polyethylene sheet is not considered, heat flux passing through droplets drastically increases with decreasing droplet diameter, i.e., the maximum heat flux passes through the smallest diameter droplets. The inclusion of the thermal resistance due to the polyethylene substrate decreases the overall heat transfer substantially, the deterioration effect being the largest in smaller sized droplets. Low thermal conductivity of the polyethylene sheet results in a large reduction in heat transfer through the smaller droplets in the present experiments.

The individual thermal resistances which constitute the overall thermal resistance are shown in Fig. 14.7b. Below a droplet diameter of about 0.5 mm, the thermal resistance due to the polyethylene sheet exceeds that of the net conductive resistance due to the droplet. Moreover, the interface heat transfer coefficient and the added curvature dependent thermal resistance due are not significant constituents in the overall thermal resistance. The curvature resistance becomes significant for extremely small droplets. For drop diameters beyond about 1.5 mm, the major resistance is due to the heat conduction within the droplet.

Figure 14.7a also shows the experimentally obtained heat flux through individual droplets. As stated in Sect. 14.2, the minimum droplet diameter that could be measured in the present experiment was of the order of 0.4 mm while the maximum

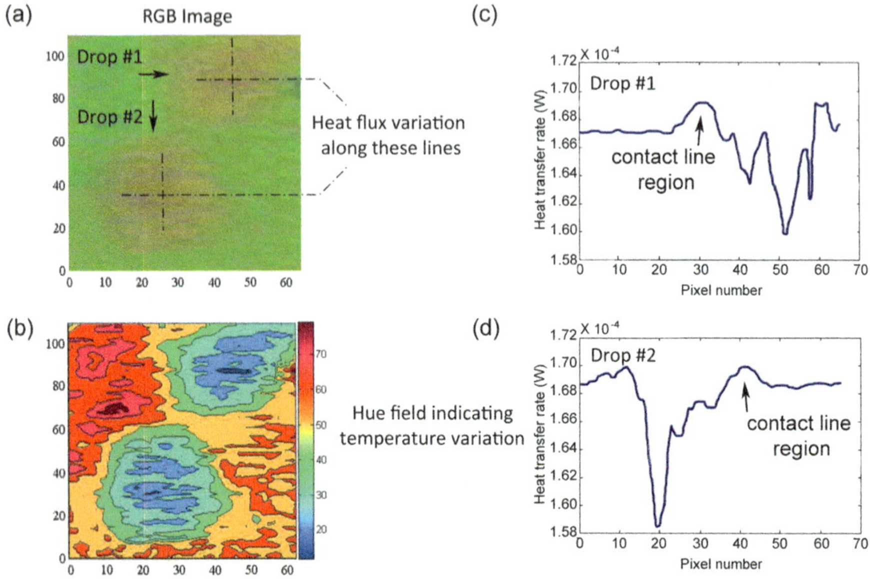

Fig. 14.6 (**a**, **b**) LCT-RGB images of two adjacent pendant water droplets, marked Drop #1 and Drop #2 below a liquid crystal sheet during dropwise condensation and their corresponding hue contour plot, at quasi steady-state. (**c**, **d**) Heat transfer rate through the mid plane passing through the two drops, as indicated along the base diameter

droplet sizes were about 4.73 mm. In this range of droplet diameters, Fig. 14.7a shows data for ten representative isolated droplets. Deviation from the predicted values becomes greater as the droplet size, and hence the thermal resistance through it, decreases. Data reduction for droplets below about 0.5 mm would not have resulted in meaningful conclusions since thermal resistance of the polyethylene sheet is dominant below this diameter. Reasonable level of accuracy is obtained with the one-dimensional heat transfer approximation for larger drops only due to rate limitation by the polyethylene substrate.

The average heat transfer coefficient estimated by using Eq. (14.6) is found to be sensitive to degree of subcooling. Specifically, for an increase in the average condensing vapor temperature, heat transfer coefficient increases from 4470 W/m^2K at 40.3 °C to 5650 W/m^2K at 41.1 and 7580 W/m^2K at 42.1 °C.

14.5 Conclusions

High resolution liquid crystal thermography permits the detection of the spatial temperature profile of individual condensing droplets. Vivid isotherms on the droplet base could be distinctly observed, when water condensed on a polyethylene sheet.

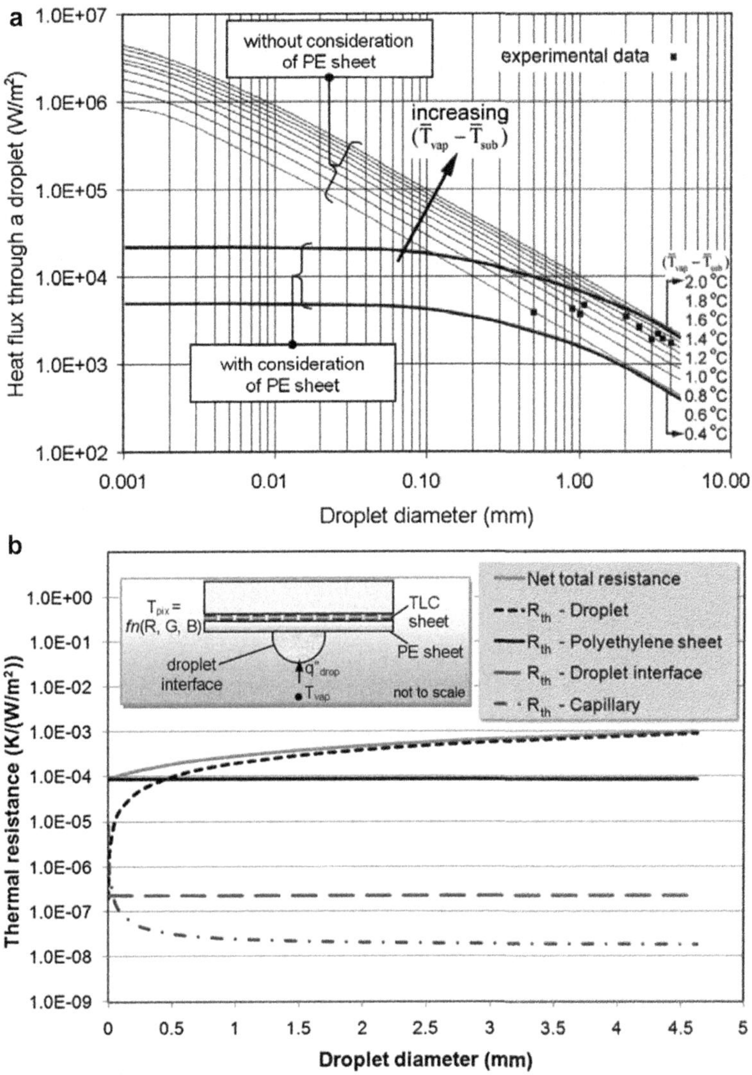

Fig. 14.7 (**a**) Heat flux variation over the base area of a drop is represented here as a function of the drop diameter. Experimental heat transfer rates have been determined from LCT data with the one-dimensional heat transfer model. The simulations shown are (**i**) without and (**ii**) with the thickness consideration of the PE substrate. (**b**) Although the highest heat flux passes through the small sized drops, the thermal resistance manifested by these drops diminishes with decreasing diameter, and the resistance due to the PE foil starts dominating, thereby thwarting heat transfer through the smaller diameter droplets

Both, normal light videography and liquid crystal thermography was simultaneously performed to estimate local and average heat transfer coefficients of condensing pendant drops with one-dimensional approximation. The three-phase contact line region of the droplet was seen to provide the path of least resistance for heat transfer. As expected, the average heat transfer rate increased with increase in subcooling and operating saturation pressures. It is adequate to image condensation patterns of drops on a surface to estimate local and average heat transfer coefficients by including all the applicable thermal resistances encountered between the condensing vapor and the subcooled substrate.

Chapter 15
Evaporation Dynamics of a Sessile Droplet on a Hydrophobic Surface

Sachin K. Singh, Mohit Gogna, Sameer Khandekar, and K. Muralidhar

Nomenclature

c	Concentration, mol/m^3
c_p	Heat capacity, J/(kg-K)
D_c	Diffusion coefficient, m^2/s
RH	Relative humidity, %
h_{fg}	Latent heat of vaporization, J/kg
k	Thermal conductivity, W/(m-K)
M_w	Molecular weight, kg/mol
m''	Evaporative flux, kg/(m^2-s)
r	Radius, m
r_c	Contact radius, m
R	Universal gas constant, J/(kg-K)
T	Temperature, K
t	Substrate thickness, m
q''	Heat flux, W/m^2
α	Thermal diffusivity, m^2/s
θ	Contact angle, °
ρ	Density, kg/m^3
σ	Surface tension, N/m

Subscripts

atm	Atmospheric pressure
sat	Saturated

S. Khandekar, K. Muralidhar, *Drop Dynamics and Dropwise Condensation on Textured Surfaces*, Mechanical Engineering Series,
https://doi.org/10.1007/978-3-030-48461-3_15

15.1 Introduction

The previous chapters discussed drop dynamics and dropwise condensation in detail. Drop motion arising from gravitational instability and coalescence were key intermediate steps during the condensation process, contributing to both shear stress and heat transfer rates. In this chapter, we examine evaporation as a contrasting physical mechanism to condensation. Even without liquid motion, evaporation rates are shown to be affected by a distribution of humidity around the liquid-gas interface, but also by interfacial cooling, wettability, and thermophysical properties of the substrate.

Evaporation of a sessile droplet placed on a solid substrate is encountered in a variety of applications. Examples can be seen in inkjet printing (Talbot et al. 2014; Wijshoff 2018), additive manufacturing (Galliker et al. 2012; Hirt et al. 2017), painting (Deegan et al. 2000; Popov 2005), hot spot cooling (Kumari and Garimella 2011), spraying of pesticides (Yu et al. 2009), DNA mapping (Akbari and Foroutan 2018; Bhar et al. 2018), virus transmission (Bhardwaj and Agarwal 2020), and thin film coatings (Kimura et al. 2003). A complete understanding of factors affecting droplet evaporation is important in improving the performance of many of these devices.

Picknett and Bexon (1977) were among the first to study sessile droplet evaporation. The authors found that evaporation of sessile droplet on solid substrate takes place in two distinct modes: constant contact radius (CCR) and constant contact angle (CCA). Assuming vapor diffusion as the only transport mechanism responsible for these modes, the authors derived a theoretical solution for evaporation rates from a simple one-dimensional differential equation model. Subsequent studies show droplet evaporation to be categorized into four modes: CCR, CCA, a mixed mode (Xu et al. 2013) that includes a gradual decrease in contact radius and contact angle, and stick-slip mode (Shanahan 1995) that shows rapid pinning and de-pinning of the contact line. Hu and Larson (2002) studied vapor diffusion during sessile droplet evaporation both numerically and analytically, and proposed correlations for the evaluation of evaporation rates. Popov (2005) derived an improved vapor diffusion-based model that can be employed to calculate both the evaporative flux and evaporation rate for a droplet over a range of contact angles ($0° < \theta < 180°$).

In addition to vapor diffusion, droplet evaporation involves several other transport mechanisms, as shown schematically in Fig. 15.1. For instance, droplet evaporation is accompanied by interface cooling due to latent heat supplied for vaporization by the bulk liquid. The resulting temperature gradients developed within the drop drive heat transfer in the solid, liquid, and gas phases. David et al. (2007) experimentally demonstrated evaporative cooling at liquid-gas interface and found that the droplet evaporation rate was limited by the thermal properties of the substrate. Dunn et al. (2009) improved upon the diffusion model by encompassing both evaporative cooling and heat conduction in all the three phases. The predicted evaporation rates were in very good agreement with experiments. Later, Saada et al. (2013) improved the model of Dunn et al. (2009) by incorporating thermal diffusion

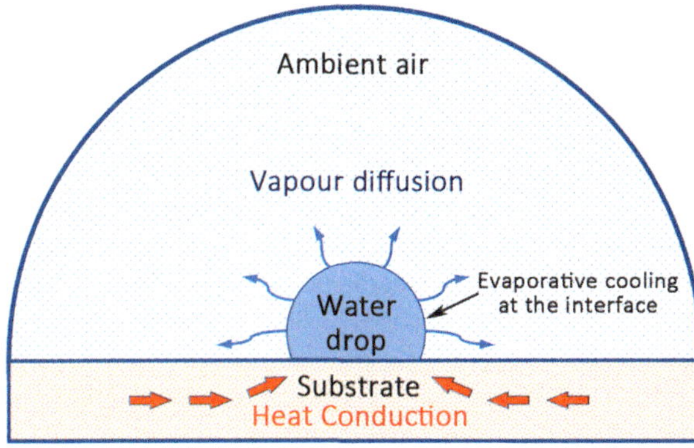

Fig. 15.1 Transport mechanisms involved in evaporation of a sessile droplet placed on solid substrate—vapor diffusion, evaporative cooling at interface and heat conduction within the substrate. The far-field is a region of constant relative humidity

in the gas phase. The non-uniform evaporation at the liquid-gas interface is accompanied by non-uniform temperatures distributed along the liquid-gas interface, leading to convective flow inside the evaporating droplet. Deegan et al. (2000) modeled convective flow inside a droplet to predict the commonly observed coffee stain effect. Thereafter, Hu and Larson (2006) experimentally and theoretically evaluated Marangoni convection in droplets of both water and organic liquids. While a strong Marangoni flow was observed in organic liquids, Marangoni flow was observed to be weak in water droplets. Sobac and Brutin (2012) experimentally investigated the thermal effect of substrate on sessile droplet evaporation in both hydrophilic and hydrophobic conditions. At ambient temperatures, the experimental results were mostly in agreement with quasi-steady, vapor diffusion-driven model. However, when the temperature of the substrate increased, the vapor diffusion models were incapable of precisely predicting the rates of droplet evaporation. Wang et al. (2015a, b) proposed a combined field approach to evaluate the effect of the underlying substrate and evaporative cooling at the liquid-gas interface. The influence of the substrate was found to be largely coupled with evaporative cooling. Later, Chandramohan et al. (2017) reported infrared thermography experiments on the effect of interfacial evaporative cooling on temperature distribution in an evaporating droplet deposited on a heated copper substrate.

In addition to the thermal properties, the evaporation dynamics of a sessile water droplet is also limited by the surface characteristics of the substrate. Shin et al. (2009) studied the evaporation dynamics of sessile droplet on hydrophobic and hydrophilic surfaces. As the hydrophobicity of the substrate became stronger, the pinning time was found to decrease, along with an increase in the evaporation time. Song et al. (2011) experimentally studied the evaporation behavior of water drops on various surfaces with a wide range of wettability. An empirical model $f(\theta) = a\theta + b$ was proposed with two empirical material constants a and b, for predicting the evolution of droplet volume during evaporation on substrates of distinct wettability. Sobac and Brutin (2011) investigated the influence of substrate wettability on the evaporation process. The experimental data were in very good agreement with Popov's diffusion model (Popov 2005) over a wide range of wettability (17° and

135°). Recently, Dash and Garimella (2013) studied the evaporation dynamics of water droplets placed on superhydrophobic substrates with negligible hysteresis. Droplet evaporation was observed to occur in the CCA mode of evaporation. However, Popov's diffusion model (Popov 2005) was found to overestimate the evaporation rate for hydrophobic substrates, as it does not consider evaporative cooling at the liquid-gas interface. Thereafter, Pan et al. (2013, 2014) developed a droplet evaporation model that investigated the effect of substrate wettability on the relevant transport mechanisms—evaporative cooling, natural convection in gas domain, along with vapor diffusion. The evaporative cooling and natural convection were found to counter each other, with gas-phase natural convection dominating for high wettability substrates and evaporative cooling dominating for low wettability substrates. To predict the shape evolution of an evaporating droplet on superhydrophobic surfaces, Li et al. (2017) developed a mathematical model based on the interplay of driving force and resistance at three-phase contact line. The droplet evolution during evaporation on superhydrophobic substrate was found to undergo the following sequence of modes: (i) CCR, (ii) CCA, (iii) CCR, and (iv) mixed mode. Around the same time, He et al. (2017) experimentally examined the evaporation dynamics of a multicomponent droplet on a chemically patterned surface. Compared to a homogeneous substrate, the chemically patterned substrate was found to enhance evaporation by stretching the contact line, along with change in evaporation modes from three ((i) CCR, (ii) CCL, and (iii) mix mode) for homogeneous to two ((i) CCR and (ii) moving contact line) for a patterned substrate. Recently, Misyura et al. (2020) experimentally investigated the evaporation of a sessile droplet placed on a hot wall with cavities. Compared to a smooth wall, the presence of cavities increased the liquid convection inside the droplet leading to enhanced droplet evaporation. Therefore, textured surfaces modify the dynamics of droplet evaporation by altering the effective thermal resistance between solid substrate and gas phase, evaporation modes, three-phase contact line motion, and convective flow inside the droplet.

Evaporation of a sessile droplet has been shown in the literature to be a complex process that is governed by various transport mechanisms. A numerical model is presented here to understand the interaction among these factors and their combined effect on the resulting droplet evaporation rates. An extended vapor diffusion model has been built in COMSOL® which incorporates evaporative cooling at the liquid-gas interface and heat conduction in solid-liquid-gas domain, in addition to vapor diffusion in the gas phase. Fluid convection in liquid and gas domains have been neglected in the model as the water droplets evaporating on an unheated substrate are known to encounter negligible convective flow within (Hu and Larson 2006; Wang et al. 2015a, b; Ristenpart et al. 2007). Conduction is the dominant mode of heat transfer during droplet evaporation on unheated substrates. An experimental setup has been built to carefully measure evaporation rates of sessile water droplets on three different substrates. Results thus obtained are used for validation of the extended vapor diffusion model. Thereafter, the model has been used to investigate the combined effect of substrate wettability and thermal properties on evaporation dynamics of a sessile water droplet.

15.2 Experimental Details

An experimental setup was constructed to carefully measure the droplet evaporation rate under controlled conditions and understand the effect of substrate wettability and thermal properties on sessile droplet evaporation. Several experiments were carried out with water droplets of 5 μl volume over copper (k = 385 W/mK, Bi ~10^{-2}) and glass (k = 1.6 W/mK, Bi ~1) substrates, with a range of wettability controlled from highly hydrophilic (θ = 22°) to highly hydrophobic (θ = 147°). Even though no convection is considered in the models of the present work, a representative Biot number (Bi = $U \cdot t/k$) has been calculated using the universal heat transfer coefficient (U, W/mK) for comparison of the heat transfer characteristics of substrates. The universal heat transfer coefficient is calculated as the ratio of average heat flux q'' at the solid-liquid interface and the difference between the average contact surface temperature and average liquid bulk temperature.

15.2.1 Experimental Apparatus

The experimental setup was constructed in such a way that the boundary conditions during droplet evaporation could be carefully controlled. The relative humidity (RH) in the test section was carefully controlled by providing an accurate mix of dry and saturated air through a humidity control loop, as shown in Fig. 15.2a (Singh et al. 2013). The humidity control loop consisted of an air compressor, a de-humidifying chamber, an insulated test section, an air hydrating mechanism, along with connecting pipes with regulatory valves. The test chamber is a large cubical volume (80 × 40 × 40 mm), as shown in Fig. 15.2b. The four longitudinal sides of the cubical were completely sealed, whereas the remaining two lateral sides were closed with the help of filter paper with 8 μm average pore size. Constant humidity conditions were maintained at the filter paper end of the test chamber by means of conditioned air supply at a constant flow rate, thus providing constant humidity (RH) Dirichlet boundary conditions. The complete setup was arranged in such a way that bulk convection is minimized inside the test cell; the vapor transport was essentially diffusive in nature. RH in the test section was measured using a GE RH sensor (least count: 0.36%) and recorded in a PC using National Instruments® (NI-4351) data acquisition board.

A small hole on top of the test section enabled insertion of an isolated micro-liter droplet (deionized, degassed, and filtered water) with the help of a micro-syringe. The drop was inserted only when constant 45% RH condition was maintained inside the test chamber. On placing a drop in the test chamber, diffusive evaporation started immediately due to concentration gradient formed between the drop interface and the ambient. It was observed that the RH in the test chamber started rising as soon as the drop was placed on the substrate. The RH rose for initial few minutes, reached a saturation value, and then dropped to initial base value towards the end. The

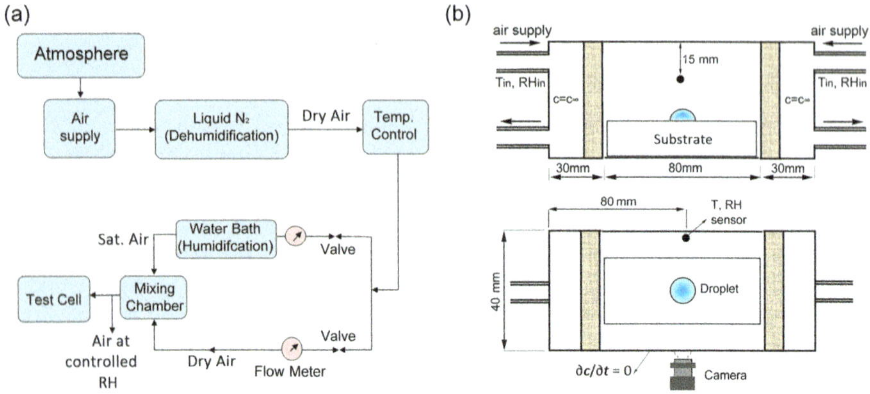

Fig. 15.2 (**a**) Air flow and humidity control loop used in the experimental setup for studying the evaporation of sessile water droplet placed on substrates of different wettability and thermal properties; (**b**) schematic diagram of the experimental setup showing the droplet evaporation chamber, location of RH sensor, and the image/data acquisition system (adapted from Singh et al. 2013)

evaporating dynamics of a sessile drop on different substrates was captured using a micro lens camera. The photographic images along with the RH data of an evaporating drop were analyzed together. These images were analyzed using ImageJ® software to measure the contact angle, and B-splines were used to fit the shape of the drop interface. Besides the equilibrium contact angle, the software was also utilized to measure the droplet height and contact radius on the substrate. These inputs are necessary to recreate the drop at different time instants on the simulation platform, which were then discretized as an evaporating sessile drop in a quasi-steady framework.

15.2.2 Substrate Preparation

Three substrates with contact angles of $\theta = 22°$, $91°$, and $147°$ were used in the experiment to understand the effect of wettability on droplet evaporation rates. Two of these substrates ($\theta = 91°$ and $147°$) were fabricated from a thin ($t = 1$ mm) copper plate. The preparation of highly hydrophobic substrate is achieved by chemically texturing the copper surface. The third high wettability substrate ($\theta = 22°$) used in experiments was a BLUE STAR® micro-glass slide. The detailed steps followed in preparation of highly hydrophobic substrates are given below.

Cleaning and Finishing of Copper Surface

The copper substrate was first surface finished by grinding with 1500 grit silicon carbide abrasive grinding paper. The finished substrate was cleaned with acetone and flushed with deionized water. Thereafter, the substrate was immersed in a solution of 20 g NaCl, 250 ml acetic acid in 1 l deionized water and placed in the furnace at 60 °C for 2 h. This was again followed by cleaning of the substrate with acetone and flushing with deionized water. The cleaned substrate was sonicated in ethanol bath for 20 min. The substrate was then rinsed with deionized water and dried with compressed nitrogen gas. This substrate had a contact angle of 91° with water. The SEM image of the hydrophobic copper substrate (in Fig. 15.3a–c) shows a smooth surface with microscale ridges, attributed to surface finishing/milling process.

Preparation of Hydrophobic Surface Using Ammonium Persulfate

The copper substrate cleaned and finished (contact angle, $\theta = 91°$) as per the procedure described above was immersed in an aqueous solution of HCl (4.0 mol/ m^3) for 60 s. The substrate was then sonicated in ethanol bath for 20 min, immersed in a solution of 2.5 M NaOH for 15 min and subsequently in 0.13 M $(NH_4)_2S_2O_8$ for 120 min at room temperature. The substrate was thereafter rinsed with deionized water and dried with compressed nitrogen gas. This substrate was immersed in 20 g/ L ethanol solution of steric acid for 24 h. The substrate was then heated in vacuum chamber at 60 °C for 1 h. The substrate thus prepared, gave a contact angle of $147° \pm 2°$. Fine randomly oriented fern-like structures (Fig. 15.3e, f) on the scale of 1–2 μm are seen at 5000× magnification through a Scanning Electron Microscope (SEM). The SEM image of the micro-glass slide shown in Fig. 15.3g, h reveals a very smooth surface with only sub-micron deformities.

15.3 Simulation Details

As noted, evaporation of liquid droplet is a complex process, governed by various transport mechanisms such as vapor diffusion, evaporative cooling, and heat conduction between the droplet, substrate, and the ambient surrounding. Furthermore, it is also affected by the wettability and thermal properties of the substrate. To analyze the combined effect of substrate wettability and thermal properties on the droplet evaporation dynamics, two axisymmetric models in *r-z* coordinates have been built using COMSOL® platform. *Model #1,* referred here as the simple vapor diffusion model (Hu and Larson 2002), solves only the vapor diffusion equation in the gas domain. *Model #2,* to be called the extended vapor diffusion model, solves for evaporative cooling and heat conduction in solid-liquid-gas domain, in addition to the vapor diffusion in gas domain. The simple vapor diffusion model on the

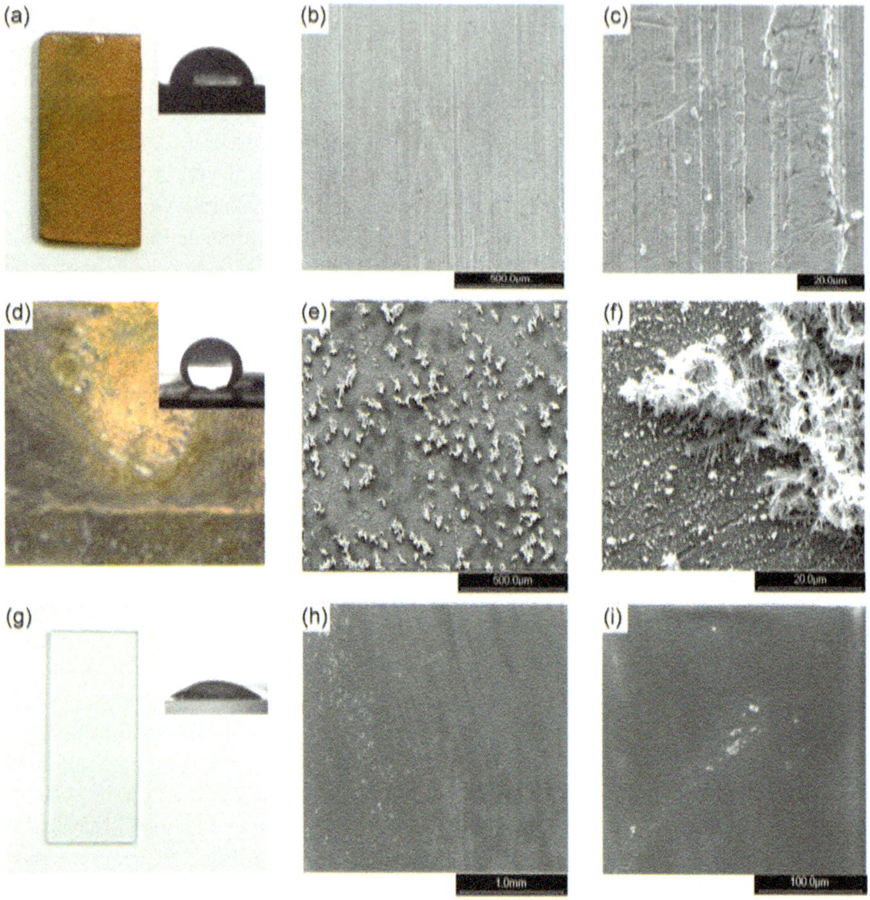

Fig. 15.3 Substrates with varying wettability used in the study of sessile droplet evaporation. (**a**) Polished copper ($\theta = 91°$, Bi ~10^{-2}), (**d**) chemically treated copper ($\theta = 147°$, Bi ~10^{-2}), and (**g**) glass slide ($\theta = 22°$, Bi ~1). Images (**b**, **c**), (**e**, **f**), and (**h**, **i**) show the corresponding SEM images for polished copper, chemically treated copper and glass slide. Inset images show the initial shape of 5 μl water droplet when deposited on the respective substrates

simulation platform utilizes only the transport of diluted species (TDS) physics, whereas the extended vapor diffusion model utilizes both the transport of diluted species physics along with heat transfer (fluid and solid) physics. The characteristic timescale for vapor diffusion (R_i^2/D_c, where R_i is the initial radius of droplet) is 1.75, 0.067, and 0.043 s, for droplet radius of 6.76, 1.33, and 1.06 mm, on 22°, 91°, and 147° substrates, respectively. Since, the timescale of droplet evaporation is significantly larger than the timescale of the transport mechanisms incorporated in simulation, the present model utilizes a quasi-steady assumption for droplet evaporation, i.e., time-independent study, correcting only for the drop shape.

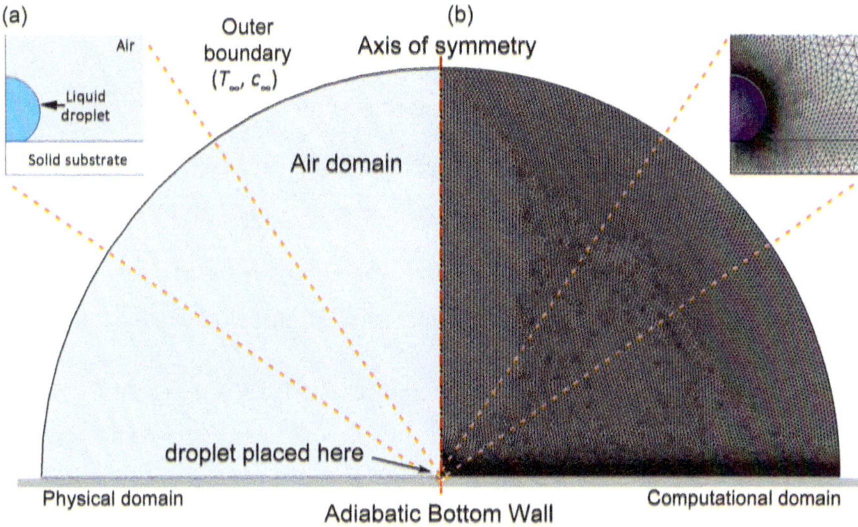

Fig. 15.4 (**a**) Geometry considered for simulation along with the outer boundary conditions used in COMSOL®; (**b**) triangular mesh used in the simulation domain. The mesh near liquid-gas and solid-gas interface is much finer than near outer boundary. The contact angle of the droplet shown in inset is 147°

The simulation considers a water droplet of a fixed initial volume (5 μl), evaporating on hydrophilic (22°), hydrophobic (91°) and highly hydrophobic (147°) substrates, respectively. The droplet evaporates in a controlled environment with ambient temperature and relative humidity set at 298.15 K and 45%, respectively. The droplet volume, liquid-solid contact angles, substrate materials and ambient conditions used in simulations are similar to those used/observed in experiments, so that direct comparison can be made between the simulations and the experiments. The simulation geometry with the boundary conditions applied to evaporating droplet, along with triangular mesh used for 147° contact angle droplet, is shown in Fig. 15.4, as an example.

The governing equations of evaporation of a single drop are

Gas domain diffusion equation $$\nabla \cdot (-D_c \nabla c) = 0 \tag{15.1}$$

Saturated vapor concentration $$c_{sat} = \frac{p_{sat}}{RT} \tag{15.2}$$

Solid, liquid and gas domains energy equation $$\nabla \cdot (-k \nabla T) = 0 \tag{15.3}$$

Table 15.1 Thermodynamic and thermophysical properties used in the simulations

Properties of water (at 298.15 K)	
Latent heat of vaporization, h_{fg}	2448 (kJ/kg)
Diffusivity coefficient of water vapor in air, D_c	2.61×10^{-5} (m^2/s)
Saturation concentration, c_{sat}	1.2741 (mol/m^3)
Relative humidity, RH	0.45
Vapor molecular weight, M_w	18.015 (g/mol)

$$\text{Liquid–gas interface evaporative cooling} \quad -n\cdot(-k\nabla T) = h_{fg}n\cdot(-D_c\nabla c) \tag{15.4}$$

The temperature and heat fluxes across all other interfaces in the geometry shown in Fig. 15.1 are assumed to be continuous

$$T_1 = T_2 \text{ and } k_1\left(\vec{n}_1\cdot\nabla T\right) = k_2\left(\vec{n}_2\cdot\nabla T\right) \tag{15.5}$$

The saturation pressure p_{sat} (appearing in Eq. 15.2) at temperature T (in Kelvin) is calculated by

$$p_{sat}(T) = 610.7 \times 10^{7.5f(T)} \quad [\text{Pa}]$$
$$\text{where } f(T) = \frac{T - 273.15}{T - 35.85} \tag{15.6}$$

In addition to the basic properties internally available in COMSOL®, the properties listed in Table 15.1 are also used in the simulations of sessile droplet evaporation. Furthermore, thermodynamic properties of air-water vapor mixture in the gas domain for *Model #2* is available in COMSOL®, when moist air is chosen as the fluid type in the gas domain. The initial temperature across all domains and boundaries is specified as 298.15 K.

The formula for saturated vapor concentration (c_{sat}), given in Eq. (15.2), is used only in *Model #2* for specifying vapor concentration at the liquid-gas interface. This formula ensures coupling of saturated vapor concentration to the interface temperature, resulting in reduced vapor concentration when evaporating cooling takes place at the interface. For *Model #1,* which is isothermal, the vapor concentration at liquid-gas interface is specified as c_{sat} given in Table 15.1. The vapor concentration at the outer boundary is specified as RH $\times$ c_{sat}. The rate of evaporation of the liquid from the liquid-gas interface of the droplet is directly affected by the temperature of the interface. This interfacial condition, in turn, affects the vapor pressure at the interface, and hence controls the vapor diffusion process in the atmosphere. As enthalpy is removed from the droplet due to vaporization, its temperature reduces, thereby affecting the temperature of the solid substrate on which the droplet is present.

The diffusion coefficient of water vapor in dry air is a function of temperature and is often given by the expression (Hu and Larson 2002)

$$D_c(T) = 2.5 \times 10^{-4} \exp\left(-\frac{684.15}{T}\right) \tag{15.7}$$

Here, T is temperature in Kelvin and the diffusion coefficient is recovered in units of m^2/s. For the simulations of this chapter, an average value given in Table 15.1 has been incorporated.

For the numerical solution of the problem, a total of 55,000–66,000 triangular mesh elements have been used across all domains based on the droplet contact angle. A local refinement of mesh is carried out at the liquid-gas interface, such that the mesh in gas domain is much finer near the liquid-gas interface than the outer boundary. A mesh-independence study was also performed to confirm that the simulation results were insensitive to further refinement of mesh.

15.3.1 Drop Discretization

As can be seen in Fig. 15.5a–c, the general trend of temporal RH data for an evaporating drop varied exponentially during the initial stages, stabilized and then kept diminishing as the surface area of the sessile drop kept reducing with time. To understand the effectiveness of *Models 1* and *2* in predicting evaporation rates and temperature distribution at different stages of evaporation process, the RH data was discretized into different sections (labeled *S1*, *S2* ...). The droplet characteristics (contact radius, height, contact angle) obtained via ImageJ® by the analysis of the central image (for instance, the images recorded at $t = 175$, 675, 1300, 1925, and 2400 s in Fig. 15.5a) in each subsection was later used to reconstruct a similar sessile droplet for simulations. Calculations were carried out using *Models 1* and *2* described above for each reconstructed droplet profile to predict the total interfacial evaporative flux and temperature distribution around the evaporating droplet at the corresponding time instant. The droplet evaporation rates were obtained by integration of the evaporative flux along the liquid-gas interface.

15.4 Results and Discussion

To evaluate the combined effect of substrate wettability and thermophysical properties on evaporation of sessile droplet, several experiments and simulations were carried out. Here, evaporation of 5 μl water droplet was studied on substrates with different contact angles ($\theta = 91°$, 147°, and 22°) on copper (Bi ~10^{-2}) and glass (Bi ~1). Experimental results are presented below wherein the measured data is utilized to validate *Models 1* and *2* described above in predicting evaporation rates at various time instants during the evaporation process.

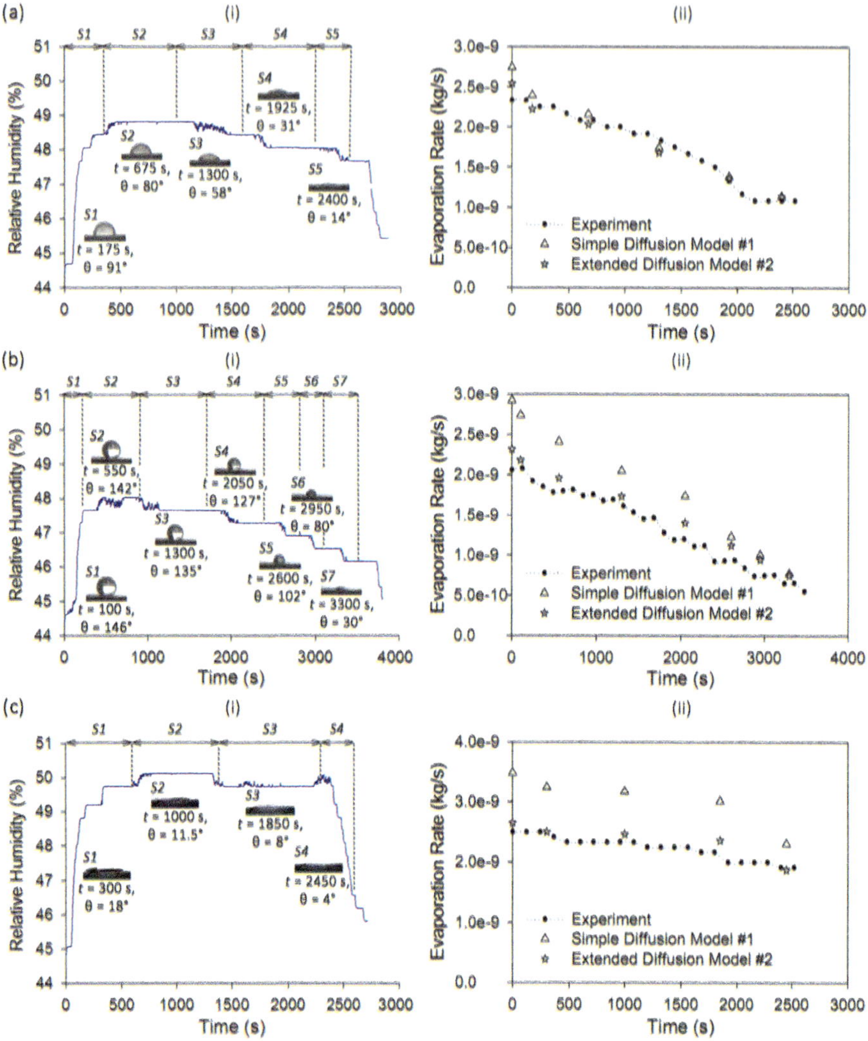

Fig. 15.5 Variation of relative humidity (RH) in the test section (left), and the corresponding transient variation of evaporation rate (right) when a 5 μl water droplet is placed on individual substrates. (**a**) Polished copper ($\theta = 91°$, Bi ~10^{-2}), (**b**) chemically treated copper ($\theta = 147°$, Bi ~10^{-2}), and (**c**) glass slide ($\theta = 22°$, Bi ~1). The evaporation phenomenon is studied by splitting the RH curve into smaller subsections and evaluating evaporation kinetics for the droplet shape corresponding to each subsection. The results compare the simulated evaporation rates from the simple vapor diffusion-only model and extended vapor diffusion model (incorporating vapor diffusion, heat conduction, and evaporative cooling) with experimentally measured evaporation rates

15.4.1 Validation of Simulation with Experiments

Droplet Evaporation on Polished Copper Substrate ($\theta = 91°$, Bi ~10^{-2})

The evaporation rate of a 5 μl sessile water droplet on a polished copper substrate was carefully measured in a controlled environment. The resulting droplet evolution and evaporation rates are shown in Fig. 15.5a, (i) and (ii), respectively. The droplet is observed to be in the constant contact angle (CCA) mode during evaporation. From Fig. 15.5a-(ii), it is seen that the evaporation rate gradually decreases with time, as the droplet evaporation is accompanied by a progressive decrease in droplet surface area and its footprint radius. On comparison of experimentally measured evaporation rates with numerically predicted evaporation rates at different time instants, it is found that the simple vapor diffusion model is in reasonable agreement of the droplet evaporation rate for the high thermal conductivity hydrophobic (91°) substrate. On incorporating the evaporative cooling and heat conduction to the diffusion model, the resulting model gives an improved prediction of the droplet evaporation rate.

Droplet Evaporation on a Superhydrophobic Copper Substrate ($\theta = 147°$, Bi ~10^{-2})

On placing of a 5 μl water droplet on highly hydrophobic copper substrate, it is found to initially evaporate in CCA mode of evaporation for about 400 s, followed by a mixed mode of evaporation for the rest of the time (Fig. 15.5b-(i)). In Fig. 15.5b-(ii), it is clear that the simple vapor diffusion model greatly overestimates droplet evaporation rates for hydrophobic surfaces. In contrast, the extended vapor diffusion model gives very good prediction of evaporation rates, as it considers the interfacial evaporative cooling and resulting reduction in interfacial vapor concentration. The overestimation of vapor diffusion model with respect to the extended model decreases with time as the droplet contact angle decreases, leading to a reduction in evaporative cooling as well.

Droplet Evaporation on Glass Substrate ($\theta = 22°$, Bi ~1)

The evaporation of water droplet on glass substrate follows a constant contact radius (CCR-pinned) mode of evaporation (Fig. 15.5c-(i)). On evaluation of the evaporation rate during sessile droplet evaporation on a low thermal conductivity substrate (glass), the simple vapor diffusion mode of evaporation is found to greatly overestimate the evaporation rate (Fig. 15.5c-(ii)). However, the fully coupled extended model provides excellent prediction of droplet evaporation rates, as it also considers the accompanying heat conduction and evaporative cooling in the solid-liquid-gas domain.

Therefore, the one-equation, simple vapor diffusion model is limited in its capacity to predict sessile droplet evaporation for low thermal conductivity and low wettability substrates. The extended vapor diffusion model given by Eqs. (15.1)–(15.6) is superior in this respect.

15.4.2 Distribution of Evaporative Mass Flux over the Drop

The simple vapor diffusion model of evaporation is found to predict droplet evaporation rates reasonably well and matches the extended model, if the substrate is highly wetting and has relatively high thermal conductivity. However, if the substrate has low thermal conductivity, the simple vapor diffusion model predicts a higher droplet evaporation rate than the extended vapor diffusion model, even for high wettability substrates. These trends are clearly seen in Fig. 15.5.

In addition to overall evaporation rates, the spatial distribution of evaporation flux and temperature at the interface have also been evaluated for evaporation of a 5 μl water droplet on various substrates (Fig. 15.6). Uniformly, the evaporative flux is higher towards the contact line and is smaller at the apex of the drop. The simple vapor diffusion model predicts the highest and lowest evaporation rates at the three-phase contact line for low and high contact angle droplets, respectively. However, when interfacial evaporative cooling and substrate heat conduction are incorporated into the diffusion model, evaporative flux is found to be the highest at the contact line for evaporation of water droplet on all substrates, regardless of their wettability. This is because the temperature at the liquid-gas interface is also the highest at the contact line due to heat conduction from the substrate. From Fig. 15.6, it is seen that evaporative flux decreases with a reduction in the thermal conductivity of the substrate.

15.5 Closure

Evaporation rate of a single droplet is examined in this chapter from simulation neglecting fluid motion and experiments. The simple vapor diffusion model is shown to be inadequate. Including interface cooling and conjugate heat transfer to the substrate are seen to be important. An extended vapor diffusion model is developed in COMSOL® to evaluate the evaporation rate of a sessile droplet on substrates of distinct wettability and thermophysical properties. The transient evaporation process is modeled as multi-stage steady-state process, where the droplet geometry during the transient is taken from experimental data. The numerical model incorporates the vapor diffusion process, cooling of the bulk liquid droplet due to evaporative flux at the liquid-gas interface, and heat conduction in the solid-liquid-gas domain to temperature differentials created during the evaporative process. The

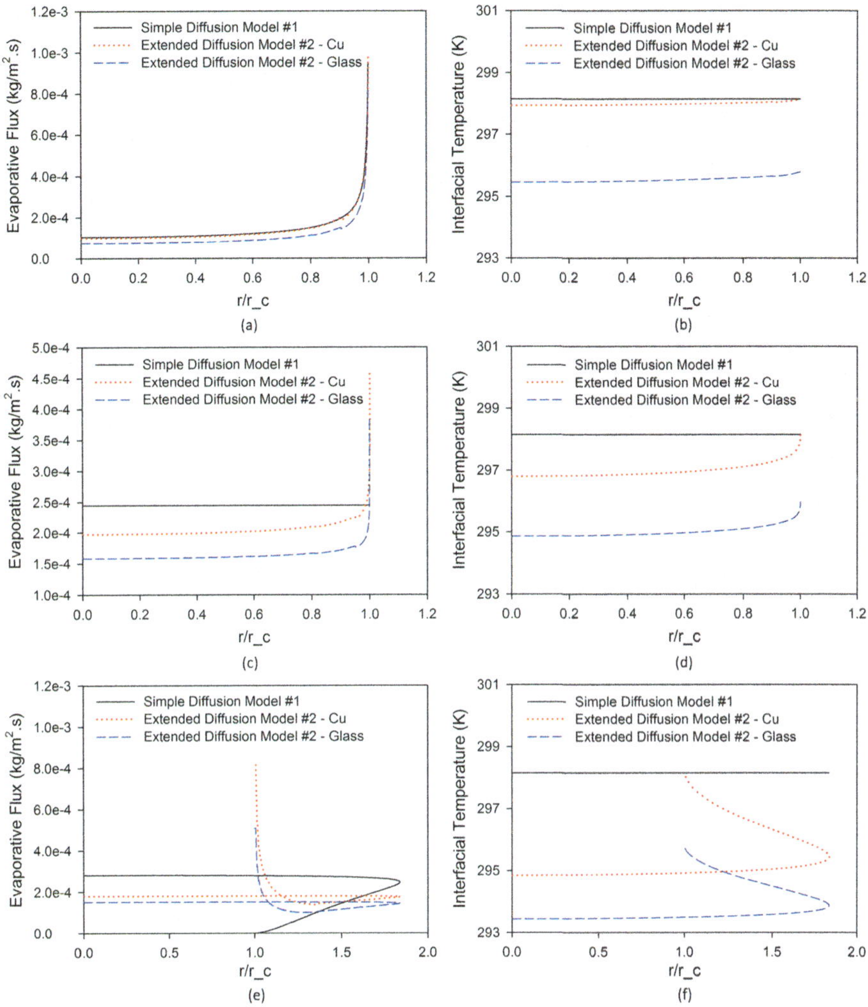

Fig. 15.6 Variation of evaporative flux and temperature along the liquid-gas interface of an evaporating sessile droplet on substrates of varying wettability (**a**, **b**) $\theta = 22°$, (**c**, **d**) $\theta = 91°$, and (**e**, **f**) $\theta = 147°$. Evaporative flux and interfacial temperature have been calculated for copper and glass substrates, considering both the simple vapor diffusion model and the extended model

numerical solution is seen to be in good agreement with experiments. From the numerical study, the following conclusions can be drawn:

(a) The simple vapor diffusion model over-predicts the droplet evaporation rate, especially on highly hydrophobic substrates. This is because an increase in the equilibrium contact angle decreases the contact area available for heat conduction from the substrate to the droplet decreases, resulting in much higher evaporative cooling at the liquid-gas interface. As a result, the evaporative flux is smaller than that predicted by the vapor diffusion model.

(b) An overestimation of evaporation rate from the simple vapor diffusion model is compounded for low thermal conductivity substrates such as glass. The droplet does not cool rapidly enough at the wall, leading to further lowering of the droplet temperature and a reduced evaporative mass flux.

(c) Evaporative flux is non-uniformly distributed around the drop surface, being low over the top surface and particularly high around the three-phase contact line.

Chapter 16
Closing Remarks and Prospects

Sameer Khandekar and K. Muralidhar

16.1 Droplet Statics

Dropwise condensation is a multiscale process; small-scale processes impact the macroscopically observed drop distribution and heat transfer rates. The smallest drop that is nucleated over the substrate arises from a balance of internal pressure and drop curvature superimposed over the thermodynamic phase behavior. Subsequently, it grows by direct condensation of vapor over the liquid-vapor interface. In this connection, static considerations of a liquid drop resting over a textured surface are discussed in the text. The governing equation arising from force balance is the Young-Laplace equation, with surface characteristics specified in terms of the contact angle distribution around the three-phase contact line of the drop. Methods of solving this equation, mainly for the drop shape, are discussed. The contact line and the contact angle are shown to be important factors in fixing the drop shape.

16.2 Droplet Spreading

For a given volume of the drop and prescribed surface properties and inclination, the drop attains an equilibrium shape. Of course, the shape tends to become a part of a sphere for increasing levels of hydrophobicity and diminishing contact hysteresis. Any departure from equilibrium leads to drop spreading, namely a process in which fluid velocity arises from non-equilibrium conditions. Spreading itself is a multiscale process in time during which droplet oscillations may be observed, but is ultimately followed by a viscous dissipation controlled, asymptotic decay of the transience. The spreading process is intricate and is rich in details. In the text, a contact line model that is free of modeling approximations is presented. It is shown to match experiments quite well.

S. Khandekar, K. Muralidhar, *Drop Dynamics and Dropwise Condensation on Textured Surfaces*, Mechanical Engineering Series,
https://doi.org/10.1007/978-3-030-48461-3_16

16.3 Coalescence Dynamics

The coalescence model adopted in most studies approximates the nuances of the overall merger process, by requiring that it occurs instantaneously, moving from one equilibrium shape to the next. Preliminary experiments highlight the subtleties of the process and reveal complex flow patterns including oscillations of the free surface and large instantaneous wall shear stresses and wall heat fluxes. Specifically, drop coalescence commences when two droplets approach each other and contact either at the three-phase contact line or above the surface, depending on the degree of hydrophobicity of the surface. A tiny liquid bridge is immediately formed, induced by the van der Waals forces. The coalescence process gets initiated by the extra surface energy released in the process. The difference in internal pressure between the two drops drives fluid motion. Immediately afterward, the coalescence process tends to get limited by viscous and inertia forces. In typical water droplets, free surface oscillation can last ~20–40 ms, depending on their size, substrate orientation, and thermophysical properties. Long-term relaxation can occur over 40–100 ms. Rapid transients in the early stage of coalescence will induce large shear stresses over the substrate, further accompanied by enhancement of the wall heat fluxes. To address this issue, experiments on coalescence of sessile and pendant drops were carried out on a hydrophobic surface. The phenomenon was imaged using a high-speed camera. The length, velocity, and timescales of coalescence were determined and compared with analytical estimates. This data was subsequently integrated with the dropwise condensation model to incorporate finite time effects in the simulation.

16.4 Dropwise Condensation

Dropwise condensation, the primary application of interest, is vapor-to-liquid phase change in the form of discrete drops on or underneath horizontal and inclined substrates. The process is hierarchical in the sense that it occurs over a wide range of length and timescales. A mathematical model of dropwise condensation underneath textured surfaces, horizontal (with or without wettability gradient) and inclined, is reported. The model starts from the formation of drops at the atomic scale at randomized nucleation sites and follows its growth by direct condensation and coalescence, till the drop is large enough to fall off or slide away. The atomic model shows that the largest stable cluster size in the atomic scale matches the minimum drop radius estimated from thermodynamic considerations. The drops of minimum radius are insensitive to surface texturing and do not provide controllability at large length and timescales. In the model, nucleation sites are randomly distributed over the substrate. Growth rate at each nucleation site is derived, on the basis that vapor condenses on the free surface of the drop, and releases latent heat, that is transferred through the liquid drop to the cold substrate. The stability criterion is developed as a force balance equation at the level of a drop. Transport parameters of a sliding drop are determined using a CFD model and presented in the form of correlations. Fluids considered are water, and liquid metals such as mercury and sodium, representing a wide range of Prandtl numbers. Performing simulation of the complete cycle of dropwise condensation, the spatio-temporal distribution of drops is obtained, from which local and area-averaged heat transfer rates, as a function of time are predicted.

An experimental study of water vapor condensation underneath a chemically textured substrate is carried out for validation of the complete dropwise condensation model. Substrate preparation involves coating the glass surface using chemical vapor deposition of silane molecules. The spatio-temporal drop distribution recorded during the experiment and observed in simulation underneath an inclined chemically textured substrate show fair to good agreement. Heat transfer rates are also validated against experiment data of water vapor and mercury available in the literature. Specific conclusions arrived at in the present study are listed below.

16.4.1 Drop Instability

The critical drop radius at which commencement of sliding takes place is a function of the thermophysical properties of the fluid, inclination of the substrate and contact angle hysteresis. Fluids with higher surface tension show larger size at instability. Reduction in contact angle hysteresis reduces the critical size, for a given angle of inclination.

16.4.2 Modeling Fluid Motion Inside a Moving Drop

During motion, a circulation pattern is set up within the drop volume. The center of the circulation pattern moves towards the solid surface at higher Reynolds numbers. Pressure and wall shear stress are nearly uniform at the base of the drop, except at the periphery, where large gradients prevail. Heat transfer in drops of high Prandtl numbers is characterized by the appearance of thermal boundary layers. Temperature distribution across the drop shows large gradients near the walls while temperature inversion is seen in the core. At lower Prandtl numbers, diffusive transport governs heat transfer rates and a near-linear variation of the temperature profile is obtained.

16.4.3 Macroscopic Modeling

The overall condensation model includes the effect of contact angle, hysteresis, inclination of the substrate, thermophysical properties of fluids, nucleation site density, degree of subcooling, saturation temperature, promoter layer thickness, and wettability gradient. Simulation for various fluids and substrate inclinations shows the following trends.

1. Dropwise condensation is necessarily a quasi-cyclic process from which the average drop size, distribution, and cycle time as well as overall heat transfer coefficient and wall shear can be computed.
2. Two distinct phases of droplet growth are observed: growth due to direct condensation and growth primarily due to coalescence.
3. Increase of static contact angle (decrease in wettability of the substrate) reduces the droplet area coverage. Reduction of coverage is also observed by increasing the substrate inclination.
4. Decrease in wettability results in earlier fall-off (horizontal substrate) and earlier slide-off (inclined substrate).
5. The critical radius of droplet at which commencement of sliding takes place is a function of the thermophysical properties of the fluid, inclination of the substrate and contact angle hysteresis. Fluids with higher surface tension show larger critical radius. Reduction in contact angle hysteresis reduces the critical radius of the droplet for a given angle of inclination.
6. Inclining the substrate results in larger number of small drops and hence in higher heat transfer coefficient.
7. Heat transfer coefficient increases with an increase in the degree of subcooling, saturation temperature, and is a strong function of the Prandtl number.
8. Providing wettability gradient serves the purpose of passively destabilizing drops in a manner similar to inclined surfaces in a gravity field. It results in a larger number of small drops and hence will lead to a higher average heat transfer coefficient.
9. Nucleation sites density is an uncertain parameter that can be determined only indirectly but has a definite effect on the heat transfer coefficient. A high nucleation density leads to frequent drop mergers and instability followed by fresh nucleation, thus reducing the average drop size. Hence, it leads to a high overall heat transfer coefficient.

16.4.4 Water Versus Bismuth

The dropwise condensation model has been first simulated with water, and the condensation characteristics are then compared with that of bismuth. Differences arise mainly from those in thermophysical properties. Relative to water, liquid bismuth has a higher density, higher thermal conductivity, higher surface tension with vapor, higher viscosity but a much smaller latent heat of vaporization, thus generating several contrasts with it. Cycle time, minimum radius of drop, holdup, and vapor-liquid interfacial resistance of liquid bismuth are larger than water for both horizontal and vertical surfaces. Maximum radius of drop, average radius of drop, conductance resistance, and average heat transfer coefficient of liquid bismuth are smaller than water for both horizontal and vertical surfaces. As a result, wall heat flux based on direct condensation as well as coalescence is substantially larger for water than liquid bismuth for both horizontal and vertical surfaces. The average heat transfer coefficient in bismuth is smaller than for water. Fluctuations in heat fluxes in water are small but frequent while in bismuth, the local peaks arising from drop instability are larger and occasional. Wall shear stresses generated during coalescence as well as sliding of drops are larger for liquid bismuth than water for both horizontal and vertical surfaces.

16.4.5 Surface Texturing

For laboratory-scale experiments on dropwise condensation, small patches of a surface can be chemically textured to generate the required level of hydrophobicity. Such techniques have been summarized in the text. Tools required for surface characterization in terms of equilibrium contact angle and contact angle hysteresis are also presented. For larger surface areas, surface patterning such as micro-grooves and nanoscale pillars are recommended. Texturing tubular surfaces is an ongoing part of research.

16.4.6 Measurement of Wall Heat Flux

Wall heat fluxes during dropwise condensation are large and show considerable variation in both space and time. Even on a scale of a single drop, the variation can go from small values at the center to a large value along the three-phase contact line. Heat fluxes will be large under drops that have just nucleated and small under those subject to incipient instability. An experiment involving liquid crystal thermography for measuring heat flux distribution under a single drop is discussed in the text. Since the space-averaged heat flux can also exhibit strong time dependence, an inverse methodology for estimation of the instantaneous heat flux from single thermocouple data is also discussed.

16.5 Future Work

Opting for dropwise condensation in advanced engineering systems involves not only an understanding of the fundamental thermo-fluidic transport phenomena but also the microscale issues associated with the individual drops and the surface characteristics of the substrate. Hence, future research should address the following concerns:

1. Surface hydrophobicity is essential for realizing the dropwise condensation cycle. However, forces generated during spreading, coalescence, and movement of drops lead to surface damage. Hence, chemical texturing is limited by factors such as leaching that tend to peal out surface coatings. In this context, pillared or patterned surfaces are to be preferred.
2. Creating physical textures by patterning that will ensure sustained hydrophobicity is a long-standing problem in surface engineering. The connection of a surface pattern with contact angle and hysteresis is unresolved. Specifically, the modeling of drop-level dynamics such as spreading and coalescence and the dropwise condensation cycle for a pillared surface are topics of importance.
3. It is desirable that shear stresses experienced by the wall be small while wall heat fluxes are large. While models are helpful in determining these quantities, experimental validation is important. However, experimental determination of local and instantaneous wall shear stress and heat transfer coefficient is a challenge. The statistical nature of droplet distribution in the ensemble further contributes to the intricacy of analysis and interpretation.
4. The mathematical model presented captures the major constituents of dropwise condensation process quite satisfactorily. There are local discrepancies, however. Looking at the experiments closely, the three-dimensional geometry of the growing drops, highly localized three-phase contact line motion and the 3D dynamics of coalescence need deeper investigation.
5. Information gathered on dropwise condensation calls for its utilization in applications as diverse as water harvesting from atmospheric moist air and distillation of precious metals.
6. Discussion on the behavior of small liquid drops will be useful in designing microfluidic switches, bio-MEMS devices, lab-on-chip, and electro-locomotion of liquid drops on dielectric surfaces.

References

Aarts, D. G. A. L., Lekkerkerker, H. N. W., Guo, H., Wegdam, G. H., & Bonn, D. (2005). Hydrodynamics of droplet coalescence. *Physical Review Letters, 95*(16), 164503.

Abdulhussein, A. T., Kannarpady, G. K., Wright, A. B., Ghosh, A., & Biris, A. S. (2016). Current trend in fabrication of complex morphologically tunable superhydrophobic nano scale surfaces. *Applied Surface Science, 384*, 311–332.

Abu-Orabi, M. (1998). Modeling of heat transfer in dropwise condensation. *International Journal of Heat and Mass Transfer, 41*, 81–87.

Agrawal, N., Prabhakar, A., & Das, S. K. (2015). Hydrogen distribution in nuclear reactor containment during accidents and associated heat and mass transfer issues - A review. *Heat Transfer Engineering, 36*, 859–879.

Ahsan, M. S., Dewanda, F., Lee, M. S., Sekita, H., & Sumiyoshi, T. (2013). Formation of superhydrophobic soda-lime glass surface using femtosecond laser pulses. *Applied Surface Science, 265*, 784–789.

Ajaev, V. S. (2012). *Interfacial fluid mechanics: A mathematical modeling approach*. New York: Springer.

Alkemade, P. F. A. (2006). Propulsion of ripples on glass by ion bombardment. *Physical Review Letters, 96*, 107602.

Amar, J. G., Popescu, M. N., & Family, F. (1999). Self-consistent rate equation approach to nucleation and growth in point/extended island models of 1-D homoepitaxy. In *Proceedings of the Material Research Society* (Vol. 570(3)).

Andrieu, C., Beysens, D. A., Nikolayev, V. S., & Pomeau, Y. (2002). Coalescence of sessile drops. *Journal of Fluid Mechanics, 453*, 427–438.

Annapragada, S. R., Garimella V. S., & Murthy Y. J. (2010). Experimental characterization of droplet motion on inclined hydrophobic surfaces. In *Proceedings of the 9th International ISHMT-ASME Heat and Mass Transfer Conference, Mumbai, India*, Paper #343 (pp. 1298–1302).

Annapragada, S., Murthy, J., & Garimella, S. (2012). Droplet retention on an incline. *International Journal of Heat and Mass Transfer, 55*(5–6), 1457–1465.

Anzenberg, E., Perkinson, J. C., Madi, C. S., Aziz, M. J., & Ludwig, K. F., Jr. (2012). Nanoscale surface pattern formation kinetics on germanium irradiated by Kr+ ions. *Physical Review B, 86*, 245412.

Arkhipov, V. A., Bondarchuk, S., Usanina, A., & Shrager, G. (2015). Influence of the viscosity of a liquid on the dynamics of spreading of its drop. *Journal of Engineering Physics and Thermophysics, 88*, 42–51.

S. Khandekar, K. Muralidhar, *Drop Dynamics and Dropwise Condensation on Textured Surfaces*, Mechanical Engineering Series,
https://doi.org/10.1007/978-3-030-48461-3

Arya, M., Khandekar, S., Pratap, D., & Ramakrishna, S. A. (2016). Pool boiling of water on nano structured micro wires at sub atmospheric conditions. *Heat and Mass Transfer, 52*(9), 1725–1737.

ASTM, Committee E20. (1993). *Manual on the use of thermocouples in temperature measurement*. ASTM ISBN #0-8031-1466-4.

Attia, M. H., Cameron, A., & Kops, L. (2002). Distortion in thermal field around inserted thermocouples in experimental interfacial studies, part 4: End effect. *Journal of Manufacturing Science and Engineering, 124*, 135–145.

Aursand, E., & Ytrehus, T. (2019). Comparison of kinetic theory evaporation models for liquid thin-films. *International Journal of Multiphase Flow, 116*, 67–79.

Azehara, H., Kasanuma, Y., Ide, K., Hidaka, K., & Tokumotona, H. (2008). Distinct chemical contrast in adhesion force images of hydrophobic-hydrophilic patterned surfaces using multiwalled carbon nanotube probe tips. *Japanese Journal of Applied Physics, 47*, 3594–3599.

Badam, V. K., Kumar, V., Durst, F., & Danov, K. (2007). Experimental and theoretical investigations on interfacial temperature jumps during evaporation. *Experimental Thermal and Fluid Science, 32*(1), 276–292.

Bakulin, N. V., Ivanovskii, M. N., Sorokin, V. P., Subbotin, V. I., & Chulkov, B. A. (1967). Phase and diffusion resistance in the condensation of an alkali metal. *Journal of Atomic Energy, 22*(5), 413–415.

Baldacchini, T., Carey, J. E., Zhou, M., & Mazur, E. (2006). Superhydrophobic surfaces prepared by microstructuring of silicon using a femtosecond laser. *Langmuir, 22*, 4917–4919.

Bansal, G. D., Khandekar, S., & Muralidhar, K. (2009). Measurement of heat transfer during dropwise condensation of water on polyethylene. *Nanoscale and Microscale Thermophysical Engineering, 13*, 184–201.

Baojin, Q., Li, Z., Hong, X., & Yan, S. (2011). Experimental study on condensation heat transfer of steam on vertical titanium plates with different surface energies. *Experimental Thermal and Fluid Science, 35*, 211–218.

Bartelt, M. C., Tringides, M. C., & Evans, J. W. (1993). Island size scaling in surface deposition processes. *Physics Review B, 47*, 13891–13894.

Barth, T. J., & Jespersen, D. C. (1989). *The design and application of upwind schemes on unstructured meshes*. AIAA Paper #0366 (Vol. 89).

Battaglia, J.-L., Cois, O., Puigsegur, L., & Oustaloup, A. (2001). Solving an inverse heat conduction problem using a non-integer identified model. *International Journal of Heat and Mass Transfer, 44*, 2671–2680.

Battoo, N. K., Sikarwar, B. S., Khandekar, S., & Muralidhar K. (2010). Mathematical simulation of dropwise condensation exposed to vapor flux. In *Proceedings of the 9th International ISHMT-ASME Heat and Mass Transfer Conference, Mumbai, India* (Vol. 346, pp. 1330–1336).

Bauerle, D. W. (1996). *Laser processing and chemistry* (2nd ed.). Berlin: Springer. ISBN #978-3-642-17613-5.

Beck, J. V., & Woodbury, K. A. (2016). Inverse heat conduction problem: Sensitivity coefficient insights, filter coefficients, and intrinsic verification. *International Journal of Heat and Mass Transfer, 97*, 578–588.

Beck, J. V., Blackwell, B., & Clair, C. R. S. (1985). *Inverse heat conduction: Ill posed problems*. New York: Wiley. ISBN #0-471-083194.

Bentley, P. D., & Hands, B. A. (1978). The condensation of atmospheric gases on cold surfaces. *Royal Society London A, 359*, 319–343.

Bergman, T. L., Incropera, F. P., DeWitt, D. P., & Lavine, A. S. (2011). *Fundamentals of heat and mass transfer* (7th ed.). Hoboken, NJ: Wiley. ISBN #13978-0470-50197-9.

Berthier, J. (2008). *Microdrops and digital microfluidics* (pp. 75–179). Norwich, CT: William Andrew Inc..

Beysens, D. (2006). Dew nucleation and growth. *Comptes Rendus Physique, 7*, 1082–1100.

Bhushan, B., & Jung, Y. C. (2008). Wetting, adhesion and friction of superhydrophobic and hydrophilic leaves and fabricated micro/nanopatterned surfaces. *Journal of Physics: Condensed Matter, 20*(22), 225010.

Bhushan, B., & Jung, Y. C. (2011). Natural and biomimetic artificial surfaces for superhydrophobicity, self-cleaning, low adhesion, and drag reduction. *Progress in Materials Science, 56*, 1–108.

Bhushan, B., Jung, Y. C., & Koch, K. (2009). Micro-, nano- and hierarchical structures for superhydrophobicity, self-cleaning and low adhesion. *Philosophical Transactions of Royal Society A, 367*, 1631–1672.

Bhutani, G. (2007). *Imaging hanging drops on inclined textured surfaces.* Master's thesis, IIT Kanpur, India.

Bhutani, G., Muralidhar, K., & Khandekar, S. (2013). Determination of apparent contact angle and shape of a static pendant drop on a physically textured inclined surface. *Interfacial Phenomena and Heat Transfer, 1*(1), 29–49.

Biance, A.-L., Clanet, C., & Quéré, D. (2004). First steps in the spreading of a liquid droplet. *Physical Review E, 69*, 016301.

Bird, J. C., Mandre, S., & Stone, H. A. (2008). Short-time dynamics of partial wetting. *Physical Review Letters, 100*, 234501.

Bird, R.B., Stewart, W.E., and Lightfoot, E.N., Transport Phenomena, John Wiley & Sons (2002).

Bisetto, A., Torresin, D., Tiwari, M. K., Del Col, D., & Poulikakos, D. (2014). Dropwise condensation on superhydrophobic nanostructured surfaces: Literature review and experimental analysis. *Journal of Physics: Conference Series, 501*, 012028.

Bizi-Bandoki, P., Valette, S., Audouard, E., & Benayoun, S. (2013). Time dependency of the hydrophilicity and hydrophobicity of metallic alloys subjected to femtosecond laser irradiations. *Applied Surface Science, 273*, 399–407.

Blackman, L. C. F., Dewar, M. J. S., & Hampson, H. (1957). An investigation of compounds promoting the dropwise condensation of stem. *Applied Chemistry, 7*, 160–157.

Blake, T. D. (2006). The physics of moving wetting lines. *Journal of Colloid and Interface Science, 299*, 1–13.

Bloch, P. E., Smargiassi, E., Car, R., Laks, D. B., Andreoni, W., & Pantelides, S. T. (1993). First principle calculations self-diffusion constants in silicon. *Physics Review Letters, 70*, 2435–2438.

Bonner, R. (2009). Condensation on surfaces with graded hydrophobicity. In *ASME Summer Heat Transfer Conference, San Francisco, CA, USA.*

Bonner, R. W., III (2010). Dropwise condensation life testing of self-assembled monolayers. In *Proceedings of the International Heat Transfer Conference (IHTC14), Washington, DC, USA.*

Boreyko, J. B., & Chen, C. H. (2009). Self-propelled dropwise condensate on superhydrophobic surfaces. *Physical Review Letters, 103*, 184501–184504.

Bormashenko, E. (2015). Progress in understanding wetting transitions on rough surfaces. *Advances in Colloid and Interface Science, 222*, 92–103.

Brakke, K. (1992). The surface evolver. *Experimental Mathematics, 1*(2), 141–165.

Braun, R. J., Murray, B. T., Boettinger, W. J., & McFadden, G. B. (1995). Lubrication theory for reactive spreading of a thin prop. *Physics of Fluids, 7*, 1797–1810.

Brenn, G., Valkovska, D., & Danov, K. D. (2001). The formation of satellite droplets by unstable binary drop collisions. *Physics of Fluids, 13*(9), 2463–2477.

Briscoe, B. J., & Galvin, K. P. (1991a). Growth with coalescences during condensation. *Physical Review A, 43*, 1906–1918.

Briscoe, B. J., & Galvin, K. P. (1991b). The sliding of sessile and pendent droplets the critical condition. *Journal of Colloid and Interface Science, 52*, 219–229.

Briscoe, B. J., Williams, D. R., & Galvin, K. P. (2005). Condensation on hydrosol modified polyethylene. *Colloids and Surfaces A, 264*, 101–105.

Brown, R., Orr, F., Jr., & Scriven, L. (1980). Static drop on an inclined plate: Analysis by the finite element method. *Journal of Colloid and Interface Science, 73*(1), 76–87.

Brune, H. (1998). Microscopic view of epitaxial metal growth: Nucleation and aggregation. *Surface Science Reports, 31*(3), 121–229.

Brune, H., Bales, G. S., Jacobsen, J., Boragno, C., & Kern, K. (1999). Measuring surface diffusion from nucleation island densities. *Physical Review B, 60*, 5991–6006.

Bansal, G. D., Khandekar, S., & Muralidhar, K. (2009). Measurement of Heat Transfer during Dropwise Condensation of Water on Polyethylene, Nanoscale and Microscale *Thermophysical Engineering, 13*(3), 184–201.

Burnside, B. M., & Hadi, H. A. (1999). Digital computer simulation of dropwise condensation from equilibrium droplet to detectable size. *International Journal of Heat and Mass Transfer, 42*, 3137–3146.

Calvert, P. (2001). Inkjet printing for materials and devices. *Chemistry of Materials, 13*(10), 3299–3305.

Carey, V. P. (2008). *Liquid-vapor phase-change phenomena* (2nd ed., pp. 45–472). New York: Taylor and Francis Group LLC.

Cassie, A. B. D., & Baxter, S. (1944). Wettability of porous surfaces. *Transactions of the Faraday Society, 40*, 546–551.

Castrejón-Pita, J. R., Betton, E. S., Kubiak, K. J., Wilson, M. C. T., & Hutchings, I. M. (2011). The dynamics of the impact and coalescence of droplets on a solid surface. *Biomicrofluidics, 5*(1), 014112.

Çengel, Y. A., & Ghajar, A. J. (2017). *Heat and mass transfer: Fundamentals and applications.* New York: McGraw Hill Education. ISBN #978-9339223199.

Cha, T. G., Yi, G. W., Moon, M. W., Lee, K. R., & Kim, H. Y. (2010). Nanoscale patterning of micro-textured surfaces to control superhydrophobic robustness. *Langmuir, 26*(11), 8319–8326.

Chandramohan, A., Weibel, J. A., & Garimella, S. V. (2017). Spatiotemporal infrared measurement of interface temperatures during water droplet evaporation on a nonwetting substrate. *Applied Physics Letters, 110*, 1–4, Paper #041605.

Chang, C.-W., Liu, C.-H., & Wang, C.-C. (2018). Review of computational schemes in inverse heat conduction problems. *Smart Science, 6*, 94–103.

Chaterjee, S., Bhattacharya, S., Maurya, S. K., Srinivasan, V., Khare, K., & Khandekar, S. (2017). Surface wettability of an atomically heterogeneous system and the resulting intermolecular forces. *Euro Physics Letters (EPL), 118*, 68006.

Chatterjee, J. (2001). A criterion for buoyancy induced drop detachment based on an analytical approximation of the drop shape. *Colloids and Surfaces A: Physicochemical and Engineering Aspects, 178*, 249–263.

Chatterjee, J. (2002). Critical Eötvös numbers for buoyancy-induced oil drop detachment based on shape analysis. *Advances in Colloid and Interface Science, 98*, 245–283.

Chatterjee, A., Derby, M. M., Peles, Y., & Jensen, M. K. (2014). Enhancement of condensation heat transfer with patterned surfaces. *International Journal of Heat and Mass Transfer, 71*, 675–681.

Chaudhury, M. K., & Whiteside, G. M. (1992). How to make water run uphill. *Science, 256*, 1539–1541.

Chen, L. Y., Baldan, M. R., & Ying, S. C. (1996). Surface diffusion in the low-friction limit: Processes. *Physics Review B, 54*, 8856–8861.

Chen, L. H., Chen, C. Y., & Lee, Y. L. (1999). Nucleation and growth of clusters in the process of vapor deposition. *Surface Science, 429*, 150–160.

Chen, C. H., Cai, Q., Tsai, C., & Chen, C. L. (2007). Dropwise condensation on superhydrophobic surfaces with two-tier roughness. *Applied Physics Letters, 90*, 173108–173111.

Chen, L., Liang, S., Yan, R., Cheng, Y., Huai, X., & Chen, S. (2009). N-octadecanethiol self-assembled monolayer coating with microscopic roughness for dropwise condensation of steam. *Journal of Thermal Science, 18*(2), 60–165.

Chang, C.-W., Liu, C.-H., & Wang, C.-C. (2017). Review of computational schemes in inverse heat conduction problems, *Smart Science, 6*(1):1–10.

Chen, H., Frankel, J. I., & Keyhani, M. (2018). Two-probe calibration integral equation method for nonlinear inverse heat conduction problem of surface heat flux estimation. *International Journal of Heat and Mass Transfer, 121*, 246–264.

Childs, P. R. N., Greenwood, J. R., & Long, C. A. (1999). Heat flux measuring techniques. *Proceedings of the Institute of Mechanical Engineers, 213*, 655–677.

Cho, H. J., Preston, D. J., Zhu, Y., & Wang, E. N. (2016). Nanoengineered materials for liquid-vapor phase-change heat transfer. *Nature Reviews Materials, 2*, 16092.

Chowdhury, A., & Bhattacharjee, S. (2011). Localized subsurface modification of materials using micro-low energy multiple ion beamlets. *AIP Advances, 1*, 042150.

Chowdhury, A., & Bhattacharjee, S. (2013). Experimental investigation of change in sheet resistance and Debye temperatures in metallic thin films due to low-energy ion beam irradiation. *Journal of Physics D: Applied Physics, 46*, 435304.

Citakoglu, E., & Rose, J. W. (1968a). Dropwise condensation some factors influencing the validity of heat-transfer measurements. *International Journal of Heat and Mass Transfer, 11*, 523–537.

Citakoglu, E., & Rose, J. W. (1968b). Dropwise condensation the effect of surface inclination. *International Journal of Heat and Mass Transfer, 12*, 645–451.

Cognard, J. (1984). Adhesion to gold: A review. *Gold Bulletin, 17*, 131–139.

Collier, J. G., & Thome, J. R. (1994). *Convective boiling and condensation* (3rd ed.). Oxford: Oxford University Press. ISBN #978-0198562962.

Collier, J. G., Thome, J. R., (1996). Convective Boiling and Condensation, 3rd ed., Oxford Engineering Science Series, Clarendon Press, ISBN-10: 0198562969.

Cox, R. (1986). The dynamics of the spreading of liquids on a solid surface: Part I—Viscous flow. *Journal of Fluid Mechanics, 168*, 169–194.

Cox, R. (1998). Inertial and viscous effects on dynamic contact angles. *Journal of Fluid Mechanics, 357*, 249–278.

Cras, J. J., Rowe-Tait, C. A., Nivens, D. A., & Ligler, F. S. (1999). Comparison of chemical cleaning methods of glass in preparation for silanization. *Biosensors and Bioelectronics, 14*, 683–688.

Cui, M., Yang, K., Xu, X.-L., Wang, S.-D., & Gao, X.-W. (2016). A modified Levenberg–Marduardt algorithm for simultaneous estimation of multi-parameters of boundary heat flux by solving transient non-linear inverse heat conduction problems. *International Journal of Heat and Mass Transfer, 97*, 908–916.

Daniel, S., Chaudhury, M. K., & Chen, J. C. (2001). Fast drop movements resulting from the phase-change on a gradient surface. *Science, 291*, 633–636.

Das, A. K., & Das, P. K. (2009). Simulation of drop movement over an inclined surface using smoothed particle hydrodynamics. *Langmuir, 25*, 11459–11466.

Das, A. K., Kilty, H. P., & Marto, P. J. (2000a). Dropwise condensation of steam on horizontal corrugated tubes using an organic self-assembled monolayer coating. *Journal of Enhanced Heat Transfer, 7*(2), 109–123.

Das, A. K., Kilty, H. P., & Marto, P. J. (2000b). The use of an organic self-assembled monolayer coating to promote dropwise condensation of steam on horizontal tubes. *ASME Journal of Heat Transfer, 122*(2), 278–286.

Dash, S., & Garimella, S. V. (2013). Droplet evaporation dynamics on a superhydrophobic surface with negligible hysteresis. *Langmuir, 29*, 10785–10795.

Date, A. W. (1996). Complete pressure correction algorithm for solution of incompressible Navier–Stokes equations on a non-staggered grid. *Numerical Heat Transfer, Part B, 29*, 441–458.

Date, A. W. (2003). Fluid dynamical view of pressure check boarding problem and smoothing pressure correction on meshes with collocated variables. *International Journal of Heat and Mass Transfer, 46*, 4885–4898.

Date, A. W. (2005a). *Introduction to computational fluid dynamics*. New York: Cambridge University Press.

Date, A. W. (2005b). Solution of transport equations on unstructured meshes with cell cantered collocated variables, Part I: Discretization. *International Journal of Heat and Mass Transfer, 48*, 1117–1127.

David, S., Sefiane, K., & Tadrist, L. (2007). Experimental investigation of the effect of the thermal properties of the substrate in the wetting and evaporation of sessile drops. *Colloids and Surfaces A: Physiochemical and Engineering Aspects, 298*, 108–114.

De Coninck, J., & Blake, T. (2008). Wetting and molecular dynamics simulations of simple liquids. *Annual Review of Materials Research, 38*, 1–22.

de Gennes, P.-G. (1985). Wetting: Static and dynamics. *Review of Modern Physics, 57*, 827–863.

de Gennes, P.-G., Brochard-Wyart, F., & Quéré, D. (2004). *Capillarity and wetting phenomena: Drops, bubbles, pearls, waves*. New York: Springer. ISBN #978-0387005928.

de Gennes, P.-G., Brochard-Wyart, F., & Quéré, D. (2013). *Capillarity and wetting phenomena: Drops, bubbles, pearls, waves*. New York: Springer Science & Business Media. ISBN #978-0-38721656-0.

De Marco, C., Gaidukeviciute, A., Kiyan, R., Eaton, S. M., Levi, M., Osellame, R., Chichkov, B. N., & Turri, S. (2012). A new perfluoropolyether-based hydrophobic and chemically resistant photoresist structured by two-photon polymerization. *Langmuir, 29*, 426–431.

Deegan, R. D., Bakajin, O., Dupont, T. F., Huber, G., Nagel, S. R., & Witten, T. A. (2000). Contact line deposit in an evaporating drop. *Physical Review E, 62*, 756–765.

Denial, S., Chaudhury, M. K., & Chen, J. C. (2001). Fast drop movements resulting from the phase-change on a gradient surface. *Science, 291*, 633–636.

Dietz, C., Rykaczewski, K., Fedorov, A. G., & Joshi, Y. (2010). Visualization of droplet departure on a superhydrophobic surface and implications to heat transfer enhancement during dropwise condensation. *Applied Physics Letters, 97*, 033104–033103.

Diggle, J., Downie, T., & Goulding, C. (1968). Anodic oxide films on aluminium. *Chemical Reviews, 69*, 365–405.

Diller, T. E. (1993). Advances in heat flux measurements. *Advances in Heat Transfer, 23*, 279–368.

Dimitrakopoulos, P., & Higdon, J. J. L. (1999). On the gravitational displacement of the three-dimensional fluid droplets from inclined solid surfaces. *Journal of Fluid Mechanics, 395*, 181–209.

Duchemin, L., Eggers, J., & Josserand, C. (2003). Inviscid coalescence of drops. *Journal of Fluid Mechanics, 487*, 167–178.

Dugas, V., Broutin, J., & Souteyrand, E. (2005). Droplet evaporation study applied to DNA chip manufacturing. *Langmuir, 21*(20), 9130–9136.

Dunn, G. J., Wilson, S. K., Duffy, B. R., David, S., & Sefiane, K. (2009). The strong influence of substrate conductivity on droplet evaporation. *Journal of Fluid Mechanics, 623*, 329–351.

Dussan, E. (1976). Moving contact line-slip boundary-condition. *Journal of Fluid Mechanics, 77*, 665–684.

Dussan, E. B. (1979). On the spreading of liquid on solid surfaces: Static and dynamic Contact Lines. *Annual Review of Fluid Mechanics, 11*, 371–400.

Dussan, E. B. (1985). On the ability of drops or bubbles to stick to non-horizontal surface of solids. *Journal of Fluid Mechanics, 151*, 1–20.

Dussan, E. B., & Chow, R. T. (1983). On the ability of drops or bubbles to stick to non-horizontal surfaces of solids. *Journal of Fluid Mechanics, 137*, 1–29.

Eddi, A., Winkels, K. G., & Snoeijer, J. H. (2013). Short time dynamics of viscous drop spreading. *Physics of Fluids, 25*, 013102.

Eggers, J. (1998). Coalescence of spheres by surface diffusion. *Physical Review Letters, 80*, 2634–2637.

Eggers, J., Lister, J. R., & Stone, H. A. (1999). Coalescence of liquid drops. *Journal of Fluid Mechanics, 401*, 293–310.

Eiswirth, R. T., Bart, H.-J., Ganguli, A. A., & Kenig, E. Y. (2012). Experimental and numerical investigation of binary coalescence: Liquid bridge building and internal flow fields. *Physics of Fluids, 24*(6), 062108.

ElSherbini, A. I., & Jacobi, A. M. (2004a). Liquid drops on vertical and inclined surfaces—I: An experimental study of drop geometry. *Journal of Colloid and Interface Science, 273*, 556–565.
ElSherbini, A. I., & Jacobi, A. M. (2004b). Liquid drops on vertical and inclined surfaces—II: An experimental study of drop geometry. *Journal of Colloid and Interface Science, 273*, 566–575.
ElSherbini, A. I., & Jacobi, A. M. (2006). Retention forces and contact angles for critical liquid drops on non-horizontal surfaces. *Journal of Colloid and Interface Science, 299*, 841–849.
Enright, R., Miljkovic, N., Alvarado, J. L., Kim, K., & Rose, J. W. (2014). Dropwise condensation on micro- and nanostructured surfaces. *Nanoscale and Microscale Thermophysical Engineering, 18*, 223–250.
Erb, R. A. (1965). Promoting permanent dropwise condensation. *Industrial and Engineering Chemistry, 57*, 49–52.
Erb, R. A. (1973). Dropwise condensation on gold. *Gold Bulletin, 6*, 2.
Erb, R. A., & Thelen, E. (1965). Dropwise condensation on hydrophobic metal and metal-sulfide surfaces. In *149th National Meeting and Symposium of the American Chemical Society, Detroit, MI, USA*.
Erb, R. A., & Thelen, E. (1966). *Dropwise condensation characteristics of permanent hydrophobic system*. U.S. Department of Interior, R and D Report #184 (pp. 54–57).
Eucken, A. (1937), Energie und stoffaustausch an grenzflachen, *Naturwissenschaften, 25*, 209–219.
Extrand, C. W. (2016). Origins of wetting. *Langmuir, 32*, 7697–7706.
Extrand, C. W., & Kumagai, Y. (1995). Liquid drop on an inclined plane: The relation between contact angles, drop shape and retentive forces. *Journal of Colloid and Interface Science, 170*, 515–521.
Fahrni, R., Schmid, J., Klippel, M., & Frangi, A. (2018). Investigation of different temperature measurement designs and installations in timber members as low conductive material. In *SiF 2018—The 10th International Conference on Structures in Fire, Belfast, UK.*
Fang, X., Li, B., Petersen, E., Seo, Y.-S., Samuilov, V. A., Chen, Y., Sokolov, J. C., Shew, C.-Y., & Rafailovich, M. H. (2006). Drying of DNA droplets. *Langmuir, 22*(14), 6308–6312.
Fang, C., Hidrovo, C., Wang, F., Eaton, J., & Goodson, K. (2008). 3-D numerical simulation of contact angle hysteresis for microscale two phase flow. *International Journal of Multiphase Flow, 34*, 690–705.
Feng, J., Qin, Z., & Yao, S. (2012). Factors affecting the spontaneous motion of condensate drops on superhydrophobic copper surfaces. *Langmuir, 28*, 6067–6075.
Frezzotti, A., Gibelli, L., Lockerby, D. A., & Sprittles, J. E. (2018). Mean-field kinetic theory approach to evaporation of a binary liquid into vacuum. *Physical Review Fluids, 3*(5), 054001.
Frink, N. T., Paresh, P., & Shahyar, P. (1991). *A fast-upwind solver for the Euler equations on three dimensional unstructured meshes* (Vol. 91). AIAA Paper #0102.
Furmidge, C. G. (1962). The sliding drop on solid surfaces and a theory for spray retention. *Journal of Colloid Science, 17*, 309–324.
Furstner, R., Barthlott, W., Neinhuis, C., & Walzel, P. (2005). Wetting and self-cleaning properties of artificial superhydrophobic surfaces. *Langmuir, 21*, 956–961.
Gao, L., & McCarthy, T. J. (2006). Contact angle hysteresis explained. *Langmuir, 22*, 6234–6237.
Garg, S. K., Dutta, D. P., Ghatak, J., Thakur, I., Khare, K., Kanjilal, D., & Som, T. (2016). Tunable wettability of Si through surface energy engineering by nanopatterning. *Royal Society of Chemistry Advances, 6*, 48550–48557.
Genzer, J., & Efimenko, K. (2000). Creating long-lived super hydrophobic polymer surface through mechanically assembled monolayer. *Science, 290*, 2130–2133.
Gerasimov, D. N., & Yurin, E. I. (2018). *Kinetics of evaporation* (Vol. 68). Cham: Springer.
Glassford, A. P. M. (1992). Practical model for molecular containment deposition kinetics. *Journal of Thermophysics and Heat Transfer, 6*(4), 656–664.
Glicksman, R. L., & Hunt, W. A. (1972). Numerical simulation of dropwise condensation. *International Journal of Heat and Mass Transfer, 15*, 2251–2269.
Gogna, M. (2016). *Effect of surface texture on evaporation of a sessile water droplet*. Ph.D. thesis, Department of Mechanical Engineering, Indian Institute of Technology Kanpur, India.

Goldstein, R. J., Ibele, W. E., Patankar, S. V., Simon, T. W., Kuehn, T. H., Strykowski, P. J., Tamma, K. K., Heberlein, J. V. R., Davidson, J. H., Bischof, J., Kulacki, F. A., Kortshagen, U., Garrick, S., & Srinivasan, V. (2006). Heat transfer—a review of 2003 literature. *International Journal of Heat and Mass Transfer, 49*, 451–534.

Gonzalez, R. C., Woods, R. E., & Eddins, S. L. (2004). *Digital image processing using MATLAB*. Hoboken, NJ: Pearson Education Inc..

Gorman, C. B., Biebuyck, H. A., & Whitesides, G. M. (1995). Control of the shape of liquid lenses on a modified gold surface using an applied electrical potential across a self-assembled monolayer. *Langmuir, 11*, 2242–2246.

Gose, E., Mucciordi, A. N., & Baer, E. (1967). Model for dropwise condensation on randomly distributed sites. *International Journal of Heat and Mass Transfer, 10*, 15–22.

Graham, C. (1969). *The limiting heat transfer mechanism of dropwise condensation*. Ph.D. thesis, Massachusetts Institute of Technology, Cambridge, MA, USA.

Graham, C., & Griffith, P. (1973). Dropwise size distribution and heat transfer in dropwise condensation. *International Journal of Heat and Mass Transfer, 16*, 337–346.

Graham, P. J., Farhangi, M. M., & Dolatabadi, A. (2012). Dynamics of droplet coalescence in response to increasing hydrophobicity. *Physics of Fluids, 24*, 112105.

Grand, N. L., Daerr, A., & Limit, L. (2005). Shape and motion of drops sliding down an inclined plane. *Journal of Fluid Mechanics, 541*, 293–315.

Greenspan, H. P. (1978). On the motion of a small viscous droplet that wets a surface. *Journal of Fluid Mechanics, vol., 84*(1), 125–143.

Griffith, P. (1985). Dropwise condensation. In W. P. Rohsenow, J. P. Hartnett, & E. N. Ganic (Eds.), *Handbook of heat transfer fundamentals* (2nd ed., pp. 37–49). New York: McGraw-Hill.

Griffith, P., & Lee, M. S. (1967). The effect of surface thermal properties and finish on dropwise condensation. *International Journal of Heat and Mass Transfer, 10*, 697–707.

Grischke, M., Trojan, K., & Dimigen, H., (1994). Deposition of low energy coating with DLC-like properties. In *Proceedings of the 11th Conference on High Vacuum, Interfaces and Thin Films* (pp. 433–436).

Gu, Y., Liao, Q., Zhu, X., & Wang, H. (2005). Dropwise condensation heat transfer coefficient on the horizontal surface with gradient surface energy. *Journal of Engineering Thermophysics, 26* (5), 820–822.

Gunjan, M. R., Somwanshi, P., Agrawal, A., Khandekar, S., & Muralidhar, K. (2015). Recoil of drops during coalescence on superhydrophobic surfaces. *Interfacial Phenomena and Heat Transfer, 3*, 203–220.

Hallcrest. (2008). *Handbook of thermochromic liquid crystal technology*. Hallcrest Publication. Retrieved from http://www.hallcrest.com/

Hampson, H. & özisik, N. (1952). An investigation into the condensation of steam. Proc. Inst. Mech. Eng. 1, Ser. B, p. 282–293.

Hansen, H., Redinger, A., Messlinger, S., Stoian, G., Rosandi, Y., & Urbassek, H. M. (2006). Mechanisms of pattern formation in grazing-incidence ion bombardment of Pt (111). *Physical Review B, 73*, 235414.

Hao, P., Lv, C., Yao, Z., & He, F. (2010). Sliding behavior of water droplet on superhydrophobic surface. *A Letter Journal Exploring (EPL), 90*, 66003, 1–6.

Hashimoto, H., & Kotake, S. (1995). In-situ measurement of clustering process near condensate. *Thermal Science and Engineering, 3*, 37–43.

Hatamiya, S., & Tanaka, H. (1986). A study on the mechanism of dropwise condensation (1st report, measurement of heat-transfer coefficient of steam at low pressures). *Transactions of the Japan Society of Mechanical Engineers Series B, 52*(476), 1828–1833.

Hay, J. L., & Hollingsworth, D. K. (1998). Calibration of micro-encapsulated liquid crystals using hue angle and a dimensionless temperature. *Experimental Thermal and Fluid Sciences, 18*, 251–257.

Hoar, T., & Mott, N. (1959). A mechanism for the formation of porous anodic oxide films on aluminum. *Journal of Physics and Chemistry of Solids, 9*, 97–99.

Hoffman, J. D., & Frankel, S. (2001). *Numerical methods for engineers and scientists* (2nd ed.). New York: CRC Press. ISBN #978-0-82470443-8.
Höhmann, C., & Stephan, P. (2002). Microscale temperature measurement at an evaporating liquid meniscus. *Experimental Thermal and Fluid Sciences, 26*, 157–162.
Houser, J. E., & Hebert, K. R. (2009). The role of viscous flow of oxide in the growth of self-ordered porous anodic alumina films. *Nature Materials, 8*, 415–420.
Hsieh, C. T., Chen, W. Y., & Wu, F. L. (2008). Fabrication and super-hydrophobicity of fluorinated carbon fabrics with micro/nano-scaled two-tier roughness. *Carbon, 46*, 1218–1224.
Hu, H., & Larson, R. G. (2002). Evaporation of a sessile droplet on a substrate. *Journal of Physical Chemistry B, 106*(6), 1334–1344.
Hu, H., & Larson, R. G. (2006). Marangoni effect reverses coffee-ring depositions. *Journal of Physical Chemistry B, 110*(14), 7090–7094.
Huang, H., Liang, D., & Wetton, B. (2004). Computation of a moving drop/bubble on a solid surface using a front-tracking method. *Communications in Mathematical Science, 2*, 535–552.
Huh, C., & Mason, S. G. (1977). Effect of surface roughness on wetting (theoretical). *Journal of Colloids and Interface Science, 60*, 11–37.
Huh, C., & Scriven, L. E. (1971). Hydrodynamic model of steady movement of a solid/liquid fluid contact line. *Journal of Colloid and Interface Science, 35*, 85–101.
Irving, J. and Kirkwood, J. (1950). The statistical mechanical theory of transport processes. (iv) the equations of hydrodynamics. The *Journal of Chemical Physics, 18*(6):817–829.
Israelachvili, J. N. (2011). *Intermolecular and surface forces*. New York: Academic Press. ISBN #978-0-12391927-4.
Ivanovskii, M. N., Subbotin, V. I., & Milovanov, Y. V. (1967). Heat transfer with dropwise condensation of mercury vapor. *Teploenergetika, 14*, 81–85.
Izumi, M., Kumagai, S., Shimada, R., Yamakawa, N. (2004). Heat transfer enhancement of dropwise condensation on a vertical surface with round shaped grooves, Exp. Thermal Fluid Sci. *28*(2) 243–248.
Jagdheesh, R., Pathiraj, B., Karatay, E., Romer, G. R. B. E., & Huis Veld, A. J. (2011). Laser-induced nanoscale superhydrophobic structures on metal surfaces. *Langmuir, 27*, 8464–8469.
Jain, V., & Nema, S. K. (2007). Deposition of super-hydrophobic nanostructure teflon-like coating using expanding plasma arc. *Applied Surface Science, 253*, 5462–5466.
Jakob, M. (1936). Heat transfer in evaporation and condensation. *Mechanical Engineers, 58*, 643–660.
Jani, M. A. M., Losic, D., & Voelcker, N. H. (2013). Nanoporous anodic aluminum oxide: Advances in surface engineering and emerging applications. *Progress in Materials Science, 58*, 636–704.
Jessensky, O., Müller, F., & Gösele, U. (2003). Self-organized formation of hexagonal pore arrays in anodic alumina. *Applied Physics Letters, 72*(9), 1173–1175.
Jiang, L., & Tsai, H. L. (2003). Femtosecond laser ablation: Challenges and opportunities. In *Proceedings of the NSF Workshop on Research Needs in Thermal, Aspects of Material Removal* (pp. 163–177).
Joanny, J. F., & de Gennes, P. G. (1984). A model for contact angle hysteresis. *Journal of Chemical Physics, 81*, 552.
Kalinin, Y., Berejnov, V., & Thorne, R. (2009). Contact line pinning by microfabricated patterns: Effects of microscale topography. *Langmuir, 25*(9), 5391–5397.
Kananeh, B., Rausch, M. H., Fröba, A. P., & Leipertz, A. (2006). Experimental study of dropwise condensation on plasma-ion implanted stainless steel tubes. *International Journal of Heat and Mass Transfer, 49*, 5018–5026.
Kapur, N., & Gaskell, P. H. (2007). Morphology and dynamics of droplet coalescence on a surface. *Physical Review Letters, 97*, 056315, 1–4.
Kasi, A. K., Kasi, J. K., Afzulpurkar, N., Bohez, E., Tuantranont, A., & Mahaisavariya, B. (2010). Novel anodic aluminum oxide (AAO) nanoporous membrane for wearable hemodialysis device. In *Third International Conference on Communications and Electronics (ICCE)* (pp. 98–101).

Kast, W. (1965) Theoretische und experimentelle Untersuchung der Wärmeübertragung bei Tropfenkondensation. Fortschrittsber. VDI—Z., Reihe 3, Nr. 6, Düsseldorf.

Kazemi, M. A., Nobes, D. S., & Elliott, J. A. (2017). Experimental and numerical study of the evaporation of water at low pressures. *Langmuir, 33*(18), 4578–4591.

Keller, F., Hunter, M. S., & Robinson, D. L. (1953). Structural features of oxide coatings on aluminum. *Journal of the Electrochemical Society, 100*, 411–419.

Keller, A., Fritzsche, M., Ogaki, R., Bald, I., Facsko, S., Dong, M., Kingshott, P., & Besenbacher, F. (2011). Tuning the hydrophobicity of mica surfaces by hyperthermal Ar ion irradiation. *Journal of Chemical Physics, 134*, 104705.

Khakshouri, S., & Duffy, D. M. (2009). Influence of electronic effects on the surface erosion of tungsten. *Physical Review B, 80*, 035415.

Khandekar, S., & Muralidhar, K. (2013). *Dropwise condensation on inclined textured surfaces*. New York: Springer. ISBN #978-1-4614-8446-2.

Khandekar, S., & Muralidhar, K. (2014). *Dropwise condensation on inclined textured surfaces* (Springer briefs in applied sciences and technology). New York: Springer.

Kietzig, A.-M., Hatzikiriakos, S. G., & Englezos, P. (2009). Patterned superhydrophobic metallic surfaces. *Langmuir, 25*, 4821–4827.

Kim, S., & Kim, K. J. (2011). Dropwise condensation modeling suitable for superhydrophobic surfaces. *ASME Journal of Heat Transfer, 133*(8), 081502, 1–8.

Kim, H., Lee, H., & Kang, B. H. (2002). Sliding of drops down an inclined solid surface. *Journal of Colloid Science, 247*, 372–382.

Kim, W.-S., Jin, J.-H., & Bae, B.-S. (2006). Low adhesive force of fluorinated sol-gel hybrid materials for easy de-moulding in a UV-based nano-imprint process. *Nanotechnology, 17*(5), 1212.

Kim, W., Kim, D., Park, S., Lee, D., Hyun, H., & Kim, J. (2018). Engineering lotus leaf-inspired micro- and nanostructures for the manipulation of functional engineering platforms. *Journal of Industrial and Engineering Chemistry, 61*, 39–52.

Kimura, M., Misner, M. J., Xu, T., Kim, S. H., & Russell, T. P. (2003). Long-range ordering of di-block copolymers induced by droplet pinning. *Langmuir, 19*(23), 9910–9913.

Kirkwood, J. and Buff., F. (1949). The statistical mechanical theory of surface tension. The Journal of Chemical Physics, 17(3):338–343.

Koch, G., Zhang, D. C., & Leiertz, A. (1997). Condensation of steam on the surface of hard coated copper discs. *Heat and Mass Transfer, 32*, 149–297.

Koch, G., Kraft, K., & Leipertz, A. (1998a). Parameter study on the performance of dropwise condensation. *Revue Générale de Thermique, 37*, 539–548.

Koch, G., Zhang, D., & Leipertz, A. (1998b). Study of plasma enhanced CVD coated material to promote dropwise condensation. *International Journal of Heat and Mass Transfer, 41*(13), 1899–1900.

Kokya, B. A., & Kokya, T. A. (2008). Proposing a formula for evaporation measurement from saltwater resources. *Hydrological Processes: An International Journal, 22*(12), 2005–2012.

Kotake, S. (1998). Molecular clusters. In C.-L. Tien, A. Majumdar, & F. M. Gerner (Eds.), *Microscale energy transport* (pp. 167–185). Washington, DC: Taylor and Francis.

Krasovitski, B., & Marmur, A. (2005). Drops down the hill: Theoretical study of limiting contact angles and the hysteresis range on a tilted plate. *Langmuir, 21*(9), 3881–3885.

Krischer, S., & Grigull, U. (1971). Microscopic study of dropwise condensation. *Wärme-und Stoffübertragung, 4*, 48–59.

Kryukov, A. P., Levashov, V. Y., & Pavlyukevich, N. V. (2014). Condensation coefficient: Definitions, estimations, modern experimental and calculation data. *Journal of Engineering Physics and Thermophysics, 87*(1), 237–245.

Kryokov, A.P. and Levashov, V.Y. (2011), About evaporation-condensation coefficients on the vapor liquid interface of high thermal conductivity matters, Int. *J. Heat and Mass Transfer, Vol. 54*, 3042–3048.

Kumar, T., Singh, U. B., Kumar, M., Ojha, S., & Kanjilal, D. (2014). Tuning of ripple patterns and wetting dynamics of Si(100) surface using ion beam irradiation. *Current Applied Physics, 14*(3), 312–317.

Kumari, N., & Garimella, S. V. (2011). Characterization of the heat transfer accompanying electrowetting or gravity induced droplet motion. *International Journal of Heat and Mass Transfer, 54*, 4037–4050.

Kundu, P., & Cohen, I. (2004). *Fluid mechanics*. San Diego, CA: Elsevier Academic Press. ISBN #978-0-12381399-2.

Lan, Z., Ma, X. Z., Zhang, Y., & Zhou, X. D. (2009). Theoretical study of dropwise condensation heat transfer: Effect of the liquid-solid surface free energy difference. *Journal of Enhanced Heat Transfer, 16*, 61–71.

Lara, J. R., & Holtzapple, M. T. (2011). Experimental investigation of dropwise condensation on hydrophobic heat exchangers. Part II: Effect of coatings and surface geometry. *Desalination, 280*, 363–369.

Larmour, I. A., Bell, S. E. J., & Saunders, G. C. (2007). Remarkably simple fabrication of superhydrophobic surfaces using electroless galvanic deposition. *Angewandte Chemie International Edition, 46*, 710–1712.

Lau, K. K. S., Bico, J., Teo, K. B. K., Chowilla, M., Amaratunga, G. A. J., Milne, W. I., McKinley, G. H., & Gleason, K. K. (2003). Superhydrophobic carbon nanotube forests. *Nano Letters, 3*, 1701–1705.

Lauga, E., Brenner, M., & Stone, H. (2007). *Microfluidics: The no-slip boundary condition* (Handbook of experimental fluid mechanics) (pp. 1219–1240). Berlin: Springer.

Lawal, A., & Brown, R. A. (1982). The stability of an inclined pendent drop. *Journal of Colloid and Interface Science, 89*, 332–345.

Le Fevre, E. J., & Rose, J. W. (1964). Heat-transfer measurement during dropwise condensation of steam. *International Journal of Heat and Mass Transfer, 7*, 272–273.

Le Fevre, E. J., & Rose, J. W. (1965). An experimental study of heat transfer by dropwise condensation. *International Journal of Heat and Mass Transfer, 8*, 1117–1133.

Le Fevre, E. J., & Rose J. W. (1966). A theory of heat transfer by dropwise condensation. In *Proceedings of the 3rd International Heat Transfer Conference, Chicago* (Vol. 2, pp. 362–375).

Leach, R. N., Stevens, F., Langford, S. C., & Dickinson, J. T. (2006). Dropwise condensation: Experiments and simulations of nucleation and growth of water drops in a cooling system. *Langmuir, 22*, 8864–8872.

Leal, L. G. (2007). *Advanced transport phenomena: Fluid mechanics and convective transport processes*. Cambridge: Cambridge University Press. ISBN #978-0-52117908-9.

Lee, Y. L., & Maa, J. R. (1991). Nucleation and growth of condensate clusters on solid surfaces. *Journal of Materials Science, 26*, 6068–6072.

Lee, K. H., & Wong, C. C. (2009). Decoupling two-step anodization in anodic aluminum oxide. *Journal of Applied Physics, 106*, 1–3.

Lee, L. Y., Fang, T. H., Yang, Y. M., & Maa, J. R. (1998). The enhancement of dropwise condensation by wettability modification of solid surface. *International Communications in Heat and Mass Transfer, 25*(8), 1095–1103.

Lee, W., Ji, R., Gosele, U., & Nielsch, K. (2006). Fast fabrication of long range ordered porous alumina membranes by hard anodization. *Nature Materials, 5*, 741–747.

Lee, W., Schwirn, K., Steinhari, M., Pippel, E., Scholz, R., & Gosele, U. (2008). Structural engineering of nanoporous anodic aluminum oxide by pulse anodization of aluminum. *Nature Nanotechnology, 3*, 234–239.

Legendre, D., & Maglio, M. (2013). Numerical simulation of spreading drops. *Colloids and Surfaces A: Physicochemical and Engineering Aspects, 432*, 29–37.

Leger, L., & Joany, J. F. (1977). Liquid spreading. *Reports on Progress in Physics, 57*, 431–487.

Leipertz, A. (2010). Dropwise condensation (section J3). In *VDI heat atlas* (pp. 933–938). Berlin: Springer.

Leipertz, A., & Choi, K. H. (2000). Dropwise condensation on ion-implanted metallic surfaces. In *Proceedings of the 3rd European Thermal Sciences Conference* (pp. 917–921).

Leipertz, A., & Fröba A. P. (2006). Improvement of condensation heat transfer by surface modification. In *Proceedings of the 7th ASME, Heat and Mass Transfer Conference, IIT Guwahati, India* (Vol. K7, pp. k85–k99).

Leipertz, A., & Fröba, A. P. (2008). Improvement of condensation heat transfer by surface modifications. *Heat Transfer Engineering, 29*(4), 343–356.

Lenz, P., & Lipowsky, R. (1998). Morphological transitions of wetting layer on structured surfaces. *Physical Review Letters, 80*(9), 1920–1998.

Lexmond, A. S., & Geld, C. W. M. (2005). The effect of plate thickness, surface tension and fluid flow on detachment of drop from a plate. *Experimental Thermal and Fluid Science, 29*, 813–819.

Li, W., & Amirfazli, A. (2007). Microtextured superhydrophobic surfaces: A thermodynamic analysis. *Advances in Colloid and Interface Science, 132*, 51–68.

Li, A. P., Muller, F., Birner, A., Nielsch, K., & Gosele, U. (1998a). Hexagonal pore arrays with a 50–420 nm interpore distance formed by self-organization in anodic alumina. *Journal of Applied Physics, 84*(11), 6023–6026.

Li, F., Zhang, L., & Metzger, R. M. (1998b). On the growth of highly ordered pores in anodized aluminum oxide. *Chemistry of Materials, 10*, 2470–2480.

Liao, Q., Wang, H., Zhu, X., & Li, M. (2006). Liquid droplet movement on horizontal surface with gradient surface energy. *Science in China Series E: Technological Sciences, 49*(6), 733–741.

Liao, Q., Zhu, X., Xing, S., & Wang, H. (2008). Visualization study on coalescence between pair of water drops on inclined surfaces. *Experimental Thermal and Fluid Science, 32*(8), 1647–1654.

Liao, J., Ling, Z., Li, Y., & Hu, X. (2016). The role of stress in the self-organized growth of porous anodic alumina. *ACS Applied Materials and Interfaces, 8*, 8017–8023.

Liu, X., & Cheng, P. (2015). Dropwise condensation theory revisited: Part II. Droplet nucleation density and condensation heat flux. *International Journal of Heat and Mass Transfer, 83*, 842–849.

Liu, Y., Chen, X., & Xin, J. H. (2006). Super-hydrophobic surfaces from a simple coating method: A bionic nanoengineering approach. *Nanotechnology, 17*, 3259–3263.

Liu, T., Mu, C., Sun, X., & Xia, S. (2007). Mechanism study on formation of initial condensate droplets. *AIChE Journal, 53*(4), 1050–31055.

Lo, H. Y., Liu, Y., & Xu, L. (2017). Mechanism of contact between a droplet and an atomically smooth substrate. *Physical Review, 7*, 021036.

Löhle, S., Fuchs, U., Digel, P., Hermann, T., & Battaglia, J.-L. (2013). Analyzing inverse heat conduction problems by the analysis of the system impulse response. *Inverse Problems in Science and Engineering, 22*, 297–308.

Long, J., Fan, P., Gong, D., Jiang, D., Zhang, H., Li, L., & Zhong, M. (2015a). Superhydrophobic surfaces fabricated by femtosecond laser with tunable water adhesion: From lotus leaf to rose petal. *ACS Applied Materials and Interfaces, 7*, 9858–9865.

Long, J., Zhong, M., Fan, P., Gong, D., & Zhang, H. (2015b). Wettability conversion of ultrafast laser structured copper surface. *Journal of Laser Applications, 27*, S29107.

Ludviksson, V., & Lightfoot, E. N. (1968). Deformation of advancing menisci. *AIChE Journal, 14*, 674–677.

Ma, X., & Wang, B. (1999). Life-time test of dropwise condensation on polymer-coated surfaces. *Heat Transfer-Asian Research, 28*(7), 551–558.

Ma, X., Tao, B., Chen, J., Xu, D., & Lin, J. (2000a). Dropwise condensation heat transfer of steam on a polytethefluoroethylene film. *Journal of Thermal Science, 10*(3), 247–253.

Ma, X., Rose, J. W., Xu, D., Lin, J., & Wang, B. (2000b). Advances in dropwise condensation heat transfer: Chinese research. *Chemical Engineering Journal, 78*, 78–93.

Ma, X., Chen, J., Xu, D., Lin, J., Ren, C., & Long, Z. (2002). Influence of processing conditions of polymer film on dropwise condensation heat transfer. *International Journal of Heat and Mass Transfer, 45*, 3405–3411.

Ma, X. H., Zhou, X. D., Lan, Z., Li, Y. M., & Zhang, Y. (2008). Condensation heat transfer enhancement in the presence of non-condensable gas using the interfacial effect of dropwise condensation. *International Journal of Heat and Mass Transfer, 51*, 1728–1737.

Ma, X., Wang, S., Lan, Z., Peng, B., Ma, H. B., & Cheng, P. (2012). Wetting mode evolution of steam dropwise condensation on superhydrophobic surface in the presence of non-condensable gas. *ASME Journal of Heat Transfer, 134*, 021501–021509.

Maa, J. R. (1978). Drop-size distribution and heat flux of dropwise condensation. *Chemical Engineering Journal, 16*, 171–176.

Madi, C. S., Davidovitch, B., George, H. B., Norris, S. A., Brenner, M. P., & Aziz, M. J. (2008). Multiple bifurcation types and the linear dynamics of ion sputtered surfaces. *Physical Review Letters, 101*, 246102.

Majumdar, A., & Mezic, I. (1999). Instability of ultra-thin water film and the mechanism of droplet formation on hydrophobic surfaces. *Journal of Heat Transfer, 121*, 964–970.

Mannion, P. T., Magee, J., Coyne, E., O'Connor, G. M., & Glynn, T. J. (2004). The effect of damage accumulation behavior on ablation thresholds and damage morphology in ultrafast laser micromachining of common metals in air. *Applied Surface Science, 233*, 275–287.

Marek, R., & Straub, J. (2001). Analysis of the evaporation coefficient and the condensation coefficient of water. *International Journal of Heat and Mass Transfer, 44*(1), 39–53.

Martin, D. S., Cole, R. J., & Weightman, P. (2005). Effects of ion bombardment on the optical and electronic properties of Cu(110). *Physical Review B, 72*, 035408.

Marto, P. J., Looney, D. J., & Rose, J. W. (1986). Evaluation of organic coating for the promotion of dropwise condensation of steam. *International Journal of Heat and Mass Transfer, 29*, 1109–1117.

Masuda, H., & Fukuda, K. (1995). Ordered metal nanohole arrays made by a two-step replication of honeycomb structures of anodic alumina. *Science, 268*, 1466–1468.

Masuda, H., Hasegwa, F., & Ono, S. (1997). Self-ordering of cell arrangement of anodic porous alumina formed in sulphuric acid solution. *Journal of the Electrochemical Society, 144*, L127–L130.

Masuda, H., Yada, K., & Osaka, A. (1998). Self-ordering of cell configuration of anodic porous alumina with large-size pores in phosphoric acid solution. *Japanese Journal of Applied Physics, 37*, L1340–L1342.

Masuda, H., Asoh, H., Watanabe, M., Nishio, K., Nakao, M., & Tamamura, T. (2001). Square and triangular nanohole array architectures in anodic alumina. *Advanced Materials, 13*(3), 189–192.

McCarthy, M., Enright, R., Gerasopoulos, K., Culver, J., Ghodssi, R., & Wang, E. N. (2010). Biomimetic superhydrophobic surfaces using viral nano templates for self-cleaning and dropwise condensation. In *Proceedings of the Solid-State Sensor Actuator and Micro-system Workshop, Hilton Head, SC, USA*.

McCormick, J. L., & Baer, E. (1963). On the mechanism of heat transfer in dropwise condensation. *Journal of Collide Science, 18*, 208–216.

McCormick, J. L., & Westwater, J. W. (1965). Nucleation sites for dropwise condensation. *Chemical Engineering Science, 20*, 1021–1031.

McCoy, B. J. (2000). Vapor nucleation and droplet growth: Cluster distribution kinetics for open and closed systems. *Journal of Colloid and Interface Science, 228*, 64–72.

Meakin, P. (1992). Steady state behavior in a model for droplet growth, sliding and coalescence: The final stage of dropwise condensation. *Physica A, 183*, 422–438.

Menchaca-Rocha, A., Martínez-Dávalos, A., Núñez, R., Popinet, S., & Zaleski, S. (2001). Coalescence of liquid drops by surface tension. *Physical Review E, 63*(4), 046309.

Mikic, B. B. (1969). On the mechanism of dropwise condensation. *International Journal of Heat and Mass Transfer, 12*, 1311–1315.

Miljkovic, N., Enright, R., & Wang, E. N. (2012). Effect of droplet morphology on growth dynamics and heat transfer during condensation on superhydrophobic nanostructured surfaces. *ACS Nano, 6*, 1776–1785.

Mills, A. F., & Seban, R. A. (1967). The condensation coefficient of water. *International Journal of Heat and Mass Transfer, 10*, 1815–1827.

Milne, A. J. B., & Amirfazli, A. (2012). The Cassie equation: How it is meant to be used. *Advances in Colloid and Interface Science, 170*, 48–55.

Mistry, A. (2014). *Generalized Lagrangian model for drop spreading on textured surfaces*. M. Tech. thesis, Indian Institute of Technology Kanpur, India.

Mistry, A., & Muralidhar, K. (2015a). Axisymmetric model of drop spreading on a horizontal surface. *Physics of Fluids, 27*, 092103.

Mistry, A., & Muralidhar, K. (2015b). Non-thermal spreading of liquid drops: Effect of fluid convection, Keynote lecture. In *Asian Society for Computational Heat Transfer and Fluid Flow—ASCHT2015, Busan, Republic of Korea*.

Mistry, A., & Muralidhar, K. (2018). Spreading of a pendant liquid drop underneath a textured substrate. *Physics of Fluids, 30*, 042104.

Miwa, M., Nakajima, A., Fujishima, A., Hashimoto, K., & Watanabe, T. (2000). Effects of the surface roughness on sliding angles of water droplets on superhydrophobic surfaces. *Langmuir, 16*, 5754–5760.

Moradi, S., Kamal, S., Englezos, P., & Hatzikiriakos, S. G. (2013). Femtosecond laser irradiation of metallic surfaces: Effects of laser parameters on super-hydrophobicity. *Nanotechnology, 24*, 415302.

Mori, K., Fujita, N., Horie, H., More, S., Miyashita, M., & Matsuda, M. (1991). Heat transfer promotion of aluminum-brass cooling tube by surface treatment with triazinethiols. *Langmuir, 7*, 1161–1166.

Moumen, N., Subramanian, R., & McLaughlin, J. B. (2006). Experiments on the motion of drops on a horizontal solid surface due to wettability gradient. *Langmuir, 22*, 2682–2690.

Mu, C., Pang, J., & Liu, T. (2008). Effect of surface topography of material on nucleation site density of dropwise condensation. *Chemical Engineering Science, 63*, 874–880.

Muwanga, R., & Hassan, I. (2006). Local heat transfer measurements in microchannels using liquid crystal thermography: Methodology development and validation. *ASME Journal of Heat Transfer, 128*, 617–627.

Nakajima, A., Hashimoto, K., & Watanabe, T. (2001). Recent studies on super-hydrophobic films. *Monatschefte für Chemie, 132*, 31–41.

Nam, Y., Seo, D., Lee, C., & Shin, S. (2015). Droplet coalescence on water repellant surfaces. *Soft Matter, 11*(1), 154–160.

Narhe, R., Beysens, D., & Nikolayev, V. S. (2004). Contact line dynamics in drop coalescences and spreading. *Langmuir, 20*, 1213–1221.

Narhe, R., Beysens, D., & Nikolayev, V. S. (2005a). Dynamics of drop coalescence on a surface: The role of initial conditions and surface properties. *International Journal of Thermophysics, 26*, 8593–8597.

Narhe, R., Beysens, D., & Nikolayev, V. S. (2005b). Dynamics of drop coalescence on a surface: The role of initial conditions and surface properties. *International Journal of Thermophysics, 26* (6), 1743–1757.

Narhe, R., Beysens, D., & Nikolayev, V. S. (2006). Contact line dynamics in drop coalescence and spreading. *Langmuir, 20*, 1213–1221.

Narhe, R. D., Beysens, D. A., & Pomeau, Y. (2008). Dynamic drying in the early-stage coalescence of droplets sitting on a plate. *Europhysics Letters, 81*(4), 46002.

Necmi, S., & Rose, J. W. (1977). Heat-transfer measurements during dropwise condensation of mercury. *International Journal of Heat and Mass Transfer, 20*, 877–880.

Neeumaqnn, A. W., Abdelmessihm, A. H., & Hameed, A. (1978). The role of contact angles and contact angles hysteresis in dropwise condensation heat transfer. *International Journal of Heat and Mass Transfer, 21*, 947–953.

Nepomnyashchy, A. A., & Simanovskii, I. B. (2004). Influence of thermocapillary effect and interfacial heat release on convective oscillations in a two-layer system. *Physics of Fluids, 16* (4), 1127–1139.

Nielsch, K., Choi, J., Schwirn, K., Wehrspohn, R. B., & Gosele, U. (2002). Self-ordering regimes of porous alumina: The 10% porosity rule. *Nano Letters, 2*, 677.

Niknejad, J., & Rose, J. W. (1984). Comparisons between experiment and theory for dropwise condensation. *International Journal of Heat and Mass Transfer, 20*, 2253–2257.

Nilsson, M. A., & Rothstein, J. P. (2011). The effect of contact angle hysteresis on droplet coalescence and mixing. *Journal of Colloid and Interface Science, 363*(2), 646–654.

Niu, D., Guo, L., Hu, H. W., & Tang, G. H. (2017). Dropwise condensation heat transfer model considering the liquid-solid interfacial thermal resistance. *International Journal of Heat and Mass Transfer, 112*, 333–342.

Oh, J. (2010). *Porous anodic aluminum scaffolds: Formation mechanisms and applications*. Ph.D. thesis, Department of Materials Science and Engineering, Massachusetts Institute of Technology, Cambridge, MA.

Oh, J., Birbarah, P., Foulkes, T., & Yin, S. L. (2017). Jumping droplet electronics hot-spot cooling. *Applied Physics Letters, 110*, 123107.

Onda, T., Shibuichi, S., Satoh, N., & Tsujii, K. (1996). Super-water-repellent fractal surfaces. *Langmuir, 12*, 2125–2127.

Ondarçuhu, T. (1995). Total or partial pinning of a droplet on a surface with chemical discontinuity. *Journal of Physics II France, 5*, 227–241.

Öner, D., & McCarthy, T. J. (2000). Ultra-hydrophobic surfaces: Effects of topography length scales on wettability. *Langmuir, 16*, 7777–7782.

Oura, K., Lifshits, V. G., Saranin, A. A., Zotov, A. V., & Katayama, M. (2003). *Surface science* (1st ed.). Berlin: Springer.

Özisik, M. N., & Orlande, H. R. B. (2000). *Inverse heat transfer: Fundamentals and applications*. New York: Taylor and Francis. ISBN #1-56032-838-X.

Pan, Y., & Suga, K. (2005). Numerical simulation of binary liquid droplet collision. *Physics of Fluids, 17*(8), 082105.

Pan, Z., Dash, S., Weibel, J. A., & Garimella, S. V. (2013). Assessment of water droplet evaporation mechanisms on hydrophobic and superhydrophobic substrates. *Langmuir, 29*, 15831–15841.

Pan, Z., Weibel, J. A., & Garimella, S. V. (2014). Influence of surface wettability on transport mechanisms governing water droplet evaporation. *Langmuir, 30*(32), 9726–9730.

Park, J., & Moon, J. (2006). Control of colloidal particle deposit patterns within picoliter droplets ejected by ink-jet printing. *Langmuir, 22*(8), 3506–3513.

Parkhutik, V., & Shershulsky, V. (1992). Theoretical modelling of porous oxide growth on aluminum. *Journal of Physics D: Applied Physics, 25*, 1258–1263.

Paulsen, J. D., Burton, J. C., & Nagel, S. R. (2011). Viscous to inertial crossover in liquid drop coalescence. *Physical Review Letters, 106*(11), 114501.

Paulsen, J. D., Burton, J. C., Nagel, S. R., Appathurai, S., Harris, M. T., & Basaran, O. A. (2012). The inexorable resistance of inertia determines the initial regime of drop coalescence. *Proceedings of the National Academy of Sciences, 109*(18), 6857–6861.

Peng, X. F., Liu, D., Lee, D. J., Yan, Y., & Wang, B. X. (2000). Cluster dynamics and fictitious boiling in micro-channels. *International Journal of Heat and Mass Transfer, 43*, 4259–4265.

Peterson, A. C., & Westwater, J. W. (1966). *Dropwise condensation of ethylene and glycol*. Chemical engineering progress symposium series 62 (Vol. 64, pp. 135–142).

Phan, H. T., Caney, N., Marty, P., Colasson, S., & Gavillet, J. (2009). Surface wettability control by nanocoating: The effects on pool boiling heat transfer and nucleation mechanism. *International Journal of Heat and Mass Transfer, 52*, 5459–5471.

Picknett, R. G., & Bexon, R. (1977). The evaporation of sessile or pendant drops in still air. *Journal of Colloid and Interface Science, 61*(2), 336–350.

Pierce, E., Carmona, F. J., & Amirfazli, A. (2008). Understanding of sliding and contact angles results in tilted plate experiments. *Colloids and Surfaces A: Physicochemical and Engineering Aspects, 323*, 73–82.

Poinern, G. E., Ali, N., & Fawcett, D. (2011). Progress in nano-engineered anodic aluminum oxide membrane development. *Materials, 4*, 487–526.

Polikarpov, A. P., Graur, I. A., Gatapova, E. Y., & Kabov, O. A. (2019). Kinetic simulation of the non-equilibrium effects at the liquid-vapor interface. *International Journal of Heat and Mass Transfer, 136*, 449–456.

Popov, Y. O. (2005). Evaporative deposition patterns: Spatial dimensions of the deposit. *Physical Review E, 71*, 1–17.

Pozrikidis, C. (1997). *Introduction to theoretical and computational fluid dynamics*. New York: Oxford University Press. ISBN #0-19-509320-8.

Pozrikidis, C. (2009). *Fluid dynamics: Theory, computation, and numerical simulation* (2nd ed., pp. 237–250). Boston, MA: Springer.

Prabhu, N. K., Fernades, P., & Kumar, G. (2009). Effect of substrate surface roughness on the wetting behavior of vegetable oils. *Materials and Design, 30*(2), 297–305.

Pratap, D. (2015). *Anisotropic and hyperbolic metamaterials in the cylindrical geometry*. Ph.D. thesis, Department of Physics, Indian Institute of Technology Kanpur, India.

Pratap, V., Moumen, N., & Subramanian, R. (2008). Thermocapillary motion of a liquid drop on a horizontal solid surface. *Langmuir, 24*, 2185–5193.

Pukha, V. E., Mikhailov, I. F., Drozdov, A. N., & Fomina, L. P. (2005). Dependence of the condensation coefficient of bismuth on the energy of particles deposited from an ion beam on silicon substrates. *Physics of the Solid State, 47*, 595–598.

Punetha, M., & Khandekar, S. (2017). A CFD based modelling approach for predicting steam condensation in the presence of non-condensable gases. *Nuclear Engineering and Design, 324*, 280–296.

Qi, B., Wei, J., Zhang, L., & Xu, H. (2015). A fractal dropwise condensation heat transfer model including the effects of contact angle and drop size distribution. *International Journal of Heat and Mass Transfer, 83*, 259–272.

Qian, J., & Law, C. K. (1997). Regimes of coalescence and separation in droplet collision. *Journal of Fluid Mechanics, 331*, 59–80.

Qin, T. and Grigoriev, R.O. (2015), The effect of noncondensables on buoyancy-thermocapillary convectiuon of volatile fluids in confined geometries, Int. *J. Heat and Mass Transfer, 90*, 678–688

Quéré, D. (2005). Non-sticking drops. *Reports on Progress in Physics, 68*, 2495–2532.

Quéré, D. (2008). Wetting and roughness. *Annual Review of Materials Research, 38*, 71–99.

Rajendra, A., Parmar, B. J., Sharma, A. K., Bhojraj, H., Nayak, M. M., & Rajanna, K. (2005). Hard anodization of aluminum and its application to sensorics. *Surface Engineering, 21*, 193–197.

Ramiasa-MacGregor, M., Mierczynska, A., Sedeva, R., & Vasilev, K. (2016). Tuning and predicting the wetting of nanoengineered material surface. *Nanoscale, 8*, 4635–4642.

Ramos, S. M. M., Charlaix, E., Benyagoub, A., & Toulemonde, M. (2003). Wetting on nanorough surfaces. *Physical Review E, 67*, 031604.

Ratsch, C., & Venables, J. A. (2003). Nucleation theory and the early stages of thin film growth. *Journal Vacuum Society Technology, A21*(5), s96–s109.

Ratsch, C., & Zangwill, A. (1994). Saturation and scaling of epitaxial island densities. *Physics Review Letters, 72*, 3194–3197.

Ratsch, C., Seitsonen, A. P., & Scheffler, M. (1997). Strain dependence of surface diffusion: Ag on Ag (111) and Pt (111). *Physics Review B, 55*, 6750–6753.

Rausch, M. H., Fröba, A. P., & Leipertz, A. (2007). Dropwise condensation on plasma-ion implanted aluminum surface. *International Journal of Heat and Mass Transfer, 51*, 1061–1070.

Rausch, M. H., Fröba, A. P., & Leipertz, A. (2008). Dropwise condensation heat transfer on an ion implanted aluminum surfaces. *International Journal of Heat and Mass Transfer, 51*, 1061–1070.

Rausch, M. H., Leipertz, A., & Fröba, A. P. (2010a). Dropwise condensation of steam on ion-implanted titanium surfaces. *International Journal of Heat and Mass Transfer, 53*, 423–430.

Rausch, M. H., Leipertz, A., & Fröba, A. P. (2010b). On the mechanism of dropwise condensation on ion-implanted metallic surface. *ASME Journal of Heat Transfer, 132*, 94503, 1–3.

Razavi, S., Koplik, J., & Kretzschmar, I. (2014). Molecular dynamics simulations: Insight into molecular phenomena at interfaces. *Langmuir, 30*(38), 11272–11283.

Ren, S. L., Yang, S. R., Wang, J. Q., Liu, W. M., & Zhao, Y. P. (2004). Preparation and tribological studies of stearic acid self-assembled monolayers on polymer-coated silicon surface. *Chemistry of Materials, 16*, 428–434.

Rio, E., Daerr, A., Andreotti, B., & Limat, L. (2005). Boundary conditions in the vicinity of a dynamic contact Line: Experimental investigation of viscous drops sliding down an inclined plane. *Physical Review Letters, 94*, 024503, 1–4.

Ristenpart, W. D., McCalla, P. M., Roy, R. V., & Stone, H. A. (2006). Coalescence of spreading droplets on a wettable substrate. *Physical Review Letters, 97*(6), 064501.

Ristenpart, W. D., Kim, P. G., Domingues, C., Wan, J., & Stone, H. A. (2007). Influence of substrate conductivity on circulation reversal in evaporating drops. *Physical Review Letters, 99*, 1–4. Paper #234502.

Rose, J. W. (1972). Dropwise condensation of mercury. *International Journal of Heat and Mass transfer, 15*, 1431–1434.

Rose, J. W. (1976). Further aspects of dropwise condensation theory. *International Journal of Heat and Mass Transfer, 19*, 1363–1370.

Rose, J. W. (1978a). The effect of surface thermal conductivity on dropwise condensation heat transfer. *International Journal of Heat and Mass Transfer, 21*, 80–81.

Rose, J. W. (1978b). Effect of conductivity tube material on heat transfer during dropwise condensation of steam. *International Journal of Heat and Mass Transfer, 21*, 835–840.

Rose, J. W. (1981). Dropwise condensation theory. *International Journal of Heat and Mass Transfer, 24*, 191–194.

Rose, J. W. (1998). Condensation heat transfer fundamentals. *Transactions of the AIChE Journal, 76A*, 143–152.

Rose, J. W. (2002). Dropwise condensation theory and experiment: A review. *Proceedings of the Institution of Mechanical Engineers, Part A: Journal of Power and Energy, 216*, 115–128.

Rose, J. W. (2004). Surface tension effects and enhancements of condensation heat transfer. *Chemical Engineering Research and Design, 82*, 419–429.

Rose, J. W., & Glicksman, L. R. (1973). Dropwise condensation—The distribution of drop sizes. *International Journal of Heat and Mass Transfer, 16*, 411–425.

Rose, J. W., Tanasawa, I., & Utaka, Y. (1999). Dropwise condensation. In *Handbook of phase-change: Boiling and condensation* (pp. 581–594). Boca Raton, FL: Taylor and Francis.

Rykaczewski, K. (2012). Microdroplet growth mechanism during water condensation on superhydrophobic surfaces. *Langmuir, 28*, 7720–7729.

Rykaczewski, K., Scott, J. H. J., Rajauria, S., Chinn, J., Chinn, A. M., & Jones, W. (2011). Three dimensional aspects of droplet coalescence during dropwise condensation on superhydrophobic surfaces. *Soft Matter, 7*, 8749–8752.

Saada, M. A., Chikh, S., & Tadrist, L. (2013). Evaporation of a sessile drop with pinned or receding contact line on a substrate with different thermophysical properties. *International Journal of Heat and Mass Transfer, 58*, 197–208.

Sadhal, S. S., Ayyaswamy, P. S., & Chung, J. N. (1997). *Transport phenomena with drops and bubbles* (Mechanical engineering series) (1st ed., pp. 218–230). New York: Springer.

Sakai, M., & Hashimoto, A. (2007). Image analysis system for evaluating sliding behavior of a liquid droplet on a hydrophobic surface. *Review of Scientific Instruments, 78*, 045105–045109.

Sakai, M., Hashimoto, A., Yoshida, N., Suzuki, S., Kameshima, Y., & Nakajima, A. (2006). Direct observation of internal fluidity in a water droplet during sliding on hydrophobic surfaces. *Langmuir, 22*, 4906–4909.

Santos, M., & White, J. (2011). Theory and simulation of angular hysteresis on planar surfaces. *Langmuir, 27*(24), 14868–14875.

Schmidt, E., Schurig, W., & Sellschopp, W. (1930). Versuche über die condensation von wasserdampf in film-und tropfenform, Forsch. *Ingenieurwes, 1*(2), 53–63.

Schmuki, P., Erickson, L. E., & Lockwood, D. J. (1998). Light emitting micropatterns of porous Si created at surface defects. *Physical Review Letters, 80*, 4060.

Schrader, M. E. (1984). Wettability of clean metal surfaces. *Journal of Colloid and Interface Science, 100*, 372–380.

Sellier, M., & Trelluyer, E. (2009). Modeling the coalescence of sessile droplets. *Biomicrofluidics, 3*(2), 022412.

Sellier, M., Nock, V., & Verdier, C. (2011). Self-propelling, coalescing droplets. *International Journal of Multiphase Flow, 37*(5), 462–468.

Shanahan, M. E. R. (1995). Simple theory of stick-slip wetting hysteresis. *Langmuir, 11*(3), 1041–1043.

Shi, F., Shim, Y., & Amar, J. G. (2005). Island-size distribution and capture numbers in three-dimensional nucleation: Comparison with mean-field behavior. *Physical Review B, 71*, 245411–245416.

Shibuichi, S., Onda, T., Satoh, N., & Tsujii, K. (1996). Super-water-repellent fractal surfaces. *Journal of Physical Chemistry, 100*, 19512–19617.

Shikhmurzaev, Y. D. (1997). Spreading of drops on solid surfaces in a quasi-static regime. *Physics of Fluids, 9*, 266.

Shin, D. H., Lee, S. H., Jung, J.-Y., & Yoo, J. Y. (2009). Evaporating characteristics of sessile droplet on hydrophobic and hydrophilic surfaces. *Microelectronic Engineering, 86*, 1350–1353.

Shin, S., Seo, J., Han, H., Kang, S., Kim, H., & Lee, T. (2016). Bio-inspired extreme wetting surfaces for biomedical applications. *Materials, 9*, 116.

Shishkova, I. N., Kryukov, A. P., & Levashov, V. Y. (2017). Study of evaporation-condensation problems: From liquid through interface surface to vapor. *International Journal of Heat and Mass Transfer, 112*, 926–932.

Sikarwar, B. S. (2013). *Modelling dropwise condensation underneath inclined textured surfaces*. Ph.D. dissertation, Indian Institute of Technology Kanpur, India.

Sikarwar, B. S., Battoo, N. K., Khandekar, S., & Muralidhar, K. (2010a). Dropwise condensation underneath chemically textured surfaces: Simulation and experiments. *Journal of Heat Transfer, 133*, 021501.

Sikarwar, B. S., Muralidhar, K., & Khandekar, S. (2010b). Flow and heat transfer in a pendant liquid drop sliding on an inclined plane. In *Proceedings of the 9th International ISHMT-ASME Heat and Mass Transfer Conference, Mumbai, India*, Paper #345 (pp. 1322–1329).

Sikarwar, B. S., Muralidhar, K., & Khandekar, S. (2010c). Flow and thermal fields in a pendant droplet moving on lyophobic surface. In *Proceedings of the 14th International IHTC-ASME Heat and Mass Transfer Conference, Washington, DC, USA*, Paper #IHTC14-22520 (p. 169).

Sikarwar, B. S., Battoo, N. K., Khandekar, S., & Muralidhar, K. (2011). Dropwise condensation underneath chemically textured surfaces: Simulation and experiments. *ASME Journal of Heat Transfer, 133*(2), 021501, 1–15.

Sikarwar, B. S., Khandekar, S., Agrawal, S., Kumar, S., & Muralidhar, K. (2012). Dropwise condensation studies on multiple scales. *Heat Transfer Engineering, 33*(4–5), 301–341.

Sikarwar, B. S., Khandekar, S., & Muralidhar, K. (2013a). Mathematical modelling of dropwise condensation on textured surfaces. *Sadhana-Academy Proceedings in Engineering Sciences, 38*, 1135–1171.

Sikarwar, B. S., Khandekar, S., & Muralidhar, K. (2013b). Simulation of flow and heat transfer in a liquid drop sliding underneath a hydrophobic surface. *International Journal of Heat and Mass Transfer, 57*(2), 786–811.

Singh, S. K., Khandekar, S., Pratap, D., & Ramakrishna, S. A. (2013). Wetting dynamics and evaporation of sessile droplets on nano-porous alumina surfaces. *Colloids and Surfaces A: Physicochemical and Engineering Aspects, 432*, 71–81.

Singh, S. K., Yadav, M. K., & Khandekar, S. (2016a). Measurement issues associated with surface mounting of thermopile heat flux sensors. *Applied Thermal Engineering, 114*, 1105–1113.

Singh, U. B., Yadav, R. P., Pandey, R. K., Agarwal, D. C., Pannu, C., & Mittal, A. K. (2016b). Insight mechanisms of surface structuring and wettability of ion-treated Ag thin films. *Journal of Physical Chemistry C, 120*, 5755–5763.

Singh, S. K., Yadav, M. K., Sonawane, R., Khandekar, S., & Muralidhar, K. (2017). Estimation of time-dependent wall heat flux from single thermocouple data. *International Journal of Thermal Sciences, 115*, 1–15.

Smith, E. R., Theodorakis, E. R., Craster, P. R., & Matar, O. (2018). Moving contact lines: Linking molecular dynamics and continuum-scale modeling. *Langmuir, 34*(42), 2501–12518.

Snoeijer, J. H., & Andreotti, B. (2013). Moving contact lines: Scales, regimes, and dynamical transitions. *Annual Review of Fluid Mechanics, 45*, 269–292.

Sobac, B., & Brutin, D. (2011). Triple-line behaviour and wettability controlled by nanocoated substrates: Influence on sessile drop evaporation. *Langmuir, 27*(24), 14999–15007.

Sobac, B., & Brutin, D. (2012). Thermal effects of the substrate on water droplet evaporation. *Physical Review E, 86*, 1–10. Paper #021602.

Sommers, A. D., & Jacobi, A. M. (2006). Creating micro-scale surface topology to achieve anisotropic wettability on an aluminum surface. *Journal of Micromechanics and Micro Engineering, 16*, 1571–1578.

Sommers, A. D., & Jacobi, A. M. (2008). Wetting phenomena on micro-grooved aluminum surfaces and modeling of the critical droplet size. *Journal of Colloid and Interface Science, 328*, 402–411.

Somwanshi, P., Muralidhar, K., & Khandekar, S. (2018). Coalescence dynamics of sessile and pendant liquid drops placed on a hydrophobic surface. *Physics of Fluids, 30*, 092103.

Sondag-Huethorst, J. A. M., & Fokkink, L. J. G. (1994). Potential-dependent wetting of electroactive ferrocene-terminated alkanethiolate monolayers on gold. *Langmuir, 10*, 4380–4387.

Song, Y., Xu, D., Lin, J., & Tsian, S. (1991). A study on the mechanism of dropwise condensation. *International Journal of Heal Mass Transfer, 34*(11), 2827–2831.

Song, T., Lan, Z., Ma, X., & Bai, T. (2009). Molecular clustering physical model of steam condensation and the experimental study on the initial droplet size distribution. *International Journal of Thermal Sciences, 48*, 2228–2236.

Song, H., Lee, Y., Jin, S., Kim, H.-Y., & Yoo, J. Y. (2011). Prediction of sessile drop evaporation considering surface wettability. *Microelectronic Engineering, 88*, 3249–3255.

Sprittles, J. E., & Shikhmurzaev, Y. D. (2012). Coalescence of liquid drops: Different models versus experiment. *Physics of Fluids, 24*(12), 122105.

Stalder, A. F., Kulik, G., Sage, D., Barbieri, L., & Hoffmann, P. (2006). A snake-based approach to accurate determination of both contact points and contact angles. *Colloids and Surfaces A: Physicochemical and Engineering Aspects, 286*(1–3), 92–103.

Stasiek, J., Stasiek, A., Jewartowski, M., & Collins, M. W. (2006). Liquid crystal thermography and true-color digital image processing. *Optics and Laser Technology, 38*, 243–256.

Stephan, K. (1992). *Heat transfer in condensation and boiling* (1st ed.). New York: Springer. ISBN #978-3642524592.

Stroscio, J. S., & Pierce, D. T. (1994). Scaling of diffusion mediated island growth in iron-on-iron homoepitaxy. *Physics Review B, 49*, 8522–8525.

Stylianou, S. A., & Rose, J. W. (1975). Drop-to-filmwise condensation transition: Heat transfer measurements for ethanediol. *International Journal of Heat and Mass Transfer, 97*, 72–98.

Stylianou, S. A., & Rose, J. W. (1980). Dropwise condensation on surface having different thermal conductivities. *ASME Journal of Heat Transfer, 102*, 477–482.

Stylianou, S. A., & Rose, J. W. (1983). Drop-to-filmwise condensation transition: Heat transfer measurements for ethandiol. *International Journal of Heat and Mass Transfer, 26*(5), 747–760.

Subramanian, R., Moumen, N., & McLaughlin, J. B. (2005). The motion of a drop on a solid surface due to a wettability gradient. *Langmuir, 21*, 11844–11849.

Sui, Y., Ding, H., & Spelt, P. D. (2014). Numerical simulations of flows with moving contact lines. *Annual Review of Fluid Mechanics, 46*, 97–119.

Sukhatme, S. P., & Rohsenow, W. M. (1966). Heat transfer during film condensation of a liquid metal vapor. *ASME Journal of Heat Transfer, 88*, 19–28.

Suzuki, S., Nakajima, A., Sakai, M., Song, J., Yoshida, N., Kameshima, Y., & Okada, K. (2006). Sliding acceleration of water droplets on a surface coated with fluoroalkysline and octadecyltrimethoxysilane. *Surface Science, 600*, 2214–2219.

Suzuki, S., Nakajima, A., Sakai, M., Song, J., Yoshida, N., Kameshima, Y., & Okada, K. (2008). Slipping and rolling ratio of sliding acceleration for a water droplet sliding on fluoroalkylsilane coating of different roughness. *Chemistry Letters, 37*, 58–59.

Swartz, M. M., & Yao, S. C. (2017). Experimental study of turbulent natural-convective condensation on a vertical wall with smooth and wavy film interface. *International Journal of Heat and Mass Transfer, 113*, 943–960.

Ta, D. V., Dunn, A., Wasley, T. J., Kay, R. W., Stringer, J., Smith, P. J., Connaughton, C., & Shephard, J. D. (2015). Nanosecond laser textured superhydrophobic metallic surfaces and their chemical sensing applications. *Applied Surface Science Part A, 357*, 248–254.

Takeyama, T., & Shimizu S. (1974). On the transition of dropwise-film condensation. In *Proceedings of the 5th International Heat Transfer Conference* (Vol. 3, pp. 274–278).

Tambe, N. S., & Bhushan, B. (2005). Nanotribological characterization of self-assembled monolayers deposited on silicon and aluminium substrates. *Nanotechnology, 16*, 1549.

Tanaka, H. (1975a). A theoretical study of dropwise condensation. *Journal of Heat Transfer, 97*(1), 97–103.

Tanaka, H. (1975b). Measurement of drop-size distribution during transient dropwise condensation. *Journal of Heat Transfer, 97*, 341–346.

Tanasawa, I. (1991). Advance in condensation heat transfer. In J. P. Hartnett, T. F. Irvine, & I. Y. Cho (Eds.), *Advances in heat transfer* (Vol. 21, pp. 56–136). San Diego, CA: Academic Press.

Tanasawa, I., & Utaka, Y. (1983). Measurement of condensation curves for dropwise condensation of steam at atmospheric pressure. *Journal of Heat Transfer, 1*(5), 633–638.

Tanasawa, I., Ochiiai, J., Utaka, Y., & Enya, S. (1974). Dropwise condensation. In *11th Japanese Heat Transfer Symposium* (Vol. 229).

Tanasawa, I., Ochiai, J., Utaka, Y., & Enya, S. (1976). Experimental study on dropwise condensation (Effect of departing drop size on heat-transfer coefficient). *Transactions of the JSME, 42* (361), 2846–2853.

Tang, C., Zhang, P., & Law, C. K. (2012). Bouncing, coalescence, and separation in head-on collision of unequal-size droplets. *Physics of Fluids, 24*(2), 022101.

Tanner, L. H. (1979). The spreading of silicone oil drops on horizontal surfaces. *Journal of Physics D: Applied Physics, 12*, 1473.

Tanner, D. W., Potter, C. J., Pope, D., & West, D. (1965). Heat transfer in dropwise condensation: part II—Surface chemistry. *International Journal of Heat and Mass Transfer, 8*, 427–436.

Tanner, D. W., Pope, D., Potter, C. J., & West, D. (1968). Heat transfer in dropwise condensation at low stem pressure in the absence of non-condensable gas. *International Journal of Heat and Mass Transfer, 11*, 181–190.

Thoroddsen, S. T., Takehara, K., & Etoh, T. G. (2005). The coalescence speed of a pendent and a sessile drop. *Journal of Fluid Mechanics, 527*, 85–114.

Thoroddsen, S. T., Qian, B., Etoh, T. G., & Takehara, K. (2007). The initial coalescence of miscible drops. *Physics of Fluids, 19*, 072110.

Tian, Y., Wang, X. D., & Peng, X. F. (2004). Analysis of surface inside metastable bulk phase during gas-liquid phase transition. *Journal of Engineering Thermophysics (Chinese), 25*, 100–102.

Tien, C. L., Majumdar, A., & Gerner, M. (1998). *Microscale energy transport* (pp. 167–183). Washington, DC: Taylor and Francis.

Tiwari, G. N., & Sahota, L. (2017). *Advanced solar-distillation systems: Basic principles, thermal modeling and its application*. Singapore: Springer.

Trimmer, W. S. N. (1989). Microrobots and micromechanical system. *Sensors and Actuators, 19*, 267–287.

Tryggvason, G., Scardovelli, R., & Zaleski, S. (2011). *Direct numerical simulations of gas-liquid multiphase flows*. Cambridge: Cambridge University Press. ISBN #9780521782401.

Tsuruta, T. (1993). Constriction resistance in dropwise condensation. In *Proceedings of the ASME Engineering Foundation Conference on Condensation and Condenser Design* (pp. 109–170).

Tsuruta, T., & Tanaka, H. (1991). A theoretical study on the constriction resistance in dropwise condensation. *International Journal of Heat and Mass Transfer, 34*(11), 2779–2786.

Tsuruta, T., Tanaka, H., & Togashi, S. (1991). Experimental verification of constriction resistance theory in dropwise condensation heat transfer. *International Journal of Heat and Mass Transfer, 34*(11), 2787–2796.

Ucar, I. O., & Erbil, H. Y. (2012). Dropwise condensation rate of water breath figures on polymer surfaces having similar surface free energies. *Applied Surface Science, 259*, 515–523.

Umur, A., & Griffith, P. (1965). Mechanism of dropwise condensation. *ASME Journal of Heat Transfer, 87*, 275–282.

Utaka, Y., Saito A., Ishikawa, H., & Yanagida, H. (1987). Transition from dropwise condensation to film condensation of propylene glycol, ethylene glycol, and glycerol vapors. In *Proceedings of the 2nd ASME-JSME Thermal Engineering Conference* (Vol. 4, pp. 377–384).

Utaka, Y., Kubo, R., & Ishii, K. (1994). Heat transfer characteristics of condensation of camphor on a lyophobic Surface. In *Proceedings of the 10th International Heat Transfer Conference* (Vol. 3, pp. 401–406).

Van der Geld, C. W. M., Ganzevles, F. L. A., Simons, C. T. P. F., & Weitz, F. (2001). Geometry adaptations to improve the performance of compact polymer condensers. *Transactions of the IChemE (Part A), 79*, 357–362.

Vemuri, S., & Kim, K. J. (2006). An experimental and theoretical study on the concept of dropwise condensation. *International Journal of Heat and Mass Transfer, 49*, 649–657.

Vemuri, S., Kim, K. J., Wood, B. D., Govindaraju, S., & Bell, T. W. (2006). Long term testing for dropwise condensation using self-assembled monolayer coating of n-octadecyl mercaptan. *Applied Thermal Engineering, 26*, 421–429.

Venables, J. A. (2000). *Introduction to surface and thin film processes* (1st ed., pp. 144–165). Cambridge: Cambridge University Press. ISBN-10 #0521785006.

Venables, J. A., & Brune, H. (2002). Capture numbers in the presence of repulsive adsorbate interactions. *Physical Review B, 66*, 195404–195419.

Voinov, O. V. (1976). Hydrodynamics of wetting. *Fluid Dynamics, 11*, 714–721.

Wang, X. D., Tian, Y., & Peng, X. F. (2003). Self-aggregation of vapor-liquid phase transition. *Progress in Natural Science (Chinese), 13*, 281–286.

Wang, H., Liao, Q., Zhu, X., Li, J., & Tian, X. (2010). Experimental studies of liquid droplet coalescence on the gradient surface. *Journal of Superconductivity and Novel Magnetism, 23*(6), 1165–1168.

Wang, F.-C., Yang, F., & Zhao, Y.-P. (2011). Size effect on the coalescence-induced self-propelled droplet. *Applied Physics Letters, 98*, 053112.

Wang, M., Liu, Y., & Yang, H. (2012). A unified thermodynamic theory for the formation of anodized metal oxide structures. *Electrochimica Acta, 62*, 424–432.

Wang, S., Liu, K., Yao, X., & Jiang, L. (2015a). Bioinspired surfaces with superwettability: New insight on theory, design, and applications. *Chemical Reviews, 115*, 8230–8293.

Wang, Y., Ma, L., Xu, X., & Luo, J. (2015b). Combined effects of underlying substrate and evaporative cooling on the evaporation of sessile liquid droplets. *Soft Matter, 11*, 5632–5640.

Watson, R. G. H., Birt, D. C. P., Honour, C. W., & Ash, B. W. (1962). The promotion of dropwise condensation by Montan wax I: Heat transfer measurements. *Journal of Applied Chemistry, 12*, 539–546.

Wayner, P. C. (1993). Spreading of a liquid film with a finite contact angle by the evaporation/condensation process. *Langmuir, 9*, 294–299.

Weeks, M. (2007). *Digital signal processing using MATLAB and wavelets*. Hingham, MA: Infinity Science Press LLC.

Wen, R., Ma, X., Lee, Y.-C., & Yang, R. (2018). Liquid-vapor phase-change heat transfer on functionalized nanowired surfaces and beyond. *Joule, 2*, 2307–2347.

Wenzel, H., Versuche über Tropfenkondensation. Allg. Wärmetech. 8 (1957) 33–39.

Wenzel, R. N. (1936). Resistance of solid surfaces to wetting by water. *Industrial and Engineering Chemistry Research, 28*, 988–994.

Welch, J. F. and Westwater, J. W. (1961). Microscopic study of dropwise condensation. Int. Dev. Heat Transfer 2 302–309.

White, F. M. (2015). *Fluid mechanics*. New York: McGraw-Hill. ISBN #978-0-07-352934-9.

Wilkins, D., & Bromley, L. (1973). Dropwise condensation phenomena. *AIChE Journal, 19*, 839–845.

Wilmshurst, R., & Rose J. W. (1970). Dropwise condensation-further heat-transfer measurements. In *Proceedings of the 4th International Heat Transfer Conference* (Vol. 4, pp. 1–4).

Wilmshurst, R., & Rose J. W. (1974). Dropwise and filmwise condensation of aniline, ethandiol, and nitrobenzene. In *Proceedings of the 5th International Heat Transfer Conference* (Vol. 3, pp. 269–273).

Winkels, K. G., Weijs, J. H., Eddi, A., & Snoeijer, J. H. (2012). Initial spreading of low-viscosity drops on partially wetting surfaces. *Physical Review E, 85*, 055301.

Wilmshurst, R. (1979), "Heat transfer during dropwise condensationof steam, ethane 1,2 diol, aniline and nitrobenzene," PhD thesis, London University, UK.

Wu, W. H., & Maa, J. R. (1976). On the heat transfer in dropwise condensation. *Chemical Engineering Journal, 12*, 225–231.

Wu, Y., Yang, C., & Yuan, X. (2001). Drop distribution and numerical simulation of dropwise condensation heat transfer. *International Journal of Heat and Mass Transfer, 44*, 4455–4464.

Wu, M., Cubaud, T., & Ho, C. M. (2004). Scaling law in liquid drop coalescence driven by surface tension. *Physics of Fluids, 16*, L51–L54.

Wu, B., Zhou, M., Li, J., Ye, X., Li, G., & Cai, L. (2009). Superhydrophobic surfaces fabricated by microstructuring of stainless steel using a femtosecond laser. *Applied Surface Science, 256*(1), 61–66.

Xie, J., Xu, J., Shang, W., & Zhang, K. (2018). Dropwise condensation on superhydrophobic nanostructure surface, part II: Mathematical model. *International Journal of Heat and Mass Transfer, 127*, 1170–1187.

Xin, B., & Hao, J. (2010). Reversibly switchable wettability. *Chemical Society Reviews, 39*, 769–782.

Xing, X. Q., Butler, D. L., Ng, S. H., Wang, Z., Danyluk, S., & Yang, C. (2007). Simulation of droplet formation and coalescence using lattice Boltzmann-based single-phase model. *Journal of Colloid and Interface Science, 311*(2), 609–618.

Xu, W., Leeladhar, R., Kang, Y. T., & Choi, C.-H. (2013). Evaporation kinetics of sessile water droplet on micro pillared super hydrophobic surfaces. *Langmuir, 29*, 6032–6041.

Yadav, M. K. (2019). Steam condensation studies towards understanding post-severe nuclear accident scenarios. Ph.D. dissertation, Indian Institute of Technology Kanpur, India.

Yadav, M. K., Khandekar, S., & Sharma, P. K. (2016). An integrated approach to steam condensation studies inside reactor containments: A review. *Nuclear Engineering and Design, 300*, 181–209.

Yadav, M. K., Singh, S. K., Parwez, A., & Khandekar, S. (2018). Inverse models for transient wall heat flux estimation based on single and multi-point temperature measurements. *International Journal of Thermal Sciences, 124*, 307–317.

Yan, Y. Y., Gao, N., & Barthlott, W. (2011). Mimicking natural superhydrophobic surfaces and grasping the wetting process: A review on recent progress in preparing superhydrophobic surfaces. *Advances in Colloid and Interface Science, 169*, 80–105.

Yang, C. X., Wang, L. G., Yuan, X. G., & Ma, C. F. (1998). Dropwise condensation as typical fractal growth process. *Journal of Aerospace Power, 13*(3), 272–276.

Yarin, A. (2006). Drop impact dynamics: Splashing, spreading, receding, bouncing. *Annual Review of Fluid Mechanics, 38*, 159–192.

Yeh, S.-I., Fang, W.-F., Sheen, H.-J., & Yang, J.-T. (2013). Droplets coalescence and mixing with identical and distinct surface tension on a wettability gradient surface. *Microfluidics and Nanofluidics, 14*, 785–795.

Yong, J., Yang, Q., Chen, F., Zhang, D., Bian, H., Ou, Y., Si, J., Du, G., & Hou, X. (2013). Stable superhydrophobic surface with hierarchical mesh-porous structure fabricated by a femtosecond laser. *Applied Physics A: Materials Science and Processing, 111*, 243–249.

Yoshida, N., Abe, Y., Shigeta, H., Nakajima, A., Ohsaki, K., & Watanable, T. (2006). Sliding behavior of droplet on flat polymer surface. *Journal of American Chemical Society, 128*, 743–747.

Young, T. (1805). An essay on the cohesion of fluids. *Philosophical Transactions of Royal Society of London, 95*, 65–87.

Ytrehus, T., & Østmo, S. (1996). Kinetic theory approach to interphase processes. *International Journal Multiphase Flow, 22*(1), 133–155.

Yu, Y., Zhu, H., Frantz, J. M., Reding, M. E., Chan, K. C., & Ozkan, H. E. (2009). Evaporation and coverage area of pesticides droplets on hairy and waxy leaves. *Biosystems Engineering, 104*(3), 324–334.

Zhang, D. C., Lin, Z. Q., & Lin, J. F. (1986). New materials for dropwise condensation. In *Proceedings of the 8th International Heat Transfer Conference* (Vol. 4, pp. 1677–1682).

Zhang, F. H., Li, E. Q., & Thoroddsen, S. T. (2009). Satellite formation during coalescence of unequal size drops. *Physical Review Letters, 102*, 104502.

Zhang, D., Chen, F., Yang, Q., Yong, J., Bian, H., Ou, Y., Si, J., Meng, X., & Hou, X. (2012). A simple way to achieve pattern-dependent tunable adhesion in superhydrophobic surfaces by a femtosecond laser. *ACS Applied Materials and Interfaces, 4*, 4905–4912.

Zhang, Z., Xu, S., & Ren, W. (2014). Derivation of a continuum model and the energy law for moving contact lines with insoluble surfactants. *Physics of Fluids, 26*, 062103.

Zhao, H., & Beysens, D. (1995). From droplet growth to film growth on a heterogeneous surface: Condensation associated with a wettability gradient. *Langmuir, 11*(2), 627–634.

Zhao, H., & Beysens, D. (1996). From droplet growth to film growth on a heterogeneous surface: Condensation associated with a wettability gradient. *Langmuir, 11*, 627–634.

Zhao, Q., & Burnside, B. M. (1994). Dropwise condensation of steam on ion-implanted condenser surfaces. *Heat Recovery Systems and CHP, 14*, 525–534.

Zhao, Q., Zhang, D. C., Zhu, X. B., Xu, D. Q., Lin, Z. Q., & Lin, J. F. (1990). Industrial application of dropwise condensation. In *Proceedings of the 9th International Heat transfer Conference* (Vol. 4, pp. 391–394).

Zhao, Q., Zhang, D. C., & Lin, J. F. (1991). Surface materials with dropwise condensation mode by ion implantation technology. *International Journal of Heat and Mass Transfer, 34*, 2833–2835.

Zhao, Q., Zhang, D. C., Lin, J. F., & Wang, G. M. (1996). Dropwise condensation on L-B film surface. *Chemical Engineering and Processing, 35*, 473–477.

Zhao, X., Blunt, M. J., & Yao, J. J. (2010). Pore-scale modeling: Effects of wettability on waterflood oil recovery. *Journal of Petroleum Science and Engineering, 71*, 169–178.

Zhu, X., Wang, H., Liao, Q., Ding, Y. D., & Gu, Y. B. (2009). Experiments and analysis on self-motion behaviors of liquid droplets on gradient surfaces. *Experimental Thermal and Fluid Sciences, 33*(6), 947–954.

Zhu, G., Yao, J., Zhang, L., Sun, H., Li, A., & Shams, B. (2016). Investigation of the dynamic contact angle using a direct numerical simulation method. *Langmuir, 32*(45), 11736–11744.

Ziberi, B., Frost, F., Hoche, T., & Rauschenbach, B. (2005). Ripple pattern formation on silicon surfaces by low-energy ion-beam erosion: Experiment and theory. *Physical Review B, 72*, 235310.

Index

S. Khandekar, K. Muralidhar, *Drop Dynamics and Dropwise Condensation on Textured Surfaces*, Mechanical Engineering Series,
https://doi.org/10.1007/978-3-030-48461-3

H

I

K

The manufacturer's authorised representative in the EU is Springer Nature Customer Service Centre GmbH, Europaplatz 3, 69115 Heidelberg, Germany. If you have any concerns regarding our products, please contact ProductSafety@springernature.com

Printed and bound by CPI Group (UK) Ltd, Croydon, CR0 4YY
30/07/2026
02181743-0002